Exocyclic DNA Adducts in Mutagenesis and Carcinogenesis

International Agency for Research on Cancer

The International Agency for Research on Cancer (IARC) was established in 1965 by the World Health Assembly, as an independently financed organization within the framework of the World Health Organization. The headquarters of the Agency are at Lyon, France.

The Agency conducts a programme of research concentrating particularly on the epidemiology of cancer and the study of potential carcinogens in the human environment. Its field studies are supplemented by biological and chemical research carried out in the Agency's laboratories in Lyon, and, through collaborative research agreements, in national research institutions in many countries. The Agency also conducts a programme for the education and training of personnel for cancer research.

The publications of the Agency are intended to contribute to the dissemination of authoritative information on different aspects of cancer research. Information about IARC publications and how to order them is available via the Internet at: http://www.iarc.fr/

Cover illustration:
Duplex DNA catalyses the chemical rearrangement of the exocyclic malondialdehyde deoxyguanosine adduct 3-(2´-deoxy-β-D-erythro-pentofuranosyl)-pyrimido[1,2*a*]purin-10(3*H*)one (M_1G).

Left panel: A C (green) : M_1G (red) base pair located within a $(CG)_3$ iterated repeat found in the *hisD3052* gene of a frameshift tester strain of *Salmonella typhimurium.*

Right panel: The cytosine opposite the M_1G lesion quantitatively catalyses the conversion of the exocyclic adduct to the corresponding N^2-(3-oxo-1-propenyl)-dG derivative. Structure of the C (green) : N^2-(3-oxo-1-propenyl)-dG (red) base pair located within the $(CG)_3$ iterated repeat as determined by NMR spectroscopy.

Work of Hui Mao, Nathalie Schnetz-Boutaud, Jason P. Weisenseel, Lawrence J. Marnett and Michael P. Stone, Departments of Chemistry and Biochemistry, Center in Molecular Toxicology, and Vanderbilt Cancer Center, Vanderbilt University, Nashville, TN 37235, USA.

INTERNATIONAL AGENCY FOR RESEARCH ON CANCER
WORLD HEALTH ORGANIZATION

Exocyclic DNA Adducts in Mutagenesis and Carcinogenesis

Edited by
B. Singer and H. Bartsch

IARC Scientific Publications No. 150

International Agency for Research on Cancer
Lyon, France
1999

Published by the International Agency for Research on Cancer,
150 cours Albert-Thomas, 69372 Lyon cedex 08, France

Distributed by Oxford University Press, Walton Street, Oxford OX2 6DP, UK
(fax: +44 1865 267782) and in the USA by Oxford University Press, 2001 Evans Road, Carey,
NC 27513 (fax: +1 919 677 1303). All IARC publications can also be ordered directly from IARC*Press*
(fax: +33 04 72 73 83 02; e-mail: press@iarc.fr).

IARC Library Cataloguing in Publication Data

Exocyclic DNA Adducts in Mutagenesis and Carcinogenesis/ editors, B. Singer, H. Bartsch

(IARC scientific publications; 150)

1. DNA Adducts
2. Mutagenesis
3. Neoplasms
I. Singer, B. (Beatrice), 1922–
II. Bartsch, H. III. Series

ISBN 92 832 2150 8 (NLM Classification: W1)
ISSN 0300-5085

Printed in France

Contents

Contributors

S. Akasaka
Osaka Prefectural Institute of Public Health
1-3-69 Nakamichi-Higashinari-ku
Osaka 537
Japan

S. Ambs
Megabios Corporation
863A Mitten Road
Burlingame, CA 94010
United States

J.P. Aston
Toxicology Department
Shell International Chemicals BV
Shell Research and Technology Centre
Amsterdam
PO Box 38000
NL-1030 BN Amsterdam
The Netherlands

A. Barbin
Unit of Gene-Environment Interactions
International Agency for Research on Cancer
150 cours Albert Thomas
69372 Lyon Cedex 08
France

T. Barlow
Department of Chemistry
University of Edinburgh
Edinburgh, Scotland
United Kingdom

H. Bartsch
Division of Toxicology and Cancer Risk Factors
German Cancer Research Centre (DKFZ)
Im Neuenheimer Feld 280
D-69210 Heidelberg
Germany

A.K. Basu
Department of Chemistry
University of Connecticut
Storrs, CT 06269
United States

S.A. Bingham
MRC Dunn Nutrition Centre
Cambridge
United Kingdom

M.S. Bogdanffy
Haskell Laboratories
E.I. du Pont de Nemours & Co.
PO Box 50
1090 Elkton Road
Newark, DE 19714
United States

S. Boivin-Angele
International Agency for Research on Cancer
150 cours Albert Thomas
69372 Lyon Cedex 08
France

K.J. Breslauer
Department of Chemistry
Rutgers, The State University of New Jersey
Piscataway, NJ 08854-8087
United States

T. Brown
Department of Chemistry
University of Southampton
Southampton, SO17 1BJ
United Kingdom

Budiawan
Department of Chemistry
Faculty of Sciences
University of Indonesia
Jakarta
Indonesia

J. Cadet
Département de Recherche Fondamentale sur la Matière Condensée
Laboratoire 'Lésions des Acides Nucléiques'
17 avenue des Martyrs
F-38054 Grenoble Cedex 9
France

V.M. Carvalho
Departamento de Bioquimica
Instituto de Quimica
Universidade de São Paulo
CP 26077
CEP 05599-970 São Paulo
Brazil

A. Chenna
Donner Laboratory
Lawrence Berkeley National Laboratory
University of California
Berkeley, CA 94720
United States

F.-L. Chung
Division of Carcinogenesis and Molecular Epidemiology
American Health Foundation
Valhalla, NY 10595
United States

J.M. Cieśla
Polish Academy of Sciences
Institute of Biochemistry and Biophysics
ul. Pawinskiego 5a
02-106 Warszawa
Poland

A.E. Crane
Shell Research Ltd
Sittingbourne Research Centre
Sinttingbourne, Kent ME9 8AG
United Kingdom

W. Cui
Department of Chemistry
University of Missouri
Columbia, MO 65211
United States

D. Cullinan
Department of Pharmacological Sciences
State University of New York at Stony Brook
Stony Brook, NY 11794-8651
United States

E.J. Deag
MRC Toxicology Unit
Hodgkin Building
University of Leicester
Leicester LE1 9HN
United Kingdom

P. Di Mascio
Departamento de Bioquimica
Instituto de Quimica
Universidade de São Paulo
CP 26077
CEP 05599-970 São Paulo
Brazil

T. Douki
Département de Recherche Fondamentale sur la Matière Condensée
Laboratoire 'Lésions des Acides Nucléiques'
17 avenue des Martyrs
F-38054 Grenoble Cedex 9
France

E. Eder
Department of Toxicology
Versbacher Strasse 9
D-97078 Würzburg
Germany

M. Eisenberg
Department of Pharmacological Sciences
State University of New York at Stony Brook
Stony Brook, NY 11794-8651
United States

J.-L. Fang
Division of Biochemical Toxicology
National Center for Toxicological Research
Jefferson, AR 72079-9502
United States

C.A. Gelfand
Department of Chemistry
Rutgers, The State University of New Jersey
Piscataway, NJ 08854-8087
United States

P. Goelzer
Department of Chemistry
University of Missouri
Columbia, MO 65211
United States

B.T. Golding
Department of Chemistry
Newcastle University
Newcastle-upon-Tyne NE1 7RU
United Kingdom

E. Griech
MRC Toxicology Unit
Hodgkin Building
University of Leicester
Leicester LE1 9HN
United Kingdom

A.P. Grollman
Department of Pharmacological Sciences
State University of New York at Stony Brook
Stony Brook, NY 11794-8651
United States

F.P. Guengerich
Center in Molecular Toxicology
Department of Biochemistry
Vanderbilt University School of Medicine
Nashville, TN 37232-0146
United States

A. Ham
Pathology Department
CB # 7525
University of North Carolina
Chapel Hill, NC 27599
United States

B. Hang
Donner Laboratory
Lawrence Berkeley National Laboratory
University of California
Berkeley, CA 94720
United States

C.C. Harris
Laboratory of Human Carcinogenesis
Building 37, Room 2C01
National Cancer Institute
37 Convent Drive MSC 4255
Bethesda, MD 20892-4255
United States

S.S. Hecht
University of Minnesota Cancer Center
Box 806 Mayo
420 Delaware Street SE
Minneapolis, MN 55455
United States

K. Hemminki
Molecular Epidemiology Unit
CNT, Department of Biosciences at Novum
Karolinska Institute
S-141 57 Huddinge
Sweden

S. Holt
Curriculum in Toxicology
CB # 7270
University of North Carolina
Chapel Hill, NC 27599
United States

R. Hughes
MRC Dunn Nutrition Centre
Cambridge
United Kingdom

S.P. Hussain
Laboratory of Human Carcinogenesis
Building 37, Room 2C01
National Cancer Institute
37 Convent Drive MSC 4255
Bethesda, MD 20892-4255
United States

F. Johnson
Department of Pharmacological Sciences
State University of New York at Stony Brook
Stony Brook, NY 11794-8651
United States

A. Kim
Curriculum in Toxicology
CB # 7270
University of North Carolina
Chapel Hill, NC 27599
United States

P. Kowalczyk
Polish Academy of Sciences
Institute of Biochemistry and Biophysics
ul. Pawinskiego 5a
02-106 Warszawa
Poland

L. Kronberg
Department of Organic Chemistry
Åbo Akademi University
Biskopsgatan 8
FIN-20500 Åbo/Turku
Finland

S. Langouët
INSERM U456
Faculté des Sciences Pharmaceutiques et Biologiques
2, avenue du Professeur Léon Berard
35043 Rennes Cedex
France

J. Laval
Groupe 'Reparation des lésions Radio- et Chimio-Induites',
UMR 1772 CNRS
Institut Gustave Roussy
94805 Villejuif Cedex
France

F. Le Curieux
Department of Organic Chemistry
Åbo Akademi University
Biskopsgatan 8
FIN-20500 Åbo/Turku
Finland

C. Leuratti
MRC Toxicology Unit
Hodgkin Building
University of Leicester
Leicester LE1 9HN
United Kingdom

D. Li
Department of Clinical Investigation
The University of Texas M.D. Anderson Cancer Center
Houston, TX 77030
United States

Q. Li
Department of Pharmacology & Molecular Toxicology
University of Massachusetts Medical School
Worcester, MA 01655-0126
United States

R.N. Loeppky
Department of Chemistry
University of Missouri
Columbia, MO 65211
United States

D.B. Ludlum
Department of Pharmacology & Molecular Toxicology
University of Massachusetts Medical School
Worcester, MA 01655-0126
United States

J. Mäki
Department of Organic Chemistry
Åbo Akademi University
Biskopsgatan 8
FIN-20500 Åbo/Turku
Finland

M.-J. Marion
INSERM, Unité 271
151 cours Albert Thomas
69424 Lyon Cedex 03
France

L.J. Marnett
A.B. Hancock Jr Memorial Laboratory for Cancer Research
Center in Molecular Toxicology and the Vanderbilt Cancer Center
Department of Biochemistry and Chemistry
Vanderbilt University School of Medicine
Nashville, TN 37232
United States

A.J. Marrogi
Laboratory of Human Carcinogenesis
Building 37, Room 2C01
National Cancer Institute
37 Convent Drive MSC 4255
Bethesda, MD 20892-4255
United States

H. Maruenda
Pontificia Universidad Catolica del Peru
Seccion Quimica
Apartado Postal 1761
Lima 100
Peru

Z. Matijasevic
Department of Pharmacology & Molecular Toxicology
University of Massachusetts Medical School
Worcester, MA 01655-0126
United States

W.G. McGregor
Department of Microbiology
Michigan State University
Lansing, MI 48824
United States

J.M. McNulty
Department of Chemistry
University of Connecticut
Storrs, CT 06269
United States

M.H.G. Medeiros
Departamento de Bioquimica
Instituto de Quimica
Universidade de São Paulo
CP 26077
CEP 05599-970 São Paulo
Brazil

A.N. Mican
Center in Molecular Toxicology
Department of Biochemistry
Vanderbilt University School of Medicine
Nashville, TN 37232-0146
United States

E.J. Morinello
Curriculum in Toxicology
CB # 7270
University of North Carolina
Chapel Hill, NC 27599
United States

M. Moriya
Department of Pharmacological Sciences
State University of New York at Stony Brook
Stony Brook, NY 11794-8651
United States

M. Müller
Abteilung Arbeits-und-Sozialmedizin der Georg-August-Universität Göttingen
Waldweg 37
D-37073 Göttingen
Germany

T. Munter
Department of Organic Chemistry
Åbo Akademi University
Biskopsgatan 8
FIN-20500 Åbo/Turku
Finland

J. Nair
Division of Toxicology and Cancer Risk Factors
German Cancer Research Centre (DKFZ)
Im Neuenheimer Feld 280
D-69210 Heidelberg
Germany

R.G. Nath
Covance Laboratories
9200 Leesburg Pike
Vienna, VA 22182
United States

M.J.M. Nivard
Medical Genetics Centre South-West Netherlands
Department of Radiation Genetics and Chemical Mutagenesis
Leiden University Medical centre
Wassenaarseweg 72
NL-2300 RA Leiden
The Netherlands

J.E. Ocando
Division of Carcinogenesis and Molecular Epidemiology
American Health Foundation
Valhalla, NY 10595
United States

J. Onuki
Département de Recherche Fondamentale sur la Matière Condensée
Laboratoire 'Lésions des Acides Nucléiques'
17 avenue des Martyrs
F-38054 Grenoble Cedex 9
France

G.A. Pandya
American Health Foundation
Valhalla, NY 10595
United States

M. Park
Department of Chemistry
University of Missouri
Columbia, MO 65211
United States

M. Persmark
Glaxo Wellcome
Room 2.2134, 5 Moore Drive
Research Triangle Park, NC 27709
United States

J.P. Plastaras
Department of Biochemistry
Vanderbilt University School of Medicine
Nashville, TN 37232
United States

G.E. Plum
Department of Chemistry
Rutgers, The State University of New Jersey
Piscataway, NJ 08854-8087
United States

D. Pluskota
Department of Organic Chemistry
Åbo Akademi University
Biskopsgatan 8
FIN-20500 Åbo/Turku
Finland

D. Potter
Shell Research Ltd
Sittingbourne Research Centre
Sinttingbourne, Kent ME9 8AG
United Kingdom

A. Ranasinghe
Department of Environmental Sciences and Engineering
CB # 7400
University of North Carolina
Chapel Hill, NC 27599
United States

J. Sági
Donner Laboratory
Lawrence Berkeley National Laboratory
University of California
Berkeley, CA 94720
United States

C. de los Santos
Department of Pharmacological Sciences
State University of New York at Stony Brook
Stony Brook, NY 11794-8651
United States

M. Saparbaev
Groupe 'Reparation des lésions Radio- et Chimio-induites',
UMR 1772 CNRS
Institut Gustave Roussy
94805 Villejuif Cedex
France

N. Scheller
Department of Environmental Sciences and Engineering
CB # 7400
University of North Carolina
Chapel Hill, NC 27599
United States

D. Schuler
Department of Toxicology
Versbacher Strasse 9
D-97078 Würzburg
Germany

D.E.G. Shuker
MRC Toxicology Unit
Hodgkin Building
University of Leicester
Leicester LE1 9HN
United Kingdom

B. Singer
Donner Laboratory
Lawrence Berkeley National Laboratory
University of California
Berkeley, CA 94720
United States

R. Singh
MRC Toxicology Unit
Hodgkin Building
University of Leicester
Leicester LE1 9HN
United Kingdom

R. Sjöholm
Department of Organic Chemistry
Åbo Akademi University
Biskopsgatan 8
FIN-20500 Åbo/Turku
Finland

J.J. Solomon
Nelson Institute of Environmental Medicine
New York University School of Medicine
57 Old Forge Road
Tuxedo, NY 10987
United States

J.A. Swenberg
Department of Environmental Sciences and Engineering
CB # 7400
University of North Carolina
Chapel Hill, NC 27599
United States

B. Tudek
Polish Academy of Sciences
Institute of Biochemistry and Biophysics
ul. Pawinskiego 5a
02-106 Warszawa
Poland

P. Upadhyaya
University of Minnesota Cancer Center
Box 806 Mayo
420 Delaware Street SE
Minneapolis, MN 55455
United States

P. Upton
Department of Environmental Sciences and Engineering
CB # 7400
University of North Carolina
Chapel Hill, NC 27599
United States

E.W. Vogel
Medical Genetics Centre South-West Netherlands
Department of Radiation Genetics and Chemical Mutagenesis
Leiden University Medical centre
Wassenaarseweg 72
NL-2300 RA Leiden
The Netherlands

M. Wang
University of Minnesota Cancer Center
Box 806 Mayo
420 Delaware Street SE
Minneapolis, MN 55455
United States

W.P. Watson
Toxicology Department
Shell International Chemicals BV
Shell Research and Technology Centre Amsterdam
PO Box 38000
NL-1030 BN Amsterdam
The Netherlands

K. Yang
Molecular Epidemiology Unit
CNT, Department of Biosciences at Novum
Karolinska Institute
S-141 57 Huddinge
Sweden

Q. Ye
Department of Chemistry
University of Missouri
Columbia, MO 65211
United States

L. Zhang
Division of Carcinogenesis and Molecular Epidemiology
American Health Foundation
Valhalla, NY 10595
United States

Foreword

It is now commonly agreed that carcinogens of diverse chemical structure bind covalently to DNA and that the resulting DNA adducts, if not repaired, lead to mutations, a critical step in the initiation of many neoplasias. As already evidenced at a conference held at IARC in 1984 (organized by B. Singer and H. Bartsch; IARC Scientific Publications No. 70), interest has been directed to a particular class of genotoxic chemicals or metabolites that can react like bifunctional alkylating agents with nucleic acid bases to form one or more additional ring systems, i.e. exocyclic DNA-base adducts. While the previous focus was on environmental, industrial and chemotherapeutic agents to which humans are known to be exposed, the renewed interest in exocyclic adducts in the 1990s was triggered by the finding that they can also be formed through endogenous processes and reactions involving lipid peroxidation-derived bifunctional aldehydes and hydroxyalkenals. With the development of ultrasensitive detection methods, background levels of various exocyclic adducts have subsequently been detected and quantified in DNA of tissues of unexposed rodents and humans. In studies of animal models and by analysis of human cancer-prone tissues, it was shown that the background levels can be increased by several pathological conditions and cancer risk factors, such as life style, chronic infections and inflammation. This finding suggests that exocyclic adducts, with other oxidative DNA damage, could play a role in the multistage process of human carcinogenesis, in which persistent oxidative stress is increasingly recognized as a driving force towards malignancy. For these reasons, the use of exocyclic adducts as biomarkers offers a promising tool in studies of cancer etiology and prevention, particularly for human neoplasias in which the causative factors and mechanisms are still poorly understood.

Because of the growing interest in this field, a second international conference was held at the Deutsches Krebsforschungzentrum (DKFZ) in Heidelberg in September 1998, which attracted world experts in the domain. The presentations which resulted in this volume of IARC Scientific Publications comprise a comprehensive treatise on the current state of the art and scientific information on exocyclic DNA adducts. The volume includes sections on ultrasensitive detection methods, formation from exogenous and endogenous sources, DNA repair, physical chemical approaches to structural elucidation, use as biomarkers and their role in mutagenesis and carcinogenesis.

Financial support for the meeting was provided by the Deutsche Forschungsgemeinschaft, Bonn, Germany, the National Cancer Institute, Bethesda, United States, the National Institute for Environmental Health Sciences, Research Triangle Park, United States, IARC, Lyon, France, and the DKFZ, Heidelberg, Germany. All of the contributions are gratefully acknowledged.

We should like to thank the organizers, B. Singer, Donner Laboratory, University of California, Berkeley, United States, and H. Bartsch, Division of Toxicology and Cancer Risk Factors, DKFZ, Heidelberg, Germany. The organizers are indebted to N. Frank (the meeting secretary), J. Nair (DKFZ) and M. Medina (University of California) for their unstinting efforts towards the success of both the meeting and the publication.

Harald zur Hausen
Chairman, Deutsches Krebsforschungzentrum

Paul Kleihues
Director, IARC

Acknowledgements

The editors are indebted to Michael Medina for his dedication and expert editorial assistance throughout the genesis of this book. We also thank Dr Norbert Frank, the meeting secretary, and Dr Jagadeesan Nair from the DKFZ who both greatly assisted in the preparation and local arrangements for running a successful conference.

Abbreviations

In order to render the text more accessible to the reader by standardizing the abbreviations used in this publication, the editors adopted a system that corresponds most closely to the international system currently in use which is applicable to DNA constituents and adducts.

Th abbreviations used for nucleic acid constituents are summarized in the table. It was considered that designation of 3´ and 5´ monophosphates is desirable in the context of ^{32}P-postlabelling in order to avoid ambiguity, but that the form dNpdN could be used in the case of dinucleotides as the position of the phosphodiester is clear.

Abbreviations for nucleic acid constituents

Constituent	Adenine	Guanine	Cytosine	Thymine
Base	Ade	Gua	Cyt	Thy
Nucleoside	Ado	Guo	Cyd	Thd
2´-Deoxynucleoside	dA	dG	dC	dT
2´-Deoxynucleoside-3´ (or 5´) monophosphate	dA3´(5´)p	dG3´(5´)p	dC3´(5´)p	dT3´(5´)p
2´-Deoxynucleoside-3´,5´-diphosphate	dA3´5´pp	dG3´5´pp	dC3´5´pp	dT3´5´pp
2´-Deoxynucleotide in DNA sequence	A	G	C	T

Two systems of abbreviation are commonly used for DNA adducts.

Abbreviations derived from the 'core' nucleoside with an added group

position		alkyl group		nucleoside
X	–	name	–	dN

e.g., 7-methyl-dG, O^6-ethyl-dG

Note that the convention in nucleic acid nomenclature is to abbreviate substitution on ring atoms by numbers: 1, 7, 8, etc.; designation of the atom, N or C, is usually unnecessary. Substitutions on exocyclic atoms, however, take a superscript; thus, N^2, N^6, O^6, etc.

Abbreviations derived from the addition of the nucleoside to the adducting molecule

Nucleoside		position		alkyl group
dN	—	Y	—	name

e.g., dG-(C)8-MeIQx

The choice of abbreviation is governed by the IUPAC name of the adduct. Thus, for example, O^6-(2-chloroethyl)-2´-deoxyguanosine becomes O^6-CE-dG, wheras *N*-(2´-deoxyguanosin-8-yl)-2-amino-1-methyl-6-phenylimidazo[4,5-*b*]pyridine becomes dG-(C)8-PhIP. As indicated above, designation of ring atoms is optional.

Exocyclic DNA Adducts in Mutagenesis and Carcinogenesis
B. Singer & H. Bartsch
IARC Scientific Publications No. 150
International Agency for Research on Cancer, Lyon, 1999

Keynote address: Exocyclic adducts as new risk markers for DNA damage in man

H. Bartsch

Background levels of exocyclic DNA adducts detected by ultrasensitive methods in tissues from unexposed humans and rodents arise from endogenous lipid peroxidation products such as *trans*-4-hydroxy-2-nonenal, crotonaldehyde and malondialdehyde. The levels of DNA adducts in rodent and human tissues and leukocytes were found to be highly variable and to be affected by lifestyle, the dietary intake of antioxidants and the type and amount of fatty acids and persistent chronic infections or inflammations, in which nitric oxide is often over-produced. Limited evidence suggests that etheno–DNA adducts play a role not only in vinyl chloride- and urethane-induced tumorigenesis but, together with other exocyclic lesions, also in several human cancers in which persistent oxidative stress leads to malignancy by increasing mutation rates and genomic instability. Therefore, promutagenic exocyclic adducts appear to be promising markers in molecular epidemiological studies for identifying endogenous sources of DNA damage and resulting oxidative modifications in cancers with poorly defined etiology and mechanisms and in intervention studies to assess the protective effects of antioxidants against cancer and, possibly, neurodegenerative diseases.

Introduction

Of the exocyclic DNA adducts, etheno (ε) bases (for structures, see Figure 1) have been the most widely studied over the last 25 years, as this class of DNA lesion is formed by many genotoxic carcinogens (reviewed by Bartsch *et al.*, 1994).

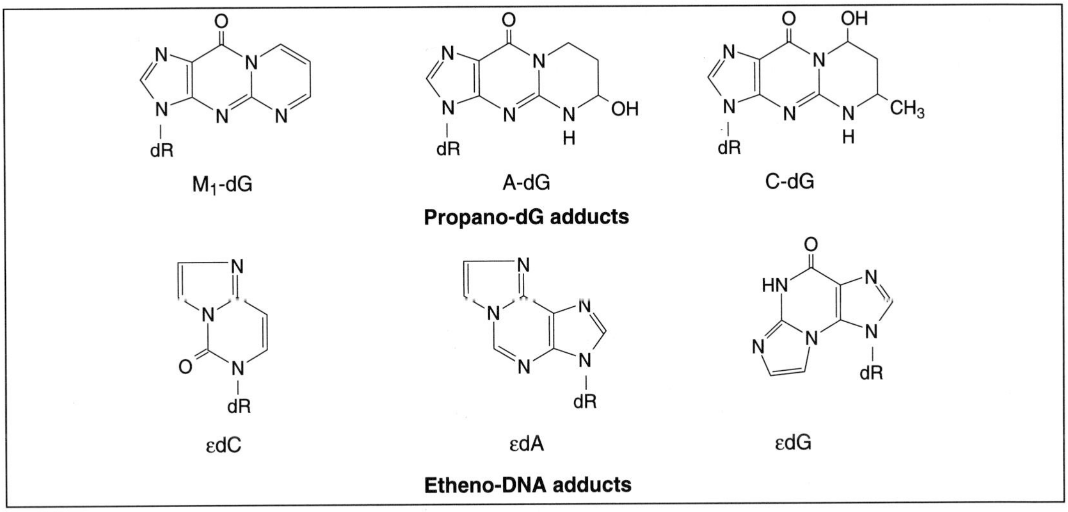

M₁-dG A-dG C-dG

Propano-dG adducts

εdC εdA εdG

Etheno-DNA adducts

Figure 1. Chemical structures of exocyclic propano adducts derived by reaction of malonaldehyde (M_1-dG), crotonaldehyde (C-dG), acrolein (A-dG) or etheno adducts derived from *trans*-4-hydroxy-2-nonenal *via* its epoxy derivative (N^2,3-εdG, εdA, εdC)

Etheno bases were first described by Kochetkov *et al.* (1971), who identified them as fluorescent analogues in biochemical studies and probes for nucleic acid structures (Leonard, 1992). Interest in etheno lesions was renewed in 1975 (for a chronicle of important landmarks, see Tables 1–3), when it was found that they were generated *in vitro* by the vinyl chloride metabolites chloroethylene oxide and 2-chloroacetaldehyde and by the multi-organ and multi-species carcinogen urethane *via* their oxirane intermediates. The results of replication and transcription fidelity assays with etheno-modified oligo- and polynucleotides established that 1,N^6-ethenodeoxyadenosine (εdA) and 3,N^4-ethenodeoxycytidine (εdC) have miscoded base-repairing properties and could thus be involved in the mutagenic and carcinogenic effects of vinyl chloride (IARC, 1979; Barbin *et al.*, 1981; Hall *et al.*, 1981; Spengler & Singer, 1981; Table 2). The promutagenic properties of etheno bases were established by primer extension assays *in vitro* and by site-specific mutagenesis in *Escherichia coli* and mammalian cells (reviewed by Bartsch *et al.*, 1994). Etheno adducts were found to be repaired by mammalian cell extracts (Oesch *et al.*, 1986, 1994), and the enzymes involved were later characterized as specific DNA glycosylases (Matijasevic *et al.*, 1992; Dosanjh *et al.*, 1994; Saparbaev *et al.*, 1995; Hang *et al.*, 1996, 1997, 1998). The substitution found at G:C or A:T base pairs in vinyl chloride-induced tumours led to the hypothesis that three promutagenic etheno-DNA adducts (N^2,3-ε-deoxyguanosine [N^2,3-εdG], εdC, εdA) are the initiating lesions (IARC, 1979; Barbin & Bartsch, 1986).

The main reason for the renewed interest in exocyclic DNA lesions in the 1990s was the development of ultrasensitive detection methods (Eberle *et al.*, 1989; Fedtke *et al.*, 1990; Guichard *et al.*, 1993; Nair *et al.*, 1995; Chung *et al.*, 1996) notably for etheno- and propano-DNA adducts, which made it possible to study the formation of exocyclic adducts in experimental animals and humans. Using gas chromatography–mass spectrometry, high-performance liquid chromatography with radioimmunoassay or immunoaffinity/^{32}P-postlabelling (IPPA) assays to detect background levels of exocyclic etheno adducts in tissues from unexposed humans and rodents (Table 3), Chung *et al.* (1989) and Fedtke *et al.* (1990) reported the presence of exocyclic base adducts in rat and human liver tissues, although the identities of the adducts were not vigorously confirmed at

Table 1. 25 Years of research on etheno (ε) DNA-base adducts: Mechanistic highlights and sources of formation

Observation	Reference
First synthesis of εA, εC	Kochetkov *et al.* (1971)
εdA-DNA adducts by vinyl chloride metabolites	Barbin *et al.* (1975)
εA, εC in RNA of vinyl chloride-exposed rodents	Laib & Bolt (1977, 1978)
εdA, εdC, εdG in rodent liver and lung DNA after exposure to vinyl chloride	Green & Hathway (1978); Eberle *et al.* (1989); Fedtke *et al.* (1990)
εdA and εdC in RNA (DNA) by urethane in mice	Ribovich *et al.* (1982); Fernando *et al.* (1996)
ε Adducts from halooxiranes and vinyl carbamate epoxide	Guengerich *et al.* (1981); Guengerich & Kim (1991); Guengerich (1992)

Table 2. 25 Years of research on etheno (ε) DNA-base adducts: Miscoding and promutagenic properties

Observation	Reference
ε Adducts as miscoding lesions	Barbin *et al.* (1981); Hall *et al.* (1981); Spengler & Singer (1981)
ε Adducts in structural perturbation of DNA/RNA	Leonard (1984, 1992)
Repair of ε adducts by cell extracts	Oesch *et al.* (1986)
ε Adducts as promutagenic lesions	Basu *et al.* (1993); Moriya *et al.* (1994); Pandya & Moriya (1996)
εdG, εdC, εdA-compatible mutations in *ras*, *p53* of angiosarcoma induced in rat or human liver by vinyl chloride	Marion *et al.* (1991); Hollstein *et al.* (1994)
Repair of *N*[2],3-ethano/G by *E. coli* DNA glycosylase	Habraken *et al.* (1991)
Repair of ε adducts by *E. coli*, rat and human DNA glycosylases	Matijasevic *et al.* (1992); Dosanjh *et al.* (1994); Saparbaev *et al.* (1995)

Table 3. Background levels of exocyclic DNA adducts in rodent and human tissues as detected by ultrasensitive methods

DNA lesion (detection method)	Reference
1,*N*[2]-Propano-dG (CdG*) in rat liver (^{32}P)	Chung *et al.* (1989)
εdG* in rat liver (GC–MS)	Fedtke *et al.* (1990)
M_1-dG in human liver (GC–MS)	Chaudhary *et al.* (1994)
M_1-dG in human leukocytes and breast (^{32}P)	Vaca *et al.* (1995)
C-dG, A-dG in human and rodent tissues (HPLC–^{32}P)	Nath & Chung (1994)
εdA, εdC in human and rodent tissues (IPPA)	Bartsch *et al.* (1994); Nair *et al.* (1995); Guichard *et al.* (1996)
εdG in human liver (LC–MS, GC–MS)	Scheller *et al.* (1995); Swenberg *et al.* (1995)

* Identity not vigorously confirmed

^{32}P, ^{32}P-post-labelling; ε, etheno; GC–MS, gas chromatography–mass spectrometry; HPLC, high-performance liquid chromatography; IPPA, immunopurification–^{32}P-post-labelling; LC, liquid chromatography

that time. In 1994, unequivocal identification of the malondialdehyde-derived deoxyguanosine (M_1-dG) adduct was reported by Chaudhary *et al.* (1994) in human liver. The same adduct was later also found in human breast and leukocytes by Vaca *et al.* (1995). Using more specific and ultrasensitive methods by coupling with ^{32}P-postlabelling, Nath and Chung (1994) and Nath *et al.* (1994, 1996) reported the presence of background levels of propano adducts, and our group (Bartsch *et al.*, 1994; Nair *et al.*, 1995; Guichard *et al.*, 1996) found background levels of etheno adducts in DNA of various human and rodent tissues; Swenberg and coworkers (Scheller *et al.*, 1995; Swenberg *et al.*, 1995) later confirmed the presence of N^2,3-εdG in human liver by mass spectrometric techniques.

These findings suggested an endogenous pathway for the formation of exocyclic adducts *via* lipid peroxidation products, e.g. from 4-hydroxyalkenals, crotonaldehyde and malondialdehyde (Figure 2). Much effort was expended (Marnett & Burcham, 1993) to elucidate (i) the sources, toxico-

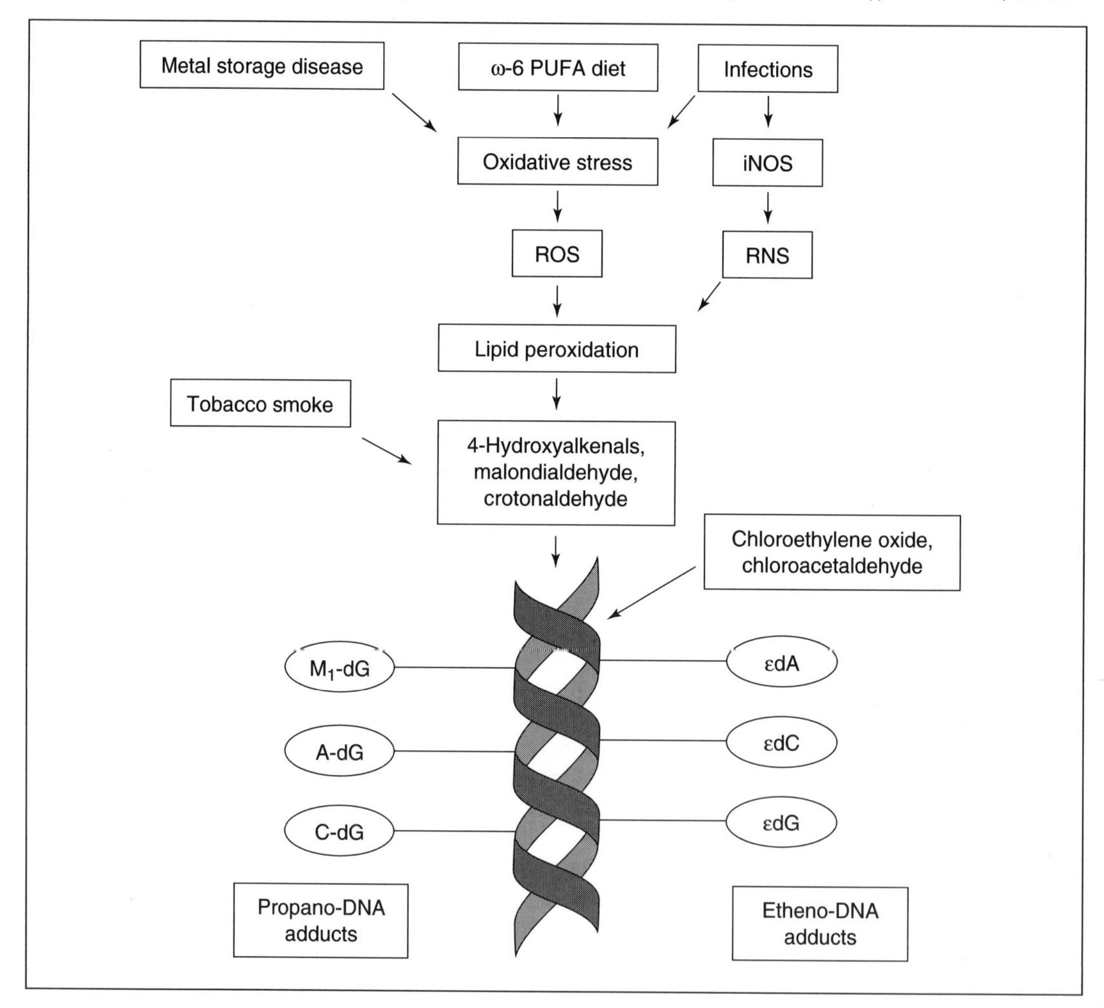

Figure 2. A proposed scheme of carcinogenic factors leading to oxidative stress-induced reactive oxygen (ROS) and nitrogen (RNS) species; these can trigger lipid peroxidation that yields dialdehydes and alkenals which form exocyclic DNA-base damage.

PUFA, polyunsaturated fatty acid; iNOS, inducible nitric oxide synthase

logical significance and genetic consequences of these background DNA lesions, (ii) the kinds of exogenous factors or physiopathological conditions that lead to their increase and (iii) whether these adducts, when generated through oxidative stress-mediated reactions, play a causative role in human carcinogenesis or in neurodegenerative diseases. In this paper, I briefly summarize some milestones in knowledge of DNA adducts, their sources and mechanisms of formation; some recent results demonstrating that inflammatory conditions, often associated with chronic infections and overproduction of nitric oxide, increase exocyclic DNA base damage in affected tissues; and a general hypothetical scheme in which persistent oxidative stress, for which etheno adducts seem to be key markers, acts as the ultimate driving force to cellular malignancy. For completeness, the reader is referred to other articles, especially those of Nair and Chung *et al.* in this volume.

Lipid peroxidation products as a source of endogenous DNA damage

The chemical structures of the main exocyclic adducts most often referred to are shown in Figure 1. The five-membered exocyclic ring is present in three etheno–DNA adducts known to be formed *in vivo*; the reaction of malonaldehyde, crotonaldehyde and acrolein with dG or dA leads to the addition of a six-membered exocyclic ring system. Evidence that background levels of exocyclic adducts are derived from DNA-reactive lipid peroxidation products is now supported by several findings, summarized in Table 4: etheno and propano adducts are formed (i) *in vitro* in the presence of DNA bases and α,β-unsaturated aldehydes such as *trans*-4-hydroxy-2-nonenal (Chung & Hecht, 1983; Winter *et al.*, 1986) *via* their epoxides (Sodum & Chung, 1989, 1991) or fatty acid hydroperoxides (Chen & Chung, 1996), all of which are lipid peroxidation products; and (ii) under conditions that increase lipid peroxidation *in vitro* in experimental systems, such as the addition of bivalent iron or cumene hydroperoxide to substrates for lipid peroxidation like arachidonic acid, a polyunsaturated fatty acid (El Ghissassi *et al.*, 1995). Treatment of rats with carbon tetrachloride (Chaudhary *et al.*, 1994) or an iron overload (Nair *et al.*, 1999) together with ethanol and carbon tetrachloride resulted in the formation of

Table 4. Sources and mechanisms of formation of background exocyclic DNA adducts in rodent and human tissues

Observation	Reference
Substituted propano-dG from enals	Winter *et al.* (1986)
1,N^2-εdG from hydroynonenal and its epoxide	Sodum & Chung (1988, 1991)
ε Adducts from enals by autoxidation	Chen & Chung (1994)
M_1-dG increased by CCl_4 in rodent liver	Chadhary *et al.* (1994)
εA, εC from rat liver microsomes/ω–6 polyunsaturated fatty acid by lipid peroxidation	El Ghissassi *et al.* (1995)
εA, εG from fatty acid peroxides and hydroxynonenal	Chen & Chung (1996)
εA, εC, M_1-dG in human leukocytes after high ω–6 polyunsaturated fatty acid diet	Vaca *et al.* (1995); Nair *et al.* (1997)
M_1-dG, dAdo adducts in human breast	Wang *et al.* (1996)
εA, εC in rat and human tissues with excess Fe and Cu storage	Nair *et al.* (1996, 1998b)
εA, εC after nitric oxide overproduction by inducible nitric oxide synthase	Nair *et al.* (1998a)

increased levels of M_1-dG and etheno adducts, respectively, in the livers of treated rodents. In animals and humans *in vivo*, conditions that induce oxidative stress or lipid peroxidation, like overproduction of nitric oxide (Nair *et al.*, 1998a) and certain genetic metal storage diseases such as Wilson disease and primary haemochromatosis (Nair *et al.*, 1998b), increased the formation of exocyclic DNA adducts in the affected tissues. Owing to accumulation of Cu and Fe in the livers of these patients and $OH^{\bullet -}$ radical production by Haber-Weiss and Fenton-type reactions, the increase in etheno adduct levels in the liver was highly significant and was positively correlated with the hepatic copper concentration. The same was seen in Long Evans cinnamon (LEC) rats, a model for Wilson disease in humans (Nair *et al.*, 1996). Further supportive evidence for lipid peroxidation-derived pathways of exocyclic adduct formation comes from the finding of increased M_1-dG adduct levels in leukocyte DNA of volunteers with a high dietary intake of ω–6 polyunsaturated fatty acids (Fang *et al.*, 1996). Interestingly, malonaldehyde-derived DNA adducts have also been detected in normal breast tissue and in tissue adjacent to tumours in breast cancer patients (Vaca *et al.*, 1995; Wang *et al.*, 1996). The levels of etheno adducts in leukocyte DNA were also drastically elevated in women on a high ω–6 polyunsaturated fatty acid (linoleic acid) diet (Nair *et al.*, 1997).

Ultrasensitive detection methods and background levels of etheno–DNA adducts

The evidence summarized in Table 3 indicates that etheno adducts are generated by reaction of DNA bases with lipid peroxidation products, as depicted in Figure 3, involving the following steps: As exemplified for linoleic acid, lipid peroxidation produces fatty acid hydroperoxides and then the reactive *trans*-4-hydroxy-2-nonenal, a major product *in vivo*, which can be oxidized by microsomes of fatty acid hydroperoxides to form its epoxide intermediate. This bifunctional agent attacks the nitrogen atom in DNA bases to form the etheno ring in cytidine, adenosine and guanosine. The three etheno adducts shown in Figure 1 have been detected *in vivo*. We have analysed εdC and εdA by a highly specific, sensitive method involving immunoaffinity chromatography coupled with ^{32}P-postlabelling (IPPA), which was developed by our group (Guichard *et al.*, 1993; Nair *et al.*, 1995). The monoclonal antibodies used were developed in collaboration with Professor M. Rajewsky, University of Essen, Germany. The IPPA method was validated against a radioimmunoassay method to yield a coefficient of variation of 9–26% (Guichard *et al.*, 1996). As its detection limit is 5 etheno adducts per 10^{10} normal nucleotides, it is suitable for analysing background levels in human samples, requiring 25–50 μg DNA.

Use of this method unambiguously and quantitatively revealed the existence of background levels of εdA and εdC in tissues from unexposed rodents and humans (Bartsch *et al.*, 1994; Nair *et al.*, 1995; Guichard *et al.*, 1996; Fernando *et al.*, 1996; Nair *et al.*, 1999). In adult Sprague-Dawley rats fed a standard pelleted diet, background levels were found in all tissues examined, being relatively low in liver (molar ratio of etheno base/10^8 parent bases, 0.02–10) and higher in lungs, kidneys and circulating lymphocytes (molar ratios, 0.7–13) (Guichard *et al.*, 1996). εdA and εdC were also detected in liver and lung DNA from unexposed control mice of various strains at levels of 0.3–1.4 x 10^8 in liver and lungs, depending on the strain. Preliminary data also suggested that the background levels in liver DNA were affected by the type of animal diet: higher levels of εdA and εdC were found in the livers of adult $B6C3F_1$ mice that had been fed a purified diet (AIN76) than in those fed a nonpurified, natural diet (NIH-07). AIN76 has been associated with a higher spontaneous incidence of liver neoplasms in BALB/c mice than the natural diet. This difference has been attributed to cancer-protective antioxidants present in the NIH-07 diet, which could also inhibit lipid peroxidation and formation of endogenous etheno adducts (Fernando *et al.*, 1996). More studies are warranted to elucidate the role of etheno adducts in spontaneous neoplasia of the liver and the protective effect of antioxidants.

Analysis of 19 liver DNA samples from humans with unknown exposure showed the presence of εdA and εdC residues in the range of ≤ 0.05 to 4 adducts per 10^8 parent bases (Nair *et al.*, 1995, 1998b, 1999). Liver DNA samples from two cases of sudden infant death and one of infant primary haemochromatosis had low or undetectable levels of etheno-modified DNA, suggesting that etheno

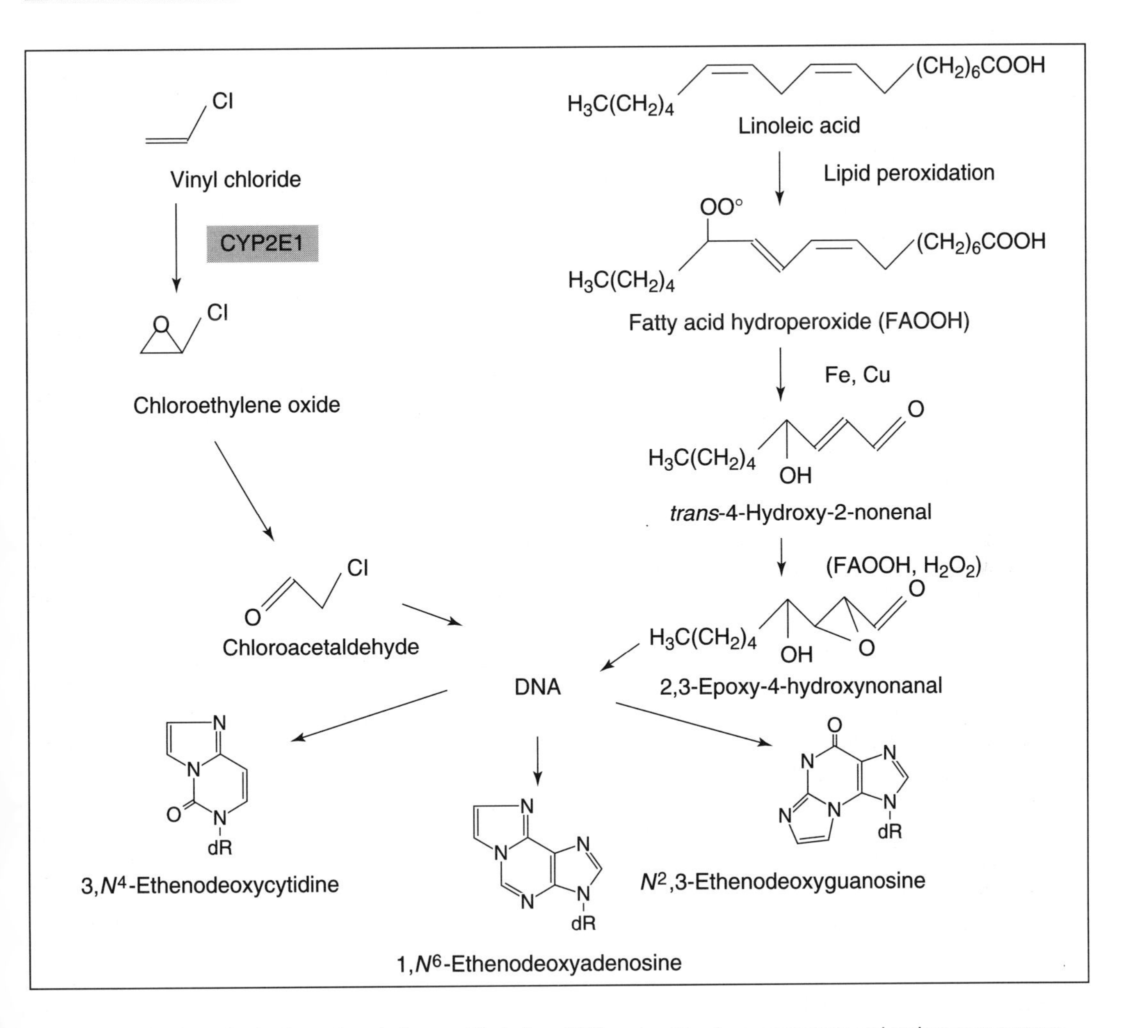

Figure 3. Mechanisms for the formation of etheno adducts from DNA nucleosides from exogenous and endogenous sources, as exemplified for vinyl chloride metabolites and *trans*-4-hydroxy-2-nonenal derived by lipid peroxidation from linoleic acid, an ω–6 polyunsaturated fatty acid

dR, deoxyribose

adducts may accumulate with age and/or increase with long-term exposure to lipid peroxidation products. An age-dependent increase in etheno adduct levels was confirmed in LEC rats (Nair *et al.*, 1996). As in human livers, comparable but variable εdA and εdC levels were detected in DNA of asymptomatic human oesophageal and pancreatic tissues and in leukocytes from male and female volunteers on a low ω–6 polyunsaturated fatty acid diet. Except in one report (Swenberg *et al.*, 1995), the N^2,3-εG levels quantified so far by mass spectrometric methods in unexposed human and rat liver DNA were comparable and only about 1.5 to 2 times higher than the εdA and εdC values detected by IPPA (Yen *et al.*, 1996; Scheller *et al.*, 1995).

Etheno adduct formation as a consequence of overproduction of nitric oxide and chronic inflammatory conditions

Overproduction of nitric oxide is catalysed by inducible nitric oxide synthase under pathological conditions of chronic infection and inflammation, both of which are known risk factors for several human cancers. The DNA damage caused by nitric oxide includes the formation of oxidized DNA bases from OH• and peroxynitrite, the latter being generated by reaction of nitric oxide and superoxide anion when formed concomitantly (Liu & Hotchkiss, 1995; Tamir *et al.*, 1996). Although nitric oxide-associated DNA damage has been hypothesized to occur *in vivo* (Ohshima & Bartsch, 1994; Tamir & Tannenbaum, 1996), much of the experimental evidence has not yet been generated. Peroxynitrite-induced oxidative stress would be expected to induce lipid peroxidation-derived DNA modifications such as etheno and propano adducts, and this was recently shown to occur in a mouse model as a consequence of overproduction of nitric oxide *in vivo* (Nair *et al.*, 1998a).

The toxicological response to excess nitric oxide production was investigated in SJL mice, which produce a high concentration of nitric oxide after injection of RcsX (pre-B lymphoma) cells (Gal *et al.*, 1996). Three groups of mice were used (Figure 4): the first served as controls; in the second group, each mouse received a single intraperitoneal injection of RcsX cells, and the third additionally received N^G-methyl-*L*-arginine, an inhibitor of inducible nitric oxide synthase. After 14 days, organs were collected and spleen DNA was analysed for etheno adducts. Injection of RcsX cells led to a 50-fold increase in nitric oxide production within 14 days, and the etheno adduct levels were six times higher in the spleen DNA of mice that had received RcsX cells than in that of controls, whereas mice initially given the inhibitor of inducible nitric oxide synthase had adduct levels that were not significantly different from those of controls. Surprisingly, no concomitant increase in 7,8-dihydro-8-oxo-2′-deoxyguanosine (8-oxodG) was detected in spleen DNA from the mice given RscX cells, which might have been expected, as peroxynitrite is a strong oxidizing agent. As the rate of degradation of 8-oxodG by peroxynitrite to ring-open products appears to be faster than its rate of generation from guanosine (Uppu *et al.*, 1996; Niles *et al.*, 1998), it may not always be a good marker for oxidative stress-induced DNA damage. When peroxynitrite is involved, more stable secondary oxidation products such as exocyclic adducts appear to be more useful markers.

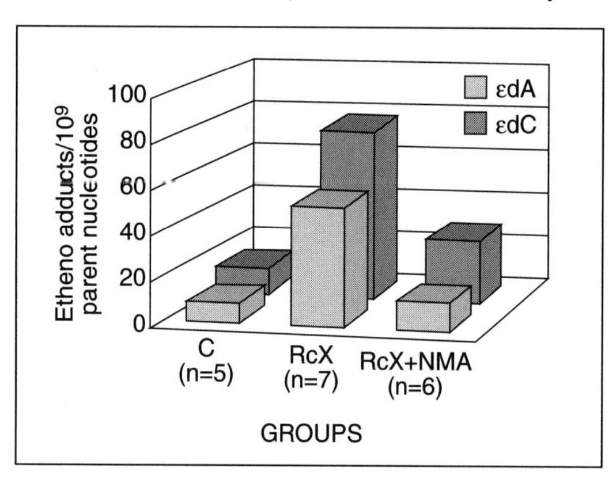

Figure 4. Etheno adducts (εdA and εdC per 10^9 nucleotides) in spleen DNA of SJL mice injected with RcsX cells or RcsX cells plus N^G-methyl-L-arginine (NMA), an inhibitor of inducible nitric oxide synthase

C, control mice; numbers per group in parentheses; data extracted from Nair *et al.* (1998a)

These results demonstrated for the first time that increased formation of etheno–DNA adducts is associated with nitric oxide overproduction *in vivo*. This suggests that exocyclic and oxidative DNA damage play a role in the development of human cancers associated with chronic viral, bacterial and parasitic infections and/or cancers that have an inflammatory component in their etiopathogenesis. Table 5 lists examples of cancer-prone human tissues in which the hypothesis could be tested. Interim results of on-going studies in this area are described briefly below.

The synergistic hepatocarcinogenic effects of hepatitis B viral infection and dietary aflatoxin (AFB_1) are well documented in man. In a study of the interaction of viral and chemical carcinogenesis in a woodchuck model (Bannasch *et al.*, 1995), one group was infected with the woodchuck hepatitis virus, which is closely related to the human virus; a second additionally received AFB_1 in the diet for life, and a third received AFB_1 only. The woodchuck hepatitis virus carriers had a high incidence of hepatocellular carcinoma, and the addition of AFB_1 to the diet

Table 5. Viral, bacterial and parasitic oncogenesis and and inflammation-associated neoplasia

In the affected tissues and cells, persistent oxidative stress or overproduction of nitric oxide could lead to accumulation of exocyclic DNA adducts that act as a driving force to malignancy. This hypothesis, supported by the author's work on hepatitis virus-induced neoplasia and on colonic polyps from patients with familial adenomatous polyposis, can now be tested in a wider range of human neoplasias, as listed

Herpes virus-associated neoplasia
- Burkitt lymphoma
- Nasopharyngeal carcinoma
- Kaposi sarcoma

Retrovirus-associated neoplasia
- Adult T-cell leukaemia/lymphoma

Papillomavirus-associated neoplasia
- Cervical and anal cancers

Hepatitis virus-associated neoplasia
- Hepatocellular carcinoma

Bacteria- and parasite-associated neoplasia
- *Helicobacter pylori* and gastric cancer
- *Salmonella typhimurium/paratyphium* and gall-bladder cancer
- *Opisthorchis viverrini* (liver fluke) and cholangiocarcinoma
- *Schistosoma haematobium* and urinary bladder cancer

Chronic inflammation-associated neoplasia
- Crohn disease and bowel cancer
- Ulcerative colitis and bowel cancer
- Pancreatitis and pancreas cancer
- Cirrhosis and liver cancer
- Asbestosis and lung cancer
- Familial adenomatous polyposis and colon cancer

resulted in an earlier appearance and a higher incidence of liver tumours. No such tumours were found in animals that received only AFB_1 or no treatment. Preliminary analyses of εdA and εdC levels in the liver DNA of groups of two to four animals taken randomly from each of the four groups revealed that only woodchucks with both chronic active hepatitis and hepatocellular carcinoma had increased etheno–DNA modifications in non-tumorous parts of their liver tissue (J. Nair *et al.*, unpublished data). High εdA and εdC levels thus seemed to predispose individual woodchucks to the development of hepatitis and tumours. We have developed an immunoprecipitation-high-performance liquid chromatography–fluorescence method for quantifying εdA in human urine with high sensitivity (Nair *et al.*, 1998c) so that the hypothesis can be tested in humans infected with hepatitis B virus.

A second study relates to patients with familial adenomatous polyposis (FAP). FAP is an autosomal, dominantly inherited disease due to mutations in the *APC* gene, resulting in the development of multiple adenomatous polyps in the colon of affected individuals, who invariably develop carcinomas (Lynch *et al.*, 1992). In such patients, the cyclooxygenase gene *COX-2* is upregulated in colorectal adenoma tissue (Eberhart *et al.*, 1994), and the activity of xanthine oxidase, which generates free radicals, is increased (Spigelman *et al.*, 1995). In a pilot study, we analysed DNA from asymptomatic epithelial colon, colonic polyp tissue from FAP patients and colorectal cancer tissue for etheno adducts. Despite the small number of FAP patients available, we observed a statistically significant (about twofold) increase in εdA and εdC adduct levels in the polyps of FAP patients in comparison with controls, whereas the adduct levels were lower in colon cancer tissue (Schmid *et al.*, in preparation). A plausible explanation for this increase in DNA damage could be the chronic production of lipid peroxidation products in this precancerous lesion by COX-2 and/or xanthine oxidase up-regulation, which may contribute to the progression of polyps to malignancy.

Biological significance of etheno adducts

The hypothesis that etheno adducts play a causal role in the initiation and progression of carcinogenesis is supported by the following observations:

(i) The promutagenic properties of etheno bases were established in studies based on steady-state kinetics or primer extension assays *in vitro* and site-specific mutagenesis in *E. coli* and in mammalian cells (reviewed by Bartsch *et al.*, 1994). The three etheno bases produced mainly base-pair substitution mutations: εdA can lead to AT→GC transitions and to AT→TA and AT→CG transversions (Basu *et al.*, 1993; Pandya & Moriya, 1996); εdC can cause CG→AT transversions and CG→TA transitions (Palejwala *et al.*, 1993; Moriya *et al.*, 1994); and N^2,3-εdG can lead to GC→AT transitions (Cheng *et al.*, 1991). The mutagenic efficiency of εdA and εdC was higher in mammalian (simian kidney) cells than in *E. coli* (Pandya & Moriya, 1996; Shibutani *et al.*, 1996). These promutagenic properties of etheno bases strongly implicate them in the initiation of carcinogenesis by vinyl chloride (Barbin *et al.*, 1981), urethane (ethyl carbamate) (Miller & Miller, 1983) and other etheno adduct-forming chemicals.

(ii) Increased levels of etheno bases have been demonstrated in target tissues for carcinogenesis in rodents exposed to ethyl carbamate and its metabolites or to vinyl chloride (Fernando *et al.*, 1996). After exposure of rats to vinyl chloride, an apparent persistence of these adducts was observed in hepatic DNA. During prolonged exposure, εdA and εdC accumulated in the liver, the major target organ, whereas only εdC accumulated in kidneys and lungs, organs which are less susceptible (Guichard *et al.*, 1996).

(iii) The types of *ras* and *p53* gene base-pair substitutions observed in liver tumours induced by vinyl chloride in rats are consistent with the promutagenic properties of etheno bases (Froment *et al.*, 1994; Barbin *et al.*, 1997). Human angiosarcomas of the liver associated with exposure to vinyl chloride showed a characteristic mutation spectrum in the Ki-*ras* and *p53* genes (Marion *et al.*, 1991; Hollstein *et al.*, 1994). The substitutions found at G:C or A:T base pairs in these tumours were compatible with the involvement of promutagenic etheno–DNA adducts (N^2,3-εdG, εdC, εdA) as the initiating lesions.

(iv) The finding that etheno adducts are recognized by specific repair enzymes further confirms the fact that these genotoxic DNA lesions are recognized by cellular defense mechanisms. Etheno bases are repaired efficiently by mammalian DNA glycosylases *in vitro* (Matijasevic *et al.*, 1992; Dosanjh *et al.*, 1994; Saparbaev *et al.*, 1995; Hang *et al.*, 1996, 1997) and by crude extracts from rat (Oesch *et al.*, 1986) or human (Oesch *et al.*, 1994) cells, εdC being removed by a human mismatch-specific thymine-DNA glycosylase (Hang *et al.*, 1998; Saparbaev & Laval, 1998). Either because of inhibition of these repair enzymes or their lack of expression in certain cell types, some of these miscoding etheno lesions were found to accumulate or persist in the DNA of certain rodent tissues and cells (Swenberg *et al.*, 1992; Guichard *et al.*, 1996).

(v) On the basis of our observations on εdA and εdC levels in experimental and human tumorigenesis in several organs (liver, colon, pancreas and skin investigated so far; the latter two not shown), in which etheno adducts either accumulated with age (LEC rat), were increased at preneoplastic stages (FAP) or paralleled the progression from benign to malignant disease stages (LEC rat, woodchuck liver), we propose the general scheme outlined in Figure 5. Thus, oxidative stress is a major event in malignancy, and etheno adducts are among the key markers for an increasing load of oxidative DNA damage, which, in turn, increases mutation rates, cell proliferation and genomic instability. If true, this scheme could apply to a number of human tumours that are caused by infectious agents or persistent inflammatory processes (listed in Table 5).

(vi) Lastly, the work of Guichard *et al.* (1996; Fernando *et al.*, 1996) indicated the possibility of concurrent formation of etheno bases through lipid peroxidation, either during or after exposure to exogenous carcinogens. This implies that agent-specific DNA adduct formation may occur after exposure to genotoxic carcinogens and, particularly at high doses, may be accompanied by secondary DNA lesions related to induction of oxidative stress or lipid peroxidation which cause genetic alterations resulting in disease progression. This could be one of the reasons why agent-specific genetic fingerprints of the primary carcinogen are sometimes not found in cancer-relevant genes of the tumour but often have a complex mutational profile, possibly caused by modification of DNA bases by oxidation and lipid peroxidation (Hollstein *et al.*, 1998).

Conclusions and perspectives

With the advent of ultrasensitive, specific detection methods for exocyclic DNA adducts, new insights can be gained in experimental and human studies into the mechanisms involved in human cancers with poorly defined etiology. Non-invasive methods for monitoring excreted etheno-deoxynucleosides in humans as a potential marker for lipid peroxidation-related DNA damage occurring in the body, such as the method developed by our group (Nair *et al.*, 1998c; Nair, this volume), will obviously be helpful.

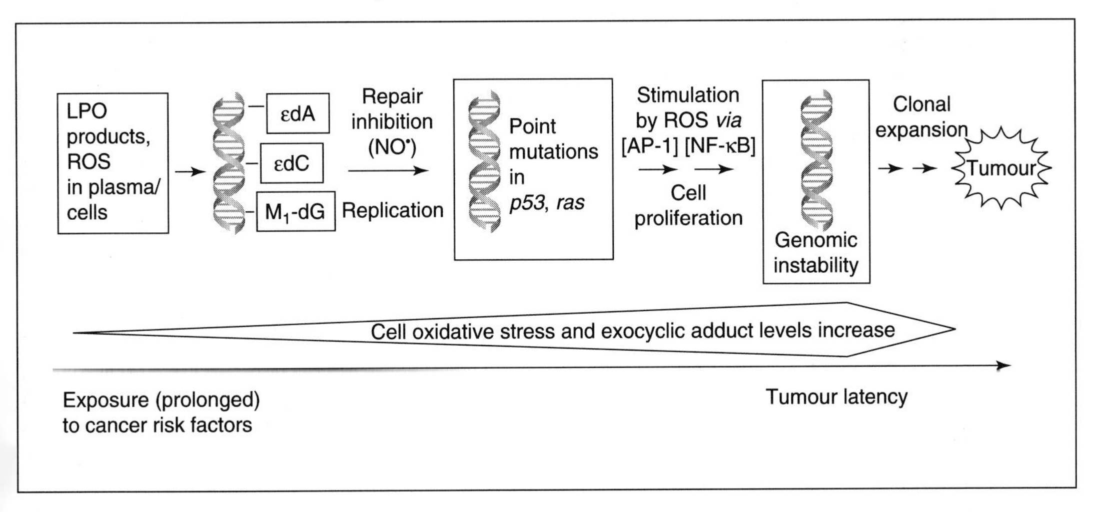

Figure 5. Hypothetical scheme for the role of cumulative oxidative stress in promotion and progression of multistage carcinogenesis, in which exocyclic DNA adducts are lead markers for oxidative DNA-base damage

The model is supported by observations in Long-Evans cinnamon rats and in human tissues including liver and colon from cancer-prone patients; for more details, see text. LPO, lipid peroxidation; ROS, reactive oxygen species; NO, nitric oxide

In closing, I quote a statement by Shapiro (1969), who first reported cyclic nucleic adducts formed from glyoxal: 'As additional reactions of related compounds with nucleic acids are discovered, the biological importance of this class of compounds will undoubtedly grow.' Numerous presentations in this volume give proof that his prediction made 30 years ago was right.

Acknowledgements

The author's research in this area was in part supported by EU contracts ENV4-CT970505 and STEP-CT910145. The author wishes to acknowledge the contributions of J. Nair, A. Barbin, Y. Guichard and K. Schmid and collaboration with P. Bannasch, M. Mutanen, M. Nagao, D. Phillips, S. Tannenbaum, C. Vaca, G. Winde and G. Wogan. The skilled secretarial assistance by G. Bielefeldt is particularly acknowledged.

References

Bannasch, P., Khoshkhou, N.I., Hacker, H.J., Radaeva, S., Mrozek, M., Zillmann, U., Kopp-Schneider, A., Haberkorn, U., Elgas, M., Tolle, T., Roggendorf, M. & Toshkov, I. (1995) Synergistic hepatocarcinogenic effect of hepadnaviral infection and dietary aflatoxin B1 in woodchucks. *Cancer Res.* 55, 3318–3330

Barbin, A. & Bartsch, H. (1986) Mutagenic and promutagenic properties of DNA adducts formed by vinyl chloride metabolites. In: *The Role of Cyclic Nucleic Acid Adducts in Carcinogenesis and Mutagenesis* (Singer, B. & Bartsch, H., eds), IARC Sci. Publ. 70, Lyon, IARC, 345–358

Barbin, A., Brésil, H., Croisy, A., Jacquignon, P., Malaveille, C., Montesano, R. & Bartsch, H. (1975) Liver-microsome-mediated formation of alkylating agents from vinyl bromide and vinyl chloride. *Biochem. Biophys. Res. Commun.* 67, 596–603

Barbin, A., Bartsch, H., Leconte, P. & Radman, M. (1981) Studies on the miscoding properties of 1,N^6-ethenoadenine and 3,N^4-ethenocytosine, DNA reaction products of vinyl chloride metabolites, during in vitro DNA synthesis. *Nucleic Acids Res.* 9, 375–387

Barbin, A., Froment, O., Boivin, S., Mariona, M.-J., Belpoggi, F., Maltoni, C. & Montesano, R. (1997) p53 Gene mutation pattern in rat liver tumours induced by vinyl chloride. *Cancer Res.* 57, 1695–1698

Bartsch, H., Barbin, A., Marion, M.J., Nair, J. & Guichard, Y. (1994) Formation, detection, and role in carcinogenesis of ethenobases in DNA. *Drug Metab. Rev.* 26, 349–371

Basu, A.K., Wood, M.L., Niedernhofer, L.J., Ramos, L.A. & Essigmann, J.M. (1993) Mutagenic and genotoxic effects of three vinyl chloride-induced DNA lesions: 1,N^6-ethenoadenine, 3,N^4-ethenocytosine, and 4-amino-5-(imidazol-2-yl)imidazole. *Biochemistry* 32, 12793–12801

Chaudhary, A.K., Nokubo, M., Reddy, G.R., Yeola, S.N., Morrow, J.D., Blair, I.A. & Marnett, L.J. (1994) Detection of endogenous malondialdehyde-deoxyguanosine adducts in human liver. *Science* 265, 1580–1582

Chen, H.J. & Chung, F.L. (1994) Formation of etheno adducts in reactions of enals via autoxidation. *Chem. Res. Toxicol.* 7, 857–860

Chen, H.-J.C. & Chung, F.-L. (1996) Epoxidation of *trans*-4-hydroxy-2-nonenal by fatty acid hydroperoxides and hydrogen peroxide. *Chem. Res. Toxicol.* 9, 306–312

Cheng, K.C., Preston, B.D., Cahill, D.S., Dosanjh, M.K., Singer, B. & Loeb, L.A. (1991) The vinyl chloride DNA derivative N^2,3-ethenoguanine produces G → A transitions in *Escherichia coli*. *Proc. Natl Acad. Sci. USA* 88, 9974–9978

Chung, F.-L. & Hecht, S.S. (1983) Formation of cyclic 1,N^2-adducts by reaction of deoxyguanosine with α-acetoxy-*N*-nitrosopyrrolidine,4-(carbethoxynitrosamino)butanal and crotonaldehyde. *Cancer Res.* 43, 1230–1235

Chung, F.-L., Young, R. & Hecht, S.S. (1989) Detection of 1,N^2-propandeoxyguanosine adducts in DNA of rats treated with *N*-nitrosopyrrolidine and mice treated with crotonaldehyde. *Carcinogenesis* 10, 1291–1297

Chung, F.-L., Chen, H.-J.C. & Nath, R.G. (1996) Lipid peroxidation as a potential endogenous source for the formation of exocyclic DNA adducts. *Carcinogenesis* 17, 2105–2111

Dosanjh, M.K., Chenna, A., Kim, E., Fraenkel Conrat, H., Samson, L. & Singer, B. (1994) All four known cyclic adducts formed in DNA by the vinyl chloride metabolite chloroacetaldehyde are released by a human DNA glycosylase. *Proc. Natl Acad. Sci. USA* 91, 1024–1028

Eberhart, C.E., Coffey, R.J., Radhika, A., Giardiello, F.M., Ferrenbach, S. & DuBois, R.N. (1994) Up-regulation of cyclooxygenase 2 gene expression in human colorectal adenomas and adenocarcinomas. *Gastroenterology* 107, 1183–1188

Eberle, G., Barbin, A., Laib, R.J., Ciroussel, F., Thomale, J., Bartsch, H. & Rajewsky, M.F. (1989) 1,N^6-Etheno-2′-deoxyadenosine and 3,N^4-etheno-2′-deoxycytidine detected by monoclonal antibodies in lung and liver DNA of rats exposed to vinyl chloride. *Carcinogenesis* 10, 209–212

El Ghissassi, F., Barbin, A., Nair, J. & Bartsch, H. (1995) Formation of 1,N^6-ethenoadenine and 3,N^4-ethenocytosine by lipid peroxidation products and nucleic acid bases. *Chem. Res. Toxicol.* 8, 278–283

Fang, J.-L., Vaca, C.E., Valsta, L.M. & Mutanen, M. (1996) Determination of DNA adducts of malonaldehyde in humans: Effects of dietary fatty acid composition. *Carcinogenesis* 17, 1035–1040

Fedtke, N., Boucheron, J.A., Walker, V.E. & Swenberg, J.A. (1990) Vinyl chloride-induced DNA adducts. II: Formation and persistence of 7-(2′-oxoethyl)guanine and N^2,3-ethenoguanine in rat tissue DNA. *Carcinogenesis* 11, 1287–1292

Fernando, R.C., Nair, J., Barbin, A., Miller, J.A. & Bartsch, H. (1996) Detection of 1,N^6-ethenodeoxyadenosine and 3,N^4-ethenodeoxycytidine by immunoaffinity/^{32}P-postlabelling in liver and lung DNA of mice treated with ethyl carbamate (urethane) or its metabolites. *Carcinogenesis* 17, 1711–1718

Froment, O., Boivin, S., Barbin, A., Bancel, B., Trépo, C. & Marion, M.J. (1994) Mutagenesis of *ras* proto-oncogenes in rat liver tumours induced by vinyl chloride. *Cancer Res.* 54, 5340–5345

Gal, A., Tamir, S., Tannenbaum, S.R. & Wogan, G.N. (1996) Nitric oxide production in SJL mice bearing the RcsX lymphoma: A model for in vivo toxicological evaluation of NO. *Proc. Natl Acad. Sci. USA* 93, 11499–11503

Green, T. & Hathway, D.E. (1978) Interactions of vinyl chloride with rat liver DNA *in vivo*. *Chem.-biol. Interactions* 22, 211–224

Guengerich, F.P. (1992) Roles of vinyl chloride oxidation products 2-chlorooxirane and 2-chloroacetaldehyde in the in-vitro formation of etheno adduct of nucleic acid bases. *Chem. Res. Toxicol.* 5, 2–5

Guengerich, F.P. & Kim, D.H. (1991) Enzymatic oxidation of ethyl carbamate to vinyl carbamate and its role as an intermediate in the formation of 1,N^6-ethenoadenosine. *Chem. Res. Toxicol.* 4 413–421

Guengerich, F.P., Mason, P.S., Stott, W.T., Fox, T.R. & Watanabe, P.G. (1981) Role of 2-haloethylene oxides and 2-haloacetaldehydes derived from vinyl bromide and vinyl chloride in irreversible binding to protein and DNA. *Cancer Res.* 41, 4391–4398

Guichard, Y., Nair, J., Barbin, A. & Bartsch, H. (1993) Immunoaffinity clean-up combined with ^{32}P-postlabelling analysis of 1,N^6-ethenoadenine and 3,N^4-ethenocytosine in DNA. In: *Postlabelling Methods for Detection of DNA Adducts* (Phillips, D.H., Castegnaro, M. & Bartsch, H., eds), IARC Sci. Publ. 124, Lyon, IARC, 263–269

Guichard, Y., El Ghissassi, F., Nair, J., Bartsch, H. & Barbin, A. (1996) Formation and accumulation of DNA ethenobases in adult Sprague-Dawley rats exposed to vinyl chloride. *Carcinogenesis* 17, 1553–1559

Habraken, Y., Carter, C.A., Sekiguchi, M. & Ludlum D.B. (1991) Release of N^2,3-ethanoguanine from haloethylnitrosourea-treated DNA by *Escherichia coli* 3-methyladenine DNA glycosylase II. *Carcinogenesis* 12, 1971–1973

Hall, J.A., Saffhill, R., Green, T. & Hathway, D.E. (1981) The induction of errors during in vitro DNA synthesis following chloroacetaldehyde-treatment of poly(dA-dT) and poly(dC-dG) templates. *Carcinogenesis* 2, 141–146

Hang, B., Chenna, A., Rao, S. & Singer, B. (1996) 1,N^6-Ethenoadenine and 3,N^4-ethenocytosine are excised by separate human DNA glycosylases. *Carcinogenesis* 17, 155–157

Hang, B., Singer, B., Margison, G.P. & Elder, R.H. (1997) Targeted deletion of alkylpurine-DNA-*N*-glycosylase in mice eliminates repair of 1,N^6-ethenoadenine and hypoxanthine but not 3,N^4-ethenocytosine or 8-oxoguanine. *Proc. Natl Acad. Sci. USA* 94, 12869–12874

Hang, B., Medina, M., Fraenkel-Conrat, H. & Singer, B. (1998) A 55-kDa protein isolated from human cells shows DNA glycosylase activity toward 3,N^4-ethenocytosine and G/T mismatch. *Proc. Natl Acad. Sci. USA* 95, 13561–13566

Hollstein, M., Marion, M.J., Lehman, T., Welsh, J., Harris, C.C., Martel Planche, G., Kusters, I. & Montesano, R. (1994) p53 Mutations at A:T base pairs in angiosarcomas of vinyl chloride-exposed factory workers. *Carcinogenesis* 15, 1–3

Hollstein, M., Moeckel, G., Hergenhahn, M., Spiegelhalder, B., Keil, M., Werle-Schneider, G., Bartsch. H. & Brickmann, J. (1998) On the origins of tumor mutations in cancer genes: Insights from the *p53* gene. *Mutat. Res.* 405, 145–154

IARC (1979) *Some Monomers, Plastics and Synthetic Elastomers, and Acrolein.* IARC Monographs on the Evaluation of Carcinogenic Risks to Humans, Vol. 19. Lyon, 377–438

Kochetkov, N.K., Shibaev, V.N. & Kost, A.A. (1971) New reaction of adenine and cytosine derivatives, potentially useful for nucleic acid modifications. *Tetrahedron Lett.* 22, 1993–1996

Laib, R.J. & Bolt, H.M. (1977) Alkylation of RNA by vinyl chloride metabolites *in vitro* and *in vivo*: Formation of 1,N^6-ethenoadenosine. *Toxicology* 8, 185–195

Laib, R.J. & Bolt, H.M. (1978) Formation of 3,N^4-ethenocytidine moieties in RNA by vinyl chloride metabolites *in vitro* and *in vivo. Arch. Toxicol.* 39, 235–240

Leonard, N.J. (1984) Etheno-substituted nucleotides and coenzymes: Fluorescence and biological activity. *CRC Crit. Rev. Biochem.* 15, 125–199

Leonard, N.J. (1992) Etheno-bridged nucleotides in structural diagnosis and carcinogenesis. *Biochem. Mol. Biol.* 3, 273–297

Liu, R.H. & Hotchkiss, J.H. (1995) Potential genotoxicity of chronically elevated nitric oxide: A review. *Mutat. Res.* 339, 73–89

Lynch, H.T., Watson, P., Smyrk, T.C., Lanspa, S.J., Borman, B.M., Boland, C.R., Lynch, J.F. Cavalieri, R.J., Leppert, M., White, R., Sidransky, D. & Vogelstein, B. (1992) Colon cancer genetics. *Cancer* 70, 1300–1312

Marion, M.J., Froment, O. & Trépo, C. (1991) Activation of Ki-ras gene by point mutation in human liver angiosarcoma associated with vinyl chloride exposure. *Mol. Carcinog.* 4, 450–454

Marnett, L.J. & Burcham, P.C. (1993) Endogenous DNA adducts: Potential and paradox. *Chem. Res. Toxicol.* 6, 778–785

Matijasevic, Z., Sekiguchi, M. & Ludlum, D.B. (1992) Release of N^2,3-ethenoguanine from chloroacetaldehyde-treated DNA by *Escherichia coli* 3-methyladenine DNA glycosylase II. *Proc. Natl Acad. Sci. USA* 89, 9331–9334

Miller, J.A. & Miller, E.C. (1983) The metabolic activation and nucleic acid adducts of naturally-occurring carcinogens: Recent results with ethyl carbamate and the spice flavors safrole and estragole. *Br. J. Cancer* 48, 1–15

Moriya, M., Zhang, W., Johnson, F. & Grollman, A.P. (1994) Mutagenic potency of exocyclic DNA adducts: Marked differences between *Escherichia coli* and simian kidney cells. *Proc. Natl Acad. Sci. USA* 91, 11899–11903

Nair, J., Barbin, A., Guichard, Y. & Bartsch, H. (1995) 1,N^6-Ethenodeoxyadenosine and 3,N^4-ethenodeoxycytine in liver DNA from humans and untreated rodents detected by immunoaffinity/^{32}P-postlabeling. *Carcinogenesis* 16, 613–617

Nair, J., Sone, H., Nagao, M., Barbin, A. & Bartsch, H. (1996) Copper-dependent formation of miscoding etheno-DNA adducts in the liver of Long Evans cinnamon (LEC) rats developing hereditary hepatitis and hepatocellular carcinoma. *Cancer Res.* 56, 1267–1271

Nair, J., Vaca, C.E., Velic, I., Mutanen, M., Valsta, L.M. & Bartsch, H. (1997) High dietary ω–6 polyunsaturated fatty acids drastically increase the formation of etheno-DNA base adducts in white blood cells of female subjects. *Cancer Epidemiol. Biomarkers Prev.* 6, 597–601

Nair, J., Gal, A., Tamir, S., Tannenbaum, S., Wogan, G. & Bartsch, H. (1998a) Etheno adducts in spleen DNA of SJL mice stimulated to overproduce nitric oxide. *Carcinogenesis* 19, 2081–2084

Nair, J., Carmichael, P.L., Fernando, R.C., Phillips, D.H., Strain, A.J. & Bartsch, H. (1998b) Lipid peroxidation-induced etheno-DNA adducts in liver of patients with the genetic metal storage disorders Wilson's disease and primary hemochromatosis. *Cancer Epidemiol. Biomarkers Prev.* 7, 435–440

Nair, J., Hofmann, I. & Bartsch, H. (1998c) Detection of 1,N^6-ethenodeoxyadenosine in human urine by immunoaffinity–HPLC–fluorescence. *Proc. AACR* 39, 333

Nair, J., Barbin, A., Velic, I. & Bartsch, H. (1999) Etheno DNA-base adducts from endogenous reactive species. *Mutat. Res.* 124, 59–69

Nath, R.G. & Chung, F.-L. (1994) Detection of exocyclic 1,N^2-propanodeoxyguanosine adducts as common DNA lesions in rodents and humans. *Proc. Natl Acad. Sci. USA* 91, 7491–7495

Nath, R.G., Chen, H.-J.C., Nishikawa, A., Young-Sciame, R. & Chung, F.-L. (1994) A ^{32}P-postlabeling method for simultaneous detection and quantification of exocyclic etheno and propano adducts in DNA. *Carcinogenesis* 15, 979–984

Nath, R.G., Ocando, J.E. & Chung, F.-L. (1996) Detection of 1,N^2-propanodeoxyguanosine adducts as potential endogenous DNA lesions in rodent and human tissues. *Cancer Res.* 56, 452–456

Niles, J.C., Burney, S., Wisnok, J.S. & Tannenbaum, S.R. (1998) Characterization of the products from the reaction of 8-oxodeoxyguanosine with peroxynitrite. *Proc. AACR* 39, 372

Oesch, F., Adler, S., Rettelbach, R. & Doerjer, G. (1986) Repair of etheno DNA adducts by *N*-glycosylases. In: *The Role of Cyclic Nucleic Acid Adducts in Carcinogenesis and Mutagenesis* (B. Singer & H. Bartsch, eds.) IARC. Sci. Publ. 70. Lyon, IARC, 373–379.

Oesch, F., Weiss, C.M. & Klein, S. (1994) Use of oligonucleotides containing ethenoadenine to study the repair of this DNA lesion. Determination of individual and collective repair activities. *Arch. Toxicol.* 68, 358–363

Ohshima, H. & Bartsch, H. (1994) Chronic infections and inflammatory processes as cancer risk factors: Possible role of nitric oxide in carcinogenesis. *Mutat. Res.* 305, 253–264

Palejwala, V.A., Rzepka, R.W., Simha, D. & Humayun, M.Z. (1993) Quantitative multiplex sequence analysis of mutational hot spots. Frequency and specificity of mutations induced by a site-specific ethenocytosine in M13 viral DNA. *Biochemistry* 32, 4105–4111

Pandya, G.A. & Moriya, M. (1996) 1,N^6-Ethenodeoxyadenosine, a DNA adduct highly mutagenic in mammalian cells. *Biochemistry* 35, 11487–11492

Ribovich, M.L., Miller, J.A., Miller, E.C. & Timmins, L.G. (1982) Labeled 1,N^6-ethenoadenosine and 3,N^4-ethenocytidine in hepatic RNA of mice given [ethyl-1,2-^{3}H]- or [ethyl-1-^{14}C]-ethyl carbamate (urethane). *Carcinogenesis* 3, 539–546

Saparbaev, M. & Laval, J. (1998) 3,N^4-Ethenocytosine, a highly mutagenic adduct, as a primary substrate for *Escherichia coli* double-strand uracil-DNA glycosylase and human mismatch-specific thymine–DNA glycosylase. *Proc. Natl Acad. Sci. USA* 95, 8508–8513

Saparbaev, M., Kleibl, K. & Laval, J. (1995) *Escherichia coli, Saccharomyces cerevisiae,* rat and human 3-methyladenine DNA glycosylases repair 1,N^6-ethenoadenine when present in DNA. *Nucleic Acids Res.* 23, 3750–3755

Scheller, N., Sangaiah, R., Ranasinghe, A., Amarnath, V., Gold, A. & Swenberg, A. (1995) Synthesis of (4,5,6,8-$^{13}C_4$) guanine, a reagent for the production of internal standards of guanyl DNA adducts. *Chem. Res. Toxicol.* 8, 333–337

Shapiro, R. (1969) Reactions with purines and pyrimidines. *Ann. N.Y. Acad. Sci.* 163, 624–630

Shibutani, S., Suzuki, N., Matsumoto, Y. & Grollman, A.P. (1996) Miscoding properties of 3,N^4-etheno-2′-deoxycytidine in reactions catalyzed by mammalian DNA polymerases. *Biochemistry* 35, 14992–14998

Sodum, R.S. & Chung, F.L. (1988) 1,N^2-Ethenodeoxyguanosine as a potential marker for DNA adduct formation by *trans*-4-hydroxy-2-nonenal. *Cancer Res.* 48, 320–323

Sodum, R.S. & Chung, F.-L. (1989) Structural characterization of adducts formed in the reaction of 2,3-epoxy-4-hydroxynonanal with deoxyguanosine. *Chem. Res. Toxicol.* 2, 23–28

Sodum, R.S. & Chung, F.L. (1991) Stereoselective formation of *in vitro* nucleic acid adducts by 2,3-epoxy-4-hydroxynonanal. *Cancer Res.* 51, 137–143

Spengler, S. & Singer, B. (1981) Transcriptional errors and ambiguity resulting from the presence of 1,N^6-ethenoadenosine or 3,N^4-ethenocytidine in polyribonucleotides. *Nucleic Acids Res.* 9, 365–373

Spigelman, A.D., Farmer, K.C.R., Oliver, S., Nugent, K.P., Bennett, P.N., Notarianni, L.J., Dobrocky, P. & Phillips, R.K.S. (1995) Caffeine phenotyping of cytochrome P4501A2, N-acetyltransferase, and xanthine oxidase in patients with familial adenomatous polyposis. *Gut* 36, 251–254

Swenberg, J.A., Fedtke, N., Cirousel, F., Barbin, A. & Bartsch, H. (1992) Etheno adducts formed in DNA of vinyl chloride exposed rats are highly persistent in liver. *Carcinogenesis* 13, 727–729

Swenberg, J.A., La, D.K., Scheller, N.A. & Wu, K-Y. (1995) Dose–response relationships for carcinogens. *Toxicol. Lett.* 82/83, 751–756

Tamir, S. & Tannenbaum, S.R. (1996) The role of nitric oxide (NO) in the carcinogenic process. *Biochim. Biophys. Acta* 1288, F31–36

Tamir, S., Burney, S. & Tannenbaum, S.R. (1996) DNA damage by nitric oxide. *Chem. Res. Toxicol.* 9, 821–827

Uppu, R.M., Cueto, R., Squadrito, G.L., Salgo, M.G. & Pryor, W.A. (1996) Competitive reactions of peroxynitrite with 2′-deoxyguanosine and 7,8-dihydro-8-oxo-2′-deoxyguanosine (8-oxodG): Relevance to the formation of 8-oxodG in DNA exposed to peroxynitrite. *Free Radical Biol. Med.* 21, 407–411

Vaca, C.E., Fang, J.-L., Mutanen, M. & Valsta, L. (1995) ^{32}P-Postlabelling determination of DNA adducts of malonaldehyde in humans: Total white blood cells and breast tissue. *Carcinogenesis* 16, 1847–1851

Wang, M., Dhingra, K., Hittelman, W.N., Liehr, J.G., de Andrade, M. & Li, D. (1996) Lipid peroxidation-induced putative malondialdehyde–DNA adducts in human breast tissues. *Cancer Epidemiol. Biomarkers Prev.* 5, 705–710

Winter, C.K., Segall, H.J. & Haddon, W.F. (1986) Formation of cyclic adducts of deoxyguanosine with the aldehydes *trans*-4-hydroxy-2-hexenal and *trans*-4-hydroxy-2-nonenal *in vitro*. *Cancer Res.* 46, 5682–5686

Yen, T.Y., Christova-Gueoguieva, N.I., Scheller, N., Holt, S., Swenberg, J.A. & Charles, M.J. (1996) Quantitative analysis of DNA adduct N^2,3-ethenoguanine using liquid chromatography/electrospray ionisation mass spectrometry. *J. Mass Spectrom.* 311, 1271–1276

Exocyclic DNA Adducts in Mutagenesis and Carcinogenesis
B. Singer & H. Bartsch
IARC Scientific Publications No. 150
International Agency for Research on Cancer, Lyon, 1999

Chapter I. Occurrence and methods of detection

Chemistry and biology of DNA damage by malondialdehyde

L.J. Marnett

Malondialdehyde is a naturally occurring product of lipid peroxidation and prostaglandin biosynthesis which is mutagenic and carcinogenic. It reacts with DNA to form adducts to deoxyguanosine and deoxyadenosine. The major adduct to DNA is a pyrimidopurinone called pyrimido[1,2-*a*]purin-10(3*H*)-one (M_1G). Studies of site-specific mutagenesis indicate that M_1G is mutagenic in bacteria and is repaired by the nucleotide excision repair pathway. M_1G has been detected in liver, leukocytes, pancreas and breast from healthy human beings at levels ranging from 1 to 120 per 10^8 nucleotides. Several assays for M_1G have been described which are based on mass spectrometry, ^{32}P-postlabelling or immunochemical techniques. M_1G appears to be a major endogenous DNA adduct in human beings that may contribute significantly to cancer linked to lifestyle and dietary factors. Recent advances in the chemistry and biology of M_1G are reviewed.

Introduction

Lipid peroxidation generates a complex variety of products, many of which are reactive electrophiles. Some of these react with protein and DNA and, as a result, are toxic and mutagenic. Maldondialdehyde (MDA) appears to be the most mutagenic product of lipid peroxidation, whereas 4-hydroxynonenal is the most toxic (Figure 1) (Esterbauer *et al.*, 1990). Exocyclic deoxyguanosine adducts derived from 4-hydroxynonenal are discussed elsewhere in this volume. As MDA is also produced by the breakdown of prostaglandin endoperoxides, its formation is tied to enzymatic lipid peroxidation catalysed by cyclooxygenases (Diczfalusy *et al.*, 1977).

A third pathway has been described recently for the formation of aldehydes that is independent of lipid peroxidation but which may be important in the generation of genotoxic compounds. Hazen *et al.* (1998) have demonstrated that myeloperoxidase in the presence of chlorine generates hypochlorous acid, which oxidizes amino acids to chloramines. The latter derivatives lose chloride and carbon dioxide to form imines; the imines are hydrolysed to aldehydes (Figure 2). In the case of threonine, the hydroxyaldehyde product undergoes dehydration to form acrolein. This may be an important source of the acrolein that damages DNA endogenously in human beings and which has been assumed to arise from lipid peroxidation (Nath & Chung, 1994).

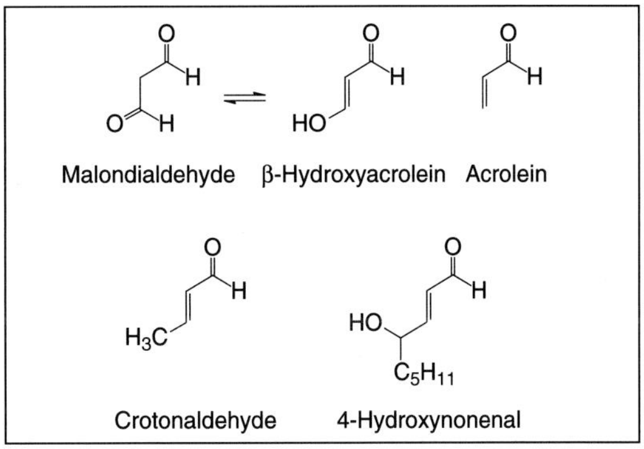

Malondialdehyde β-Hydroxyacrolein Acrolein

Crotonaldehyde 4-Hydroxynonenal

Figure 1. Structures of aldehydic products of lipid peroxidation

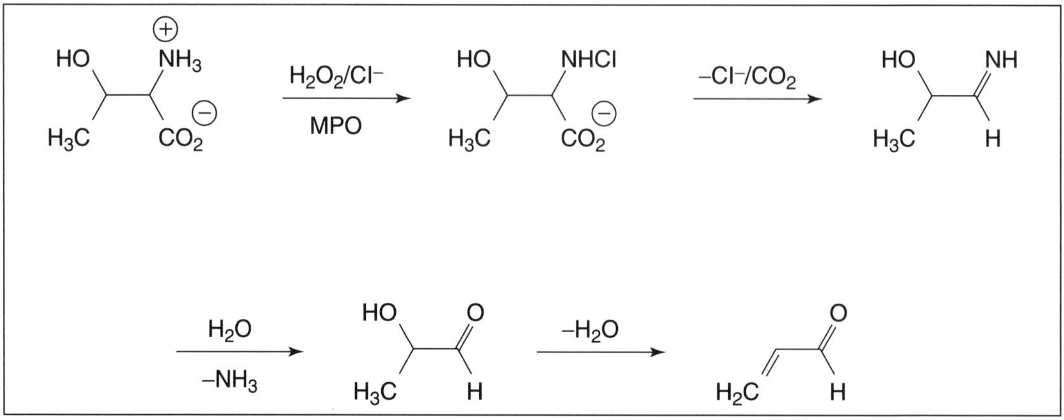

Figure 2. Oxidation of amino acids to aldehydes by myeloperoxidase

Genotoxicity of malondialdehyde

Mukai and Goldstein (1976) reported the mutagenic activity of MDA towards *Salmonella typimurium* in 1976. Since MDA is unstable, the material used for the assay was prepared by hydrolysis of tetraethoxypropane. Some impurities generated during tetraethoxypropane hydrolysis are nearly 30-fold more mutagenic than MDA and account for a significant part of the activity (~50%) detected in the assay (Marnett & Tuttle, 1980); however, assay of highly purified MDA prepared by three independent means revealed that this compound is, indeed, mutagenic (Basu & Marnett, 1983). The carcinogenic activity of MDA was established in a two-year bioassay in rodents (Spalding, 1988), in which highly purified preparations of MDA induced thyroid tumours in rats but not mice.

Several other aldehydes, including some α,β-unsaturated aldehydes are mutagenic in *Salmonella* if steps are taken to minimize their toxicity (Marnett *et al.*, 1985). These consist of adding high concentrations of glutathione shortly after administration of the aldehyde to the bacteria to strip aldehydes from cysteinyl groups of proteins, which are putative targets for their toxic action. In fact, 4-hydroxynonenal was originally identified as a highly toxic product of lipid peroxidation, and both it and its glutathione conjugate were used as chemotherapeutic agents in some small clinical trials in humans (Schauenstein *et al.*, 1977). The toxic activity of the 4-hydroxynonenal appears to be due to its reactivity with sulfhydryl proteins (Krokan *et al.*, 1985).

Reaction of malondialdehyde with DNA

MDA reacts with nucleic acid bases to form multiple adducts (Figure 3). Both of its carbonyl equivalents adduct to N^2 and N1 of deoxyguanosine (dG) with loss of two water molecules to form a pyrimidopurinone (Seto *et al.*, 1983; Marnett *et al.*, 1986). This compound is a planar aromatic molecule that is moderately fluorescent (~3% quantum yield). The condensation products with deoxyadenosine (dA) and deoxycytidine (dC) arise by addition of one of the carbonyl equivalents of MDA with the exocyclic amino groups to form an oxopropenyl deriviative (Nair *et al.*, 1984; Stone *et al.*, 1990a,b). No evidence for cyclization of these products has been obtained. Comparison of the yields of the various adducts produced in the reaction of MDA with DNA *in vitro* indicates that M_1G is the major adduct, followed by N^6-(3-oxopropenyl)deoxyadenosine (M_1dA). The amount of M_1G is approximately five times that of M_1dA. N^4-(3-Oxopropenyl)deoxycytidine (M_1dC) is formed in only trace amounts (Chaudhary *et al.*, 1996).

A complicating factor in the reaction of MDA with nucleosides is its polymerization to form dimers and trimers that also react with DNA. The dimer of MDA reacts with dG to form an oxadiazabicyclo[3:3:1]nonene derivative, whereas the trimer of MDA reacts with dA and dC to form (5*,7*-bisformyl-2**H*-3*,6*-dihydro-2*,6*-methano-1*,3*-oxazocin-3*-yl) derivatives of the exocyclic amino groups (Stone *et al.*, 1990a,b). The oligomerization of MDA is relatively slow at neutral pH, so the monomeric adducts depicted in

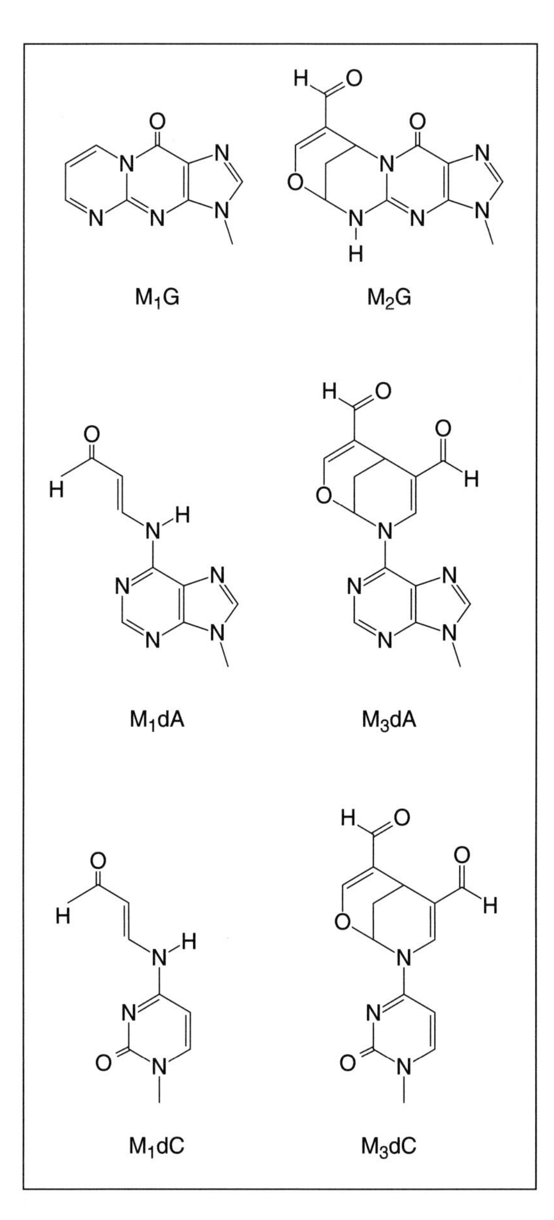

Figure 3. Malondialdehyde–DNA adducts

Figure 3 are the major products generated under physiological conditions. There may be conditions *in vitro* or possibly *in vivo*, however, in which these unusual multimeric adducts are formed. The ability of MDA to form monomeric and oligomeric adducts is a reflection of the nucleophilic reactivity of its enol functional group, whereas the electrophilic reactivity is due to its aldehyde functional group. The existence of oligomeric deoxynucleoside adducts necessitated the adoption of abbreviations for the individual adducts that designate the number of MDA units in each adduct (Figure 3).

Alternative routes to M_1G

Dedon *et al.* (1998) have recently demonstrated that M_1G (and presumably other MDA-derived adducts) can be formed independently of lipid peroxidation. Direct oxidation of DNA by agents that abstract the 4′ hydrogen atom of the sugar backbone (e.g. calicheamicin and bleomycin) initiates a cascade of reactions that lead to the formation of base propenals. Base propenals are oxopropenyl base derivatives that are structurally analogous to acroleins substituted at the β position with good-to-moderate leaving groups (Figure 4). These compounds (including M_1dA and M_1dC) transfer their oxopropenyl group to dG to form M_1G. Treatment of DNA with bleomycin or calicheamicin in the complete absence of lipid leads to the formation of

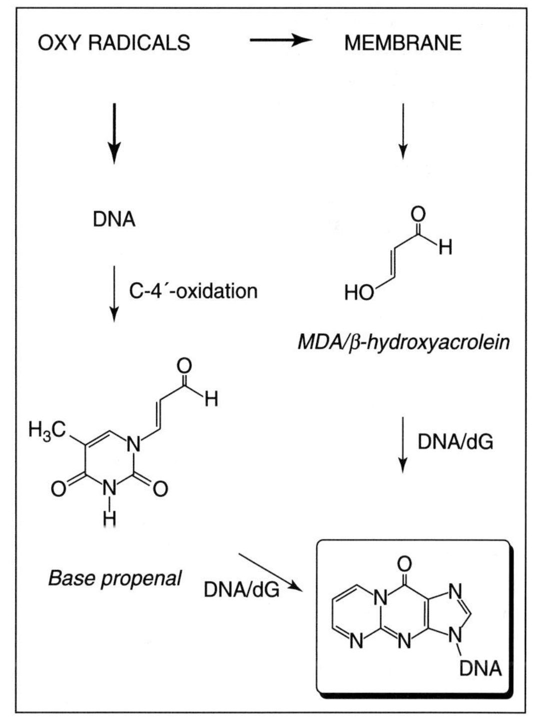

Figure 4. Reaction of oxidizing agents with DNA to form M1G

MDA, malondialdehyde; dG, deoxyguanosine

M_1G. M_1G is also formed by the direct reaction of adenine propenal with DNA. This provides another pathway linking oxidative stress to the formation of M_1G in DNA and may explain the occurrence of M_1G at high levels in certain human tissues (see below).

Mutagenic potential and repair of malondialdehyde–DNA adducts

The mutagenic potential of MDA–DNA adducts has been evaluated by random and site-specific approaches. Benamira *et al.* (1995) reacted MDA at neutral pH with a single-stranded M13 vector containing the *lacZα* gene and scored mutations to the *lacZα⁻* phenotype after replication in strains of *Escherichia coli* induced for the SOS response. Increasing concentrations of MDA led to an increase in the number of *lacZα⁻* mutations, coincident with an increase in the level of the major MDA–dG adduct, M_1G. The most common sequence changes induced by MDA were base-pair substitutions (76%). Of these, 29/68 were transversions, most of which were G→T (24/29). Transitions accounted for 39/68 of the base-pair substitutions and were comprised exclusively of C→T (22/39) and A→G (17/39). Frameshift mutations were identified in 16% of the induced mutants and were mainly single-base additions occurring in runs of reiterated bases (11/14). If one assumes that all mutations are targeted to the site of adduction, the ability of MDA to induce base-pair substitution mutations at dG, dA and dC residues coincides with its ability to form adducts at all three deoxynucleosides. No mutations were detected at dT residues, as expected from the fact that MDA does not form detectable adducts to dT.

The mutagenic potential of M_1G has been studied in site-specific experiments, which necessitated modification of the synthetic reactions used for oligodeoxynucleotide synthesis to accommodate the instability of M_1G to the base 'de-protection' conditions usually used in the last step of the synthetic pathway (Reddy & Marnett, 1995). M_1G was positioned at a defined site in a duplex M13 genome by a gapped duplex approach. Unmodified and M_1G-modified genomes containing either a cytosine or thymine opposite the lesion were transformed into repair-proficient and repair-deficient *E. coli* strains. Base-pair substitutions were quantified by hybridization analysis. Modified genomes containing a cytosine opposite M_1G resulted in roughly equal numbers of M_1G→A and M_1G→T mutations, with a few M_1G→C mutations (Fink *et al.*, 1997). The total mutation frequency was approximately 1%, which represents a 500-fold increase in mutations over that with unmodified M13MB102. Transformation of modified genomes containing a thymine opposite M_1G allowed an estimate to be made of the ability of M_1G to block replication. The (–)-strand was replicated > 80% of the time in the unadducted genome but only 20% of the time when M_1G was present. Correction of the mutation frequency for the strand bias of replication indicated that the actual frequency of mutations induced by M_1G was 18% (Fink *et al.*, 1997). The spectrum and frequency of mutations induced by M_1G were very similar to those observed with a structural analogue, 1,N^2-propanodeoxyguanosine (PdG), assayed in the same system.

Repair of M_1G was assessed by transformation of M_1G-containing genomes into *E. coli* strains deficient in individual genes of DNA repair on the basis of the assumption that inactivation of a repair gene will increase the half-life of the adduct and increase its mutagenicity in site-specific experiments. No effect on mutation frequency was observed when the cells were defective in formamidopyrimidine glycosylase or 3-methyladenine glycosylase. These two glycosylases have been shown previously to participate in the repair of 8-oxodG and exocyclic adducts, respectively (Fink *et al.*, 1997; Johnson *et al.*, 1997). Approximately a fourfold increase in mutation frequency was observed when M_1G-containing genomes were transformed into cells deficient in nucleotide excision repair (either *uvrA⁻* or *uvrB⁻*) (Fink *et al.*, 1997). Increases in mutation frequency of similar magnitude were observed with PdG-containing genomes. The ability of PdG (and by analogy M_1G) to be removed by reconstituted bacterial and mammalian nucleoside excision repair complexes was confirmed by a series of experiments *in vitro* (Johnson *et al.*, 1997). PdG appears to be removed more efficiently by the mammalian nucleotide excision repair complex than the bacterial complex.

Occurrence of M_1G in genomic DNA

Postlabelling analysis

The high mutagenic potential of M_1G begs the question of whether it is formed in human beings.

As for all endogenous DNA adducts, answering this question requires unusually robust and rigorous assays because of the lack of control DNA from an untreated animal or human sample. By definition, it is untreated DNA that is analysed for 'background DNA damage'. Thus, any assay used must be both highly sensitive and very specific. It should allow verification that the adduct detected is actually what it is presumed to be. This is no simple task for any DNA adduct.

^{32}P-Postlabelling was first used for the detection and quantification of M_1G. Vaca and coworkers (1992) developed a method involving nuclease P1 treatment for enrichment to detect the adduct in mouse liver DNA. Treatment of mice with MDA (20 μmol/kg body weight) caused an increase in adduct level to $6/10^6$ nucleotides. The levels of M_1G in untreated animals were not reported. In subsequent studies, treatment of mice with 200 μmol/kg body weight MDA stimulated the formation of MDA–haemoglobin adducts in blood and M_1G in liver. The levels of M_1G declined slowly after MDA treatment, with a half-life of 12.5 days (Kautiainen *et al.*, 1993). This slow half-life for removal is interesting considering the efficient removal of M_1G and PdG by nucleotide excision repair, described above. Interestingly, in these experiments, the levels of M_1G in the liver increased significantly up to four days after administration of MDA.

Wang and Liehr 1995a) used ^{32}P-postlabelling to demonstrate a two- to threefold increase in the levels of M_1G in the kidneys of Syrian hamsters treated with doses of diethylstilboestrol or oestradiol that induce cancer in this organ. Wang and Liehr (1995b) also provided evidence for the existence of MDA adducts to dA as well as dG. The identities of these adducts have not yet been established, but they have been found in human leukocytes and breast cells. A value in leukocytes of 2.6 ± 1.2 adducts per 10^7 nucleotides (range, 1–5 per 10^7) was reported from a study of 26 male and female volunteers (Vaca *et al.*, 1995), and a parallel analysis of seven female volunteers indicated a level of 3.0 ± 1.3 M_1G per 10^7 nucleotides in normal breast tissue. Fang *et al.* (1996) reported a dramatic increase in the level of M_1G in the leukocytes of women (but not men) fed a diet with a high concentration of sunflower oil over that in a group fed a diet with a high concentration of rapeseed oil. The diet with high polyunsaturated fatty acid stimulated a nearly 20-fold increase in the level of M_1G, up to 28 per 10^7 nucleotides. Wang *et al.* (1996) reported that the levels of M_1G and putative MDA–dA adducts were two- to threefold higher in the normal breast tissue of women with breast cancer than in the normal tissue of women without breast cancer. The levels of M_1G in the tumour tissue of women with breast cancer were lower than those in the surrounding normal tissue.

Mass spectrometric analysis

In our laboratory, we have used gas chromatography–mass spectrometry (GC–MS) with electron capture negative chemical ionization detection (GC/EC NCI/MS) to quantify M_1G (Chaudhary *et al.*, 1994; Rouzer *et al.*, 1997). Figure 5 charts the

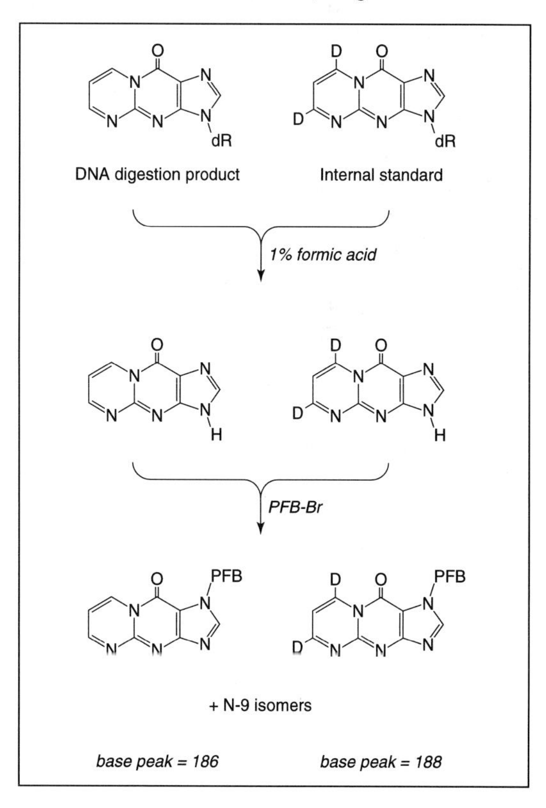

Figure 5. Steps in the analysis of M_1G by gas chromatography–mass spectrometry with electron capture negative chemical ionization detection

PBR-Br, pentafluorobenzyl bromide

course of the analytical method. Nuclear DNA is isolated, purified and digested to deoxynucleosides. During nuclease treatment, an internal standard of M_1G-deoxyribose labelled with two deuteriums is added. The mixture of deoxynucleosides is passed through an immunoaffinity column containing an immobilized monoclonal antibody raised against a periodate-cleaved M_1G-ribose conjugated to albumin (Sevilla *et al.*, 1997). Unmodified deoxynucleosides pass through the immunoaffinity column, and then the M_1G-deoxyribose is eluted with methanol. The deoxyribose residue is removed by acid hydrolysis, and the M_1G base is derivatized with pentafluoroben-

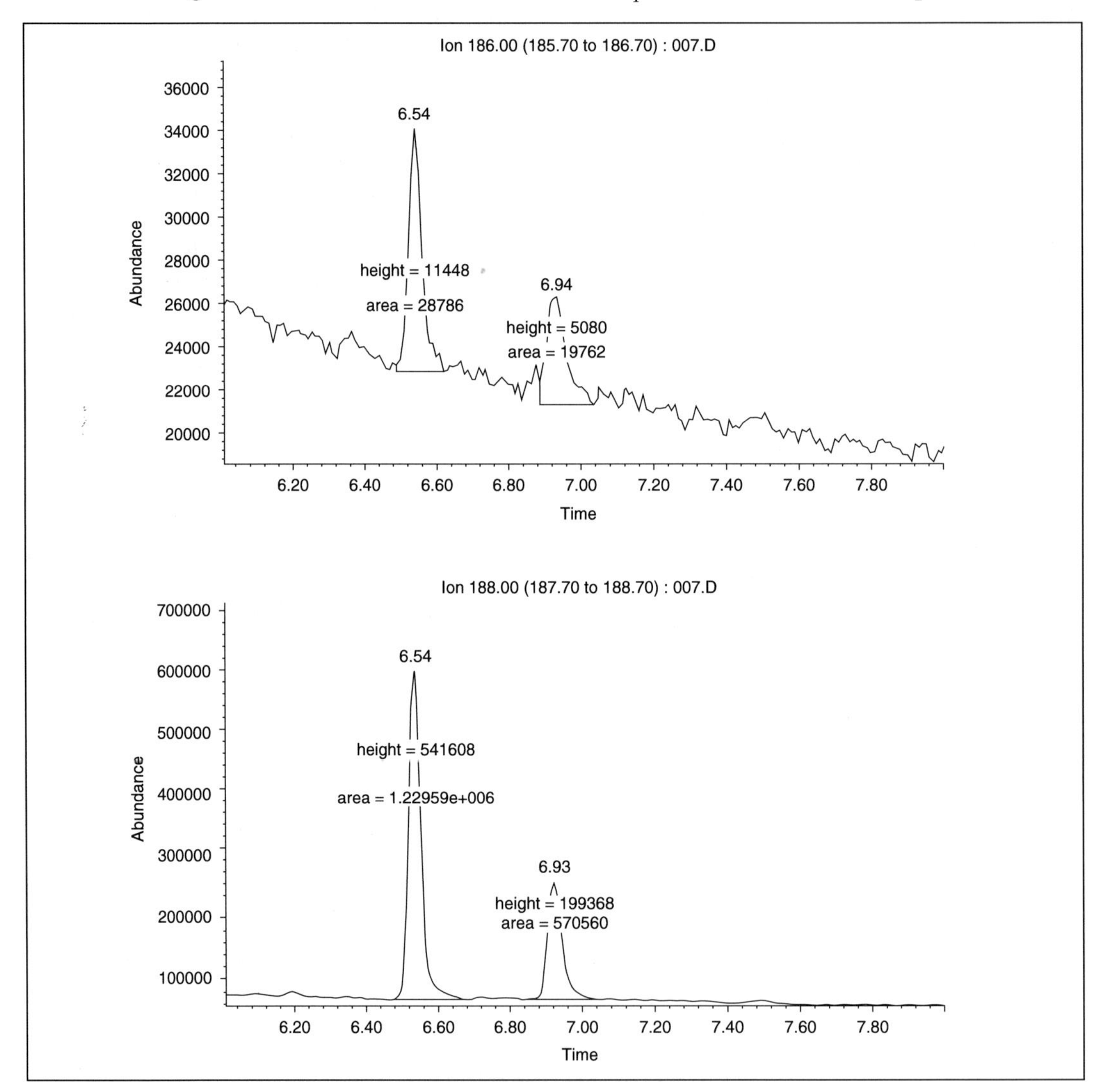

Figure 6. Selected ion monitoring profiles of pentafluorobenzyl derivatives of M_1G

The upper trace represents the channel for the material endogenously present in DNA ($m/z = 186$), whereas the lower trace represents the channel for the internal standard ($m/z = 188$).

zyl (PFB) bromide. Analysis of the PFB derivatives by GC/EC NCI/MS in the selected ion monitoring mode enables quantification by comparison of the intensity of the base peak of the endogenously occurring M_1G (m/z = 186) with that of the internal standard (m/z = 188) (Figure 6). The chromatographic trace in Figure 6 shows two separable M_1G-PFB derivatives corresponding to the N7 and N9 isomers. The GC/EC NCI/MS assays show good linearity and quantitative recovery of M_1G when oligonucleotides containing this adduct are carried through the analytical procedure (Rouzer *et al.*, 1997). The limit of sensitivity of the assay is approximately 1 M_1G per 10^8 nucleotides, starting with 1 mg of DNA. Specificity is provided by the combination of immunoaffinity chromatography, chemical derivatization, chromatographic coelution and selected ion monitoring.

Application of this technique to human liver and leukocyte DNA samples reveals that M_1G is present at levels of ~ 1 in 10^6 nucleosides and 6 in 10^8 nucleosides, respectively (Chaudhary *et al.*, 1994; Rouzer *et al.*, 1997; see Table 1). The livers were unused transplant samples, and the leukocyte samples were from human volunteers. The sample size in the latter study was rather small, but no obvious differences in M_1G levels were detected between the two smokers included in the study, and a very small but statistically significant difference was seen between men and women. A study of M_1G levels in unused human pancreas transplant samples demonstrated a bimodal distribution: the levels in approximately half the samples were below the limit of detection of the assay, and the levels in the other half were 2–50 per 10^8 nucleosides (Kadlubar *et al.*, 1998a).

As indicated above, the specificity of the GC/EC NCI/MS assay is ensured by a combination of chemical, immunochemical and spectroscopic prerequisites; however, it is possible to verify the presence of M_1G residues in DNA directly by liquid chromatography–mass spectrometry (LC–MS). A sample of human liver DNA shown to contain a high concentration of M_1G by GC/EC NCI/MS analysis was processed in the absence of an internal standard and the M_1G-containing fractions were isolated (Chaudhary *et al.*, 1995). The isolated material was then injected into a triple quadrupole LC–MS with electrospray detection. As observed with most deoxynucleosides, source-induced frag-

Table 1. Levels of M_1G in human tissues

Tissue	No. of samples	Average no. adducts/ 10^7 nucleotides (range)		Technique	Reference
Liver	6	9	5–12	GC–MS	Chaudhary *et al.* (1994)
Leukocytes	10	0.6	0.5–0.8	GC–MS	Rouzer *et al.* (1997)
Pancreas	27	3.2	< 0.1-5	GC–MS	Kadlubar *et al.* (1998b)
Leukocytes	26	2.6	1–5	^{32}P	Vaca *et al.* (1995)
Leukocytes[a]	7	0.9	0.2–2.5	^{32}P	Fang *et al.* (1996)
Leukocytes[b]	6	11	1.2–28	^{32}P	Fang *et al.* (1996)
Breast[c]	51	0.2	0.05–1.3	^{32}P	Wang *et al.* (1996)
Breast[d]	28	0.08	0.02–0.19	^{32}P	Wang *et al.* (1996)
Breast	7	3	0.7–5.6	^{32}P	Vaca *et al.* (1995)

GC–MS, gas chromatography–mass spectrometry; ^{32}P, ^{32}P-postlabelling

[a] Women eating a diet high in monounsaturated fat
[b] Women eating a diet high in polyunsaturated fat
[c] Normal tissue from breast cancer patients
[d] Normal breast tissue from non-cancer-bearing women

mentation led to loss of the deoxyribose moiety in the first quadrupole. The protonated molecular ion corresponding to M_1G (m/z = 188) was then collisionally dissociated in the second quadrupole, and the product ions analysed in the third quadrupole. The spectrum of ions generated was identical to that produced by tandem mass spectral analysis of an authentic standard of M_1G (Figure 7). This establishes unequivocally that M_1G residues are present in human liver DNA. Similar verification has been carried out on human leukocyte DNA samples, but in that case the low levels of M_1G required a mass fragmentographic approach. The offset voltage in the second quadrupole was programmed to change the fragmentation pattern of the protonated molecular ion progressively, and the expected alteration in fragmentation pattern for M_1G was observed (Rouzer *et al.*, 1997).

A concern in the analysis of endogenous DNA adducts is the possibility that they are artefactually generated during the work-up. This is especially serious when studying adducts that could be generated by oxidative processes that might occur *post mortem* or *ex vivo*. We have used various approaches to minimize this problem in the analysis of M_1G (Chaudhary *et al.*, 1994). During tissue handling and DNA isolation, large amounts of antioxidants were added to all buffers to minimize post-mortem lipid peroxidation. Interestingly, inclusion of butylated hydroxyanisole (10 mmol/L) or vitamin E (10 mmol/L) did not lower the levels of M_1G detected in liver tissue, but inclusion of vitamin C increased them by approximately twofold. We attribute this effect to the long-recognized ability of ascorbic acid to support iron-catalysed lipid peroxidation. The other approach we have used to assess possible artefactual M_1G production is the inclusion of isotopically labelled dG in the buffer used for DNA digestion (Figure 8). dG, chemically synthesized to contain four ^{13}C and one ^{15}N atoms, was added to the digestion mixture in amounts comparable to the amounts of dG in the DNA sample. If any MDA was

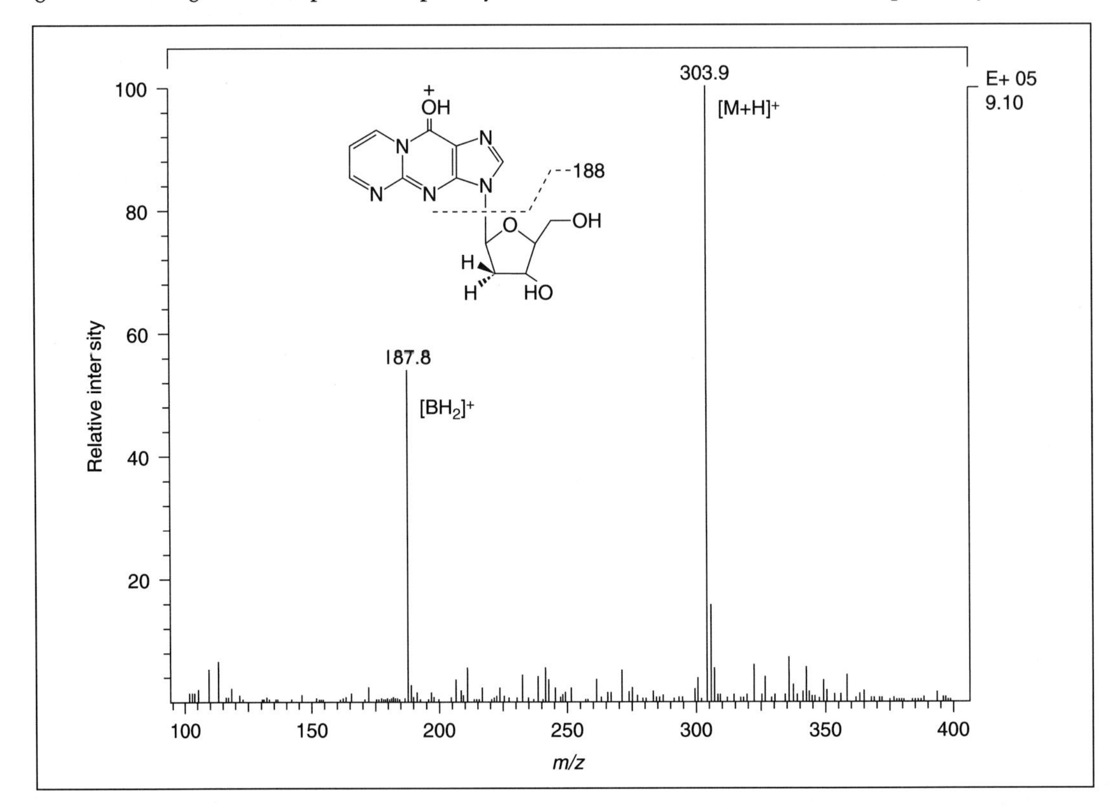

Figure 7. Mass spectrum of the collisionally dissociated m/z = 188 ion produced during analysis of a human liver DNA sample

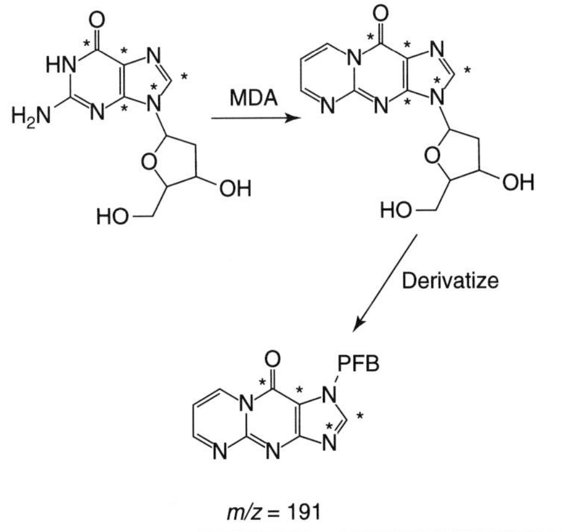

Figure 8. Isotopic probe for artefactual generation of M_1G *ex vivo*

MDA, malondialdehyde

present as a result of lipid peroxidation during tissue manipulation, it would react with the isotopically labelled dG to form M_1G that would yield a PFB derivative in GC/EC NCI/MS with a $m/z = 191$. No such material was detected in any of our analyses.

Analysis of rat tissues for M_1G by GC/EC NCI/MS reveals a level in liver (~6 per 10^7 nucleosides) comparable to that detected in human beings. The level in two-year-old rats is nearly double that in six-week-old rats. Interestingly, in testis DNA samples from the same rats the levels of M_1G were below the limit of detection of the assay (1 per 10^8 nucleosides). When rats were exposed to the lipid peroxidation stimulus carbon tetrachloride by oral intubation, the levels of M_1G in liver increased by nearly 100% (Chaudhary *et al.*, 1997).

Conclusions

The available evidence indicates that M_1G, and possibly other MDA-derived adducts, is a highly mutagenic adduct that is present in the genomic DNA of human beings (Table 1). M_1G may be produced by reaction of MDA with dG residues in DNA or by oxopropenyl transfer from base propenal to dG residues. Thus, M_1G can arise from lipid peroxidation of DNA peroxidation. The levels of M_1G are as high as 120 per 10^8 nucleosides, but considerable tissue-to-tissue variation is observed. M_1G levels appear to increase with age and the unsaturated fatty acid content of the diet. M_1G may contribute to the background of genetic mutations in human beings, but this should be established in population-based studies in large human populations.

Acknowledgements

Research in the laboratory of L.J. Marnett is supported by research and centre grants from the National Institutes of Health (CA47479, ES00267 and CA68485).

References

Basu, A.K. & Marnett, L.J. (1983) Unequivocal demonstration that malondialdehyde is a mutagen. *Carcinogenesis* 4, 331–333

Benamira, M., Johnson, K., Chaudhary, A., Bruner, K., Tibbetts, C. & Marnett, L.J. (1995) Induction of mutations by replication of malondialdehyde-modified M13 DNA in *Escherichia coli*: Determination of the extent of DNA modification, genetic requirements for mutagenesis, and types of mutations induced. *Carcinogenesis* 16, 93–99

Chaudhary, A.K., Nokubo, M., Reddy, G.R., Yeola, S.N., Morrow, J.D., Blair, I.A. & Marnett, L.J. (1994) Detection of endogenous malondialdehyde-deoxyguanosine adducts in human liver. *Science* 265, 1580–1582

Chaudhary, A.K., Nokubo, M., Oglesby, T.D., Marnett, L.J. & Blair, I.A. (1995) Characterization of endogenous DNA adducts by liquid chromatography/electrospray ionization/tandem mass spectrometry. *J. Mass Sprectrom.* 30, 1157–1166

Chaudhary, A.K., Reddy, G.R., Blair, I.A. & Marnett, L.J. (1996) Characterization of an *N*⁶-oxopropenyl-2′-deoxyadenosine adduct in malondialdehyde-modified DNA using liquid chromatography electrospray ionization tandem mass spectrometry. *Carcinogenesis* 17, 1167–1170

Dedon, P.C., Plastaras, J.P., Rouzer, C.A. & Marnett, L.J. (1998) Indirect mutagenesis by oxidative DNA damage: Formation of the pyrimidopurinone adduct of deoxyguanosine by base propenal. *Proc. Natl Acad. Sci. USA* 95, 11113–11116

Diczfalusy, U., Falardeau, P. & Hammarstrom, S. (1977) Conversion of prostaglandin endoperoxides to C17-hydroxyacids by human platelet thromboxane synthase. *FEBS Lett.* 84, 271–274

Esterbauer, H., Eckl, P. & Ortner, A. (1990) Possible mutagens derived from lipids and lipid precursors. *Mutat. Res. Rev. Genet. Toxicol.* 238, 223–233

Fang, J.L., Vaca, C.E., Valsta, L.M. & Mutanen, M. (1996) Determination of DNA adducts of malonaldehyde in humans: Effects of dietary fatty acid composition. *Carcinogenesis* 17, 1035–1040

Fink, S.P., Reddy, G.R. & Marnett, L.J. (1997) Mutagenicity in *Escherichia coli* of the major DNA adduct derived from the endogenous mutagen malondialdehyde. *Proc. Natl Acad. Sci. USA* 94, 8652–8657

Hazen, S.L., Hsu, F.F., D'Avignon, A. & Heinecke, J.W. (1998) Human neutrophils employ myeloperoxidases to convert α-amino acids to a battery of reactive aldehydes: A pathway for aldehyde generation at sites of inflammation. *Biochemistry* 37, 6864–6873

Johnson, K.A., Fink, S.P. & Marnett, L.J. (1997) Repair of propanodeoxyguanosine by nucleotide excision repair in vivo and in vitro. *J. Biol. Chem.* 272, 11434–11438

Kadlubar, F., Anderson, K., Lang, N., Thompson, P., MacLeod, S., Mikhailova, M., Chou, M., Plastaras, J., Marnett, L., Haussermann, S. & Bartsch, H. (1998a) Comparison of endogenous DNA adduct levels in human pancreas. *Proc. Am. Assoc. Cancer Res.* 39, 286

Kadlubar, F.F., Anderson, K.E., Lang, N.P., Thompson, P.A., Macleod, S.L., Chou, M.W., Mikhailova, M., Plastaras, J., Marnett, L.J., Haussermann, S., Nair, J., Velie, I. & Bartsch, H. (1998b) Comparison of endogenous DNA adduct levels in human pancreas. *Mutat. Res.* 405, 125–133

Kautianen, A., Vaca, C.E. & Granath, F. (1993) Studies on the relationship between hemoglobin and DNA adducts of malonaldehyde and their stability *in vivo*. *Carcinogenesis* 14, 705–708

Krokan, H., Grafstrom, R.C., Sundqvist, K., Esterbauer, H. & Harris, C.C. (1985) Cytotoxicity, thiol depeletion and inhibition of O^6-methylguanine-DNA methyltransferase by various aldehydes in cultured human bronchial fibroblasts. *Carcinogenesis* 6, 1755–1759

Marnett, L.J. & Tuttle, M.A. (1980) Comparison of the mutagenicities of malondialdehyde and side products formed during its chemical synthesis. *Cancer Res.* 40, 276–282

Marnett, L.J., Hurd, H.K., Hollstein, M.C., Levin, D.E., Esterbauer, H. & Ames, B.N. (1985) Naturally occurring carbonyl compounds are mutagens in *Salmonella* tester strain TA104. *Mutat. Res.* 148, 25–34

Marnett, L.J., Basu, A.K., O'Hara, S.M., Weller, P.E., Rahman, A.F.M.M. & Oliver, J.P. (1986) Reaction of malondialdehyde with guanine nucleosides: Formation of adducts containing oxadiazabicyclononene residues in the base-pairing region. *J. Am. Chem. Soc.* 108, 1348–1350

Mukai, F.H. & Goldstein, B.D. (1976) Mutagenicity of malondialdehyde, a decomposition product of peroxidized polyunsaturated fatty acids. *Science* 191, 868–869

Nair, V., Turner, G.A. & Offerman, R.J. (1984) Novel adducts from the modification of nucleic acid bases by malondialdehyde. *J. Am. Chem. Soc.* 106, 3370–3371

Nath, R.G. & Chung, F.-L. (1994) Detection of exocyclic 1,N^2-propanodeoxyguanosine adducts as common DNA lesions in rodents and humans. *Proc. Natl Acad. Sci.* 91, 7491–7495

Reddy, G.R. & Marnett, L.J. (1995) Synthesis of an oligonucleotide containing the alkaline labile malondialdehyde–deoxyguanosine adduct pyrimido[1,2-*a*]purin-10(3H)-one. *J. Am. Chem. Soc.* 117, 5007–5008

Rouzer, C.A., Chaudhary, A.K., Nokubo, M., Ferguson, D.M., Reddy, G.R., Blair, I.A. & Marnett, L.J. (1997) Analysis of the malondialdehyde-2'-deoxyguanosine adduct, pyrimidopurinone, in human leukocyte DNA by gas chromatography/electron capture negative chemical ionization/mass spectrometry. *Chem. Res. Toxicol.* 10, 181–188

Schauenstein, E., Esterbauer, H. & Zollner, H. (1977) Malondialdehyde. In: Schauenstein, E., Esterbauer, H. & Zollner, H., eds, *Aldehydes in Biological Systems.* London, Pion Ltd, pp. 133–140

Seto, H., Okuda, T., Takesue, T. & Ikemura, T. (1983) Reaction of malondialdehyde with nucleic acid. I. Formation of fluorescent pyrimido[1,2-*a*]purin-10(3H)-one nucleosides. *Bull. Chem. Soc. Jpn* 56, 1799–1802

Sevilla, C.L., Mahle, N.H., Eliezer, N., Uzieblo, A. O'Hara, S.M., Nokubo, M., Miller, R., Rouzer, C.A. & Marnett, L.J. (1997) Development of monoclonal antibodies to the malondialdehyde–deoxyguanosine adduct, pyrimidopurinone. *Chem. Res. Toxicol.* 10, 172–180

Spalding, J.W. (1988) *Toxicology and Carcinogenesis Studies of Malondialdehyde Sodium Salt (3-Hydroxy-2-propenal, Sodium Salt) in F344/N Rats and B6C3F1 Mice* (NTP Tech. Rep. 331). Washington DC, National Toxicology Program, pp. 5–13

Stone, K., Ksebati, M. & Marnett, L.J. (1990a) Investigation of the adducts formed by reaction of malondialdehyde with adenosine. *Chem. Res. Toxicol.* 3, 33–38

Stone, K., Uzieblo, A. & Marnett, L.J. (1990b) Studies of the reaction of malondialdehyde with cytosine nucleosides. *Chem. Res. Toxicol.* 3, 467–472

Vaca, C.E., Vodicka, P. & Hemminki, H. (1992) Determination of malonaldehyde-modified 2'-deoxyguanosine-3'-monophosphate and DNA by ^{32}P-postlabelling. *Carcinogenesis* 13, 593–599

Vaca, C.E., Fang, J.-L., Mutanen, M. & Valsta, L. (1995) ^{32}P-Postlabeling determination of DNA adducts of malonaldehyde in humans: Total white blood cells and breast tissue. *Carcinogenesis* 16, 1847–1851

Wang, M.-Y. & Liehr, J.G. (1990a) Induction by estrogens of lipid peroxidation and lipid peroxide-derived malonaldehyde–DNA adducts in male Syrian hamsters: Role of lipid peroxidation in estrogen-induced kidney carcinogenesis. *Carcinogenesis* 16, 1941–1945

Wang, M.-Y. & Liehr, J.G. (1990b) Lipid hydroperoxide-induced endogenous DNA adducts in hamsters: Possible mechanism of lipid hydroperoxide-mediated carcinogenesis. *Arch. Biochem. Biophys.* 316, 38–46

Wang, M., Dhingra, K., Hittleman, W.N., Liehr, J.G., de Andrade, M. & Li, D. (1996) Lipid peroxidation-induced putative malondialdehyde–DNA adducts in human breast tissues. *Cancer Epidemiol. Buimarkers Prev.* 5, 705–710

Formation and repair of DNA adducts in vinyl chloride- and vinyl fluoride-induced carcinogenesis

J.A. Swenberg, M.S. Bogdanffy, A. Ham, S. Holt, A. Kim, E.J. Morinello, A. Ranasinghe, N. Scheller & P.B. Upton

Vinyl chloride is a known human and animal carcinogen that induces angiosarcomas of the liver. We review here studies on the formation and repair of DNA adducts associated with vinyl chloride and vinyl fluoride in exposed and control rodents and unexposed humans. These vinyl halides induce etheno (ε) adducts that are identical to those formed after lipid peroxidation. Of these adducts, N^2,3-ethenoguanine (εG) is present in greatest amounts in tissues of exposed animals. After exposure to vinyl chloride for four weeks, εG levels attain steady-state concentrations, such that the amount of newly formed adducts equals the number of adducts that are lost each day. We report the first dosimetry of εG in rats exposed to 0, 10, 100 or 1100 ppm vinyl chloride for five days or four weeks. The number of adducts increased in a supralinear manner. Exposure to 10 ppm vinyl chloride for five days caused a two- to threefold increase in εG over that of the controls, while four weeks' exposure resulted in a fivefold increase. This was confirmed with [$^{13}C_2$]vinyl chloride and by measuring exogenous and endogenous adducts in the same animals. Exposure to 100 ppm vinyl chloride for four weeks caused a 25-fold increase in εG levels over that found in control rats, while exposure to 1100 ppm resulted in a 42-fold increase. The amount of endogenous εG was similar in liver DNA from rats and humans. A comparable response to exposure was seen in rats and mice exposed to 0, 25, 250 or 2500 ppm vinyl fluoride for 12 months. There was a very high correlation between εG levels in rat and mouse liver at 12 months and the incidence of haemangiosarcoma at two years. We were able to demonstrate that the target cell population for angiosarcoma, the nonparenchymal cells, contained more εG than hepatocytes, even though nonparenchymal cells are exposed by diffusion of vinyl halide metabolites formed in hepatocytes. The expression of *N*-methylpurine-DNA glycosylase mRNA was induced in rat liver after exposure to either 25 or 2500 ppm vinyl fluoride. When this induction was investigated in hepatocytes and nonparenchymal cells, it was found that the latter had only 20% of the *N*-methylpurine-DNA glycosylase mRNA of hepatocytes, and that only the hepatocytes had induction of this expression after exposure to vinyl fluoride. Thus, the target cells for vinyl halide carcinogenesis have much lower expression of this DNA repair enzyme, which has been associated with etheno adduct repair.

Introduction

Vinyl chloride and vinyl fluoride are important precursor chemicals used by the plastics industry. Over 14 thousand million pounds (over 6 thousand million kg) of vinyl chloride were produced in the United States in 1993, and the National Institute for Occupational Safety and Health (1990) estimated that 81 000 workers were potentially exposed to vinyl chloride daily. Since its commercialization in the 1930s, almost all of the vinyl chloride monomer produced has been used to manufacture polyvinyl chloride, a plastic used in plumbing, credit cards, automobile interiors and countless other applications. Contamination of soil and groundwater by vinyl chloride is primarily the result of the microbial degradation of trichloroethylene and perchloroethylene. Vinyl fluoride is used in smaller amounts to manufacture polyvinyl fluoride and other fluoropolymers.

Before the early 1970s, routine exposure to vinyl chloride was thought to pose little threat to humans, and a 500 ppm threshold limit value was set at that time (American Conference of Governmental Industrial Hygienists, 1991). Acute exposure to high concentrations of vinyl chloride was known to adversely affect the nervous system, causing dizziness, ataxia, inebriation, fatigue, narcosis and even death. Some chronically exposed workers developed 'vinyl chloride disease', a scleroderma-like condition of the connective tissues of the fingers, fracturing of bones at the tips of the fingers, liver damage and sometimes haematological changes and pulmonary effects.

The first report of cancer in animals linked to exposure to vinyl chloride was made by Viola *et al.* in 1971. Tumours were observed in Wistar rats exposed to 30 000 ppm vinyl chloride for 4 h/day on five days per week for at least 10 months. A very large study specifically designed to study vinyl chloride-induced carcinogenesis was initiated in the same year by Maltoni *et al.* and reported in 1981. The results indicated that as little as 50 ppm vinyl chloride were capable of inducing a spectrum of tumours in rodents; hepatic angiosarcoma was the hallmark lesion, occurring with a clear supralinear dose–response relationship. Tumour development was greatly influenced by duration of exposure, and younger animals were particularly susceptible. No significant sex differences were detected.

Also in 1974, it was announced that three workers exposed to vinyl chloride had died of hepatic angiosarcoma (Creech & Johnson, 1974). A subsequent epidemiological study involving a cohort of 1855 workers who had worked at the same plant between 1942 and 1976 revealed 10 cases of hepatic angiosarcoma (Dannaher *et al.*, 1981). The mean length of exposure of these workers was 19.9 years (range, 12–28 years), and all were believed to have been exposed to relatively high concentrations of vinyl chloride. The rarity of the tumour in unexposed humans—10% (12/126) of the hepatic angiosarcomas diagnosed in 1964–74 were in vinyl chloride workers (Falk *et al.*, 1981)—made hepatic angiosarcoma a 'sentinel tumour' for exposure to vinyl chloride. This and other epidemiological studies (Tabershaw & Gaffey, 1974; Nicholson *et al.*, 1975) firmly established vinyl chloride as a human carcinogen. Although causative relationships between exposure to vinyl chloride and other human cancers have been suggested by some reports, they have not been definitively established.

Vinyl chloride was consistently mutagenic in short-term assays, and increased numbers of chromosomal aberrations were detected in vinyl chloride workers (reviewed by Fabricant & Legator, 1981). Vinyl chloride-induced mutagenesis is dependent on oxidative metabolic activation. Although it was initially thought that the carcinogenic metabolite was chloroacetaldehyde, evidence was obtained *in vivo* and *in vitro* that its precursor, chloroethylene oxide, is primarily responsible (Gwinner *et al.*, 1983; Guengerich, 1992). Chloroethylene oxide is a short-lived epoxide that has been shown to react with guanine, cytosine and adenine *in vitro* to form alkylated derivatives, including the exocyclic etheno (ε) adducts.

Vinyl halides are activated via a saturable, CYP450 2E1-mediated mechanism (Gehring *et al.*, 1978; Guengerich *et al.*, 1991). As a consequence, the exposure–response to vinyl chloride would be expected to show a supralinear pattern. This is indeed the case with respect to several end-points, including excretion of vinyl chloride metabolites (Watanabe & Gehring, 1976), tumour frequency (Bogdanffy *et al.*, 1995) and DNA adduct formation (La & Swenberg, 1996). As a consequence, a simple linear extrapolation of the biological responses to high doses would provide an underestimate of the risk vinyl chloride poses at the low exposures typically experienced by humans. The influence of other factors, such as induction and/or saturation of DNA repair, on the ultimate exposure–response is, however, more difficult to predict. Clewell *et al.* (1995) applied pharmacokinetic modelling of metabolism of vinyl chloride in order to improve the risk assessment.

To better characterize the linear portion of the vinyl chloride exposure–response relationship, a more sensitive end-point than tumour frequency is required. An alternative approach is to base the exposure–response on that of the DNA adducts that cause the mutations assumed to underlie vinyl chloride-induced carcinogenesis. The molecular dose of such adducts integrates differences in absorption, distribution, metabolism, detoxification and DNA repair between exposures. This measurement can also reduce uncertainties arising from differences between individuals, species and routes of exposure. Sensitive techniques such as ^{32}P-postlabelling and gas chromatography–mass spectrometry (GC–MS) allow investigation of the effects of doses within an order of magnitude of common occupational exposures.

The major vinyl chloride-induced adduct, 7-(2′-oxoethyl)guanine (OEG), does not induce mutations *in vitro* (Barbin *et al.*, 1985) and is thus thought not to play a significant role in vinyl chloride-induced mutagenesis (see Figure 1 for structures). The exocyclic derivatives 1,N^6-ethenoadenine (εA), 3,N^4-ethenocytosine (εC) and N^2,3-ethenoguanine (εG), however, have miscoding potential *in vitro* and *in vivo* (Singer *et al.*, 1987; Cheng *et al.*, 1991; Mroczkowska & Kusmierek, 1991; Singer *et al.*, 1991; Basu *et al.*, 1993). The fidelity of replication or transcription has been studied *in vitro* when modified synthetic polynucleotides were used as templates. The mutagenic specificities found were subsequently confirmed *in vivo* with site-directed kinetic assays and by site-specific studies of mutagenesis. Both *in vitro* and *in vivo*, εA caused A→C transversions, εC caused C→A transversions and C→T transitions, and εG caused G→A transitions. The results of these studies were consistent with the findings of two separate studies on the mutations in tumours from vinyl chloride-exposed humans and rats. In five out of six human liver angiosarcomas associated with exposure to vinyl chloride, GC→AT transitions were observed at codon 13 of the c-Ki-*ras*-2 gene, consistent with either εG or εC mutagenic properties (Marion *et al.*, 1991). In another study, tumours from vinyl chloride-exposed rats showed G→A transitions, A→C transversions and C→T transitions in N-*ras* genes, all of which are consistent with promutagenic properties of etheno adducts (Froment *et al.*, 1994). Mutations in the *p53* gene have also been examined in angiosarcomas from vinyl chloride workers (Hollstein *et al.*, 1994) and in liver tumours from rats exposed to vinyl chloride (Barbin *et al.*, 1997).

Bogdanffy *et al.* (1995) demonstrated that vinyl fluoride was carcinogenic to rats and mice after long-term inhalation. A spectrum of tumours similar to that induced by vinyl chloride was observed at doses of 25 ppm or greater. The metabolism of vinyl fluoride also appeared to be saturable, with a plateau in urinary fluoride excretion at doses greater than 250 ppm. These parallels are not surprising in light of the structural and chemical similarities between vinyl chloride and vinyl fluoride, and mechanistic knowledge gained from the study of one is likely to be applicable to the other. No human cancers have been attributed to vinyl fluoride.

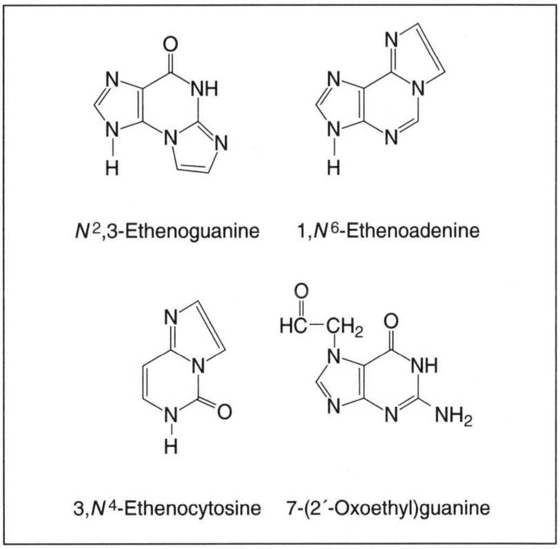

N^2,3-Ethenoguanine 1,N^6-Ethenoadenine

3,N^4-Ethenocytosine 7-(2′-Oxoethyl)guanine

Figure 1. Structures of DNA adducts resulting from exposure to vinyl chloride and vinyl fluoride

Studies on vinyl chloride-exposed rats

In initial studies on the formation and repair of vinyl chloride-induced DNA adducts, the numbers of adducts formed in rat pups and mothers exposed to 600 ppm vinyl chloride (6 h/day, five days) and the loss of these adducts over time were measured (Fedtke *et al.*, 1989). As shown in Table 1, the major DNA adduct formed was OEG. It was formed in greatest amounts in livers of rat pups, with less in kidney and livers of dams; OEG was not detected in brain DNA of pups or dams. The method used to detect OEG was high-performance liquid chromatography (HPLC) with fluorescence detection, which had a limit of detection of about 10 pmol/μmol guanine. The rate of loss of OEG was similar to that of other N7 guanine DNA adducts, with a half-life of 62 h.

Table 1. Concentrations of 7-(2′-oxoethyl)guanine (OEG) and N^2,3-ethenoguanine (εG) in rat tissue DNA measured immediately after exposure to 600 ppm vinyl chloride for 4 h/day, five days

Tissue	Adducts (pmol/μmol guanine; mean ± SD)	
	OEG	εG
Pups		
Liver	162 ± 36 (9)	1.81 ± 0.25 (5)
Lung	20 ± 7 (3)	0.21 ± 0.08 (5)
Kidney	29 ± 1 (3)	0.31 ± 0.02 (3)
Brain	< 10 (3)	< 12 (3)
Spleen	< 10 (3)	< 12 (3)
Adults		
Liver	43 ± 7 (5)	0.47 ± 0.14 (5)
Lung	20 ± 5 (3)	0.27 ± 0.03 (3)
Kidney	Not analysed	< 0.12 (3)

From Fedtke *et al.* (1990a); numbers of samples analysed in parentheses

A new GC–MS method was developed for measuring εG (Fedtke *et al.*, 1990b), which involved a synthesized [$^{13}C_4$] stable isotope as internal standard, with identical gas chromatographic properties to the analyte, which was added to the DNA at the time of mild acid hydrolysis. εG was enriched on disposable SCX columns, derivatized with α-bromo-2,3,4,5,6-pentafluorotoluene and analysed by GC–MS. The limit of detection with synthesized derivatized standards was 190 amol/μL injected solution. With this method, the amount of εG was shown to be about 1/100th that of OEG (Fedtke *et al.*, 1990a). Again, the amount formed was greatest in pup liver and lower in liver DNA from dams and other tissues (Table 1). The ratio of OEG:εG was similar in all tissues, suggesting that the DNA repair was not tissue-specific. εG was lost much more slowly than OEG, with a calculated half-life of about 30 days.

Similar studies were conducted by Ciroussel *et al.* (1990) in rats exposed to 500 ppm vinyl chloride (7 h/day, 14 days). εA and εC were enriched by HPLC and evaluated in a competitive radioimmunoassay. These tissues and the tissues from the study of Fedtke *et al.* were exchanged and analysed for OEG, εG, εA and εC (Swenberg *et al.*, 1992). OEG was lost much faster from the DNA than any of the etheno adducts (Table 2). The methods used at that time were not sensitive enough to measure etheno adducts in unexposed control rats, and the results were interpreted to mean that etheno adducts were highly persistent (Swenberg *et al.*, 1992). This interpretation was later found to be wrong, and several groups demonstrated that the etheno adducts could be repaired by DNA glycosylases (Oesch *et al.*, 1986; Matijasevic *et al.*, 1992; Dosanjh *et al.*, 1994; Hang *et al.*, 1996; Saparbaev & Laval, 1998; Hang *et al.*, 1998). Furthermore, as methods improved, it became clear that the etheno adducts were present in unexposed control animals and humans. In fact, what appeared to be persistent DNA adducts was actually the background presence of endogenous etheno adducts and a return to this level after cessation of exposure.

A collaboration between Bartsch's group at IARC and our laboratory was undertaken to better understand the rate of accumulation of etheno

Table 2. DNA adducts (per mol parent base) in rat liver after exposure to 500 or 600 ppm vinyl chloride

DNA adduct	Days after exposure				
	500 ppm	600 ppm			
	0	0	3	7	14
OEG	2.0×10^{-4}	1.6×10^{-4}	5.1×10^{-5}	1.2×10^{-5}	Not detected
εG	3.8×10^{-6}	1.8×10^{-6}	1.3×10^{-6}	8.4×10^{-7}	4.7×10^{-7}
εA	4.8×10^{-7}	9.8×10^{-7}	8.4×10^{-7}	7.1×10^{-7}	6.1×10^{-7}
εC	1.3×10^{-7}	2.1×10^{-7}	1.4×10^{-7}	1.4×10^{-7}	0.8×10^{-7}

From Ciroussel *et al.* (1990) and Swenberg *et al.* (1992)
Exposure was for 7 h/day for 14 days to 500 ppm, and 4 h/day for five days to 600 ppm.
OEG, 7-(2′-oxoethyl)guanine; ε, etheno

adducts. Using immunoaffinity chromatography with ^{32}P-postlabelling, the IARC group demonstrated accumulation of εdAdo and εdCyd in the livers of adult rats exposed to 500 ppm vinyl chloride (4 h/day) for 1, 2, 4 or 8 weeks. Both adducts accumulated, steady-state levels of about 30 fmol/μmol and 60 fmol/μmol being approximated for εA and εC in the liver, respectively, after four weeks of exposure (Guichard *et al.*, 1996). A similar time course was demonstrated for εG (Figure 2), but the number of εG adducts was 10–100-fold greater. In addition to the low-resolution GC–MS method used by Fedtke *et al.* (1990b), εG was also analysed by GC–high-resolution MS, with a resolving power of 10 000. There was no apparent loss of sensitivity, and this became the method of choice due to its greater specificity (Ranasinghe *et al.*, 1998).

One of the major goals of this research was to better understand the relationship between exposure concentration and molecular dose. All of the studies on vinyl chloride had involved high concentrations (500–600 ppm), which were known to saturate metabolism (Watanabe & Gehring, 1976). Adult rats (eight weeks of age at the start of exposure) were exposed by whole-body inhalation to 0, 10, 100 or 1100 ppm vinyl chloride for one or four weeks (6 h/day, five days per week). This exposure regimen covered the high-exposure range previously studied and extended it down to within one order of magnitude of current occupational exposure standards in most of the world. A new method for εG enrichment was used for this study (Ham & Swenberg, 1998). Immunoaffinity columns were prepared with a polyclonal antibody raised against εG, and the analyte and internal standard were derivatized with α-bromo-2,3,4,5,6-pentafluorotoluene. Preliminary data, based on three to four animals per exposure group, are shown in Figure 3.

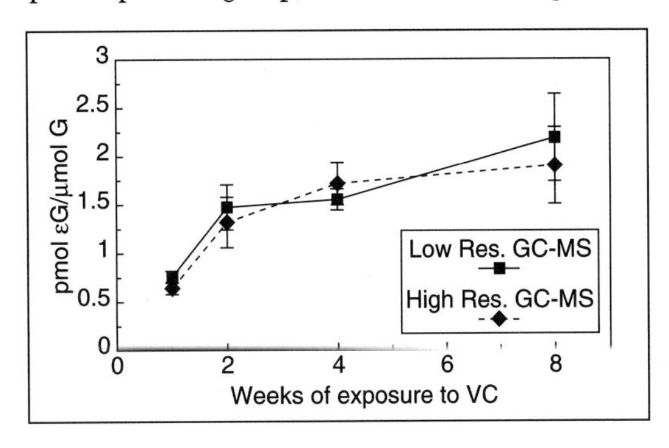

Figure 2. Quantification of N^2,3-ethenoguanine (εG) in rats exposed to vinyl chloride (VC) by low- and high-resolution gas chromatography–mass spectrometry (GC–MS)

Adult rats were exposed to 500 ppm vinyl chloride, 4 h/day for 1, 2, 4 or 8 weeks. Samples were analysed by the method of Fedtke *et al.* (1990b).

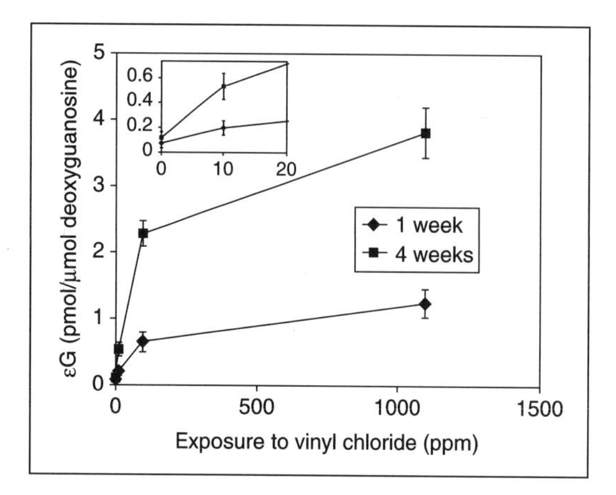

Figure 3. Quantification of N^2,3-ethenoguanine (εG) as a molecular dosimeter of vinyl chloride in rats

Inset graph depicts magnification at low exposure. Rats were exposed to 10, 100 or 1100 ppm vinyl chloride for 6 h/day, five days per week for 1 or 4 weeks. Samples were analysed by immunoaffinity chromatography followed by high-resolution gas chromatography–mass spectrometry of the corresponding pentofluorobenzyl derivatives.

Endogenous εG was detected in all control rats at an average of 90 ± 40 fmol/μmol G (Table 3). As shown in Figure 3, the molecular dose of vinyl chloride increased in a supralinear manner, with a steep slope between 0 and 100 ppm but little additional accumulation at 1100 ppm vinyl chloride. The number of εG adducts increased by threefold between one and four weeks at all exposure concentrations. It was also possible for the first time to compare the endogenous amounts of εG with those associated with exposures similar to those in the workplace. Table 3 shows the amounts of εG present in control rats and those exposed to 10 ppm vinyl chloride. When the data for four weeks were compared with those for controls, a significant difference was detected (one-tailed Student's t test, $p = 0.002$). Notably, after exposure to vinyl chloride at a concentration 10 times greater than occupational exposure standards, the background amounts of εG were increased by only 2.2-fold and fivefold after one- and four-week exposures. These are the first such data, and they provide a strong scientific basis for current occupational standards.

Since the method used to quantify εG is based on mass, we hypothesized that we could further examine the relationship between endogenous and exogenous εG by exposing rats to [$^{13}C_2$]vinyl chloride. Eight-week-old rats were exposed to 10, 100 or 1100 ppm [$^{13}C_2$]vinyl chloride for five days (6 h/day) in a nose-only exposure chamber with monitoring of endogenous εG at m/z 354.0413, [$^{13}C_2$]vinyl chloride-induced exogenous εG at m/z 356.0481 and the isotopic internal standard at m/z 360.0489. Thus, endogenous and exogenous εG could be monitored in the same animal for the first time. Preliminary data are shown in Figure 4 for animals exposed to 10, 100 or 1100 ppm [$^{13}C_2$]vinyl chloride. It can be seen that the amount of endogenous εG is about one-half that of exogenous εG induced by 10 ppm [$^{13}C_2$]vinyl chloride. This provides independent confirmation of the

Table 3. Relationship between endogenous DNA adducts in rats and exposure to 10 ppm vinyl chloride for 6 h/day

Length of exposure (days)	No. of rats	N^2,3-Ethenoguanine (fmol/μmol guanine; mean ± SD)	Fold increase
Control	6	90 ± 40	---
5	4	200 ± 50	2.2
20	3	530 ± 11	5.9

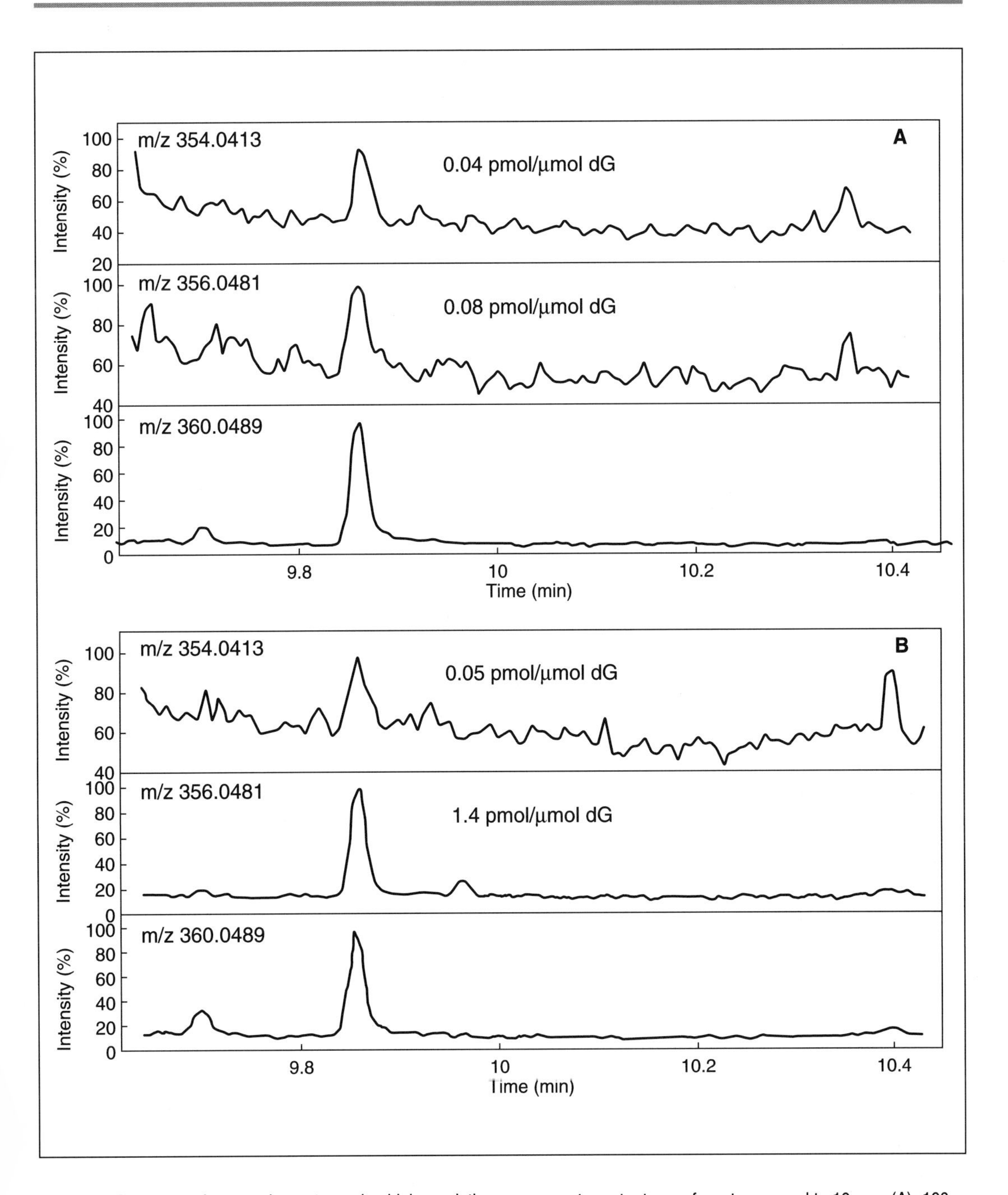

Figure 4. Representative gas chromatography–high-resolution mass spectrometry traces for rats exposed to 10 ppm (A), 100 ppm (B) or 1100 ppm (C) [$^{13}C_2$]vinyl chloride

m/z 354.0413 is endogenous N^2,3-ethenoguanine (εG), *m/z* 356.0481 is εG resulting from [$^{13}C_2$]vinyl chloride, and *m/z* 360.0489 is the internal standard. Samples were analysed by immunoaffinity chromatography followed by gas chromatography–high-resolution mass spectrometry of the corresponding pentofluorobenzyl derivatives.

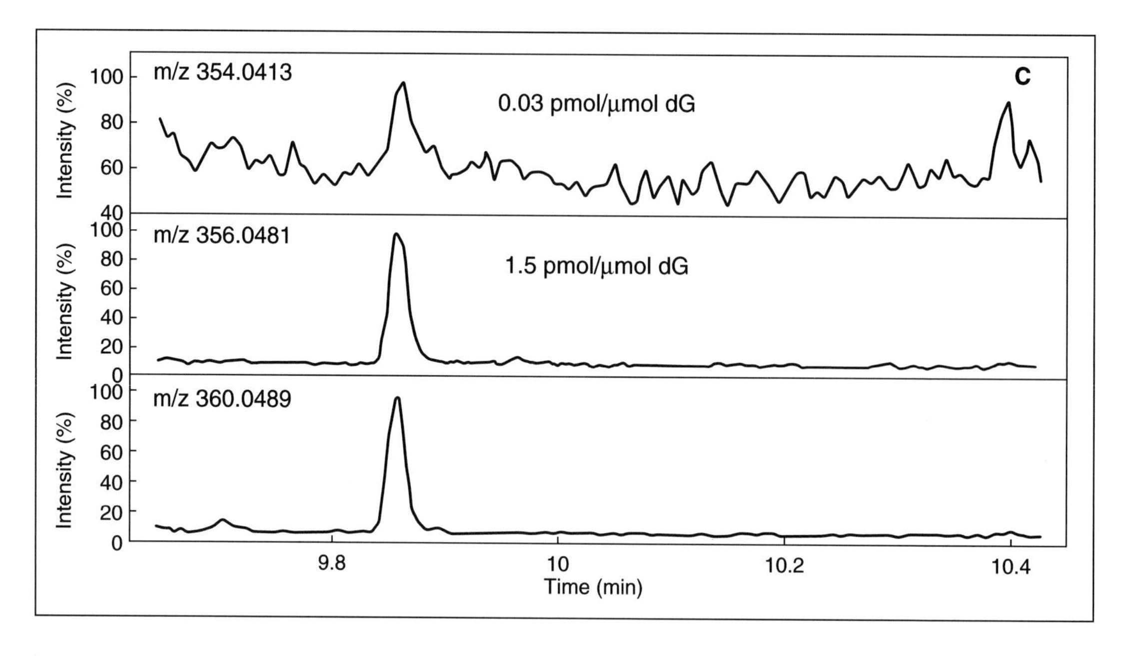

Figure 4 (Contd). Representative gas chromatography–high-resolution mass spectrometry traces for rats exposed to 10 ppm (A), 100 ppm (B) or 1100 ppm (C) [$^{13}C_2$]vinyl chloride

m/z 354.0413 is endogenous N^2,3-ethenoguanine (εG), *m/z* 356.0481 is εG resulting from [$^{13}C_2$]vinyl chloride, and *m/z* 360.0489 is the internal standard. Samples were analysed by immunoaffinity chromatography followed by gas chromatography–high-resolution mass spectrometry of the corresponding pentofluorobenzyl derivatives.

data shown in Figure 3 and Table 3. Thus, in two experiments, exposure to 10 ppm vinyl chloride for five days resulted in two to three times more εG than is present in unexposed rats. The amount of endogenous εG did not change in rats exposed to 100 or 1100 ppm [$^{13}C_2$]vinyl chloride, while the amount of exogenous εG was similar to that shown in Figure 3. Although the number of animals examined should be expanded, these data suggest that endogenous εG is repaired similarly in animals exposed to high and low concentrations of vinyl chloride and that repair is not saturated at 100 or 1100 ppm vinyl chloride. While the absolute amount of endogenous εG in the [$^{13}C_2$]vinyl chloride-exposed rats is lower than those shown in Table 3, more animals will have to be examined to determine if this difference is real. As shown below, methylpurine-DNA glycosylase is induced in livers of rats exposed to vinyl fluoride. Such induction could have a protective effect at low concentrations of exposure.

Endogenous N^2,3-ethenoguanine in human liver

Samples from 12 human livers were obtained from the Tennessee Tissue Repository and analysed with a modification of the Fedtke method (Mitro *et al.*, 1995; Swenberg *et al.*, 1995). Briefly, DNA was extracted and the stable isotope internal standard was added during mild acid hydrolysis. εG was enriched on the disposable SCX columns described by Fedtke *et al.* (1990b). After desalting, the analyte and internal standard were derivatized with 1-bromomethyl-2,3,5,6-tetrafluoro-4-(trifluoro methyl)benzene. The samples were cleaned and analysed by GC–high-resolution MS. A representative chromatogram is shown in Figure 5. The amount of εG ranged from 56 to 713 fmol/μmol guanine, with a mean of 208 ± 71 fmol/μmol guanine (Table 4). Age, sex and smoking were not significant factors, although the number of individuals sampled in these groups was small.

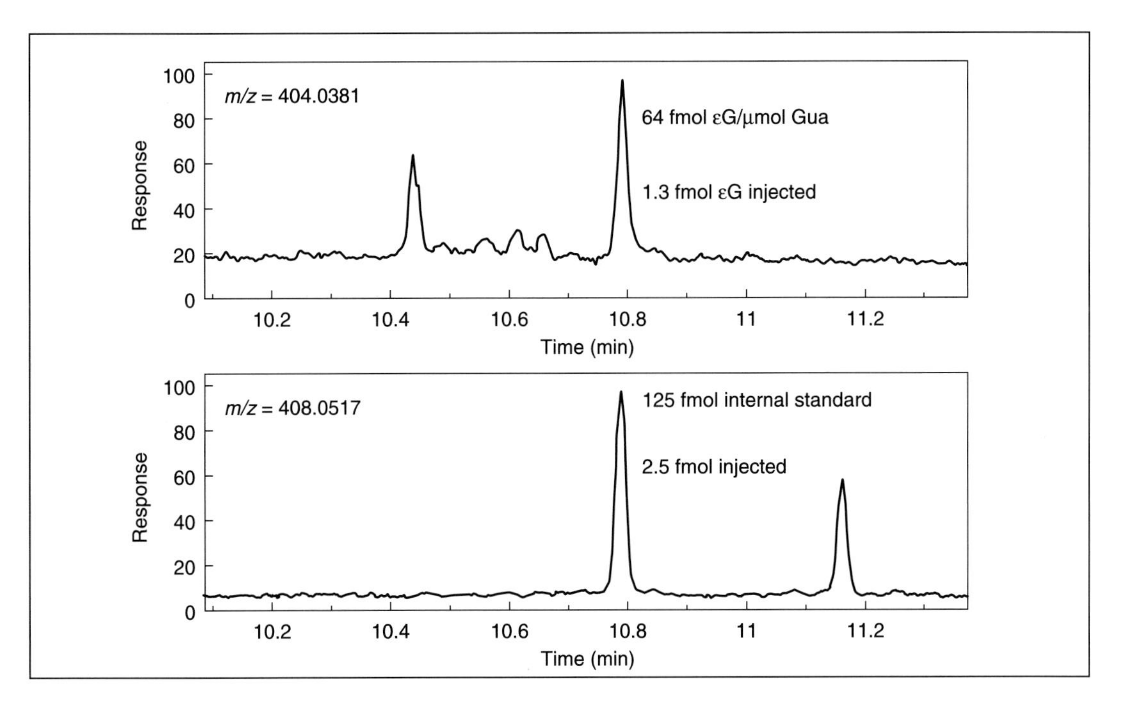

Figure 5. Representative gas chromatography–high-resolution mass spectrometry traces for N^2,3-ethenoguanine (εG) in human liver (30-year-old male)

m/z 404.0381 is εG and *m/z* 408.0517 is the internal standard. Samples were analysed by modified method of Fedtke *et al.* (1990b) with trifluoromethyltetrafluorobenzyl derivatives.

Table 4. Amounts of endogenous N^2,3-ethenoguanine (εG) in DNA from human liver samples

Age (years)	Sex	εG (fmol/μmol guanine)
8	Male	315
10	Female	56
16	Female	201
19	Male	393
25	Female	326
29	Male	358
30	Male	67
35	Male	187
39*	Male	207
45*	Male	289
51	Female	388
56	Male	713
Mean		208
SD		71

* Smoker
εG was determined by a modification of the method of Fedtke *et al.* (1990b), by analysis of trifluoromethyltetra-fluorobenzyl derivatives with high-resolution gas chromatography–mass spectrometry.

Studies on vinyl fluoride-exposed rats and mice

Molecular dose and haemangiosarcoma incidence

During the conduct of a two-year study of carcinogenicity in rats and mice, tissues were collected at a 12-month interim necropsy for analyses of OEG and εG and comparisons with the incidence of haemangiosarcoma of the liver. Exposure was to 0, 25, 250 or 2500 ppm vinyl fluoride (6 h/day, five days per week) by whole-body inhalation (Bogdanffy *et al.*, 1995). DNA was isolated from the livers of these chronically exposed animals and εG was analysed by the Fedtke procedure and GC–high-resolution MS (Swenberg *et al.*, 1995), while OEG was measured by HPLC and fluorescence detection by the method of Fedtke *et al.* (1989).

Figure 6 shows the exposure–response to vinyl fluoride as OEG in rats. The data are strongly indicative of a supralinear response due to saturation of metabolic activation, very similar to that of vinyl chloride. The number of OEG adducts did not accumulate between 2 and 52 weeks of exposure (Holt *et al.*, 1998). This result is consistent with those for other N7 guanine DNA adducts (Bedell *et al.*, 1982; Lindamood *et al.*, 1982). The number of OEG adducts in rat pups that were similarly exposed to vinyl fluoride between 10 and 21 days of age was two to three times greater than in adults. These data are similar to those shown with vinyl chloride by Fedtke *et al.* (1992b).

The molecular dose of εG showed a strongly supralinear response after 12 months of exposure to vinyl fluoride (Figure 7). Chronic exposure to 25 ppm vinyl fluoride caused a 2.5-fold increase in mice and 3.5-fold increase in rats when compared with the respective controls. The additional increase in εG associated with exposure to 250 or 2500 ppm vinyl fluoride was less than that shown for vinyl chloride. Mice had greater steady-state numbers of εG adducts than rats under similar exposure conditions. There was no difference between males and females of either species.

The incidence of haemangiosarcomas in the two-year study was reported by Bogdanffy *et al.* (1995). Mice developed nearly twice as many neoplasms (60%) as rats (30%). When the number of εG adducts for each exposure group was compared with the incidence of haemangiosarcomas, interesting species differences were noted. In mice, there was a linear relationship between the inci-

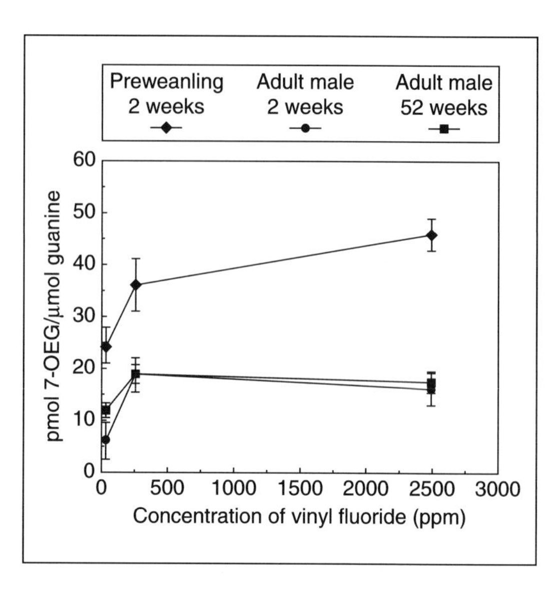

Figure 6. Exposure–response of 7-(2′-oxoethyl)guanine (OEG) to vinyl fluoride

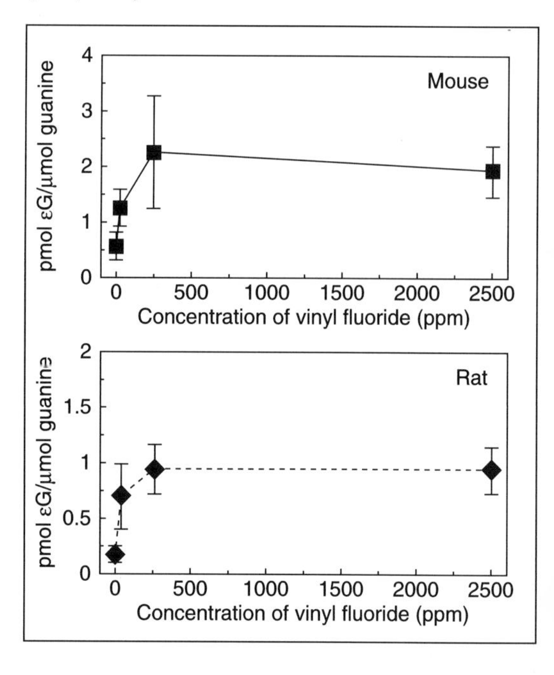

Figure 7. Molecular dose of N^2,3-ethenoguanine (εG) in mice and rats after 12 months' exposure to vinyl fluoride

dence of haemangiosarcomas and εG, while rats had a sublinear relationship at between 250 and 2500 ppm vinyl fluoride. This nonlinearity corresponded to an increase in cell proliferation demonstrated at 2500 ppm by Bogdanffy *et al.* (1990). When all of the data for rats and mice on εG and haemangiosarcomas were compared by regression analysis, a high correlation was seen (r^2 = 0.88; Figure 8). Thus, it appears that εG is highly correlated with the carcinogenesis of vinyl fluoride.

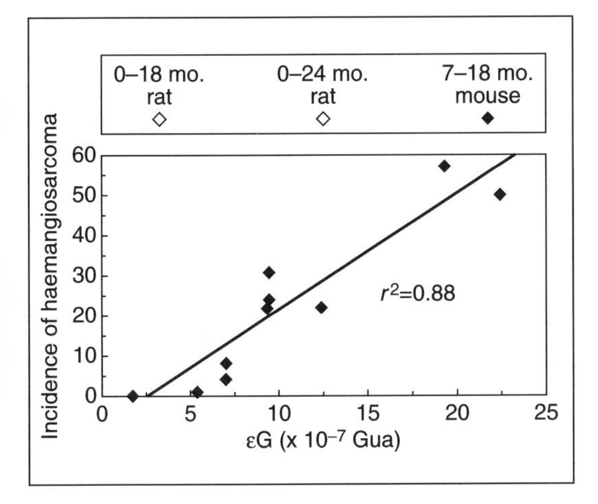

Figure 8. Correlation between N^2,3-ethenoguanine (εG) and incidence of haemangiosarcomas after exposure to vinyl fluoride

Cell type differences in formation and repair of N^2,3-ethenoguanine

Hepatic haemangiosarcomas develop from sinusoidal endothelial cells in the liver. To better approximate the target cell, adult rats were exposed to 2500 ppm vinyl fluoride for four weeks, the livers were perfused with collagenase, and nonparenchymal cells, which contain Kupffer and endothelial cells, were isolated by differential centrifugation by the methods of Bedell *et al.* (1982). The amount of εG was determined in hepatocytes and nonparenchymal cells. Vinyl fluoride is activated by cytochrome P450 2E1-mediated epoxidation, which is located in centrilobular hepatocytes. In contrast, nonparenchymal cells have little or no cytochrome P450 2E1, and essentially all metabolic activation takes place in hepatocytes. Cellular DNA would be readily exposed in hepatocytes but would be exposed in nonparenchymal cells only after diffusion from hepatocytes. Thus, initial formation would be expected to be lower in nonparenchymal cells. This was not found after four weeks of exposure, however, as nonparenchymal cells had 1.2 ± 0.9 pmol εG/µmol guanine, while hepatocytes had 0.4 pmol εG/µmol. These results suggested cell-specific differences in the repair of εG.

Dosanjh *et al.* (1994) demonstrated that purified *N*-methylpurine-DNA glycosylase can remove all four etheno–DNA adducts from DNA treated with the vinyl chloride metabolite chloroacetaldehyde. The reverse transcriptase polymerase chain reaction method of Roy *et al.* (1997) for measuring the expression of *N*-methylpurine-DNA glycosylase mRNA in cell culture was modified to quantify the expression in rat tissues (Holt *et al.*, 1998), and the effect of exposure to vinyl fluoride on gene expression was determined (Figure 9). While there was no difference between adult and preweanling control rats in whole liver, expression was induced by both 25 and 2500 ppm vinyl fluoride in both groups after 12 months. There was no difference in expression between the groups exposed to low and high concentrations.

Similar studies were conducted on hepatocytes and nonparenchymal cells of control rats and rats exposed to 2500 ppm vinyl fluoride for four weeks. The expression of *N*-methylpurine-DNA glycosy-

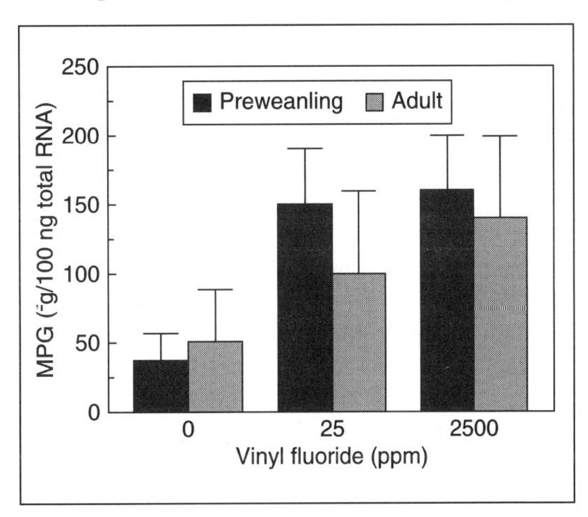

Figure 9. Effect of chronic exposure to vinyl fluoride on the expression of *N*-methylpurine-DNA glycosylase (MPG) in adult and preweanling rats

lase in nonparenchymal cells was only one-fifth that in hepatocytes (228.9 ± 36.3 versus 48.7 ± 1.6 fg per 250 ng total RNA). Furthermore, induction of this enzyme occurred only in the hepatocytes, so that the nonparenchymal cells of vinyl fluoride-exposed rats had only one-eighth the amount of *N*-methylpurine-DNA glycosylase mRNA that was present in hepatocytes. While Lefebvre *et al.* (1993) showed induction of this enzyme in cultured human cells exposed to a variety of DNA-damaging agents, this represents the first demonstration of induction of its expression *in vivo*.

Discussion

Vinyl halides are associated with the induction of hepatic haemangiosarcomas in many species of experimental animal and in humans. A total of 173 cases of haemangiosarcoma have been diagnosed in workers exposed to vinyl chloride (Lee *et al.*, 1996), all of whom were in the workforce before occupational exposure to vinyl chloride was lowered in the mid-1970s. Conversely, even though more than 20 years have elapsed, no cases of hepatic haemangiosarcoma have been reported in either European workers first exposed after 1972 or in workers in the United States first exposed after 1968 (Storm & Rozman, 1997). This is a major example of the protective effect of proper industrial hygiene. The data presented in this paper provide solid scientific evidence for the relationship between exposure to vinyl halides and endogenous εG. With two independent methods, it was shown that 10 ppm vinyl chloride resulted in a two- to threefold increase in the level of εG. Likewise, 25 ppm vinyl fluoride increased the amount of εG by 2.5–3.5 over the average amount present in unexposed rodents. As humans and rodents have similar amounts of endogenous εG, it is reasonable to assume that they would have a similar response. Simple interpolation of the data presented here suggests that current occupational standards are sufficient to protect workers. The likelihood of carcinogenic risks due to environmental exposures to parts per million is much less certain.

Our present understanding of the formation and repair of εG suggests that the epoxide of the vinyl halide is formed in hepatocytes and that it alkylates hepatocellular DNA to a greater extent than that of nonparenchymal cells but is efficiently repaired in hepatocytes. In contrast, the target cells for vinyl halide carcinogenesis are exposed to lower amounts of epoxide that have diffused from neighbouring hepatocytes, but these cells are less able to repair εG. We showed previously that nonparenchymal cells also have a higher rate of proliferation than hepatocytes (Lewis & Swenberg, 1982), so that this cell population is more likely to convert promutagenic DNA adducts into mutations. This results in greater potential for εG-related mutagenesis and carcinogenesis. It is likely that similar effects occur with the other etheno adducts, but this will require additional research.

The strength of mass spectrometric approaches for investigating endogenous and exogenous DNA damage is clear from the studies presented above. If DNA repair becomes induced or saturated, important alterations in endogenous adducts could take place. It will now be possible to determine such effects. The presence of endogenous εG in rodents and humans is thought to be associated with lipid peroxidation. Whether other factors also contribute remains to be shown. Whatever the source(s), endogenous DNA adducts are likely to play an important role in carcinogenesis. A high proportion of control animals held for two years develop neoplasia; likewise, one-third of the human population develops cancer, and known causes can account for only a portion of these cases. As endogenous DNA adducts are probably related to interindividual differences in genetics and lifestyle factors, it will be important to gain a much better understanding of their causes, extent of formation and repair.

Acknowledgements

This research was supported in part by grants from the Chemical Manufacturers' Association, the EPA/NIEHS Superfund Basic Research Program (P42-ES05948) and NIEHS Training Grants (ES07126, ES07017 and ES05779).

References

American Conference of Governmental Industrial Hygienists (1991) *Documentation of Threshold Limit Values and Biological Exposure Indices*, 6th Ed. Cincinnatti, OH

Barbin, A., Bartsch, H., Leconte, P. & Radman, M. (1985) Lack of miscoding properties of 7-(2-oxoethyl)guanine, the major vinyl chloride–DNA adduct. *Cancer Res.* 45, 2440–2444

Barbin, A., Froment, O., Boivin, S., Marion, M.-J., Belpoggi, F., Maltoni, C. & Montesano, R. (1997) p53 gene mutation pattern in rat liver tumors induced by vinyl chloride. *Cancer Res.* 57, 1695–1698

Basu, A.K., Wood, M.L., Niedernhofer, L.J., Ramos, L.A. & Essigmann, J.M. (1993) Mutagenic and genotoxic effects of three vinyl chloride-induced DNA lesions: 1,N^6-ethenoadenine, 3,N^4-ethenocytosine, and 4-amino-5-(imidazol-2-yl)imidazole. *Biochemistry* 32, 12793–12801

Bedell, M.A., Lewis, J.G., Billings, K.C. & Swenberg, J.A. (1982) Cell specificity in hepatocarcinogenesis: Preferential accumulation of O^6-methylguanine in target cell DNA during continuous exposure of rats to 1,2-dimethylhydrazine. *Cancer Res.* 42, 3079–3083

Bogdanffy, M.S., Kee, C.R., Kelly, D.P., Carakostas, M.C. & Sykes, G.P. (1990) Subchronic inhalation study with vinyl fluoride: Effects on hepatic cell proliferation and urinary fluoride excretion. *Fundam. Appl. Toxicol.* 15, 394–406

Bogdanffy, M.S., Makovec, G.T. & Frame, S.R. (1995) Inhalation oncogenicity bioassay in rats and mice with vinyl fluoride. *Fundam. Appl. Toxicol.* 26, 223–238

Cheng, K.C., Preston, B.D., Cahill, D.S., Dosanjh, M.K., Singer, B. & Loeb, L.A. (1991) The vinyl chloride DNA derivative N^2,3-ethenoguanine produces G→A transitions in *Escherichia coli*. *Proc. Natl Acad. Sci. USA* 88, 9974–9978

Ciroussel, F., Barbin, A., Eberle, G. & Bartsch, H. (1990) Investigations on the relationship between DNA ethenobase adduct levels in several organs of vinyl chloride-exposed rats and cancer susceptibility. *Biochem. Phamacol.* 39, 1109–1113

Clewell, H.J., Gentry, P.R., Gearhart, J.M., Allen, B.C. & Andersen, M.E. (1995) Considering pharmacokinetic and mechanistic information in cancer risk assessments for environmental contaminants: Examples with vinyl chloride and trichloroethylene. *Chemisphere* 31, 2561–2578

Creech, J.L. & Johnson, M.N. (1974) Angiosarcoma of the liver in the manufacture of polyvinyl chloride. *J. Occup. Med.* 16, 150–151

Dannaher, C.L., Tamburro, C.H. & Yam, L.T. (1981) Occupational carcinogenesis: The Louisville experience with vinyl chloride-associated hepatic angiosarcoma. *Am. J. Med.* 70, 279–287

Dosanjh, M.K., Chenna, A., Kim, E., Fraenkel-Conrat, H., Samson, L. & Singer, B. (1994) All four known cyclic adducts formed in DNA by the vinyl chloride metabolite chloroacetaldehyde are released by a human DNA glycosylase. *Proc. Natl Acad. Sci. USA* 91, 1024–1028

Fabricant, J.D. & Legator, M.S. (1981) Mutagenicity studies of vinyl chloride. *Environ. Health Perspectives* 41, 189–193

Falk, H., Herbert, J., Crowley, S., Ishak, K.G., Thomas, L.B., Popper, H. & Caldwell, G.G. (1981) Epidemiology of hepatic angiosarcoma in the United States: 1964–1974. *Environ. Health Perspectives* 41, 107–113

Fedkte, N., Walker, V.E. & Swenberg, J.A. (1989) Determination of 7-(2-oxoethyl)guanine and N^2,3-ethenoguanine in DNA hydrolysates by HPLC. *Arch. Toxicol.* 13S, 214–218

Fedkte, N., Boucheron, J.A., Walker, V.E. & Swenberg, J.A. (1990a) Vinyl chloride-induced DNA adducts. II: Formation and persistence of 7-(2′-oxoethyl)guanine and N^2,3-ethenoguanine in rat tissue DNA. *Carcinogenesis* 11, 1287–1292

Fedkte, N., Boucheron, J.A., Turner, M.J., Jr & Swenberg, J.A. (1990b) Vinyl chloride-induced DNA adducts. I: Quantitative determination of N^2,3-ethenoguanine based on electrophore labeling. *Carcinogenesis* 11, 1279–1285

Froment, O., Boivin, S., Barbin, A., Bancel, B., Trepo, C. & Maron, M.-J. (1994) Mutagenesis of *ras* proto-oncogenes in rat liver tumors induced by vinyl chloride. *Cancer Res.* 54, 5340–5345

Gehring, P.J., Watanabe, P.G. & Park, C.N. (1978) Resolution of dose–response toxicity data for chemicals requiring metabolic activation. *Toxicol. Appl. Pharmacol.* 44, 581–591

Guengerich, F.P. (1992) Roles of the vinyl chloride oxidation products 2-chlorooxirane and 2-chloroacetaldehyde in the *in vitro* formation of etheno adducts of nucleic acid bases. *Chem. Res. Toxicol.* 5, 2–5

Guengerich, F.P., Kim, D.H. & Iwasaki, M. (1991) Role of human cytochrome P-450 IIE1 in the oxidation of many low molecular weight cancer suspects. *Chem. Res.Toxicol.* 4, 168–179

Guichard, Y., El Ghissassi, F., Nair, J., Bartsch, H. & Barbin, A. (1996) Formation and accumulation of DNA ethenobases in adult Sprague-Dawley rats exposed to vinyl chloride. *Carcinogenesis* 17, 1553–1559

Gwinner, L.M., Laib, L.B., Filser, J.G. & Bolt, H.M. (1983) Evidence of chloroethylene oxide being the reactive metabolite of vinyl chloride towards DNA: Comparative studies with 2,2'-dichlorodiethylether. *Carcinogenesis* 4, 1483–1486

Ham, A.-J. L. & Swenberg, J.A. (1998) Immunoaffinity (IA) gas chromatography/mass spectrometry (GC/MS) method for N^2,3-ethenoguanine demonstrates importance of hydrolysis method for adduct quantitation. *Toxicol. Sci.* 1S, 180

Hang, B., Chenna, A., Rao, S. & Singer, B. (1996) 1,N^6-Ethenoadenine and 3,N^4-ethenocytosine are excised by separate human DNA glycosylases. *Carcinogenesis* 17, 155–157

Hang, B., Medina, M., Fraenkel-Conrat, H. & Singer, B. (1998) A 55-kDa protein isolated from human cells shows DNA glycosylase activity toward 3,N^4-ethenocytosine and the G/T mismatch. *Proc. Natl Acad. Sci. USA* 95, 13561–13566

Hollstein, M., Marion, M.-J., Lehman, T., Welsh, J., Harris, C.C., Martel-Planche, G., Kusters, I. & Montesano, R. (1994) p53 mutations in A:T base pairs in angiosarcomas of vinyl chloride-exposed factory workers. *Carcinogenesis* 15, 1–3

Holt, S., Scheller, N.A., Mitra, S., Bogdanffy, M.S. & Swenberg, J.A. (1998) Formation and repair of vinyl fluoride-induced DNA adducts. *Toxicol. Sci.* 1S, 180

La, D.K. & Swenberg, J.A. (1996) DNA adducts: Biological markers of exposure and potential applications to risk assessment. *Mutat. Res.* 365, 129–146

Lee, F.I., Smith, P.M., Bennett, B. & Williams, D.M.J. (1996) Occupationally related angiosarcoma of the liver in the United Kingdom 1972–1994. *Gut* 39, 312–318

Lefebvre, P., Zak, P. & Laval, F. (1993) Induction of O^6-methylguanine-DNA-methyltransferase and N^3-methyladenine-DNA-glycosylase in human cells exposed to DNA-damaging agents. *DNA Cell Biol.* 12, 233–241

Lewis, J.G. & Swenberg, J.A. (1982) The effect of 1,2-dimethylhydrazine and diethylnitrosamine on cell replication and unscheduled DNA synthesis in target and nontarget cell populations in rat liver following chronic administration. *Cancer Res.* 42, 89–92

Lindamood, C., Bedell, M.A., Billings, K.C. & Swenberg, J.A. (1982) Alkylation and de novo synthesis of liver cell DNA from C3H mice during continuous dimethylnitrosamine exposure. *Cancer Res.* 42, 4153–4157

Maltoni, C., Lefemine, G., Ciliberti, A., Cotti, G. & Carretti, D. (1981) Carcinogenicity bioassays of vinyl chloride monomer: A model of risk assessment on an experimental basis. *Environ. Health. Perspectives* 41, 3–29

Marion, M.-J., Froment, O. & Trepo, C. (1991) Activation of Ki-*ras* gene by point mutation in human liver angiosarcoma associated with vinyl chloride exposure. *Mol. Carcinog.* 4, 450–454

Matijasevic, Z., Sekiguchi, M. & Ludlum, D.B. (1992) Release of N^2,3-ethenoguanine from chloroacetaldehyde-treated DNA by *Escherichia coli* 3-methlyadenine DNA glycosylase II. *Proc. Natl Acad. Sci. USA* 89, 9331–9334.

Mitro, K.L., Scheller, N.A., Ranasinghe, A. & Swenberg, J.A. (1995) Quantitation of endogenous N^2,3-ethenoguanine in human and rat liver DNA using high resolution mass spectrometry. *Proc. Am. Assoc. Cancer Res.* 36, 142

Mroczkowska, M.M. & Kusmierek, J.T. (1991) Miscoding potential of N^2,3-ethenoguanine studied in an *Escherichia coli* DNA-dependent RNA polymerase in vitro system and possible role of this adduct in vinyl chloride-induced mutagenesis. *Mutagenesis* 6, 385–390

National Institute for Occupational Safety and Health (1990) *National Occupational Exposure Survey*. US Department of Health and Human Services, Cincinnatti, OH

Nicholson, W.J., Hammond, E.C., Seidman, H. & Selikoff, I.J. (1975) Mortality experience of a cohort of vinyl chloride-polyvinyl chloride workers. *Ann. N.Y. Acad. Sci.* 246, 225–230

Oesch, F., Adler, S., Rettelbach, R. & Doerjer, G. (1986) Repair of etheno DNA adducts by *N*-glycosylases. In: *The Role of Cyclic Nucleic Acid Adducts in Carcinogeneisis and Mutagenesis* (B. Singer & H. Bartsch, eds), IARC Sci. Publ. 70. Lyon, IARC, 373–379

Ranasinghe, A., Scheller, N., Wu, K.Y., Upton, P.B. & Swenberg, J.A. (1998) Application of gas chromatography/electron capture negative chemical ionization high-resolution mass spectrometry for analysis of DNA and protein adducts. *Chem. Res. Toxicol.* 11, 520–526

Roy, G., Roy, R. & Mitra, S. (1997) Quantitative reverse transcriptase polymerase chain reaction for measuring N-methylpurine-DNA glycosylase mRNA level in rodent cells. *Anal. Biochem.* 246, 45–51

Saparbaev, M. & Laval, J. (1998) 3,N^4-Ethenocytosine, a highly mutagenic adduct, is a primary substrate for *Escherichia coli* double-stranded uracil-DNA glycosylase and human mismatch-specific thymine-DNA glycosylase. *Proc. Natl Acad. Sci. USA* 95, 8508–8513

Singer, B., Spengler, S.J., Chavez, F. & Kusmierek, J.T. (1987) The vinyl chloride-derived nucleoside, N^2,3-ethenoguanosine, is a highly efficient mutagen in transcription. *Carcinogenesis* 8, 745–747

Singer, B., Kusmierek, J.T., Folkman, W., Chavez, F. & Dosanjh, M.K. (1991) Evidence for the mutagenic potential of the vinyl chloride induced adduct, N^2,3-etheno-deoxyguanosine, using a site-directed kinetic assay. *Carcinogenesis* 12, 745–747

Storm, J.E. & Rozman, K.K. (1997) Evaluation of alternative methods for establishing safe levels of occupational exposure to vinyl halides. *Reg. Toxicol. Pharmacol.* 44, 481–489

Swenberg, J.A., Fedtke, N., Ciroussel, F., Barbin, A. & Bartsch, H. (1992) Etheno adducts formed in DNA of vinyl chloride-exposed rats are highly persistent in liver. *Carcinogenesis* 13, 727–729

Swenberg, J.A., La, D.K., Scheller, N.A. & Wu, K.Y. (1995) Dose–response relationships for carcinogens. *Toxicol. Lett.* 82, 751–756

Tabershaw, I.R. & Gaffey, W. (1974) Mortality study of workers in the manufacture of vinyl chloride and polyvinyl chloride. *J. Occup. Med.* 16, 509–518

Viola, P.L., Bigotti, A. & Caputo, A. (1971) Oncogenic response of rat skin, lungs, and bones. *Cancer Res.* 31, 512–522

Watanabe, P.G. & Gehring, P.G. (1976) Dose-dependent fate of vinyl chloride and its possible relationship to oncogenicity in rats. *Environ. Health. Perspectives* 17, 145–152

Exocyclic DNA Adducts in Mutagenesis and Carcinogenesis
B. Singer & H. Bartsch
IARC Scientific Publications No. 150
International Agency for Research on Cancer, Lyon, 1999

Role of 1,N^2-propanodeoxyguanosine adducts as endogenous DNA lesions in rodents and humans

F.-L. Chung, L. Zhang, J.E. Ocando & R.G. Nath

Results obtained in a number of studies *in vitro* and *in vivo* support the hypothesis that short- and long-chain enals and their epoxides derived from oxidized polyunsaturated fatty acids are potential endogenous sources of cyclic propano and etheno DNA adducts. We previously reviewed the evidence from some of these studies. Here, we describe the results of our more recent studies on the role of 1,N^2-propanodeoxyguanosine adducts as endogenous DNA lesions. These studies include: the detection of distinct patterns of such adducts in various tissues of different species; the detection of long-chain *trans*-4-hydroxynonenal-derived deoxyguanosine adducts *in vivo*; the specificity of the formation of enal-derived propano adducts from ω–3 and ω–6 polyunsaturated fatty acids; and the detection of acrolein- and crotonaldehyde-derived adducts in human oral tissue DNA and their increased levels in smokers. Taken together, these studies further strengthen the hypothesis that enals produced by lipid peroxidation are the primary source for cyclic propano adducts *in vivo*, but these results cannot rule out the possible contribution of environmental and other sources. The mutagenicity of enals and their epoxides and the results of site-specific mutagenesis studies indicate that the cyclic adducts are potential promutagenic lesions; however, only circumstantial evidence is currently available for their role in carcinogenesis.

Introduction

The propano, etheno and malondialdehyde-derived adducts, named according to their newly adducted ring moeity, are perhaps the most widely studied of the cyclic DNA bases. These adducts are products of the reactions of bi-functional electrophilic compounds with DNA bases. The propano adducts are formed by reactive α,β-unsaturated aldehydes or enals, such as acrolein, crotonaldehyde and *trans*-4-hydroxy-2-nonenenal (HNE) (Chung *et al.*, 1984; Winter *et al.*, 1986; Meerman *et al.*, 1989; Eder & Hoffman, 1993; Yi *et al.*, 1997). Acrolein and crotonaldehyde are ubiquitous pollutants in the environment and also products of peroxidation of lipids, whereas the third is a unique oxidation product of ω–6 polyunsaturated fatty acids (Esterbauer *et al.*, 1991; Wu & Lin, 1995). The etheno adducts are products of reactions with chloroacetaldehyde, 1-substituted oxiranes and the epoxides of enals (Sodum & Chung, 1991; Guengerich *et al.*, 1993; Park *et al.*, 1993). The malondialdehyde-derived adduct is formed either by reaction with malondialdehyde, a known product of lipid peroxidation, or with base propenals arising from oxidized DNA (Marnett *et al.*, 1986; Dedon *et al.*, 1998).

In the past several years, we have studied acrolein-, crotonaldehyde- and HNE-derived 1,N^2-propanodeoxyguanosine (acrolein-, crotonaldehyde- and HNE-dG; Figure 1) and related cyclic adducts. We showed by a ^{32}P-postlabelling–high-performance liquid chromatography (HPLC) assay that the cyclic propano dG adducts are present in tissue DNA of humans and untreated rodents at relatively high levels (Nath & Chung, 1994; Nath *et al.*, 1994, 1996). These observations raised the possibility that the adducts are formed *via* endogenous pathways. Because peroxidation of polyunsaturated fatty acids is known to yield enals of various chain lengths, ranging from acrolein to HNE as secondary products (Esterbauer *et al.*, 1991; Wu & Lin, 1995), we proposed that membrane fatty

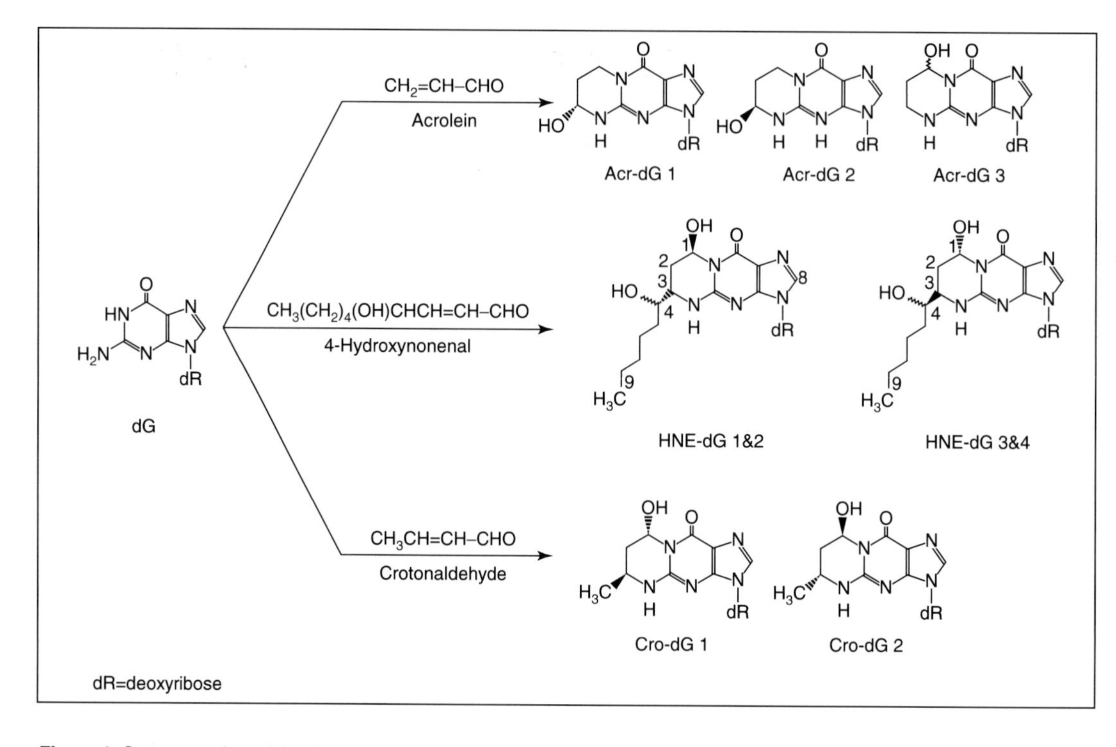

Figure 1. Structures of acrolein–deoxyguanosine (Acr-dG), crotonaldehyde–deoxyguanosine (Cro-dG) and *trans*-4-hydroxy-2-nonenenal–deoxyguanosine (HNE-dG) adducts

Acr-dG consists of three isomers, and one of them, Acr-dG 3 with a hydroxy group adjacent to N1 of guanine, is the predominant one detected *in vivo*, Cro-dG consists of two isomers, and HNE-dG consists of four isomers. The alkyl and hydroxy group of the propano ring are in *trans* configuration in Cro-dG and HNE-dG (Chung *et al.*, 1984; Yi *et al.*, 1997).

acids may be the endogenous sources of these cyclic adducts. We carried out a number of studies *in vitro* and *in vivo* to determine whether oxidation of polyunsaturated fatty acids is indeed the source. These studies show that:

- The levels of propano adducts are dramatically increased in rat liver DNA after depletion of glutathione.
- Their levels are also increased in the liver DNA of carbon tetrachloride-treated rats and of a mutant strain of Long Evans rats genetically predisposed to a higher level of lipid peroxidation.
- The cyclic propano-dG adducts of acrolein and crotonaldehyde are background lesions which appear to occur with distinct patterns in various tissues of several species, including humans.
- In addition to the propano adducts derived from the short-chain enals, the HNE-dG adducts are also present *in vivo* as background lesions.
- Under conditions of stimulated lipid peroxidation *in vitro*, acrolein-, crotonaldehyde- and HNE-dG adducts are formed specifically from arachidonic acid, linoleic acid (ω–6) or docosahexaenoic acid (ω–3).
- Acrolein- and crotonaldehyde-dG are present in human oral tissue, and their levels are increased in smokers.

Collectively, these results indicate that oxidized polyunsaturated fatty acids are an important endogenous source of propano adducts. We have reviewed some of these studies previously (Chung *et al.*, 1996). Here, we describe recent studies and discuss the possible roles of these adducts in mutagenesis and carcinogenesis.

Table 1. Species and tissues shown to contain acrolein-deoxyguanosine and crotonaldehyde–deoxyguanosine adducts and the range of total adduct levels in each species

Species	Strain	Tissue	Total adduct levels (µmol/mol guanine)
Mouse	A/J	Liver and skin	0.31–1.69
Rat	Fischer 344	Liver, mammary gland, colonic mucosa, kidney, brain, lung, prostate and leukocytes	0.11–1.12
	Wistar	Liver and urinary bladder	
Human		Liver, leukocytes, gingiva and breast	0.01–7.53[a]

[a] The highest levels were found in the gingival DNA of a smoker (Nath *et al.*, 1998)

Presence of acrolein–deoxyguanosine and crotonaldehyde–deoxyguanosine in distinct patterns in tissues and of *trans*-4-hydroxy-2-nonenenal–deoxyguanosine *in vivo*

We have found that acrolein-dG and crotonaldehyde-dG are prevalent DNA lesions in a variety of rodents and human tissues, including colon, lung, brain, prostate, kidney, mammary gland, liver and leukocytes (Nath & Chung, 1994; Nath *et al.*, 1996). Table 1 summarizes the species and tissues in which these adducts have been detected and the range of levels of acrolein-dG plus crotonaldehyde-dG in each species. These adducts are detected in each tissue in distinct patterns (Figure 2). For example, the total levels were the highest in rat brain and lung, followed by

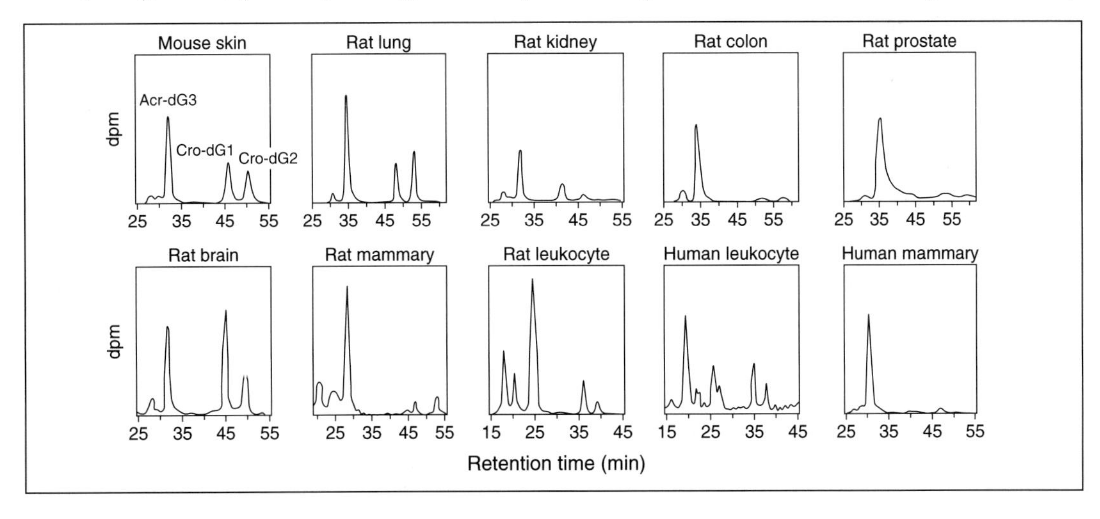

Figure 2. Presence of specific patterns of acrolein–deoxyguanosine and crotonaldehyde–deoxyguanosine adducts in rodent and human tissues in final analysis based on co-migration with the adduct standards on reverse-phase high-performance liquid chromatography (HPLC)

The retention times varied in some cases due to ageing of the HPLC columns.

colon and prostate; the adduct levels in rat kidney were the lowest of all the tissues examined so far. Rat colon, prostate and mammary gland DNA showed almost exclusively acrolein-dG3, an isomer of the acrolein-dG adduct, with very low levels of crotonaldehyde-dG. In contrast, brain and lung had relatively high levels of crotonaldehyde-dG adducts. Human tissues such as the liver and breast consistently showed much greater variation in acrolein-dG and crotonaldehyde-dG levels than mouse and rat tissues, indicating greater individual variation among humans than among inbred animals. While the significance of the tissue-specific and stereoselective pattern of these adducts in tissues is as yet unknown, these results support their endogenous origins and argue against their being artefacts. Furthermore, the levels detected in each tissue may represent a steady state as a result of constant formation and removal. Such a state of equilibrium could be altered by a number of factors, such as antioxidant status, repair activity and, possibly, fatty acid composition in the tissue.

Recently, we detected the long-chain HNE-dG adducts in human and rat tissue by a ^{32}P-postlabelling–HPLC method similar to that for acrolein- and crotonaldehyde-dG. HNE-dG adducts consist of two pairs of diastereomers of HNE-dG, 1,2 and 3,4, because of the chiral carbon in the side chain (Figure 1; Yi *et al.*, 1997). HNE-dG adducts with a long alkyl side-chain are considerably more lipophilic than acrolein-dG or crotonaldehyde-dG, and therefore solid-phase extraction can be used to enrich HNE-dG adducts in DNA hydrolysates, followed by one-dimensional thin-layer chromatography purification, and subsequent analysis by reversed-phase HPLC. The final HPLC analysis of DNA obtained from humans and rat liver or colon showed radioactive peaks due to the DNA in these tissues co-migrating with the ultraviolet standards of HNE-dG 1,2 and 3,4. In addition, the identities of these adducts were confirmed by the co-migration of the 5′-monophosphates of HNE-dG after hydrolysis by polynucleotide kinase. Unlike acrolein and crotonaldehyde, which are ubiquitous environmental pollutants, HNE is a unique product of oxidized polyunsaturated fatty acids. Thus, the detection of HNE-dG adducts indicates that peroxidation of fatty acids is likely to be an important pathway and source for their formation. While the protein adducts of HNE have been detected in tissues (Yoritaka *et al.*, 1996), our study is the first to show the presence of the HNE–DNA adducts in human and rodent tissues *in vivo*.

Cyclic propano adducts are formed from arachidonic acid, linoleic acid and docosahexaenoic acid *in vitro* under lipid peroxidation conditions

In these studies, we examined whether acrolein-dG, crotonaldehyde-dG and HNE-dG are formed when lipid peroxidation is stimulated and whether the formation of each adduct is dependent on the types of polyunsaturated fatty acids. Incubations were carried out with a mixture containing 1.5 mmol/L linoleic acid, arachidonic acid or docosahexaenoic acid and deoxyguanosine 5′-monophosphate (25 µmol or 25 µmol/L) in the presence of $FeSO_4$ (0.75 mmol/L) and NADPH (0.1 mmol/L) in 1 mL Tris-HCl buffer (0.1 mol/L, pH 7.1) under a previously described condition in which the etheno adducts of adenine and cytosine were detected with rat liver microsomes (El Ghissassi *et al.*, 1995). The incubation mixture was analysed by a sequential reversed-phase HPLC for acrolein-dG, crotonaldehyde-dG and HNE-dG. The identities of these products were confirmed by comparing the HPLC retention times and ultraviolet spectrum to those of the authentic synthetic standards. Acrolein-dG and crotonaldehyde-dG were formed predominantly from docosahexaenoic acid, whereas HNE-dG adducts were formed exclusively from arachidonic acid and linoleic acid. Figure 3 shows typical HPLC chromatograms obtained from analyses of the incubation mixture with linoleic acid. The quantities of HNE-dG adducts formed by arachidonic acid and linoleic acid were 20.6 and 3.5 nmol, respectively, whereas HNE-dG adducts were not detected in the reaction with docosahexaenoic acid; 5.4 and 0 nmol (not detectable) of acrolein-dG and crotonaldehyde-dG were obtained from linoleic acid, and 54.0 and 20.2 nmol were obtained from docosahexaenoic acid, respectively. Interestingly, acrolein-dG3, the major acrolein-dG isomer detected *in vivo*, was also the major adduct found in these reactions. Overall, these results agree with the reported enal production from various polyunsaturated fatty acids, and they clearly demonstrate the specificity of the formation of cyclic propano adducts after peroxidation of polyunsaturated fatty acids.

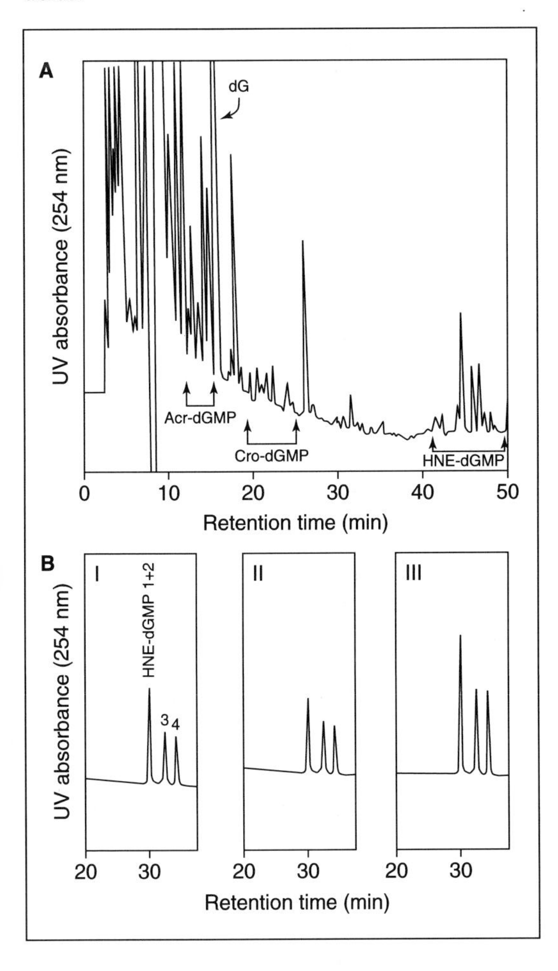

Figure 3. High-performance liquid chromatography of the incubation mixture from a reaction with linoleic acid

In (A), the chromatogram indicates the fractions corresponding to acrolein–deoxyguanosine (Acr-dG), crotonaldehyde–deoxyguanosine (Cro-dG) and *trans*-4-hydroxy-2-nonenenal–deoxyguanosine (HNE-dG) adducts; (B), chromatograms from the analysis of (i) the collected peaks corresponding to HNE-dG adducts in (A), (ii) the standard HNE-dG adducts and (iii) the HNE-dG adducts plus the standards.

Effect of depletion of reduced glutathione on acrolein–deoxyguanosine and crotonaldehyde–deoxyguanosine levels in liver DNA

Reduced glutathuone (GSH) is an endogenous antioxidant and an effective scavenger of enals. The dual function of GSH suggests that sustained depletion of this substance in tissues could increase the levels of acrolein-dG and crotonaldehyde-dG in tissue DNA, if these adducts are derived from enals generated by lipid peroxidation. We examined the effect of GSH depletion caused by L-buthionine (S,R)-sulfoximine (BSO) on the levels of these adducts in liver DNA from male Fischer 344 rats (Nath *et al.*, 1997). A group of five male rats was given BSO (30 mmol/L) in drinking-water for three weeks, and a group of eight rats was given water only. The average body weight in the BSO group was about 15% lower than that in the control group at termination. At the end of treatment, the liver from each animal was immediately processed for GSH and cyst(e)ine determination by HPLC with an electrochemical detector (Kleinman & Richie, 1995), and DNA was isolated. The hepatic GSH levels were reduced by more than 80% in the BSO-treated rats, from 4.43 to 0.72 μmol/g tissue. The mean levels of acrolein-dG3, crotonaldehyde-dG1 and crotonaldehyde-dG2 were 1.18 ± 1.03, 3.12 ± 3.26 and 2.50 ± 2.59 μmol/mol guanine in the group given BSO and 0.57 ± 0.25, 0.15 ± 0.18 and 0.16 ± 0.22 μmol/mol guanine in the control group, respectively. These increases are approximately twofold for acrolein-dG (p = 0.17) and 15–21-fold for crotonaldehyde-dG ($p < 0.05$).

The differential increase in acrolein-dG and crotonaldehyde-dG observed in the BSO-treated animals is intriguing. It could be due to various factors, including differences in the reactivity of acrolein and crotonaldehyde towards GSH and DNA or stability and repair of these adducts. It should be noted that the treated animals had significantly lowered body weights. These effects of BSO were unexpected, since the same dose did not affect the body weights of mice even though, on a body-weight basis, the mice were exposed to a higher dose of BSO than rats (Sun *et al.*, 1985). Nonetheless, this study demonstrates that tissue GSH may play an important role in protecting DNA from cyclic adduction by enals; it also supports the endogenous origin of acrolein-dG and crotonaldehyde-dG.

Detection of acrolein–deoxyguanosine and crotonaldehyde–deoxyguanosine in human oral tissue DNA and their increased levels in smokers

We recently analysed human oral tissue DNA for acrolein-dG and crotonaldehyde-dG. Since smokers generally have an extra oxidative burden due to their exposure to cigarette smoke, we investigated whether the cyclic propano-dG adduct levels are higher in the oral tissues of smokers than in those of nonsmokers (Nath *et al.*, 1998). The gingival tissue DNA from 11 smokers (four men and seven women aged 30–58 years) and 12 nonsmokers (eight men and four women aged 21–66 years old) were thus analysed by the ^{32}P-postlabelling–HPLC method. We found that the mean acrolein-dG levels (μmol/mol guanine) in smokers were significantly higher than those in nonsmokers: 1.36 ± 0.90 in smokers vs. 0.46 ± 0.26 in nonsmokers (p = 0.003). The mean crotonaldehyde-dG1 levels in smokers and nonsmokers were 0.53 ± 0.44 and 0.06 ± 0.07, respectively, corresponding to an 8.8-fold increase in smokers (p = 0.0015). The levels of crotonaldehyde-dG2 were also increased, by 5.5-fold, in smokers as compared with nonsmokers, from 0.31 ± 0.40 to 1.72 ± 1.26 μmol/mol guanine (p = 0.0014). Furthermore, the total level of cyclic adducts (acrolein-dG and crotonaldehyde-dG) in smokers was 4.4-fold greater than in nonsmokers (p = 0.0003). These results are summarized in Figure 4.

The increases in acrolein-dG and crotonaldehyde-dG levels in smokers' gingival DNA observed in this study could be due to stimulated production of endogenous enals *via* lipid peroxidation caused by the oxidants in cigarette smoke and/or direct exposure to acrolein and crotonaldehyde in the smoke. The pro-oxidant state of smokers has previously been characterized by higher plasma levels of malondialdehyde, lower GSH levels and increased 8-hydroxydeoxyguanine levels in leukocytes and lung DNA (Bridges *et al.*, 1993; Hulea *et al.*, 1995; Asami *et al.*, 1996; Yarborough *et al.*, 1996). The plasma GSH levels returned to normal when smoking was stopped

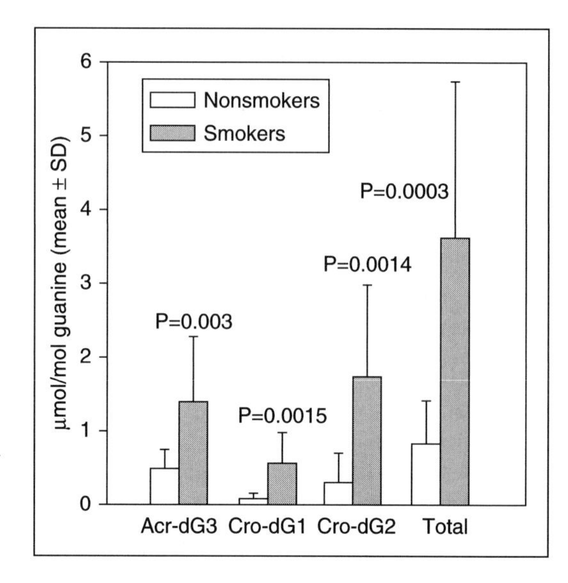

Figure 4. Levels of cyclic adducts in the oral tissue DNA of smokers and nonsmokers

Acr-dG, acrolein–deoxyguanosine; Cro-dG, crotonaldehyde–deoxyguanosine

(Lane *et al.*, 1996). As discussed above, the depletion of hepatic GSH by treatment with BSO caused a dramatic increase in the 1,N^2-propano-dG adduct levels in rat liver DNA. Interestingly, in that study, the increases in crotonaldehyde-dG in the BSO-treated rats were much more pronounced than those of acrolein-dG, a result similar to that obtained in the study in smokers. These observations lend further support to endogenous formation of acrolein-dG and crotonaldehyde-dG in smokers.

1,N^2-Propanodeoxyguanosine adducts in mutagenesis and carcinogenesis

The identification of enal-derived 1,N^2-propano-dG adducts as endogenous DNA lesions in rodent and human tissues has raised questions about their roles in mutagenesis and carcinogenesis. Although there is as yet no direct proof for their involvement in carcinogenesis, a number of studies have provided suggestive evidence.

Acrolein and crotonaldehyde are known to induce revertants in the *Salmonella* tester strains TA100 and TA104 (Marnett *et al.*, 1985), whereas

HNE was inactive in these strains (Chung *et al.*, 1993), although it induced mutation in V79 Chinese hamster cells (Cajelli *et al.*, 1987). We previously examined the mutagenicity of acrolein-dG in an immunoassay developed for acrolein-dG (Foiles *et al.*, 1989). Formation of this adduct in DNA from *S. typhimurium* tester strains TA100 and TA104 exposed to acrolein was analysed under conditions in which revertants are induced. Acrolein-dG3 was detected in a dose-dependent manner in both strains. The mutagenicity of acrolein was also measured in these strains in a liquid preincubation assay. The correlation between acrolein-dG3 levels and revertants/plate in TA100, which contains GC base pairs at the site of reversion, suggests that acrolein-dG3 is a promutagenic lesion. Acrolein also induced revertants in TA104, which has AT pairs at the site of reversion, suggesting that the adenine adducts may be involved in the mutagenicity in this strain.

Although the mutation characteristics of the etheno adducts that have been detected *in vivo*, such as 1,N^6-ethenodA, N^2,3-ethenodG and 3,N^4-ethenodC have been investigated in studies of site-specific mutagenesis (Cheng *et al.*, 1991; Palejwala *et al.*, 1991; Basu *et al.*, 1993; Pandya & Moriya, 1996), no information is available on the mutagenic property of acrolein-dG, crotonaldehyde-dG and HNE-dG. A model 1,N^2-propano-dG adduct without the hydroxy group in the propano ring has, however, been used as a substitute. This induced G→A and G→T mutations and frame-shift mutation (Benamira *et al.*, 1992; Moriya *et al.*, 1994). It is known that the outcome of studies of site-specific mutagenesis can vary with the structures of adducts, sequence context and host systems; these results, nonetheless, indicate that the model propano-dG adduct is a promutagenic lesion.

At present, there is only circumstantial evidence to support a role of acrolein-dG and crotonaldehyde-dG in carcinogenesis. The increases in adduct levels over background in rodents treated with *N*-nitrosopyrrolidine or cyclophosphamide, both of which are carcinogens known to release crotonaldehyde and acrolein, respectively, after metabolism suggest that these adducts may be important in the induction of cancer by these chemicals (Chung *et al.*, 1989; Wilson *et al.*, 1991). It should be noted, however, that other DNA alkylating agents are formed after treatment with these carcinogens, in addition to acrolein and crotonaldehyde-derived adducts. The low spontaneous liver tumour incidences in Fischer 344 rats suggests that the background propano adducts in liver may not be sufficient for tumour development during the lifetime of these animals. Further, we have shown that crotonaldehyde given in drinking-water was a weak liver carcinogen in Fischer 344 rats and gave rise to an increased formation of crotonaldehyde-dG in this tissue (Chung *et al.*, 1986, 1989). We have also shown in skin tumorigenesis in CD-1 mice and in liver tumorigenesis in newborn CD-1 mice that while the epoxide of HNE was weakly tumorigenic the parent enal was inactive in both models (Chung *et al.*, 1993). These results suggest that enals are likely to be weak tumour initiating agents. It is tempting to speculate, however, that the propano dG adducts may be important in the promotion of tumorigenesis, as this process is known to be associated with stimulated oxidative conditions.

Conclusion

The studies described here provide several lines of evidence that support the importance of enals generated by lipid peroxidation as sources for propano adducts in tissue DNA. We have shown that the formation of short-chain or long-chain propano adducts is likely to be dependent on the type of polyunsaturated fatty acids. On the basis of this information, we propose a pathway, depicted in Scheme 1, involving polyunsaturated fatty acids in the formation of acrolein-dG, crotonaldehyde-dG and HNE-dG. Further, our studies showed the presence of HNE-dG adducts in tissues. This is important because the detection of both short- and long-chain enal-derived propano adducts as background DNA lesions suggests that cyclic propano adduction is a common reaction *in vivo*, and acrolein-dG, crotonaldehyde-dG and HNE-dG may represent only a portion of these adducts in tissue DNA. Understanding of the roles of these endogenous cyclic adducts in cancer is an important and challenging area of future investigation. In this context, it is reasonable to believe that the basal levels and patterns of these promutagenic lesions in human tissues could be modi-

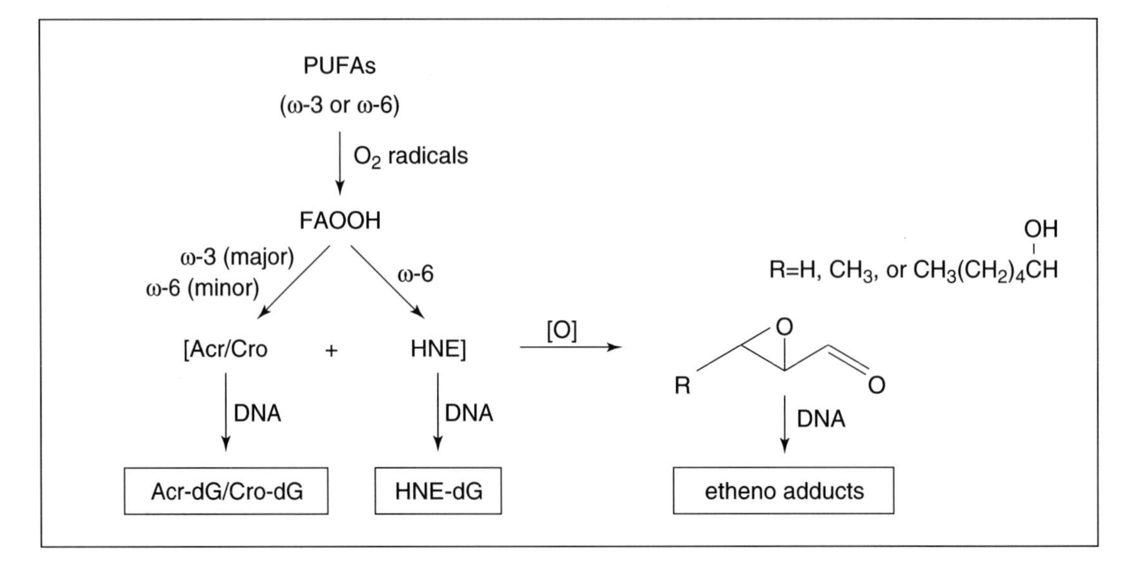

Scheme 1. Proposed endogenous pathway involving oxidation of polyunsaturated fatty acids (PUFAs) for the formation of acrolein–deoxyguanosine (Acr-dG), crotonaldehyde–deoxyguanosine (Cro-dG) and *trans*-4-hydroxy-2-nonenenal–deoxyguanosine (HNE-dG) adducts and the etheno adducts

The epoxidation of enals could be mediated by H_2O_2, fatty acid hydroperoxide (FAOOH) or autotoxidation (Chung *et al.*, 1996).

fied by a variety of genetic, environmental and dietary factors. These interactions are likely to profoundly influence the susceptibility of the hosts to carcinogens.

Acknowledgements

We are grateful to John Richie for analysing the tissue GSH and Joseph Guttenplan for providing human oral tissue DNA. These studies would not have been possible without their helpful collaboration.

References

Asami, S., Hirano, T., Yamaguchi, R., Tomioka, Y., Itoh, H. & Kasai, H. (1996) Increase of a type of oxidative DNA damage, 8-hydroxyguanine, and its repair activity in human leukocytes by cigarette smoking. *Cancer Res.*, 56, 2546–2549.

Basu, A.K., Wood, M.L., Niedernhofer, L.J., Ramos, L.A. & Essigmann, J.M. (1993) Mutagenic and genotoxic effects of three vinyl chloride-induced DNA lesions: 1,N^6-Ethenoadenine, 3,N^4-ethenocytosine, and 4-amino-5-(imidazol-2-yl)imidazole. *Biochemistry*, 32, 12793–12801

Benamira, M., Singh, U. & Marnett, L.J. (1992) Site-specific frameshift mutagenesis by a propanodeoxyguanosine adduct positioned in the (CpG)4 hot-spot of *Salmonella typhimurium* hisD3052 carried on an M13 vector. *J. Biol. Chem.*, 267, 22392–22400

Bridges, A.B., Scott, N.A., Parry, G.J. & Belch, J.J. (1993) Age, sex, cigarette smoking and indices of free radical activity in healthy humans. *Eur. J. Med.*, 2, 205–208

Cajelli, E., Ferraris, A. & Brambilla, G. (1987) Mutagenicity of 4-hydroxynonenal in V79 Chinese hamster cells. *Mutat. Res.*, 190, 169–171

Cheng, K.C., Preston, B.D., Cahill, D.S., Dosanjh, M.K., Singer, B. & Loeb, L.A. (1991) The vinyl chloride DNA derivative N^2,3-ethenoguanine produces G → A transitions in *Escherichia coli. Proc. Natl Acad. Sci. USA*, 88, 9974–9978.

Chung, F.-L., Young, R. & Hecht, S.S. (1984) Formation of cyclic 1,N^2-propanodeoxyguanosine adducts in DNA upon reaction with acrolein or crotonaldehyde. *Cancer Res.*, 44, 990–995.

Chung, F.-L., Tanaka, T. & Hecht, S.S. (1986) Induction of liver tumors in F344 rats by crotonaldehyde. *Cancer Res.*, 46, 1285–1289

Chung, F.-L., Young, R. & Hecht, S.S. (1989) Detection of cyclic 1,N^2-propanodeoxyguanosine adducts in DNA of rats treated with N-nitrosopyrrolidine and mice treated with crotonaldehyde. *Carcinogenesis*, 10, 1291–1297

Chung, F.-L., Chen, H.-J.C., Guttenplan, J.B., Nishikawa, A. & Hard, G.C. (1993) 2,3-Epoxy-4-hydroxynonanal as a potential tumor-initiating agent of lipid peroxidation. *Carcinogenesis*, 14, 2073–2077

Chung, F.-L., Chen, H.-J.C. & Nath, R.G. (1996) Lipid peroxidation as a potential endogenous source for the formation of exocyclic DNA adducts: A commentary. *Carcinogenesis*, 17, 2105–2111

Dedon, P., Plastaras, J., Rouzer, C. & Marnett, L.J. (1998) Formation of the pyrimidopurinone adduct of deoxyguanosine by base propenal, a product of oxidative DNA damage. *Proc. Am. Assoc. Cancer Res.*, 39, 286

Eder, E. & Hoffman, C. (1993) Identification and characterization of deoxyguanosine adducts of mutagenic β-alkyl substituted acrolein congeners. *Chem. Res. Toxicol.*, 6, 486–494

El-Ghissassi, F., Barbin, A., Nair, J. & Bartsch, H. (1995) Formation of 1,N^6-ethenoadenine and 3,N^4-ethenocytosine by lipid peroxidation products and nucleic acid bases. *Chem. Res. Toxicol.*, 8, 278–283

Esterbauer, H., Schaur, R.J. & H. Zollner (1991) Chemistry and biochemistry of 4-hydroxynonenal, malonaldehyde and related aldehydes. *Free Rad. Biol. Med.*, 11, 81–128

Foiles, P.G., Akerkar, S.A. & Chung, F.-L. (1989) Application of an immunoassay for cyclic acrolein deoxyguanosine adducts to assess their formation in DNA of *Salmonella typhimurium* under conditions of mutation induction by acrolein. *Carcinogenesis*, 10, 87–90

Guengerich, F.P., Persmark, M. & Humphreys, W.G. (1993) Formation of 1,N^2- and N^3,3-ethenoguanine from 2-halooxiranes: isotopic labeling studies and isolation of a hemianminal derivative of N^2-(2-oxoethyl)guanine. *Chem. Res. Toxicol.*, 6, 635–648

Hulea, S.A., Olinescu, R., Nita, S., Crocnan, D. & Kummerow, F.A. (1995) Cigarette smoking causes biochemical changes in blood that are suggestive of oxidative stress: A case–control study. *J. Environ. Pathol. Toxicol. Oncol.*, 14, 173–180

Kleinman, W.A. & Richie, J.P., Jr (1995) Determination of thiols and disulfides using high-performance liquid chromatography with electrochemical detection. *J. Chromatogr. B Biomed. Appl.*, 672, 73–80

Lane, J.D., Opara, E.C., Rose, J.E. & Behm, F. (1996) Quitting smoking raises whole blood glutathione. *Phys. Behav.*, 60, 1379–1381

Marnett, L.J., Hurd, H.K., Hollstein, M.C., Levin, D.E., Esterbauer, H. & Ames, B.N. (1985) Naturally occurring carbonyl compounds are mutagens in *Salmonella* tester strain TA104. *Mutat. Res.*, 148, 25–34

Marnett, L.J., Basu, A.K., O'Hara, S.M., Weller, P.E., Rahman, A.F.M.M. & Oliver, J.P. (1986) Reaction of malondialdehyde with guanine nucleosides: Formation of adducts containing oxadiazabicyclononene residues in the base-pairing region. *J. Am. Chem. Soc.*, 108, 1348–1350

Meerman, J.H.N., Smith, T.R., Pearson, P.G., Meier, G.P. & Nelson, S.D. (1989) Formation of cyclic 1, N^2-propanodeoxyguanosine and thymidine adducts in the reaction of the mutagen 2-bromoacrolein with calf thymus DNA. *Cancer Res.*, 49, 6174–6179

Moriya, M., Zhang, W., Johnson, F. & Grollman, A.P (1994) Mutagenic potency of exocyclic DNA adducts: Marked differences between *Escherichia coli* and simian kidney cells. *Proc. Natl Acad. Sci. USA*, 91, 11899–11903

Nath, R.G. & Chung, F.-L. (1994) Detection of exocyclic 1,N^2-propanodeoxyguanosine adducts as common DNA lesions in rodents and humans. *Proc. Natl Acad. Sci. USA*, 91, 7491–7495

Nath, R.G., Chen, H.-J.C., Nishikawa, A., Young-Sciame, R. & Chung, F.-L. (1994) A ^{32}P-postlabeling method for simultaneous detection and quantification of exocyclic etheno and propano adducts in DNA. *Carcinogenesis*, 15, 979–984

Nath, R.G., Ocando, J.E. & Chung, F.-L. (1996) Detection of 1,N^2-propanodeoxyguanosine adducts as potential endogenous DNA lesions in rodent and human tissues. *Cancer Res.*, 56, 452–456

Nath, R.G., Ocando, J.E., Richie, J.P. & Chung, F.-L. (1997) Effect of glutathione depletion on exocyclic adduct levels in the liver DNA of F344 rats. *Chem. Res. Toxicol.*, 10, 1250–1253.

Nath, R.G., Ocando, J.E., Guttenplan, J.B. & Chung, F.-L. (1998) 1,N^2-Propanodeoxyguanosine adducts: Potential new biomarkers of smoking induced DNA damage in human oral tissue. *Cancer Res.*, 58, 581–584

Palejwala, V.A., Simha, D. & Humayun, M.Z. (1991) Mechanisms of mutagenesis by exocyclic DNA adducts. Transfection of M13 viral DNA bearing a site-specific adduct shows that ethenocytosine is a highly efficient recA-independent mutagenic noninstructional lesion. *Biochemistry*, 30, 8736–8743.

Pandya, G. & Moriya, M. (1996) 1,N^6-Ethenodeoxyadenosine, a DNA adduct highly mutagenic in mammalian cells. *Biochemistry*, 35, 11487–11492

Park, K.K., Liem, A., Stewart, B.C. & Miller, J.A. (1993) Vinyl carbamate epoxide, a major strong electrophilic, mutagenic and carcinogenic metabolite of vinyl carbamate and ethyl carbamate (urethane). *Carcinogenesis*, 14, 441–450

Sodum, R.S. & Chung, F.-L. (1991) Stereoselective formation of *in vitro* nucleic acid adducts by 2,3-epoxy-4-hydroxynonanal. *Cancer Res.*, 7, 137–143

Sun, J.D., Ragsdale, S.S., Benson, J.M. & Henderson, R.F. (1985) Effects of the long-term depletion of reduced glutathione in mice administered L-buthionine (S,R)-sulfoximine, *Fundam. Appl. Toxicol.*, 5, 913–919

Wilson, V.L., Foiles, P.G., Chung, F.-L., Povey, A.C., Frank, A.A. & Harris, C.C. (1991) Detection of acrolein and crotonaldehyde DNA adducts in cultured human cells and canine peripheral blood lymphocytes by ^{32}P-postlabeling and nucleotide chromatography. *Carcinogenesis*, 12, 1483–1490

Winter, C.K., Segall, H.J. & Haddon, W.F. (1986) Formation of cyclic adducts of deoxyguanosine with the aldehydes *trans*-4-hydroxy-2-hexenal and *trans*-4-hydroxy-2-nonenal *in vitro*. *Cancer Res.*, 46, 5682–5686

Wu, H.-Y. & Lin, J.-K. (1995) Determination of aldehydic lipid peroxidation products with dabsylhydrazine by high-performance liquid chromatography. *Anal. Chem.*, 67, 1603–1612.

Yarborough, A., Zhang, Y.-J., Hsu, T.-M. & Santella, R.M. (1996) Immunoperoxidase detection of 8-hydroxydeoxyguanosine in aflatoxin B1-treated rat liver and human oral mucosal cells. *Cancer Res.*, 56, 683–688

Yi, P., Zhan, D., Samokyszyh, V.M., Doerge, D.R. & Fu, P.P. (1997) Synthesis and ^{32}P-postlabeling/HPLC separation of diastereomeric 1,N^2-(1,3-propano)-2-deoxyguanosine 3′-phosphate adducts formed from 4-hydroxy-2-nonenal. *Chem. Res. Toxicol.*, 10, 1259-1265

Yoritaka, A., Hattori, N., Uchida, K., Tanaka, M., Stadtman, E.R. & Mizuno, Y. (1996) Immunohistochemical detection of 4-hydroxynonenal protein adducts in Parkinson's disease. *Proc. Natl Acad. Sci. USA*, 93, 2696–2701

Exocyclic DNA Adducts in Mutagenesis and Carcinogenesis
B. Singer & H. Bartsch
IARC Scientific Publications No. 150
International Agency for Research on Cancer, Lyon, 1999

Lipid peroxidation-induced etheno-DNA adducts in humans

J. Nair

Increased oxidative stress and lipid peroxidation are implicated at various stages of carcinogenic processes. Recent studies have shown that reactive hydroxyalkenals derived from lipid peroxidation form the promutagenic exocyclic etheno DNA adducts 1,N^6-ethenodeoxyadenosine (εdA) and 3,N^4-ethenodeoxycytidine (εdC). A highly selective and sensitive immunoaffinity ^{32}P-postlabelling method has been developed to detect εdA and εdC, with a detection limit of about 5 adducts per 10^{10} parent nucleotides, which permitted their measurement in small amounts of human DNA. Background levels of εdA and εdC were detected in normal human tissue DNA, apparently as a result of lipid peroxidation under normal physiological conditions. High levels of εdA and εdC were found in the liver DNA of cancer-prone patients with Wilson disease or primary haemochromatosis. High dietary intake of ω–6 polyunsaturated fatty acids, which are readily oxidized to form enals, increased the εdA and εdC levels in DNA from leukocytes of women. An immunoaffinity–high-performance liquid chromatography–fluorescence method has been developed to measure εdA in human urine. Etheno DNA adducts can now be used as biomarkers to investigate the potential role of oxidative stress and lipid peroxidation in human cancers associated with certain lifestyles or chronic infections and to verify whether the levels of these adducts can be reduced by chemopreventive regimens.

Introduction

Increased oxidative stress and lipid peroxidation have been implicated in cancer and neurodegenerative diseases. A role for lipid peroxidation in tumour promotion and progression has been postulated, but only recently has it been demonstrated that DNA adducts can form reactive lipid peroxidation products. The reactive hydroxyalkenals, such as *trans*-4-hydroxy-2-nonenal (HNE), that are generated as a result of lipid peroxidation have been shown to react via further epoxidation with nucleic acids to form the DNA adducts 1,N^6-ethenodeoxyadenosine (εdA) and 1,N^2-ethenodeoxyguanosine (εdG) (Chung *et al.*, 1996). εdA, etheno-cyclicAMP and 3,N^4-ethenodeoxycytidine (εdC) were formed after oxidation of arachidonic acid and microsomal membranes in the presence of iron or cumene hydroperoxide (El Ghissassi *et al.*, 1995). The putative pathway of formation of etheno-DNA adducts from lipid peroxidation is given in Figure 1.

A sensitive immunoaffinity–^{32}P-postlabelling method for measuring εdA and εdC has been developed (Guichard *et al.*, 1993; Nair *et al.*, 1995) involving the labelling procedure and thin-layer chromatographic separation that were reported at the first international meeting on exocyclic DNA adducts (Hollstein *et al.*, 1986). The development of the method for assaying etheno-DNA adducts and its application as a biomarker for oxidative stress and lipid peroxidation in humans are described below. A specific, sensitive, non-invasive assay for εdA in urine based on immunoaffinity–high-performance liquid chromatography (HPLC)–fluorescence has also been developed to investigate the mechanism of formation and repair of these lipid peroxidation-induced ε-base adducts in humans. The mechanism appears to be complex and may be affected by dietary factors, hormonal metabolism and inflammatory process.

Materials and methods

Measurements εdA and εdC in DNA

As εdA and εdC are not stable at alkaline pH, postlabelling of these adducts as bisphosphates at pH > 8.5 is not appropriate, and a labelling procedure at near neutral pH was developed, in

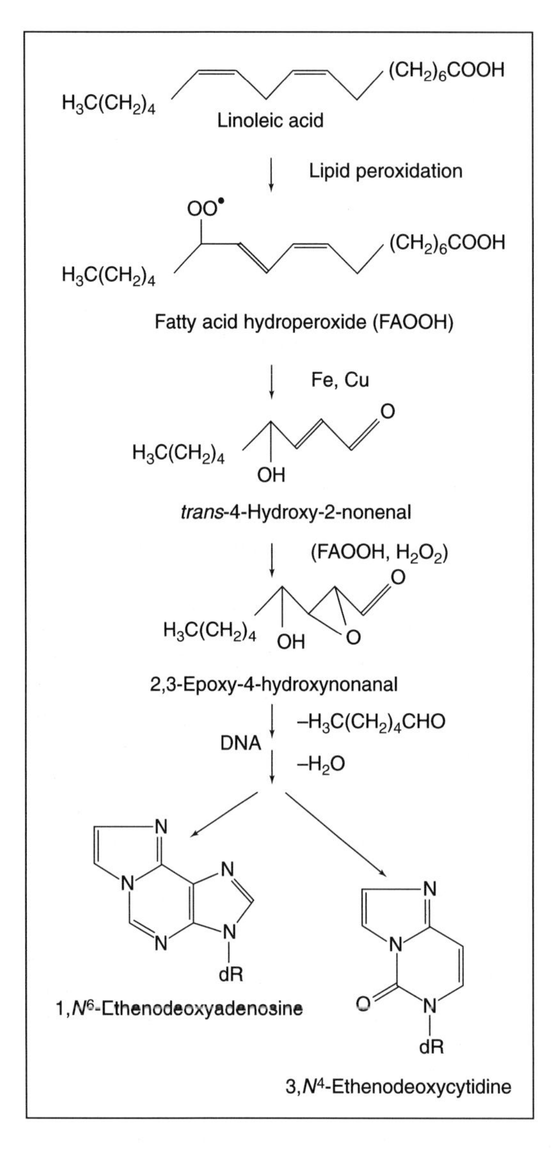

Figure 1. Putative pathways of formation of etheno adducts from lipid peroxidation of ω–6 polyunsaturated fatty acids

Linoleic acid as substrate is shown as an example.

which the kinase and phosphatase activity of T4-polynucleotide kinase (Cameron & Uhlenbeck, 1977) is used to yield the labelled 5′-monophosphate of the adduct. An internal standard deoxyuridine 3′-monophosphate was added after immunoaffinity purification of the sample in order to correct the labelling efficiency. For accurate quantification of the parent nucleotides and for quality control of the enzymatic hydrolysis, a reverse-phase HPLC step was introduced. In brief, about 25 μg of DNA were hydrolysed to nucleotide 3′-monophosphate with micrococcal nuclease (0.1 U/μg DNA) and spleen phosphodiesterase (0.01 U/μg DNA) in Tris buffer (100 mmol/L, pH 6.8) containing 20 mmol/L $CaCl_2$. For quantification of deoxyadenosine and deoxycytidine, an aliquot of hydrolysate equivalent to about 0.6 μg was injected onto a C18 column, and the normal nucleotide 3′-monophosphates were resolved by isocratic elution with 100 mmol/L ammonium formate, pH 6.8. The remainder of the sample was enriched on immunoaffinity columns prepared from the monoclonal antibody EM-1-A for εdA and EM-1-C for εdC (Guichard *et al.*, 1993). The antibodies used in this study were provided by Dr P. Lorenz, Institute for Cell Biology, University of Essen, and their characteristics have been reported (Eberle *et al.*, 1989). The dried sample was labelled with $[\gamma]^{32}$P-ATP (> 5000 Ci/mmol) with the internal standard, deoxyuridine 3′-monophosphate.

The adducts in the sample were resolved by two-directional thin layer chromatography (D1 = 1 mol/L acetic acid, pH 3.5; D2 = saturated ammonium sulfate, pH 3.5). After autoradiography, the adduct spots and the internal standard were marked and cut out, and the radioactivity was measured in a liquid scintillation counter. Typical autoradiograms obtained for standards and human liver DNA are shown in Figure 2. The absolute adduct levels were quantified against standards, and the relative adduct levels per parent nucleotides were determined from the amounts of deoxycytidine and deoxyadenosine obtained from HPLC analysis, as described previously (Nair *et al.*, 1995). The sensitivity of the method is about 5 etheno adducts per 10^{10} nucleotide from a 25-μg sample of DNA.

Measurement of ethenodeoxyadenosine in human urine

Urine samples were spiked with 1,N^6-ethenoadenosine-2,8-[^{3}H] (^{3}H-εA) as internal standard; each sample was purified by passage through an anion exchange-C18 column, followed by semi-preparatory HPLC on a C18 col-

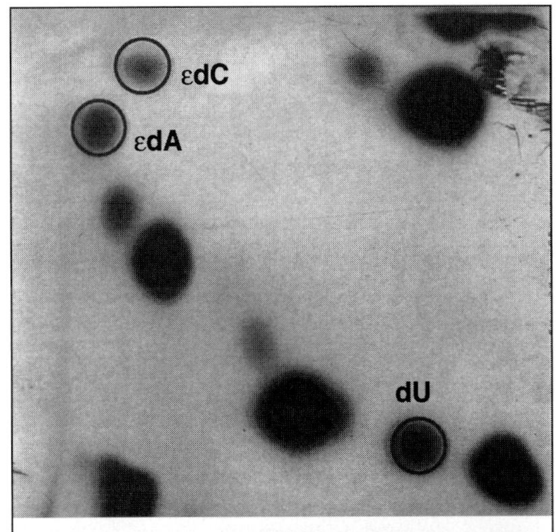

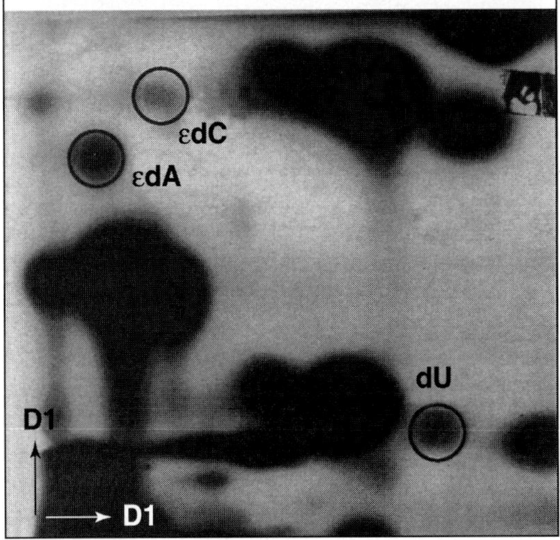

Figure 2. Autoradiograms of polyethyleneimine-cellulose maps of ^{32}P-labelled etheno adducts as 5′-monophosphates

A, standards; B, human liver DNA. εdA, 1,N^6-etheno-deoxyadenosine; εdC, 3,N^4 ethenodeoxycytidine; dU, deoxyuridine (internal standard). D1, 1 mol/L acetic acid, pH 3.5; D2, saturated ammonium sulfate, pH 3.5. The undesignated spots are normal nucleotides remaining from immunoaffinity clean-up and from trace contamination in ATP and T4-polynucleotide kinase.

umn and isocratic elution with 35 mmol/L ammonium acetate pH 6.8 (85%) and methanol (15%) and an immunoaffinity column (prepared with the monoclonal antibody EM-A-1) and analysed on a HPLC–fluorescence detector ($\lambda_{ex.}$ 230 nm, λ_{em}. 410 nm). The internal standard ^{3}H-εA and εdA were resolved by elution with ammonium phosphate 20 mmol/L, pH 5 and methanol on a gradient of 0–25 min 29% methanol, 25–35 min 30% methanol and 30–40 min 100% methanol. A representative HPLC chromatogram is shown in Figure 3. The fraction containing ^{3}H-εA was collected and counted in a liquid scintillation counter for recovery. The sensitivity of the method was 6 fmol per injection, and the overall recovery of the internal standard was 50–70%.

Results

εdA and εdC in tissue DNA

DNA isolated from human liver, leukocytes, pancreas and colon was analysed for εdA and εdC with the immunoaffinity postlabelling methods. Background levels of εdA and εdC in the range 0.05–25 x 10^8 adducts per parent nucleotides were found. The ranges of adducts in DNA from individual tissues are given in Table 1.

DNA damage caused by oxidative stress and lipid peroxidation due to copper and iron storage disorders in human liver was investigated by measuring etheno adducts in liver DNA from normal subjects and from patients with Wilson disease or primary haemochromatosis (Nair *et al.*, 1998). The mean εdA and εdC levels per 10^8 parent nucleotides in normal liver were 1.9 ± 0.5 and 2.8 ± 1.0, respectively. The mean εdA and εdC levels per 10^9 parent nucleotides in DNA from patients with Wilson disease were 6.1 ± 0.8 and 9.2 ± 3.6, respectively, which were three times higher and statistically significant ($p < 0.0001$). The levels of εdA and εdC per 10^9 parent nucleotides in DNA from patients with primary haemochromatosis were 4.7 ± 3.3 and 6.4 ± 1.2, respectively, two times higher than in the normal liver and statistically significant ($p < 0.02$). The etheno adduct levels were highly correlated with the copper content of the liver in samples from normal persons and patients with Wilson disease.

εdA and εdC were analysed in leukocyte DNA obtained from volunteers who were participating in a study to determine the effects of diets rich in linoleic and oleic acids (Valsta *et al.*, 1992). The mean levels of etheno adducts were 40 times higher in women on the high linoleic

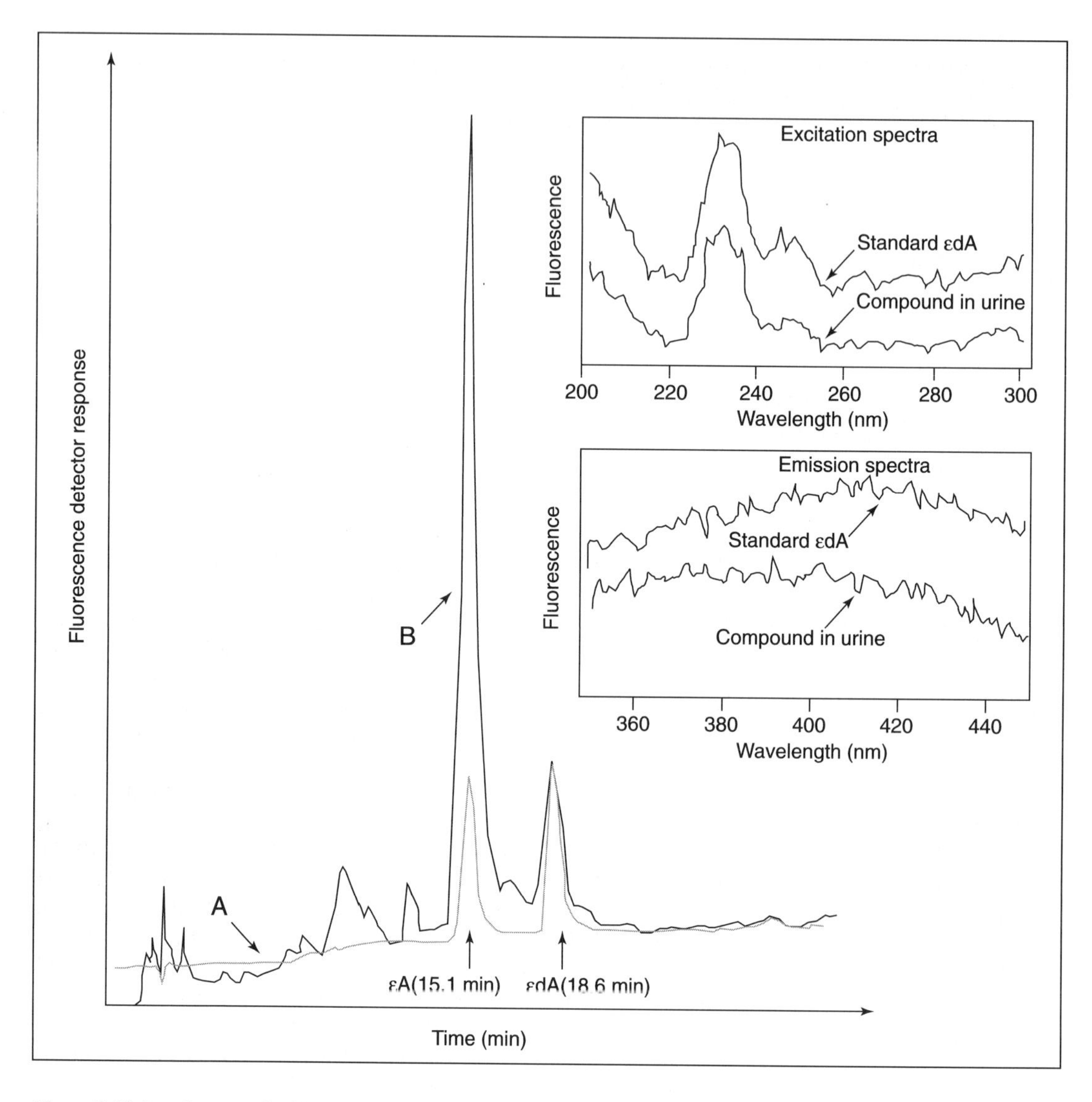

Figure 3. High-performance liquid chromatography profiles of, A, standard 1,N^6-ethenoadenosine (εA) and 1,N^6-etheno-deoxyadenosine (εdA) and, B, a urine sample

Inset: excitation and emission spectra of standard εdA and the compound in urine eluting at the same retention time

acid diet (εdA 265/10^8 dA, εdC 225/10^8 dC) than in those on the high oleic acid diet (εdA 6/10^8 dA, εdC 6/10^8 dC) but with a very large interindividual variation in lipid peroxidation-derived DNA damage. No such difference was found between two groups of male volunteers on these diets, the levels of etheno adducts being 2–7/10^8 parent nucleotides (Nair *et al.*, 1997).

1,N^6-Ethenodeoxyadenosine in human urine

εdA was analysed in spot urine samples from nine male and nine female volunteers and detected at levels ranging from 0.27 to 4.40

Table 1. Background levels of 1,N^6-ethenodeoxyadenosine (εdA) and 1,N^6-ethenodeoxycytidine (εdC) in DNA from normal human tissues

Tissue	No. of samples	Adducts /10^8 parent nucleotides	
		εdA	εdC
Liver	19	0.05–3.0	0.05–3.0
Pancreas	28	0.40–6.0	0.10–4.0
Leukocytes	14	0.30–25.0	0.40–11.0
Colon	10	0.10–9.0	0.60–2.6

fmol/μmol creatinine. The distribution of the levels is presented in Figure 4. The identity of εdA in urine was ascertained by its isolation from urine with a highly specific antibody (EM-A-1) and confirmed by comparing the chromatographic profiles with authentic standards and the fluorescence excitation emission spectra (Figure 3).

Discussion

The development of a specific, sensitive immunoaffinity ^{32}P-postlabelling method enabled us to analyse εdA and εdC in the relatively small amounts of DNA that are usually available for human studies. Both etheno adducts were detected in liver, pancreas and colon tissue and leukocytes obtained from healthy volunteers, reflecting the levels of these adducts that are formed as a result of physiological lipid peroxidation. Earlier studies in rodents had suggested that the diet could play a role in the background adduct levels detected in liver DNA (Nair *et al.*, 1999).

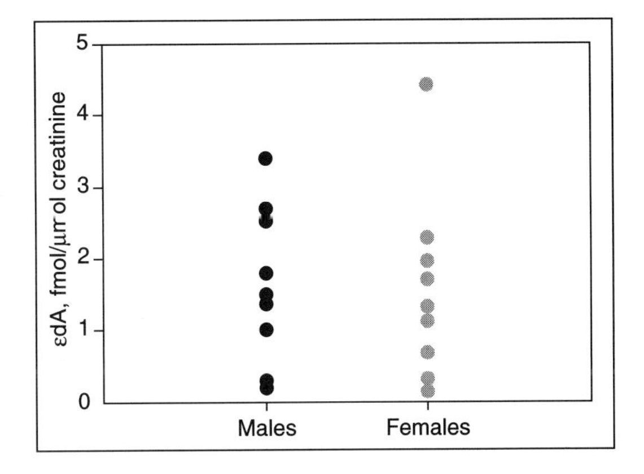

Figure 4. Distribution of levels of 1,N^6-ethenodeoxyadenosine (εdA) in the urine of nine male and nine female volunteers

Wilson disease and primary haemochromatosis are genetically determined metal storage disorders that lead to accumulation of prooxidant copper and iron ions in the liver. Enhanced levels of etheno adducts in the livers of these patients thus reflect DNA damage caused by metal storage-induced oxidative stress and lipid peroxidation. The positive correlation between adduct levels and metal ions (copper in particular) in the liver further demonstrates this association (Nair *et al.*, 1998). Studies in Long-Evans cinnamon rats, which develop primary hepatocellular carcinoma as a result of aberrant copper accumulation, as in Wilson disease, have shown that the levels of εdA and εdC peak at the time of development of hepatitis and correlate significantly with the copper content of the liver (Nair *et al.*, 1996). Copper accumulation in the livers of Wilson disease patients, if not treated by chelating therapy, leads to liver cancer (Yamaguchi *et al.*, 1993), and a 90–240-fold increase in the relative risk for primary liver cancer has been reported in patients with primary haemochromatosis (Hsing *et al.*, 1995).

The diets, prepared in a research kitchen, of the volunteers who participated in the controlled study of dietary fatty acids were similar except that the linoleic acid or oleic acid content was controlled for 25 consecutive days (Valsta *et al.*, 1992). The available information on lifestyle factors such as smoking, alcohol consumption, use of oral contraceptives and physical exercise did not explain the almost 40-fold difference in mean adduct levels in the female volunteers on the diet rich in linoleic acid. The difference could therefore be attributed to the high intake of ω-polyunsaturated fatty acids, which has been reported to result in elevated levels in membranes, implying increased lipid peroxidation (Hammer & Wills, 1979; Turpeinen *et al.*, 1995).

The observation of high etheno adduct levels only in female subjects on the linoleic acid indicates that additional factors may affect adduct formation. Lipid hydroperoxides and malondialdehyde DNA adducts were formed in male Syrian hamsters treated with oestradiol (Wang & Liehr, 1995), and induction of redox cycling by 4-hydroxylated oestradiol, a metabolite of oestradiol, was shown *in vitro* (Han & Liehr, 1995). Furthermore, a high intake of ω–6 polyunsaturated fatty acids has been shown to alter the metabolism of oestradiol (Davis *et al.*, 1993). The above observations indicate that a synergistic effect of hormonal metabolism and high linoleic acid intake may be responsible for the lipid peroxidation-induced adducts seen only in female subjects.

This is the first report of the detection of εdA in human urine, although 1,N^6-ethenoadenine was detected in the urine of untreated rats by liquid chromatography and mass spectrometry (Holt *et al.*, 1998). The εdA in urine could arise either from its formation from the deoxyadenosine nucleoside pool or as a consequence of DNA repair; however, the main pathway for repair of εdA is through 3-methyladenine DNA glycosylase (Saparbaev *et al.*, 1995; Hang *et al.*, 1996), which excises the cyclic DNA base. Whether a nucleotide excision repair pathway also operates *in vivo* to yield this adduct as the deoxynucleoside has yet to be demonstrated.

In conclusion, a sensitive, specific method developed to measure εdA and εdC was used to demonstrate their formation in DNA from human tissues and leukocytes where the levels were increased by oxidative stress and lipid peroxidation. A sensitive method has been developed to measure εdA in human urine, which, after validation, could be used to investigate the body burden of DNA damage. By using sensitive, specific methods to analyse easily available biological samples such as leukocytes and urine, various etiological factors that influence lipid peroxidation, such as high dietary fat intake, low antioxidant status and inflammatory processes, could be evaluated with etheno adducts as biomarkers.

Acknowledgements

The author wish to acknowledge the collaboration of Dr D. Phillips, Dr P. Carmichael, Dr M. Mutanen and Dr C. Vaca in the studies on human DNA, Dr A. Barbin in the development of the method and C. Ditrich, I. Kling and I. Hofmann for excellent technical assistance. This study was partly supported by EU Contract No. ENVA-CT97-0505.

References

Cameron, V. & Uhlenbeck, O.C. (1977) 3'-Phosphatase activity in T4 polynucleotide kinase. *Biochemistry*, 16, 5120–5126.

Chung, F.L., Chen, H.J. & Nath, R.G. (1996) Lipid peroxidation as a potential endogenous source for the formation of exocyclic DNA adducts. *Carcinogenesis*, 17, 2105–2111.

Davis, D.L., Bradlow, H.L., Wolff, M., Woodruff, T., Hoel, D.G. & Culver, H.A. (1993) Medical hypothesis: Xenoestrogens as preventable causes of breast cancer. *Environ. Health Perspectives*, 101, 372–377.

Eberle, G., Barbin, A., Laib, R.J., Ciroussel, F., Thomale, J., Bartsch, H. & Rajewsky, M.F. (1989) 1,N^6-Etheno-2'-deoxyadenosine and 3,N^4-etheno-2'-deoxycytidine detected by monoclonal antibodies in lung and liver DNA of rats exposed to vinyl chloride. *Carcinogenesis*, 10, 209–212.

El Ghissassi, F., Barbin, A., Nair, J. & Bartsch, H. (1995) Formation of 1,N^6-ethenoadenine and 3,N^4-ethenocytosine by lipid peroxidation products and nucleic acid bases. *Chem. Res. Toxicol.*, 8, 278–283.

Guichard, Y., Nair, J., Barbin, A. & Bartsch, H. (1993) Immunoaffinity clean-up combined with ^{32}P-postlabelling analysis of 1,N^6-ethenoadenine and 3,N^4-ethenocytosine in DNA. In: *Postlabelling Methods for Detection of DNA Adducts* (Phillips, D.H., Castegnaro, M. & Bartsch, H., eds), IARC Sci. Publ. 124, Lyon, IARC, pp. 263–269.

Hammer, C.T. & Wills, E.D. (1979) The effect of dietary fats on the composition of the liver endoplasmic reticulum and oxidative drug metabolsim. *Br. J. Nutr.*, 41, 465–475.

Han, X. & Liehr, J.G. (1995) Microsome-mediated 8-hydroxylation of guanine bases of DNA by steroid estrogens: Correlation of DNA damage by free radicals with metabolic activation to quinones. *Carcinogenesis*, 16, 2571–2574.

Hang, B., Chenna, A., Rao, S. & Singer, B. (1996) 1,N^6-Ethenoadenine and 3,N^4-ethenocytosine are excised by separate human DNA glycosylases. *Carcinogenesis*, 17, 155–157.

Hollstein, M., Nair, J., Bartsch, H., Bochner, B. & Ames, B.N. (1986) Detection of DNA base damage by ^{32}P-postlabelling: TLC separation of 5'-deoxynucleoside monophosphates. In: *The Role of Cyclic Nucleic Acid Adducts in Carcinogenesis and Mutagenesis* (Singer, B. & Bartsch, H., eds), IARC Sci. Publ. 70. Lyon, IARC, pp. 437–448.

Holt, S., Yen, T.Y., Sangaiah, R. & Swenberg, J.A. (1998) Detection of 1,N^6-ethenoadenine in rat urine after chloroethylene oxide exposure. *Carcinogenesis*, 19, 1763–1769.

Hsing, A.W., McLaughlin, J.K., Olsen, J.H., Mellemkjaer, L., Wacholder, S. & Fraumeni, J.F., Jr (1995) Cancer risk following primary hemochromatosis: A population-based cohort study in Denmark. *Int J. Cancer*, 60, 160–162.

Nair, J., Barbin, A., Guichard, Y. & Bartsch, H. (1995) 1,N^6-Ethenodeoxyadenosine and 3,N^4-ethenodeoxycytine in liver DNA from humans and untreated rodents detected by immunoaffinity/^{32}P-postlabeling. *Carcinogenesis*, 16, 613–617.

Nair, J., Sone, H., Nagao, M., Barbin, A. & Bartsch, H. (1996) Copper-dependent formation of miscoding etheno-DNA adducts in the liver of Long Evans cinnamon (LEC) rats developing hereditary hepatitis and hepatocellular carcinoma. *Cancer Res*, 56, 1267–1271.

Nair, J., Vaca, C.E., Velic, I., Mutanen, M., Valsta, L.M. & Bartsch, H. (1997) High dietary omega-6 polyunsaturated fatty acids drastically increase the formation of etheno-DNA base adducts in white blood cells of female subjects. *Cancer Epidemiol. Biomarkers Prev.*, 6, 597–601.

Nair, J., Carmichael, P.L., Fernando, R.C., Phillips, D.H., Strain, A.J. & Bartsch, H. (1998) Lipid peroxidation induced etheno-DNA adducts in the liver of patients with genetic metal storage disorders Wilson's disease and primary hemochromatosis. *Cancer Epidemiol. Biomarker Prev.*, 7, 435–440.

Nair, J., Barbin, A., Velic, I. & Bartsch, H. (1999) Etheno DNA-base adducts from endogenous reactive species, *Mutat. Res.* 424, 59–69

Saparbaev, M., Kleibl, K. & Laval, J. (1995) *Escherichia coli, Saccharomyces cerevisiae*, rat and human 3-methyladenine DNA glycosylases repair 1,N^6-ethenoadenine when present in DNA. *Nucleic. Acids Res.*, 23, 3750–3755.

Turpeinen, A.M., Alfthan, G., Valsta, L., Hietanen, E., Salonen, J.T., Schunk, H., Nyyssonen, K. & Mutanen, M. (1995) Plasma and lipoprotein lipid peroxidation in humans on sunflower and rapeseed oil diets. *Lipids*, 30, 485–492.

Valsta, L.M., Jauhiainen, M., Aro, A., Katan, M.B. & Mutanen, M. (1992) Effects of a monounsaturated rapeseed oil and a polyunsaturated sunflower oil diet on lipoprotein levels in humans. *Arterioscler. Thromb.*, 12, 50–57.

Wang, M.Y. & Liehr, J.G. (1995) Induction by estrogens of lipid peroxidation and lipid peroxide-derived malonaldehyde–DNA adducts in male Syrian hamsters: Role of lipid peroxidation in estrogen-induced kidney carcinogenesis. *Carcinogenesis*, 16, 1941–1945.

Yamaguchi, Y., Heiny, M.E. & Gitlin, J.D. (1993) Isolation and characterization of a human liver cDNA as a candidate gene for Wilson disease. *Biochem. Biophys. Res. Commun.*, 197, 271–277.

Exocyclic DNA Adducts in Mutagenesis and Carcinogenesis
B. Singer & H. Bartsch
IARC Scientific Publications No. 150
International Agency for Research on Cancer, Lyon, 1999

Detection of 1,N^6-etheno-2′-deoxyadenosine and 3,N^4-etheno-2′-deoxycytidine occurring endogenously in DNA

W.P. Watson , J.P Aston, T. Barlow, A.E. Crane, D. Potter & T. Brown

1,N^6-Etheno-2′-deoxyadenosine (εdA) and 3,N^4-etheno-2′-deoxycytidine (εdC) are DNA adducts formed by a number of genotoxic chemicals, including vinyl chloride. They are also formed endogenously in tissue DNA, probably from a reactive metabolite of lipid peroxidation. Both the qualitative and quantitative detection of endogenous adducts is important in order to place adduct formation by chemicals such as vinyl chloride in the context of this natural background level. Methods with sufficient sensitivity are therefore being developed to measure the natural background of εdA and εdC adducts. We have developed a high-performance liquid chromatography (HPLC)–^{32}P-postlabelling method to measure εdA and εdC at alkylation frequencies of 1 adduct in 10^7–10^8 nucleotides in 10-μg samples of DNA. In HPLC-^{32}P-postlabelling analysis of liver DNA from control Wistar rats, εdA and εdC were determined at levels of 1 adduct in 8.1 x 10^7 and 1 adduct in 1.8 x 10^7 nucleotides, respectively. The levels of εdA and εdC measured in liver DNA of animals exposed orally to five daily doses of 50 mg/kg body weight vinyl chloride were found by this method to be 1 adduct in 2.9 x 10^7 and 1 adduct in 1.4 x 10^7 nucleotides, respectively. In contrast, in a direct labelling study, radiolabelled εdA and εdC were not detected in liver DNA of rats exposed for 6 h by nose-only inhalation to [1,2-^{14}C]vinyl chloride at up to 45 ppm v/v. Immunochemical procedures are also being developed for recognizing etheno adducts. Thus, a monoclonal antibody raised to protein conjugates of εdC showed high selectivity in the recognition of this DNA adduct. When the antibody was immobilized on a solid support and used in an immunoenrichment procedure to purify εdC from a large excess of normal nucleotides, one εdC adduct from about 10^8 normal nucleotides could be resolved. Coupling the immunoaffinity enrichment procedure with capillary zone electrophoresis permitted the detection of approximately one εdC adduct in 3 x 10^6 nucleotides.

Introduction

Cyclic etheno nucleic acid adducts are of considerable interest because they are formed from a wide range of chemical carcinogens and are also produced endogenously in animals and man. They are of particular concern because they affect the normal Watson-Crick base pairing in DNA. Cyclic etheno adducts have been shown to be formed from vinyl chloride and its metabolites chlorooxirane (Laib & Bolt, 1977, 1978; Guichard *et al.*, 1996), vinyl carbamate (Ribovich *et al.*, 1982; Fernando *et al.*, 1996) and glycidaldehyde (Golding *et al.*, 1990, 1996; for reviews, see Leonard, 1992; Bartsch *et al.*, 1994).

Much interest in the etheno adducts 1,N^6-etheno-2′-deoxyadenosine (εdA) and 3,N^4-etheno-2′-deoxycytidine (εdC) (Figure 1) resulted from their first detection (Green & Hathway, 1978) in DNA of animals exposed to vinyl chloride, a widely used industrial chemical and human carcinogen (Maltoni *et al.*, 1982). Vinyl chloride's reactive metabolites chlorooxirane and chloroacetaldehyde alkylate DNA to produce a range of nucleic acid adducts, including *N*7-(2-oxoethyl)guanine, εdA and εdC. More recently, both εdA and εdC have been detected in tissues of animals not exposed to chemicals (Barbin *et al.*, 1993), and studies have indicated that these pro-

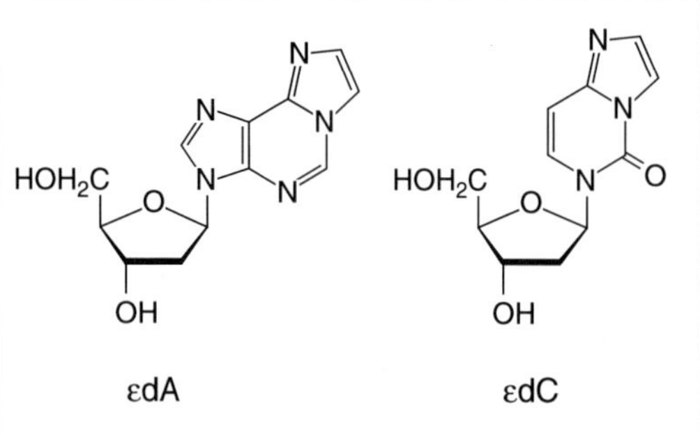

Figure 1. Structures of 1,*N*^{6}-etheno-2′-deoxyadenosine (εdA) and 3,*N*^{4}-etheno-2′-deoxycytidine (εdC)

mutagenic lesions could arise from processes involving lipid peroxidation (Barbin *et al.*, 1993; Nair *et al.*, 1995).

The most widely used procedure (Hollstein *et al.*, 1986) for the detection of these cyclic adducts and its further developed form (Nair *et al.*, 1995 and this volume) is based on the thin-layer chromatography–^{32}P-postlabelling methods of Randerath *et al.* (1981). High sensitivity is required to detect DNA adducts resulting from exposure to chemical carcinogens, because they are often formed in DNA at levels of less than 1 adduct in 10^7 nucleotides. To permit detection and quantification of adducts formed at such low levels from non-radioactive carcinogens requires selective enrichment of adducts from the unmodified nucleotides. We have developed an enrichment method for etheno adducts based on HPLC separation and have coupled this with ^{32}P-postlabelling detection (Watson & Crane, 1989; Steiner *et al.*, 1992; Watson *et al.*, 1993). These procedures permitted the detection and quantification of etheno adducts, including εdA and εdC, in the range of femtomoles in microgram amounts of DNA. We describe here the application of this approach to determine levels of εdA and εdC in livers of rats exposed to vinyl chloride and in unexposed controls.

We have also explored adduct enrichment procedures involving polyclonal antibodies which recognize etheno adducts (Wraith *et al.*, 1990) as an alternative to HPLC methods (A.E. Crane & W.P. Watson, unpublished results). In order to have an unlimited supply of antibody for preparation of adduct enrichment devices, however, a monoclonal antibody was considered preferable. Accordingly, a monoclonal antibody with high specificity for εdC was developed in the work reported here. Other groups have used this approach in the coupling of immunochemical enrichment with ^{32}P-postlabelling for the detection of etheno adducts (Young & Santella, 1988; Eberle *et al.*, 1989; Guichard *et al.*, 1993; Nair *et al.*, 1995; Misra *et al.*, 1988). Detection limits in the sub-femtomole range have been reported with ^{32}P-postlabelling detection.

Although the ^{32}P-postlabelling method is extremely sensitive, it requires the enzymatic labelling of abnormal nucleotides; for some adducts, this can result in low labelling efficiencies. The ^{32}P-postlabelling methods have been particularly useful for the detection and quantification of large, bulky adducts but in general have been less successful for low-molecular-mass adducts because of their chromatographic similarity to normal nucleotides (reviewed by Watson, 1987; Beach & Gupta, 1992; Gorelick, 1993)

In this report, we also describe a method for detecting low levels of non-radiolabelled εdC in biological samples. An immunoenrichment procedure with a new monoclonal antibody was developed and coupled with capillary zone electrophoresis to detect εdC-3′-monophosphate. A quantitative method for measuring εdC was developed by using εdC-5′,3′-bisphosphate as an internal standard, which has the advantage of eliminating the requirement for radioisotopes of high specific activity.

Materials and Methods

Chemicals

All reagents were commercially available, unless otherwise specified, and were obtained from Sigma, Aldrich or Fluka and dried or purified if necessary. The biochemical reagents were obtained from Sigma unless indicated otherwise.

Treatment of animals

Male Wistar rats aged about 10 weeks were obtained from Charles River (United Kingdom) and were acclimatized for one week, during which time they were allowed free access to laboratory food and water. The exposed group of 12 animals (average weight, 350 g) was treated on each of five consecutive days with vinyl chloride (Matheson Gas Products Ltd) as a solution in corn oil (50 mg/ml) at a dose of 50 mg/kg body weight orally from a Hamilton gas-tight syringe fitted with a short intubation tube. The treated animals were allowed free access to food and water and were killed on the sixth day by an injection of an overdose of Lethobarb. Eight untreated control animals were maintained for the same duration under the same conditions and were sacrificed at the age of 12 weeks. Livers were immediately excised from treated and control animals and were frozen in liquid nitrogen and then stored at –80 °C. Pooled liver DNA was isolated and purified by a modified version of the method of Irving and Veazy (1968).

High-performance liquid chromatography–^{32}P-postlabelling analysis of DNA adducts

The isolation of etheno adducts for ^{32}P-postlabelling analysis was based on the method described by Watson and Crane (1989). Thus, 10-µg liver DNA samples from treated and control animals were digested to nucleoside 3′-monophosphates by addition of 20 µl of 20 mmol/L sodium succinate/10 mmol/L calcium chloride (pH 6.0), 1 µl of 21 U/ml phosphodiesterase II (Sigma P 6752) and 1 µl of 50 U/ml micrococcal nuclease (Sigma N3755) and then incubated for 3 h at 37 °C. The adducted nucleoside 3′-monophosphates were isolated by HPLC on a Polymer Laboratories PLRP-S 100 Å column (5 µm, 4.6 x 250 mm) eluted isocratically (1 ml/min) with 12.5% solvent B (80% aqueous methanol) and 87.5% solvent A (100 mmol/L aqueous triethylammonium bicarbonate), and the eluant was monitored by UV absorbance (260 nm) and fluorescence detection (excitation λ_{max}, 241 nm; emission λ_{max}, 420 nm). The fractions that eluted with a retention time of 1.6–2.7 relative to that of deoxyadenosine-3′-monophosphate were collected and freeze-dried. The combined residues were dissolved in water (200 µl), applied to a pre-wetted Bond Elut CBA cartridge and eluted with water (2 x 400 µl). The eluant was evaporated to dryness in a centrifugal evaporator and stored at –80 °C until postlabelling analysis. ^{32}P-Postlabelling of fractions was carried out with 62.5 µCi [γ-^{32}ATP] and T4 polynucleotide kinase (Amersham T 2020Z), according to the procedure of Gupta (1985). Labelled 3′[5′-^{32}P]bisphosphates were 3′-dephosphorylated with nuclease P1 under conditions described by Reddy and Randerath (1986) for nucleoside-3′-monophosphates. The majority of unchanged ATP was removed on a pre-wetted Bond Elut NH_2 cartridge from which labelled adducts were eluted with 0.25 mol/L KH_2PO_4 (2 ml). ^{32}P-Labelled adducts were then analysed by co-chromatography with εdA-5′-monophosphate and εdC-5′-monophosphate by HPLC with a Hypersil ODS column (5 µm, 4.6 x 250 mm) eluted (Solvent A, 75 mmol/L KH_2PO_4; Solvent B, 80% aqueous methanol) isocratically for 5 min with Solvent B then with a gradient from 0 to 15% B for 30 min. The eluant was monitored on-line for its UV absorbance and ^{32}P activity (Raytest Ramona). Fractions were collected, and the radiolabel co-eluting with adduct reference standards was quantified by liquid scintillation counting. The values obtained were averages of four postlabelling determinations.

Immunochemical studies

Oligonucleotides were synthesized by solid-phase cyanoethyl phosphoramidite chemistry on an Applied Biosystems 394 automated DNA synthesizer and were de-protected with concentrated ammonia for 5 h at 60 °C. Oligonucleotides and compounds were purified by HPLC on a Brownlee Aquapore Octyl reversed-phase column (10 x 250 mm) with flow rate of 3 ml/min of the following: Solvent A, 0.1 mol/L NH_4OAc; buffer B, 0.1 mol/L NH_4OAc containing 25% acetonitrile (for the purification of εdC monophosphates and bisphosphates, Solvent B, 0.1 mol/L NH_4OAc with 5% acetonitrile, was used) and a gradient of 0–3 min 0% B, 3–5 min 0–20% B, 5–24 min 20–70% B, 24–26 min 70–100% B and 26–28 min 100% B.

3′-O-Succinyl εdC (5,6-dihydro-5-oxo-6β-3′-O-succinyl-2′-deoxyribofuranosyl imidazo [1,2-c]pyrimidine)

5,6-Dihydro-5-oxo-6β-D-5′-O-dimethoxytrityl-2′-deoxyribofuranosylimidazo[1,2-*c*]pyrimidine (250 mg, 460 mmol) and 4-dimethylaminopyridine (4.6 mg, 46 mmol) were co-evaporated with anhydrous pyridine (3 x 5 ml) and dissolved in anhydrous pyridine (5 ml). Succinic anhydride (60 mg, 598 mmol) was added at room temperature to a stirred solution under an argon atmosphere. After 72 h, the solution was concentrated *in vacuo* and co-evaporated with toluene (3 x 5 ml). The resulting brown gum was purified by flash-column chromatography, eluting with a 0–9% gradient of methanol in dichloromethane containing 1% triethylamine to yield 5,6-dihydro-5-oxo-6β-D-5′-*O*-dimethoxytrityl-3′-*O*-succinyl-2′-deoxyribofuranosylimidazo[1,2-*c*]pyrimidine (210 mg, 70%), Rf 0.43 (ethyl acetate:methanol, 5:1). δ_H([$^2H_6$]-dimethylsulfoxide) 9.9–9.6 (1H, s, CO_2H), 7.65 (1H, s, 2-H), 7.58–7.56 (1H, d, 7-H), 7.37–6.78 (14H, m, trityl and 3-H), 6.59–6.55 (1H, t, 1′-H), 6.27–6.25 (1H, d, 8-H), 5.48–5.46 (1H, d, 3′-H), 4.20–4.19 (1H, d, 4′-H), 3.72 (6H, s, Me), 3.52–3.41 (2H, q, 5′-H), 3.02–2.98 (4H, m, 7′-H and 6′-H) and 2.46–2.42 (2H, m, 2′-H); δ_C([$^2H_6$]-dimethylsulfoxide) 176.16 (C), 172.50 (C), 158.38 (C), 145.48 (C), 144.61 (C), 143.99 (C), 135.01 (C), 134.84 (C), 132.14 (CH), 129.81 (CH), 127.85 (CH), 127.71 (CH), 127.10 (CH), 126.83 (CH), 112.99 (CH), 112.55 (CH), 98.93 (CH), 86.86 (C), 85.24 (CH), 84.05 (CH), 74.42 (CH), 54.94 (CH_3), 44.75 (CH_2), 38.30 (CH_2), 30.24 (CH_2) and 29.79 (CH_2); λ_{max} (50% ethanol)/nm 294 (shoulder), 281 (shoulder), 272, 265 (shoulder), 257 (shoulder) and 232; υ_{max}(nujol mull)/cm^{-1} 3200–2500 (CO_2H) and 1702 (C=O and aromatic ring system); *m/z* (FAB) 654.24518 (M+1$^+$; Calc. for $C_{36}H_{36}N_3O_9$, 654.24513). The purified compound (50 mg, 77 mmol) was dissolved in acetonitrile (500 ml), and HCl solution (500 ml, 1 mol/L) was added. After undergoing stirring at room temperature for 24 h, the product was purified by reversed-phase HPLC. The appropriate peak was collected, concentrated *in vacuo* and freeze-dried to yield 3′-*O*-succinyl εdC as a white solid (14 mg, 52%), R_f 0.06 (ethyl acetate:methanol, 5:1). λ_{max} (50% ethanol)/nm 293 (shoulder), 281 (shoulder), 272, 264 (shoulder) and 257 (shoulder); *m/z* (FAB) 352.11488 (M+1$^+$; Calc. for $C_{15}H_{17}N_3O_7$, 352.11489).

εdC-3′-Monophosphate and εdC-5′,3′-bisphosphate (5,6-dihydro-5-oxo-6β-D-2′-deoxyribofuranosyl imidazo[1,2-c]pyrimidine 3′-mono and 5′,3′-bisphosphates)

εdC-3′-Monophosphate and εdC-5′,3′-bisphosphate were prepared on an automated DNA synthesizer by the incorporation of a phosphate addition monomer (P; Cruachem, 'phosphate-on' monomer) and εdC phosphoramidite into the sequences P.εdC.P.T. The ammonia de-protection step cleaved these sequences to give εdC-5′,3′-bisphosphate, which was purified by reversed-phase HPLC. The appropriate fractions were collected, desalted by gel filtration (G25 Sephadex) chromatography and freeze-dried to give εdC-5′,3′-bisphosphate: δP([$^2H_2$]-D_2O) 2.423 (2P, d, P^V); λ_{max} (H_2O)/nm 292 (shoulder), 281 (shoulder) and 272; *m/z* 412.03115 (M+1$^+$; Calc. for $C_{11}H_{15}N_3O_{10}P$, 412.03110).

Preparation of εdC conjugated proteins

Bovine serum albumin or chicken γ globulin (6 mg) and 1-ethyl-3-(3-dimethylaminopropyl) carbodiimide.HCl (13 mg, 84 mmol) were added to *N*-morpholinosulfonic acid buffer (100 mmol/L, pH 6.2, 2 ml) and warmed to 32 °C. A solution of 3′-*O*-succinyl-εdC (10 mg, 28 mmol) in dimethylsulfoxide (2 ml), also warmed to 32 °C, was added dropwise, and a slight precipitate formed. The suspension was incubated at 32 °C, and after 24 h the product was purified by gel filtration on a NAP-10 column chromatograph (Pharmacia) eluting with 50% ethanol, according to the manufacturer's specifications. The conjugate was stored in 50% ethanol at –20 °C until use.

Assay by competitive enzyme-linked immunosorbent assay

The solutions used were coating buffer (15 mmol/L sodium carbonate and 35 mmol/L sodium hydrogen carbonate at pH 9.5) and PBS/T (100 mmol/L phosphate and 50 mmol/L sodium chloride at pH 7.4 with 0.05% w/v Tween-20). 96-Well microtitre plates (Nunc) were coated with εdC-chicken γ-globulin conjugate (100 µl per well with 10 µg/ml in coating buffer) at room temperature and incubated at 4 °C overnight. The plates were washed with PBS/T (5 times), shaken, aspirated until dry and covered with plastic lids before storage at 4 °C until use. Antibody (anti-εdC) cell culture medium was applied (50 µl per well) in triplicate with a range of concentrations of inhibitor (50 µl per well in PBS/T) to the coated plates and incubated at room temperature overnight. The washings were repeated, rabbit anti-mouse antibody horseradish peroxidase conjugate (Dakopatts) (100 µl per well at 1/1000 dilution in PBS/T) was applied, and the samples were incubated at room temperature for 2 h. Washing was repeated before application of the 3,3´,4,4´-tetramethylbenzidine substrate (Sigma) (100 μl *per* well); after incubation for 30 min at room temperature, the resultant blue colour was quantified spectrophotometrically at 690 nm on an ELISA plate reader.

Immunodot–blot assays

The solutions used were PBS/T (100 mmol/L phosphate and 50 mmol/L sodium chloride at pH 7.4 with 0.05% w/v Tween-20) and blocking buffer (10% skimmed milk in PBS/T). Oligonucleotide solutions (500 nl) were spotted onto nylon membranes (Amersham Hybond), allowed to dry and irradiated at 254 nm for 5 min. Blocking buffer (20 ml) was applied, and the samples were incubated at room temperature for 30 min. The membrane was washed with PBS/T (5 times), the antibody cell culture was medium applied, and the membrane was incubated at room temperature for 3 h. Washing was repeated, rabbit anti-mouse antibody horseradish peroxidase conjugate (Dakopatts) (20 ml diluted to 1/1000 with PBS/T) was applied, and the samples were incubated at room temperature for 1 h. Washing was again repeated, 3,3'-diaminobenzidine substrate (Sigma; 10 ml) applied, and the samples were incubated at room temperature for 30 min until the colour developed.

Preparation and use of immunoaffinity chromatography gel

Antibody (anti-εdC) cell culture medium was purified by Protein-A affinity chromatography (Pierce), and the purified antibody (31.1 mg) was concentrated to 1 ml in PBS/T. The antibody was immobilized with an Immunopure Ag/Ab Immobilisation kit 4 (Pierce) according to the manufacturer's specifications to give a coupling efficiency of 37%. The concentration of antibody solution was calculated spectrophotometrically by assuming that an antibody solution of 1 mg/ml absorbs an optical density of 1.4 at 280 nm. The immunoaffinity gel was poured into a column (Pierce; 12 x 107 mm with porous polyethylene discs at the top and bottom of the gel bed) and washed with PBS/azide (20 ml). The column was equilibrated with cold PBS (3 times the column volume), and εdC-3'-monophosphate in cold PBS (25% of the column volume) was added and allowed to enter the gel bed. A further aliquot of cold PBS (5% of the column volume) was added and allowed to enter the gel. The column was capped and incubated for 1 h at room temperature, then washed with cold PBS (8 times the column volume) followed by cold water (4 times the column volume). Acetic acid (pH 2.5, 2 times the column volume) was added, and the fractions were collected (4 x 1 ml). The column was washed with PBS/azide (5 times the column volume) and stored at 4 °C for reuse.

Analysis of immunoaffinity column eluents

The fractions collected from the immunoaffinity column were freeze-dried and redissolved in distilled water (20 μl), and quantities of 10 nmol to 10 pmol εdC-5',3'-bisphosphate were added. The samples were analysed by capillary zone electrophoresis on an Applied Biosystems Model 270A Capillary Electrophoresis system with Microgel capillaries eluted with 75 mmol/L Tris-phosphate buffer containing 10% methanol at pH 7.6. On-line detection was performed by UV absorbance (272 nm), with a loading voltage of –5 kV, a running temperature of 30 °C, a running voltage of –15 kV and a sample uptake time of 5–30 s.

Results and Discussion

^{32}P-Postlabelling analyses for εdA and εdC

A HPLC ^{32}P-postlabelling procedure with an HPLC enrichment step involving multiple-column chromatography has been developed for quantitative detection at sub-femtomole sensitivity for a number of etheno cyclic adducts (Watson & Crane, 1989; Steiner *et al.*, 1992; Watson *et al.*, 1993). In the study reported here, this procedure was used to compare the amounts of εdA and εdC in liver DNA from adult Wistar rats exposed orally to vinyl chloride (5 x 50 mg/kg, 5 days) with those in similar but unexposed control animals. The ^{32}P-postlabelling analyses of DNA samples from control and exposed animals showed peaks which co-chromatographed with εdA- and εdC-5'-monophosphate. Representative chromatograms of DNA from control and exposed animals are shown in Figures 2 and 3. The amount

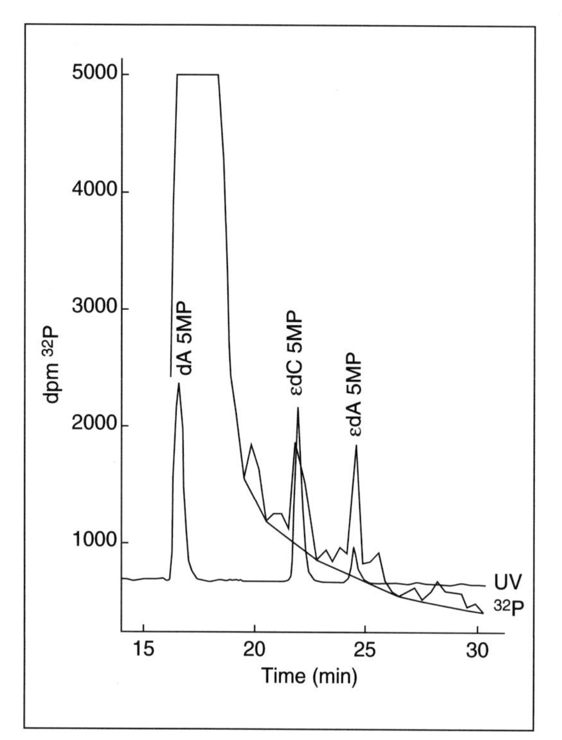

Figure 2. High-performance liquid chromatography–^{32}P-postlabelling analysis of 1,N^6-etheno-2′-deoxyadenosine (εdA) and 3,N^4-etheno-2′-deoxycytidine (εdC) in liver DNA from unexposed Wistar rats

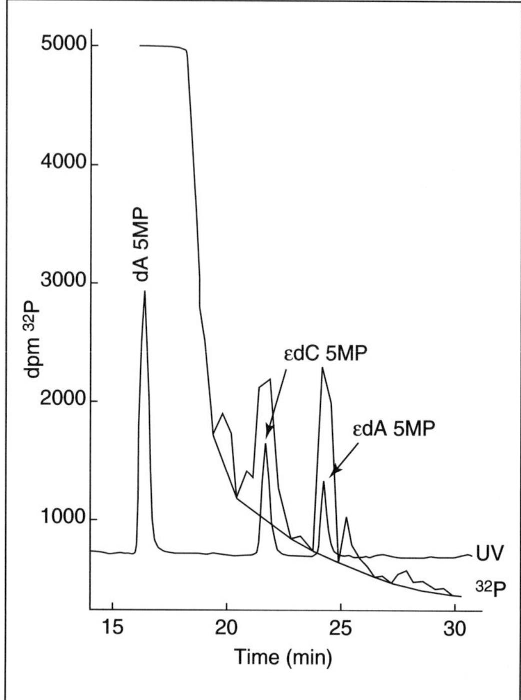

Figure 3. High-performance liquid chromatography–^{32}P-postlabelling analysis of 1,N^6-etheno-2′-deoxyadenosine (εdA) and 3,N^4-etheno-2′-deoxycytidine (εdC) in liver DNA from Wistar rats exposed to vinyl chloride (5 x 50 mg/kg body weight orally, 5 days)

of εdA, measured as εdA-5′-monophosphate, in the liver DNA of treated animals was 1 adduct per 2.9 x 10^7 nucleotides. There was a measurable background of εdA in the liver DNA of unexposed animals, at a level of 1 adduct per 8.1 x 10^7 nucleotides, i.e. about three times less than in the exposed group. The levels of εdC in liver DNA from exposed and control animals were not significantly different, the values in exposed animals being 1 adduct per 1.4 x 10^7 nucleotides and those in unexposed animals, 1 adduct per 1.8 x 10^7 nucleotides. This finding may be of significance in view of the role of εdC in mutagenesis (see below). Interestingly, in a separate radiolabelling study with [1,2-^{14}C]vinyl chloride, neither εdA nor εdC was detected (detection limit, about 1 adduct per 10^8 nucleotides in a 10-mg sample of DNA) in livers of rats exposed to vinyl chloride for 6 h at up to 45 ppm v/v (specific activity, 8.45 mCi/mmol) by nose-only inhalation (Watson *et al.*, 1991). These findings are consistent with those of other research groups, who reported substantial increases in εdA and εdC in animals exposed to vinyl chloride only at either very high or long-term exposures (Swenberg *et al.*, 1992 and this volume).

The detection of background levels of εdA and εdC in unexposed animals by an isolation and purification procedure different from those reported previously (e.g. Nair *et al.*, 1995), at levels similar to those induced by vinyl chloride at low doses, supports the view that the genotoxic risk due to exposure to low concentrations of vinyl chloride is also low. The origin of the εdA and εdC in DNA of unexposed animals is unknown but is probably related to dietary factors (see Bartsch, this volume). In the study reported here, we were unable to quantify the possible contribution from the vehicle used for oral dosing with vinyl chloride.

The role of 3,N⁴-etheno-2′-deoxycytidine in mutagenesis

Although reactive metabolites of vinyl chloride, chlorooxirane and chloroacetaldehyde, alkylate DNA to produce a range of nucleic acid adducts, the resultant mutagenicity occurs almost exclusively at cytidines (Jacobsen *et al.*, 1989), which are converted to εdC (Krzyzosiak *et al.*, 1986). When formed in DNA, the etheno bridge in εdC blocks two of the three hydrogen bonding sites of dC and so destabilizes the genetically correct base-pairing with dG (Gibson *et al.*, 1994), leading to errors in transcriptional and replicative processes. Studies on chloroacetaldehyde-modified RNA and DNA templates *in vitro* showed that RNA polymerase incorporates U and A opposite εdC (Spengler & Singer, 1981) and pol I incorporates dA and T opposite εdC (Barbin *et al.*, 1981; Hall *et al.*, 1981). Studies with chloroacetaldehyde-treated DNA transfected into *Escherichia coli* showed that 80% of the mutations at dC were of the type dC → T transitions and 20% were dC → dA transversions (Jacobsen *et al.*, 1989). The mutagenic frequency was approximately 30% for a site-specific εdC in *E. coli* (Palejawala *et al.*, 1991). The absence of significant increments in the level of εdC in animals exposed to single, low concentrations of vinyl chloride, in comparison with unexposed controls, is thus consistent with a weak mutagenic potency of vinyl chloride. This conclusion is supported by a covalent binding index of 525 for vinyl chloride, which is low in comparison with the value of 17 000 for the potent liver carcinogen aflatoxin B_1 (Lutz, 1979).

Immunoaffinity methods for 3,N⁴-etheno-2′-deoxycytidine

Antibodies have an exceptional capacity for molecular recognition and have been used extensively in immunochemical assays for protein and DNA adducts. Quantification by immunoassays is, however, dependent on the cross-reactivity of the antibody towards unknown species. When the antibodies are used principally for enrichment of adducts from a complex biological matrix, this is not a disadvantage. Immobilized antibodies have been used widely for immunoenrichment of specific adducts and members of adduct classes (Booth *et al.*, 1994). A major advantage of these immunochemical procedures when compared with conventional HPLC resolution is their relative ease and speed of use.

Preparation of antigens

A monoclonal antibody was raised against εdC conjugated to bovine serum albumin and chicken γ globulin by standard methods (Goding, 1986). The protein conjugates were prepared by addition of 3′-*O*-succinyl εdC to aqueous solutions of bovine serum albumin and chicken γ globulin in the presence of a water-soluble carbodiimide, 1-ethyl-3-(3-dimethylaminopropyl)carbodiimide. The 3′-*O*-succinyl εdC was synthesized by treatment of 5′-*O*-dimethoxytrityl εdC with succinic anhydride and subsequent acid hydrolysis of the 5′-protecting group. 5′-*O*-Dimethoxytrityl εdC was prepared by published procedures (Srivastava *et al.*, 1988). The εdC conjugated bovine serum albumin had an average ratio of εdC to carrier protein of approximately 25:1, as judged by analysis by the method of Habeeb (1966) .

Antibody screening

The antibody obtained was classed as immunoglobulin G_{2b} with an isotyping kit (Seralab). The specificity of the antibody in cell culture medium was determined by competitive ELISA with normal deoxynucleosides and the following etheno adducts: εdC, εdA and 1,N^2-ethenoguanosine (εG). The antibody showed high specificity for εdC (Table 1).

εdC was incorporated into oligonucleotides by automated DNA synthesis with εdC phosphoramidite prepared by the method of Srivastava *et al* (1994). Oligonucleotides were synthesized bearing εdC in a central position and at the 5′ end and purified by reversed-phase HPLC. These oligonucleotides and oligonucleotides devoid of εdC were fixed to nylon membranes and washed successively with cell culture medium, anti-mouse antibody horseradish peroxidase conjugate and 3,3′-diaminobenzidine substrate for immunodot–blot assay, which detected about 5 fmol of εdC when centrally positioned in DNA and 500 amol of εdC when positioned at the 5′ end. No binding was observed to the unmodified control oligonucleotide up to 5 pmol, the highest dose tested. These results indicate that the antibody had high selectivity and affinity for εdC.

Table 1. Antibody specificity to a range of competitors as judged by competitive enzyme-linked immunosorbent assay (see Materials and Methods)

Competitor	IC_{50} (pmol)
εdC	0.15
εdA	> 1 x 10^5
εG	> 1 x 10^5
dC	> 1 x 10^6
dA	> 2 x 10^5
dG	> 2 x 10^5
T	1 x 10^6

IC_{50}, concentration of inhibitor that gives 50% inhibition

The antibody was purified by protein-A affinity chromatography (Pierce) and immobilized on a solid support by periodate oxidation of carbohydrate residues situated on the Fc antibody region and subsequent reaction with hydrazide-activated agarose beads (Pierce). This method orientates the antibody binding sites distally to the solid support, allowing maximum antigen binding.

Capillary zone electrophoresis analysis

Reference standards of εdC-3′-monophosphate and εdC-3′,5′-bisphosphate were prepared on an automated DNA synthesizer with εdC phosphoramidite and 'phosphate-on' monomer (Cruachem) and purified by reversed-phase HPLC. Capillary zone electrophoresis easily resolved εdC-3′-monophosphate from εdC-3′,5′-bisphosphate, the bisphosphate eluting considerably before the monophosphate. As εdC-3′-mono and εdC-5′,3′-bisphosphate have very similar UV extinction coefficients, the 3′,5′-bisphosphate could be used as an internal standard for measuring εdC-3′-monophosphate by capillary zone electrophoresis.

The immunoaffinity gel prepared from this antibody was extremely efficient in isolating trace amounts of εdC-3′-monophosphate (1.0–0.1 nmol) from dilute solution. The nucleotide adduct could then be recovered quantitatively by elution from the column with acetic acid at pH 2.5. Elution with aqueous mixtures of organic solvents was found to damage the antibody irreversibly, so that the affinity gel could not be reused. Assay of antibody binding to εdC by ELISA over a range of pH values revealed that binding was essentially unaffected in the range pH 4.0–7.4 (Figure 4). The antibody binding was slightly reduced below pH 4.0, increased to a maximum at pH 3.0, and substantially diminished below this pH. The high binding observed at pH 3.0 was attributed to protonation of εdC at N^4. Thus, the protonated form of εdC has greater immunoreactivity than the uncharged compound.

The antibody immunoaffinity column was also effective in separating εdC-3′-monophosphate from a large excess of normal nucleotides. Defined amounts of mononucleotides were spiked with a tracer amount of εdC-3′-monophosphate and purified by the immunoenrichment procedure. Capillary zone electrophoresis analysis (Figure 5) revealed almost quantitative (> 90%) recovery of εdC-3′-monophosphate by the immunoaffinity gel in the presence of up to 500 mg of normal nucleotides; however, in the pres-

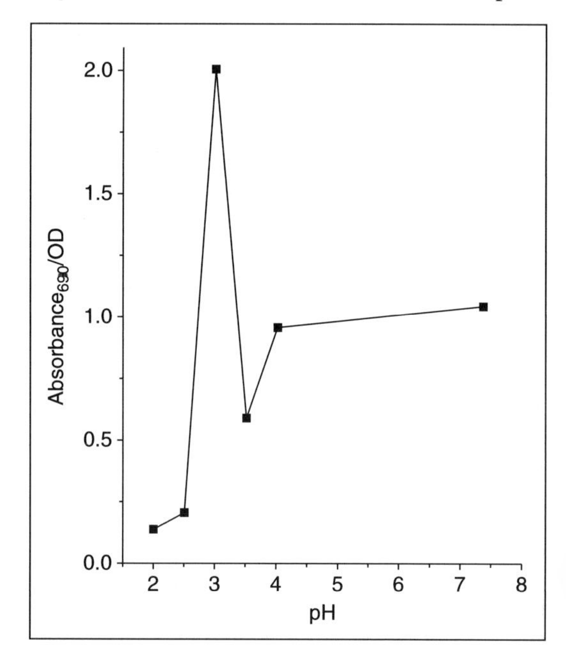

Figure 4. Immunoreactivity of antibody towards 3,N^4-etheno-2′-deoxycytidine over a range of pHs as measured by enzyme-linked immunosorbent assay

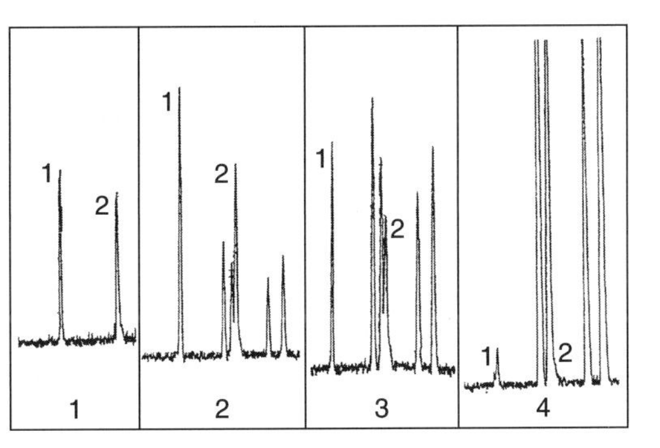

Figure 5. Capillary zone electrophoresis of recovered 3,N^4-etheno-2′-deoxycytidine-3′-monophosphate (10 pmol) from large excesses of normal nucleotides

(1) 10 mg, (2) 100 mg, (3) 500 mg, (4) 1000 mg. Recoveries were about 90% for (1), (2) and (3) and not measurable for (4). 3,N^4-Etheno-2′-deoxycytidine-3′,5′-bis phosphate (peak 1), internal standard for quantification

ence of 1000 mg of nucleotides, εdC-3′-monophosphate could not be resolved from the contaminating normal nucleotides that had bound non-specifically to the column and eluted with the analyte. The detection limits of this system were found to be in the range 1 εdC in 1.5–2.8 x 10^8 normal nucleotides. The procedure is thus sensitive in the range suitable for measuring levels of endogenous adducts.

Conclusion

When HPLC–^{32}P-postlabelling analysis was used to measure εdA and εdC in DNA from unexposed and vinyl chloride-treated rats, measurable amounts of both adducts were found in the liver DNA of both groups. The immunoenrichment–capillary zone electrophoresis procedure provides an alternative, selective, sensitive method for detection of εdC, and this type of procedure, with appropriate antibodies, should be generally applicable for the isolation and quantification of other nucleic acid adducts. εdC has a close structural relationship to the normal nucleosides, and this affects antibody selectivity and resolution by capillary zone electrophoresis from the constituents of DNA. Therefore, the detection limits for bulky lesions such as polycyclic aromatic hydrocarbon adducts which are structurally unrelated to DNA exceed those found for εdC. A significant advantage of capillary zone electrophoresis detection is, however, that there is no requirement for radioisotopes and the associated special precautions for handling high specific radioactivity.

References

Barbin, A., Bartsch, H., Leconte, P. & Radman, M. (1981) Studies on the miscoding properties of 1,N^6-ethenoadenine and 3,N^4-ethenocytidine, DNA reaction products of vinyl chloride metabolites, during in vitro DNA synthesis. *Nucleic Acids Res.* 9, 375–387.

Barbin, A., El Ghissassi, F., Nair, J. & Bartsch, H. (1993) Lipid peroxidation leads to formation of 1,N^6-ethenoadenine and 3,N^4-ethenocytosine in DNA bases. *Proc. Am. Assoc. Cancer Res.* 34, 136.

Bartsch, H., Barbin, A., Marion, M.-J., Nair, J. & Guichard, Y. (1994) Formation, detection, and role in carcinogenesis of ethenobases in DNA. *Drug Metab. Rev.* 26, 349–371.

Beach, A.C. & Gupta, R.C. (1992) Human biomonitoring and the ^{32}P-postlabeling assay. *Carcinogenesis* 13, 1053–1074.

Booth, E.D., Aston, J.P., van den Berg, P.T.M., Baan, R.A., Riddick, D.A., Wade, L.T., Wright, A.S. & Watson, W.P. (1994) Class-specific immunoadsorption purification for polycyclic aromatic hydrocarbon-DNA adducts. *Carcinogenesis* 15, 2099–2106.

Eberle, G., Barbin, A., Laib, F.C., Thomale, J., Bartsch, H. & Rajewsky, M.F. (1989) 1,N^6-Etheno-2-deoxyadenosine and 3,N^4-etheno-2′-deoxycytidine detected by monoclonal antibodies in lung and liver DNA of rats exposed to vinyl chloride. *Carcinogenesis* 10, 209–212.

Fernando, R.C., Nair, J., Barbin, A., Miller, J.A. & Bartsch, H. (1996) Detection of 1,N^6-ethenodeoxyadenosine and 3,N^4-ethenodeoxycytidine by immunoaffinity-^{32}P-postlabelling in liver and lung DNA of mice treated with ethyl carbamate (urethane) or its metabolites. *Carcinogenesis* 17, 1711–1718.

Gibson, N.J., Parkinson, J.A., Barlow, T., Watson, W.P. & Brown, T. (1994) The stability of duplex DNA containing 3,N^4-etheno-2'-deoxycytidine (εdC). A UV melting and high resolution ^{1}H NMR study. *J. Chem. Soc. Chem. Commun.* 2241.

Goding, J.W. (1986) *Monoclonal Antibodies: Principles and Practice*, New York, Academic Press.

Golding, B.T., Kennedy, G. & Watson, W.P. (1990) Structure determination of adducts from the reaction of (*R*)-glycidaldehyde and guanosine. *Carcinogenesis* 11, 865–868.

Golding, B.T., Slaich, P.K., Kennedy, G., Bleasdale, C. & Watson, W.P. (1996) Mechanisms of formation of adducts from reactions of glycidaldehyde with 2'-deoxyguanosine and/or guanosine. *Chem. Res. Toxicol.* 9 147–157.

Gorelick, N.J. (1993) Application of HPLC in the ^{32}P-postlabeling assay. *Mutat. Res.* 288, 5–18.

Green, T. & Hathway, D.E. (1978) Interactions of vinyl chloride with rat-liver DNA in vivo. *Chem.–Biol. Interactions* 22, 211–224.

Guichard, Y., Nair, J., Barbin, A. & Bartsch, H. (1993) Immunoaffinity clean-up combined with ^{32}P-postlabelling analysis of 1,*N*^{6}-ethenoadenine and 3,N^4-ethenocytosine in DNA. In: *Postlabelling Methods for Detection of DNA Adducts* (Phillips, D.H., Castegnaro, M. & Bartsch, H., eds), IARC Scientific Publications No. 124. Lyon, International Agency for Research on Cancer, pp. 263–269.

Guichard, Y., El Ghissassi, F., Nair, J., Bartsch, H. & Barbin, A. (1996) Formation and accumulation of DNA ethenobases in adult Sprague-Dawley rats exposed to vinyl chloride. *Carcinogenesis* 17, 1553–1559.

Gupta, R.C. (1985) Enhanced sensitivity of ^{32}P-postlabelling analysis of aromatic carcinogen:DNA adducts. *Cancer Res.* 45, 5656–5662.

Habeeb, A.F.S.A. (1966) Determination of free amino groups in proteins by trinitrobenzenesulfonic acid. *Anal. Biochem.* 14, 328–336.

Hall, J.A., Saffhill, R., Green, T. & Hathaway, D.E. (1981) The induction of errors during in vitro DNA synthesis following chloroacetaldehyde-treatment of poly(dA-dT) and poly(dC-dG) templates. *Carcinogenesis* 2, 141–146.

Hollstein, M., Nair, J., Bartsch, H., Bochner, B. & Ames, B.N. (1986) Detection of DNA base damage by ^{32}P-postlabelling:TLC separation of 5'-deoxynucleoside monophosphates. In: *The Role of Cyclic Nucleic acids Adducts in Carcinogenesis and Mutagenesis* (Singer, B. & Bartsch, H. eds), IARC Scientific Publications No. 70. Lyon, International Agency for Research on Cancer, pp. 437–448.

Irving, C.C. & Veazey, R.A. (1968) Isolation of deoxyribonucleic acid and ribosomal nucleic acid from rat liver. *Biochem. Biophys. Acta* 166, 246–268.

Jacobsen, J.S., Perkins, C.P., Callahan, J.T., Sambamurti, K. & Humayun, M.Z. (1989) Mechanisms of mutagenesis by chloroacetaldehyde. *Genetics* 121, 213–222.

Krzyzosiak, W.J., Wiewioroski, M. & Jaskolski, M. (1986) Chemical modification of adenine and cytosine residues with chloroacetaldehyde at the nucleoside and the tRNA level: The structural effect of chloroacetaldehyde modification. In: *The Role of Cyclic Nucleic Acid Adducts in Carcinogenesis and Mutagenesis* (Singer, B. & Bartsch, H. eds), IARC Scientific Publications No. 70. Lyon, International Agency for Research on Cancer, pp. 75–81.

Laib, R.J. & Bolt, H.M. (1977) Alkylation of RNA by vinyl chloride metabolites in vitro and in vivo: Formation of 1,N^6-etheno-adenosine. *Toxicology* 8, 185–195.

Laib, R.J. & Bolt, H.M. (1978) Formation of 3,*N*^{4}-ethenocytidine moieties in RNA by vinyl chloride metabolites in vitro and in vivo. *Arch. Toxicol.* 39, 235–240.

Leonard, N.J. (1992) Etheno-bridged nucleotides in structural diagnosis and carcinogenesis. *Chem.-Biochem. Mol. Biol.* 3, 273–297.

Lutz, W.K. (1979) In vivo covalent binding of organic chemicals to DNA as a quantitative indicator in the process of chemical carcinogenesis. *Mutat. Res.* 65, 289–356.

Maltoni, C., Lefemine, G., Ciliberti, A., Cotti, G. & Carretti, D. (1982) Vinyl chloride: A model carcinogen for risk assessment. *Environ. Sci. Res.* 25, 329–344.

Misra, R.R., Chiang, S.Y. & Swenberg, J.A. (1988) A comparison of two ultrasensitive methods for measuring 1,N^6-etheno-2′-deoxyadenosine and 3,N^4-etheno-2′-deoxycytidine. *Carcinogenesis* 15, 1647–1652.

Nair, J., Barbin, A., Guichard, Y. & Bartsch, H. (1995) 1,N^6-Ethenodeoxyadenosine and 3,N^4-ethenodeoxycytidine in liver DNA from humans and untreated rodents detected by immunoaffinity/^{32}P-postlabelling. *Carcinogenesis* 16, 613–617.

Palejawala, V.A., Simha, D. & Humayun, M.Z. (1991) Mechanisms of mutagenesis by exocyclic DNA adducts. Transfection of M13 viral DNA bearing a site-specific adduct shows that ethenocytosine is a highly efficient Rec a-independent mutagenic noninstructional lesion. *Biochemistry* 30, 8736–8743.

Randerath, K., Reddy, M.V. & Gupta, R.C. (1981) ^{32}P-Labeling test for DNA damage. *Proc. Natl Acad. Sci. USA* 78, 6126–6129.

Reddy, M.V. & Randerath, K. (1986) Nuclease P1-mediated enhancement of sensitivity of ^{32}P-labelling test for structurally diverse DNA adducts. *Carcinogenesis* 7, 1543–1551.

Ribovich, M.L., Miller, J.A., Miller, E.C. & Timmins, L.G. (1982) Labeled 1,N^6-ethenoadenosine and 3,N^4-ethenocytidine in hepatic RNA of mice given [ethyl-1,2-^{3}H or ethyl-1-^{14}C] ethyl carbamate (urethan). *Carcinogenesis* 3, 539–546.

Spengler, S. & Singer, B. (1981) Transcriptional errors and ambiguity resulting from the presence of 1,N^6-ethenoadenosine or 3,N^4-ethenocytidine in polyribonucleotides. *Nucleic Acids Res.* 9, 365–373.

Srivastava, S.C., Raza, S.K. & Misra, R. (1994) 1,N^6-Etheno deoxy and ribo adenosine and 3,N^4-etheno deoxy and ribo cytidine phosphoramidites. Strongly fluorescent structures for selective introduction in defined sequence DNA and RNA molecules. *Nucleic Acids Res.* 22, 1296–1304.

Steiner, S., Crane, A.E. & Watson, W.P. (1992) Molecular dosimetry of DNA adducts in C3H mice treated with glycidaldehyde. *Carcinogenesis* 13, 119–124.

Swenberg, J.A., Fedtke, N., Ciroussel, F., Barbin, A. & Bartsch, H. (1992) Etheno adducts formed in DNA of vinyl chloride-exposed rats are highly persistent in liver. *Carcinogenesis* 13, 727–729.

Watson, W.P. (1987) Post radiolabelling for detecting DNA damage. *Mutagenesis* 2, 319–331.

Watson, W.P. & Crane, A.E (1989) HPLC-^{32}P-postlabelling analysis of 1,N^6-ethenodeoxyadenosine and 3,N^4-ethenodeoxycytidine. *Mutagenesis* 4, 75–77.

Watson, W.P., Potter, D., Blair, D. & Wright, A.S. (1991) The relationship between alkylation of haemoglobin and DNA in Fischer 344 rats exposed to [1,2-^{14}C] vinyl chloride In: *Biomonitoring and Carcinogen Risk Assessment* (Garner, R.C., Farmer, P.B., Steele, G.T. & Wright, A.S., eds). London, Oxford University Press, pp. 421–428.

Watson, W.P., Crane, A.E. & Steiner, S. (1993) ^{32}P-Postlabelling methods for cyclic DNA adducts. In: *Postlabelling Methods for Detection of DNA Adducts* (Phillips, D.H., Castegnaro, M. & Bartsch, H., eds), IARC Scientific Publications No. 124, Lyon, International Agency for Research on Cancer, pp. 255–262.

Wraith, M.J., Watson, W.P. & Wright, A.S. (1990) Immunoassays for molecular dosimetry studies with vinyl chloride and ethylene oxide In: *Immunoassays for Trace Chemical Analysis* (Vanderlaan, M., Stanker, L.H., Watkins, B.E. & Roberts, D.W., eds). ACS Symposium Series 451. Washington DC, American Chemical Society, pp. 272–279.

Young, T.L. & Santella, R.M. (1988) Development of techniques to monitor for exposure to vinyl chloride: Monoclonal antibodies to ethenoadenosine and ethenocytidine. *Carcinogenesis* 9, 589–592.

Exocyclic DNA Adducts in Mutagenesis and Carcinogenesis
B. Singer & H. Bartsch
IARC Scientific Publications No. 150
International Agency for Research on Cancer, Lyon, 1999

Chapter II. Chemistry

Possible mechanisms of carcinogenesis after exposure to benzene

B.T. Golding & W.P. Watson

We review the history of the toxicology of benzene and consider current exposure levels, the metabolism of benzene, reactions of the metabolites with biomolecules and possible mechanisms of carcinogenesis due to benzene. Epidemiological evidence indicates a relationship between exposure to benzene and the occurrence of acute non-lymphocytic leukaemia in humans. Working groups convened by IARC and other organizations have therefore judged that there is sufficient evidence for classifying benzene as a human carcinogen. Despite much research, including numerous studies in animals, the detailed mechanism of the carcinogenicity of benzene is unknown. The significant differences in the responses of rodents and humans to benzene are not understood.

Benzene forms many metabolites, some of which are reactive towards biomolecules, but the metabolite(s) responsible for the induction of leukaemia is unknown. Candidate metabolites, either singly or in combination, include epoxides, oxepins, quinones and aldehydes, all of which are reactive towards proteins and DNA. Our studies on muconaldehydes and benzene oxide-oxepin are discussed in this context. The significance of DNA adduct formation in respect of human leukaemia is uncertain. The overall reactivity of benzene towards DNA has been shown to be very low in experimental animals, although dose-related reactivity of metabolites with DNA was observed. The lack of significant DNA reactivity is reflected in the lack of activity of benzene in short-term tests for genotoxicity; however, benzene causes oxidative stress, which can be detected as oxidative damage to DNA. Mechanisms other than DNA damage may play a role in benzene-related toxicity, e.g. reactions of benzene metabolites with essential enzymes such as topoisomerase II.

Introduction

Most chemists are aware that benzene is a human carcinogen, and its use in laboratories is therefore strictly controlled. In the nineteenth century, however, benzene was a common household degreasing agent, and it was used in the dry-cleaning industry until its inflammability, rather than its toxicity, brought about its replacement by chlorinated hydrocarbon solvents. Remarkably, benzene was one of the first cancer chemotherapeutic agents, being recommended for the treatment of leukaemia (von Koranyi, 1912). The rationale for this astonishing treatment was the leukopenia observed in some cases of benzene poisoning; however, 'benzene therapy' was short-lived because of the complex action of the substance on the blood and the ensuing toxicity (Browning, 1937).

The acute toxicity of benzene was already well known by the late nineteenth century, and measures to prevent industrial fatalities from single exposure to high levels of benzene were established by 1925. The chronic toxicity of benzene was shown to be associated with damage to blood-forming organs and a progressive, severe anaemia. In 1932, after earlier observations (Lenoir & Claude, 1897; Deloré & Bergomano, 1928), a female worker in the rubber industry was described who had been exposed to benzene for two years and who had developed myelogenous leukaemia (Weil, 1932). During the Second World War, workers at a pharmaceutical company in the United Kingdom developed vitamin C deficiency and scurvy as a result of high occupational exposures to benzene, and this led to control measures (Parke, 1996). It then took a surprisingly long time for most chemists to accept the potential hazards of benzene, and its widespread use as a solvent for chromatography and reactions continued until quite recently.

Today, the evidence for the carcinogenicity of benzene is based on extensive epidemiological data, which indicate a small excess of leukaemia cases among workers occupationally exposed repeatedly to high concentrations of benzene (reviewed by Snyder & Kocsis, 1975; CONCAWE, 1996; see below). Benzene has therefore been classified in IARC Group 1 as a human carcinogen (IARC, 1987) and is regulated as such. A link between exposure to benzene and leukaemia is seen, however, only with continuous exposure to concentrations about 100 times the current maximum exposure limit of 1 ppm. As the ambient concentrations of benzene are many orders of magnitude lower than those that have caused leukaemia in workplace settings, it is very difficult to assess quantitatively the carcinogenic risk of low exposures (see below).

Uses of and exposure to benzene

The industrial production of benzene has increased enormously during this century. For example, benzene manufacture in the United States grew from 68 x 10^6 gallons (257 x 10^6 L) in 1922 to 1362 x 10^6 gallons (5156 x 10^6 L) in 1986, and to over 2 x 10^9 gallons (7.6 x 10^9 L) in 1996 (Storck *et al.*, 1998). The substance is used as a solvent, e.g. for fats, inks, paints, plastics and rubber, and as a chemical intermediate, e.g. in the production of maleic anhydride and cyclohexane. It is a constituent of motor fuels at a level of about 2%.

Occupational exposure to benzene has been drastically reduced as exposure limits have been set ever lower. The recommended threshold limit values are currently 1–10 ppm (1 ppm = 3 mg benzene per m^3 of air). At least until the mid-1980s, several hundred thousand workers were exposed in China, where the maximum allowable concentration in the workplace was 40 mg/m^3 (Yin *et al.*, 1987a). The concentrations of benzene in ambient air in rural and suburban locations are 0.02–3 ppb (0.07–9 μg/m^3), whereas the range in urban environments is typically 0.6–6 ppb, the precise value depending on motor vehicle traffic, industrial activity, the season and the averaging time. It has been estimated that an office worker who is a non-smoker ingests on average 133 μg of benzene per day, with approximately one-third from outdoor air, one-third from travel (driving and refuelling) and the remainder from indoor air. For a typical smoker, the intake is approximately threefold higher (CONCAWE, 1996).

The risks to human health of very low concentrations of benzene have proved hard to quantify. Törnqvist and Ehrenberg (1994) attempted to estimate the risks for the Swedish population (about 8 million) due to urban air pollution and concluded that approximately five cases of leukaemia per annum will arise as a result of an average exposure to 3.6 μg/m^3 (1.2 ppb) of benzene. The CONCAWE publication (1996), however, recommended 40 ppb as a reasonable 'air quality standard' which, averaged over one year, would protect the general population from any benzene-related toxicity. In a Chinese retrospective cohort study (Yin *et al.*, 1987b; reviewed in IARC, 1987), 30 cases of leukaemia were found in a group of 28 460 occupationally exposed workers, with four cases in a reference cohort of 28 257 workers. By contemporary standards, this is regarded as an unacceptable increment in risk.

Cell types in benzene-induced carcinogenesis

The critical long-term effect of benzene in humans is acute non-lymphocytic leukaemia (reviewed by CONCAWE, 1996). This finding is based on epidemiological evidence because there is no valid animal model in which exposure to benzene can be related to leukaemia. Genotoxic and haematological effects have also been associated with exposure to benzene but are outweighed by the risk of leukaemia. The mechanism by which benzene causes leukaemia is unclear. The possible role of DNA adducts is discussed below with other factors.

Benzene is regarded as a weak carcinogen in animals, as, in contrast to humans, most of the neoplasms observed in animals exposed to benzene are of epithelial origin and occur primarily in the Zymbal gland, nasal cavity, liver and mammary tissue (US National Toxicology Program, 1986; reviewed by Huff *et al.*, 1989; Maltoni *et al.*, 1989).

In mice exposed to benzene, there was a tendency towards induction of lymphoid neoplasms, whereas in rats an increased incidence of neoplasms, which were mainly carcinomas, occurred at various sites (reviewed in IARC, 1987). Acute non-lymphocytic leukaemia or its variant, acute myelogenous leukaemia, have not been observed in experimental animals exposed to benzene.

In several studies of the effects of benzene on human health, acute myelogenous leukaemia was the dominant cell type (reviewed by CONCAWE, 1995). In a cohort of exposed workers, 82% of the leukaemia cases were acute non-lymphocytic leukaemia, which included acute myelogenous leukaemia (62% of the total) and rarer forms of acute non-lymphocytic leukaemia that also originate from the myeloid progenitor. Chronic lymphocytic leukaemia was observed in only 2% of the cases, even though this cell type is the most common in the general population. The CONCAWE rapporteurs concluded that 'acute non-lymphocytic leukaemia is linked to benzene exposure, while the other cell types are not'.

In a study aimed at determining the risks for specific cell types of leukaemia, a dose–response relationship was demonstrated for acute non-lymphocytic leukaemia but not for other cell types (Crump, 1994; Schnatter *et al.*, 1996). Other data for the same cohort did not allow accurate determination of the degree of non-linearity in the dose–response curve (Crump, 1996).

Metabolism of benzene

The metabolism of benzene is believed to play a critical role in its toxicity (for reviews, see Snyder & Kalf, 1994; Snyder & Hedli, 1996; Ross, 1996). Current knowledge is summarized in Scheme 1. Benzene metabolites are formed

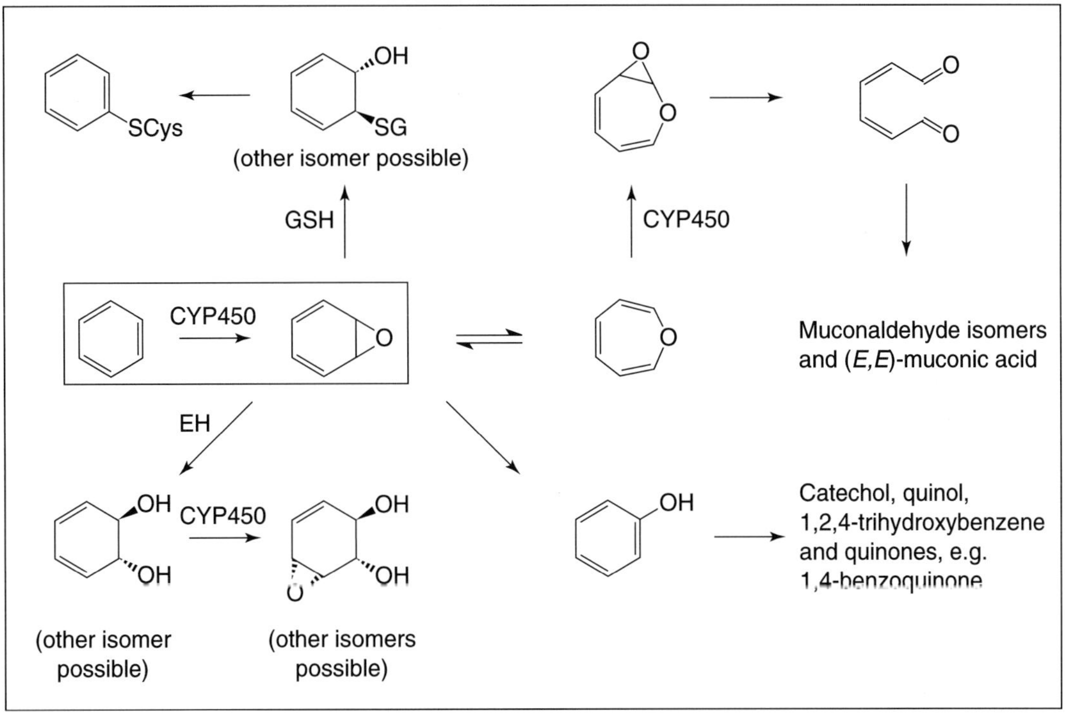

Scheme 1. Metabolism of benzene

CYP450, cytochrome P450 (principally 2E1 isozyme); GSH, glutathione (reaction possibly mediated by glutathione *S*-transferase); EH, epoxide hydrolase
Glucuronide and sulfate conjugates are also formed; some steps shown are not fully defined.

mainly by the action of cytochrome P4502E1 in the liver (Guengerich *et al.*, 1991; Gut *et al.*, 1996), and the metabolites are transported *via* the blood to the bone marrow and other organs (Scheme 2; Irons & Stillmann, 1996). The phenolic metabolites produced in the liver may be converted to quinones and semiquinones (i.e. one-electron reduction products of quinones) in the bone marrow.

The metabolites phenol (Schultzen & Naunyn, 1867), (*E,E*)-muconic acid (Jaffé, 1909; Fuchs & von Soos, 1916/17) and *S*-phenylmercapturic acid (Zbarsky & Young, 1943) were identified in remarkably early studies in the urine of animals and humans exposed to benzene. The principal human metabolite is phenol (Parke, 1996), which is formed by rearrangement of the primary metabolite benzene oxide, which is in rapid equilibrium with its valence tautomer oxepin (Vogel & Günther, 1967). The dihydroxy metabolites hydroquinone and catechol, which lead to 1,4- and 1,2-benzoquinone, respectively, may be derived from phenol. Benzene oxide-oxepin has been identified in the blood of mice, rats and humans exposed to benzene (Lindstrom *et al.*, 1997; Lovern *et al.*, 1997; Lindstrom *et al.*, 1998). Phenol is formed non-enzymatically from benzene oxide-oxepin in a reaction that is accelerated in protic media (e.g. water, with a higher rate of conversion at lower

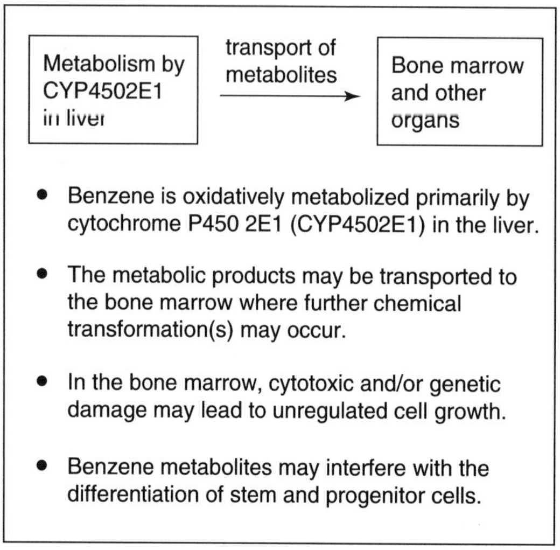

Metabolism by CYP4502E1 in liver → transport of metabolites → Bone marrow and other organs

- Benzene is oxidatively metabolized primarily by cytochrome P450 2E1 (CYP4502E1) in the liver.
- The metabolic products may be transported to the bone marrow where further chemical transformation(s) may occur.
- In the bone marrow, cytotoxic and/or genetic damage may lead to unregulated cell growth.
- Benzene metabolites may interfere with the differentiation of stem and progenitor cells.

Scheme 2. Metabolism of benzene, transport of metabolites and effects in bone marrow (see Irons & Stillman, 1996)

pHs); however, Lindstrom *et al.* (1997) demonstrated that benzene oxide-oxepin is more stable than previously recognized. Thus, it has a sufficiently long half-life in blood (7.9 min in rat blood) for it to be formed in the liver and transported to the bone marrow, where it can be further metabolized and react with macromolecules.

We have found that pure benzene oxide-oxepin is unchanged after extended periods in dry acetonitrile at 37 °C. Phenol is formed in the presence of water, its rate of formation depending on the concentration of water in acetonitrile. For example, benzene oxide-oxepin in buffered aqueous acetonitrile (1:1 v/v; pH 6.9, 37 °C) was converted into phenol with a half reaction time of approximately 9 h. In 95:5 (v/v) buffered aqueous acetonitrile at the same temperature and pH, the half reaction time for conversion into phenol was about 8 min, similar to the half-life of benzene oxide-oxepin in blood.

It has been postulated that muconaldehydes are formed *via* the further oxidation of oxepin to 2,3-epoxyoxepin (Davies & Whitham, 1977; Scheme 3), in a process that has been modelled with both mouse liver microsomes (Latriano *et al.*, 1986; Witz *et al.*, 1996) and chemical oxidants (see Scheme 4; Bleasdale *et al.*, 1997). Thus, oxidation of benzene oxide-oxepin with dimethyldioxirane at temperatures above –40 °C gave about 75% (*Z,Z*)-muconaldehyde, presumably *via* 2,3-epoxyoxepin, while one-electron oxidants (e.g. ceric ammonium nitrate) afforded primarily the (*E,Z*)-isomer (Scheme 5). Interestingly, a minor product of the oxidation with dimethyldioxirane was *sym.*-oxepin oxide, which was rearranged to (4*H*)-pyran-4-carboxaldehyde at temperatures above 10 °C. The latter compound is a masked trialdehyde, which could form DNA adducts; however, *sym.*-oxepin oxide and (4*H*)-pyran-4-carboxaldehyde have not yet been shown to be products of the metabolism of benzene.

The metabolic pathways for benzene appear to be qualitatively similar in all species studied (Henderson, 1996). There are, however, quantitative species differences in the amount of benzene metabolized, mice showing a greater overall capacity than rats or primates. There are also quantitative differences between species in the various metabolic pathways involved, mice, for example, forming relatively more hydroquinone metabolites than

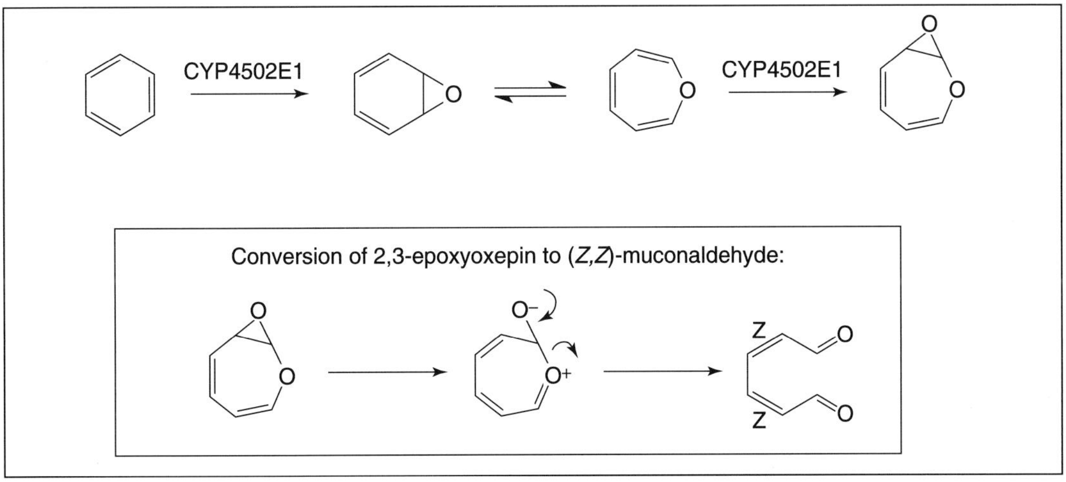

Scheme 3. Hypothetical metabolic route from benzene to (*Z,Z*)-muconaldehyde *via* 2,3-epoxyoxepin

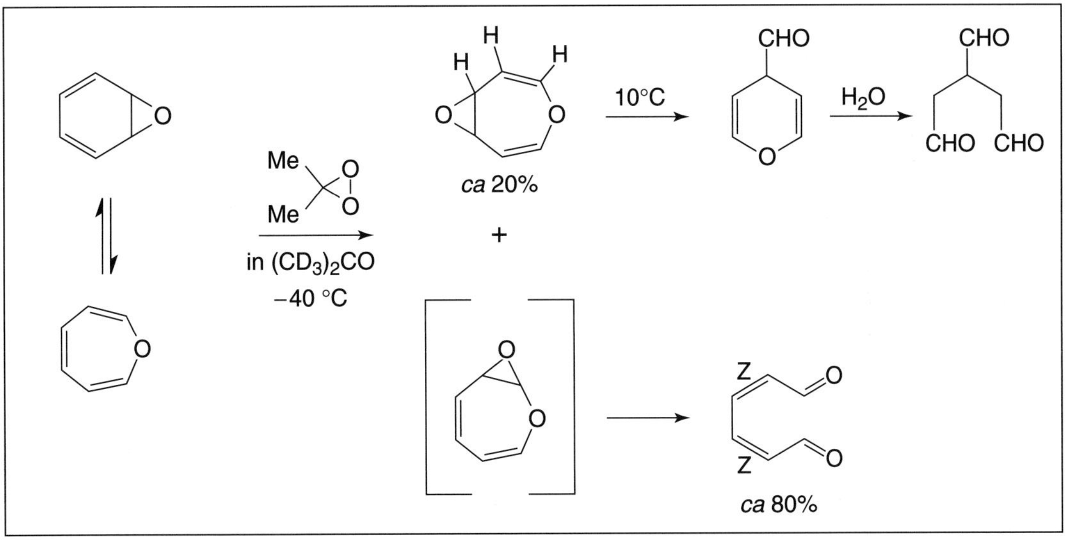

Scheme 4. Modelling of CYP450 metabolism of benzene: Oxidation of benzene oxide-oxepin with dimethyldioxirane

rats. The dose is also important. In all species studied, a higher proportion of benzene is converted to hydroquinone and ring-opened metabolites at low doses than at high doses (Rothman *et al.*, 1998).

Reactions of benzene metabolites with biomolecules

The benzene metabolites that have been shown to react (or are expected to react) with glutathione, proteins and DNA are benzene oxide-oxepin, benzoquinones, muconaldehydes and epoxides of *trans*-dihydrocatechol (for structures, see Scheme 6). These metabolites are mutagenic in some test systems (see Glatt *et al.*, 1989 concerning epoxides of *trans*-dihydrocatechol and benzoquinone; see Stark & Rastetter, 1996, concerning benzene oxide; see Bleasdale *et al.*, 1993, concerning muconaldehydes). Any or all of them may contribute to the toxicity of benzene (Snyder & Hedli, 1996).

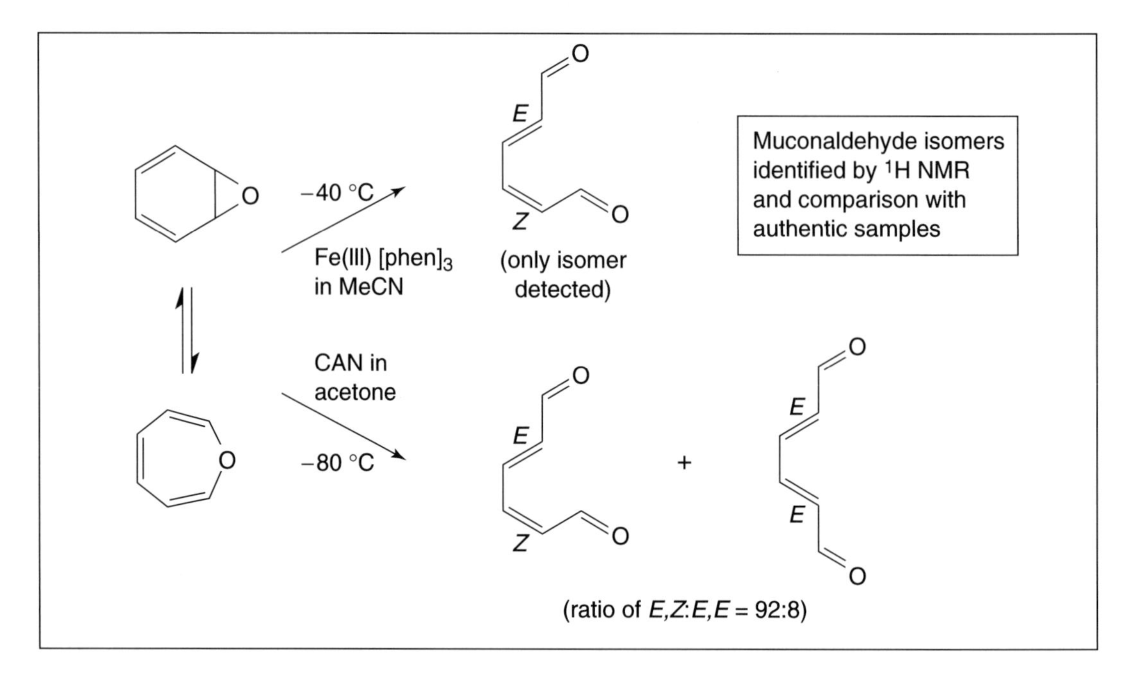

Scheme 5. One-electron oxidations of oxepin
CAN, cerium(IV) ammonium nitrate. The isolated yield of (*E,Z*)-muconaldehyde from the CAN oxidation was about 70%.

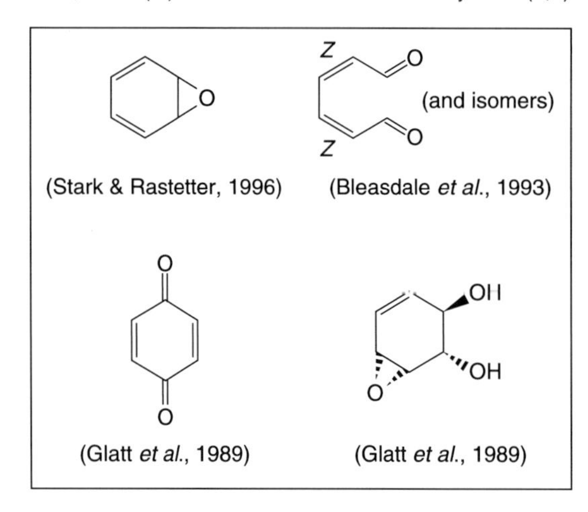

Scheme 6. Reactive metabolites of benzene, all of which form (or are expected to form) adducts with glutathione, proteins, nucleosides, DNA etc. and all of which are mutagenic in some test systems (see papers by cited authors)

We have focused on the chemistry of muconaldehydes and benzene oxide-oxepin that could be relevant to their mutagenicity. (*Z,Z*)-Muconaldehyde forms cyclic adducts with the DNA bases adenine and guanine (see below), which would disrupt Watson-Crick base pairing if formed in DNA (Bleasdale *et al.*, 1996). Reaction of benzene oxide-oxepin with DNA would be expected to release *N*7-phenylguanine; however, this adduct could not be detected in the urine of rats exposed to benzene (see below and Norpoth *et al.*, 1996). Although the reaction of benzene oxide-oxepin with glutathione presumably leads to *S*-phenylmercapturic acid, the initial adduct in this process, *S*-(2-hydroxycyclohexa-3,5-dienyl)glutathione ('pre-mercapturic acid'), and from DNA bases might primarily eliminate glutathione or the DNA base, leading to phenol (Scheme 7; L. Ehrenberg, personal communication). In a study with proton nuclear magnetic resonance, we found that addition of glutathione (2, 6 and 15 mmol/L) to benzene oxide-oxepin (15 mmol/L) in deuterium oxide-acetonitrile (95:5 v/v) buffered at pH 6.9 did not affect the rate of production of phenol, which was the only product detected. No evidence was obtained for the formation of the pre-mercapturic acid *S*-(2-hydroxycyclohexa-3,5-dienyl)glutathione.

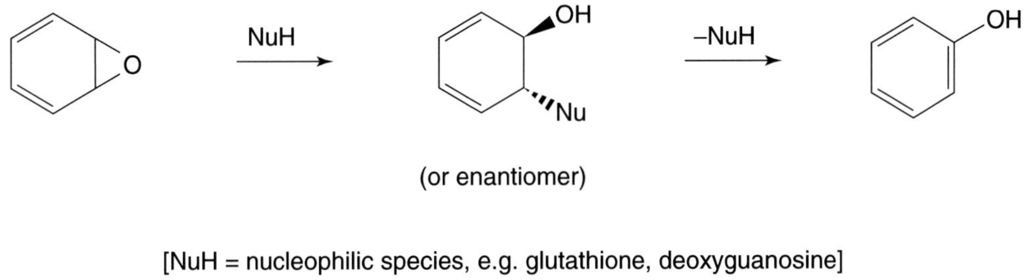

NuH → (or enantiomer) –NuH →

[NuH = nucleophilic species, e.g. glutathione, deoxyguanosine]

Scheme 7. Adduction of benzene oxide with a nucleophile; formation of phenol from the adduct

Possible mechanisms of carcinogenesis due to benzene

Mechanistic approaches based on studies of the metabolism of benzene and binding of the metabolites to macromolecules may offer the best means of understanding the genesis of human leukaemia from exposure to benzene. Such approaches are necessitated by the lack of a satisfactory animal model for studying benzene-induced leukaemia, and may allow assessment of the risks associated with exposure to benzene. Molecular mechanisms have been sought for the human carcinogenicity of benzene for many years but have not yet been found because of the complexities.

Several mechanisms have been postulated for the toxicity of benzene to bone marrow (see Smith, 1996). One is that quinones covalently bind to spindle fibre protein and thus inhibit cell replication. DNA damage could be related to the myeloproliferative disorder, which ultimately leads to acute myeloid leukaemia. Benzene metabolites could cause DNA damage by several mechanisms. Electrophilic metabolites can alkylate DNA, leading to mutation or chromosomal damage, which ultimately might be expressed as leukaemia; however, the very low levels of DNA binding of metabolites that have been observed in animals exposed to even high concentrations of benzene indicate that this mechanism alone is insufficient to explain fully the observed leukemogenic effects in humans.

Alternatively, or additionally, some metabolites may cause oxidative stress, giving rise to reactive oxygen species which may damage DNA and cause mutagenicity. Benzene metabolites may undergo redox cycling through oxidation of the reduced form of a metabolite by dioxygen to give the oxidized form of the metabolite and reactive oxygen species, e.g. hydroxyl radicals. One of the major products formed by the attack of hydroxyl radicals on DNA is 8-hydroxy-deoxyguanosine, and this has been detected in increased amounts in animals exposed to benzene (Kolachana *et al.*, 1993).

A combination of events, including damage to tubulin, histone proteins, topoisomerase II and other DNA-associated proteins, together with DNA itself, may be responsible for the consequent effects of DNA strand breaks, mitotic recombination, chromosomal translocations and aneuploidy. The occurrence of these effects in stem or progenitor cells could give rise to a leukaemic clone, which might be subject to selective growth advantage. Epigenetic effects (which cause normal cells to become tumour cells without a mutation having occurred) of benzene metabolites could be involved in the development and survival of the leukaemic clone (Smith, 1996; Irons & Stillman, 1996).

Overall, benzene does not have the characteristics of many chemical carcinogens, i.e. forming a single DNA-reactive metabolite. Thus, little binding to DNA is detected when radiolabelled benzene is administered to experimental animals (see below). In addition, benzene and most of its metabolites do not show mutagenic activity when tested in short-term tests for genotoxicity, such as the Ames *Salmonella* microsome test.

Some metabolites do, however, give rise to chromosomal damage both *in vitro* and *in vivo*. Perhaps the most likely explanation for the mechanism of action of benzene is that both genetic and epigenetic changes are necessary for its overall effects. It is also possible that promoting mechanisms are involved; however, in studies of the formation of preneoplastic foci in livers of rats treated with benzene (400 mg/kg body weight, five days per week for six weeks) benzene did not have significant promoting activity in this model of carcinogenesis (Taningher *et al.*, 1995). A current overview of the carcinogenesis of benzene is shown in Scheme 8 (see Smith, 1996).

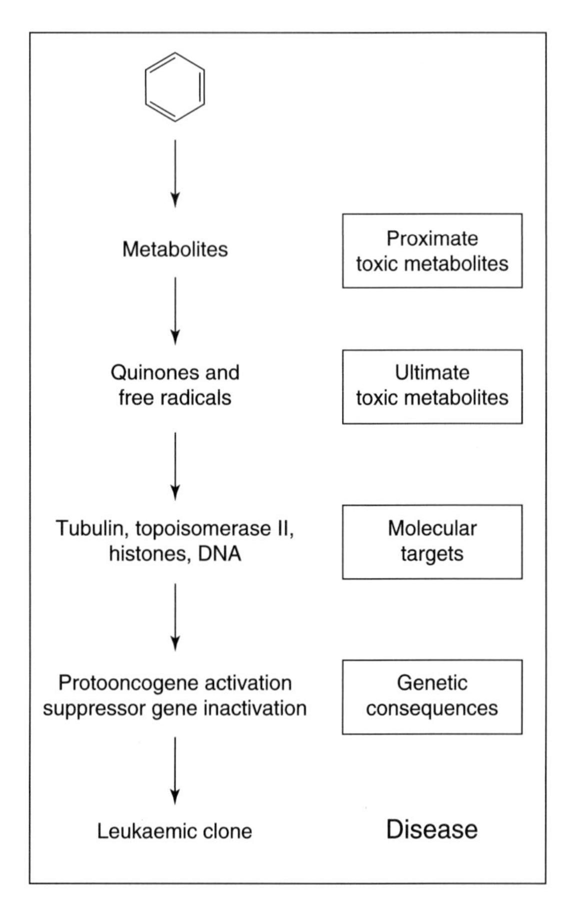

Scheme 8. Mechanism of leukaemia induction by benzene (based on Smith, 1996)

Covalent binding of metabolites to macromolecules

DNA binding

Snyder *et al* (1978) were the first to demonstrate the covalent binding of metabolites to bone marrow of mice treated *in vivo* with radiolabelled benzene. Lutz and Schlatter (1979) subsequently showed binding of low levels of radiolabel to liver DNA of rats exposed by inhalation to ^{14}C-labelled benzene. Potter and Watson (unpublished results) and others (Arfellini *et al.*, 1985) found similarly low levels of radiolabel associated with liver DNA of rats exposed to ^{14}C-benzene by intraperitoneal injection. In our studies, the total adduct formation corresponded to 1 adduct in 2 x 10^8 nucleotides, but the location of the radiolabel could not be characterized further. More recently, accelerator mass spectrometry has been used to investigate tissue distribution and DNA binding in mice exposed to very low levels of ^{14}C-benzene (Creek *et al.*, 1997). The doses spanned nine orders of magnitude, down to 700 pg/kg body weight. The DNA adduct levels were highest in the liver, peaking at 30 min; the levels peaked between 12 and 24 h in bone marrow. Dose–response assessments at 1 h indicated that the adduct levels increased linearly with dose over eight orders of magnitude up to 16 mg/kg. The levels of protein adducts, which were also measured, were one order of magnitude higher than the DNA adduct levels. The structures of the adducts were not, however, determined.

The evidence for the formation of DNA adducts *in vivo* after exposure to non-radioactive benzene, particularly in studies in which ^{32}P-postlabelling was used for adduct analysis, is contradictory (Reddy *et al.*, 1994; Bodell *et al.*, 1994, 1996; Li *et al.*, 1996). ^{32}P-Postlabelling of DNA adducts (Reddy & Randerath, 1986) is an alternative procedure for adduct determination but is quantitatively less reliable in the absence of defined reference standards. An advantage of the method is that it can be applied to non-radioactive compounds and to the analysis of DNA from human body fluids. The formation of benzene-related DNA adducts in animals exposed to non-radiolabelled benzene has been demonstrated in several studies. The first use of this technique was in a study (Bauer *et al.*, 1989) in rabbits, which showed possible formation of the hydroquinone- or 1,4-benzoquinone-derived DNA adduct 3′′-hydroxy-1,N^2-benz-2′-deoxyguanosine in DNA from the livers of animals exposed to benzene. In later work (Bodell *et al.*, 1996; Levay *et al.* 1996), nuclease P1-enhanced ^{32}P-postlabelling was used to detect one major and two minor DNA adducts in the bone marrow and leukocytes of mice treated with benzene. The DNA adducts observed were reported to be the same as those formed from the reaction of hydroquinone/1,4-benzoquinone with DNA *in vitro*. A comparison of DNA adducts from the bone-marrow DNA of mice exposed to benzene with N^2-(4-hydroxyphenyl)-2′-deoxyguanosine-3′-monophosphate showed that the principal adduct formed *in vivo* co-chromatographed with this compound. The same adduct has been observed in HL-60 cells treated with 1,4-benzoquinone (Levay *et al.*, 1993; Pathak *et al.*, 1995; Pongracz & Bodell, 1996). It has been suggested that hydroquinone is a reducing co-substrate for peroxidase enzymes and that the resultant semiquinone and 1,4-benzoquinone bind to DNA (Levay *et al.*, 1993); however, the DNA adducts in the HL-60 cells did not correspond to the benzetheno adduct 3″-hydroxy-1,N^2-benz-2′-deoxyguanosine (Levay *et al.*, 1991; Bodell *et al.*, 1994).

Substituted phenylguanines have been detected in the urine of animals exposed to benzene (Norpoth *et al.*, 1996), but *N*7-phenylguanine, an expected adduct from benzene oxide-oxepin, was not detected even with a highly sensitive immunofluorescence assay. It has been proposed that the excreted adducts were ring-hydroxylated *N*7-phenylguanines. These adducts differed from each of the authentic samples in which a phenyl group was attached to O^6, N^2 or C-8, respectively.

We have shown that (*Z,Z*)-muconaldehyde forms cyclic adducts with adenine and guanine nucleosides and deoxynucleosides (Scheme 9; Bleasdale *et al.*, 1996). These pyrrole adducts are formed under both protic and aprotic conditions, but, owing to the instability of (*Z,Z*)-muconaldehyde with respect to isomerization to the corresponding (*E,Z*)- and (*E,E*)-isomer, the yields are relatively low. The isomerization of (*Z,Z*)-muconaldehyde to the corresponding (*E,Z*)-isomer occurs either by an uncatalysed thermal reaction (half-reaction time, about 26 h in acetonitrile at 36 °C; Bleasdale *et al.*, 1996) or can be catalysed by glutathione. In the latter case, further isomerization to (*E,E*)-muconaldehyde occurs. The (*E,E*)-isomer of muconaldehyde cannot form cyclic adducts like the (*Z,Z*)-isomer, while the (*E,Z*)-isomer is much less efficient than the (*Z,Z*)-isomer in respect of this type of adduct formation. The adducts from guanine nucleosides retain the sugar component, whereas the adenine adduct undergoes depurination and the isolated adduct is an aglycone. The adenine adduct has been detected as a product of the reaction of calf thymus DNA with (*Z,Z*)-muconaldehyde *in vitro*. Although (*E,E*)-muconaldehyde does not form pyrrole adducts like (*Z,Z*)-muconaldehyde, it could form cyclic adducts analogous to those from acrolein and crotonaldehyde (Chung *et al.*, 1989).

Protein binding

As noted above, benzene metabolites have been shown to form adducts with proteins (Creek *et al.*, 1997). Haemoglobin and bone-marrow proteins have been used to study the disposition of the benzene metabolites benzene oxide-oxepin, 1,2-benzoquinone and 1,4-benzoquinone in rats and mice exposed to ^{13}C- and/or ^{14}C-benzene (Rappaport *et al.*, 1996). A high proportion of reaction occurred with cysteine residues: 38–45% with haemoglobin and 63–81% with bone-marrow protein. There were significant differences between species in the amounts of the metabolite adducts: In mice, 1,4-benzoquinone gave the highest level (10–27%) of cysteine adducts, whereas in rats benzene oxide-oxepin predominated (73% of cysteine adducts) in haemoglobin and 1,2-benzoquinone adducts (14% of cysteine adducts) predominated in bone marrow. Interestingly, high background levels of adducts of both benzoquinones were detected in unexposed animals of each species. This background, which has also been observed in humans, is believed to be derived from dietary constituents and undermines the idea that benzoquinones are the cause of the carcinogenicity of benzene (McDonald *et al.*, 1993a,b). A major adduct in haemoglobin and albumin is *S*-phenylcysteine, which could arise from benzene oxide-oxepin (Bechtold *et al.*, 1992a,b; Rappaport *et al.*, 1996).

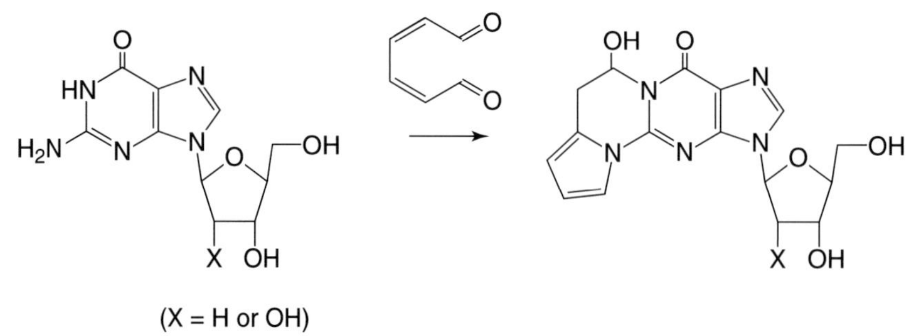

(X = H or OH)

Scheme 9. Cyclic adducts from (*Z,Z*)-muconaldehyde with deoxyguanosine or guanosine

In 0.1 mol/L aqueous phosphate buffer (pH 7.0), 35–40 °C, in darkness

Inhibition of topoisomerase II

Topoisomerase II enzymes act by creating transient breaks in both strands of DNA in order to relieve the torsional strain that occurs in DNA during its replication and transcription. A number of anticancer drugs have been identified that target topoisomerases. As inhibitors of topoisomerase II have been shown to induce leukaemia in humans, it has been proposed that inhibition of this enzyme by metabolites of benzene is responsible for the carcinogenic effects of benzene. An investigation on the inhibitory effects of the known and putative benzene metabolites phenol, 4,4′-biphenol, 2,2′-biphenol, hydroquinone, catechol, 1,2,4-benzene-triol, 1,4-benzoquinone and (*E,E*)-muconaldehyde on human topoisomerase II was carried out *in vitro* (Chen & Eastmond, 1995; Frantz *et al.*, 1996). 1,4-Benzoquinone and (*E,E*)-muconaldehyde were found to inhibit the enzyme directly, the benzoquinone being effective at a concentration about 10 times lower (about 10 µmol/L) than the muconaldehyde. All of the phenolic metabolites, phenol, 4,4′-biphenol, 2,2′-biphenol, hydroquinone, catechol and 1,2,4-benzene-triol, inhibited the enzyme in the presence of an activation system containing a peroxidase with hydrogen peroxide. The rationale for this approach was that bone marrow contains high levels of myeloperoxidase and other peroxidase enzymes. Most of the compounds tested inhibited the enzyme at concentrations below 10 µmol/L, although 2,2′-biphenol was less active.

Conclusions

Epidemiological data show that past occupational exposures to benzene were associated with a significant risk for developing leukaemia, whereas we believe that the risks due to environmental exposures may be negligible. Despite 132 years of research on the metabolism and toxicity of benzene, there is no coherent picture of how it gives rise to human leukaemia. It is obvious that only by painstaking attention to molecular mechanistic detail will it be possible to unravel this mystery.

Acknowledgements

We thank the BBSRC and Shell Research Ltd for financial support and M. Barnes, C. Bleasdale A. Campbell, K. Delaney, A. Henderson, L.D. Potter, A. Rafiq and L. Samara for contributions to this paper.

References

Arfellini, G., Grilli, S., Colacci, A., Mazzullo, M. & Prodi, G. (1985) In vivo and in vitro binding of benzene to nucleic acids and proteins of various rat and mouse organs. *Cancer Lett.* 28, 159–168.

Bauer, H., Dimitriadis, E.A. & Snyder, R. (1989) An in vivo study of benzene metabolite DNA adduct formation in liver of male New Zealand rabbits. *Arch. Toxicol.* 63, 209–213.

Bechtold, W.E., Willis, K.L., Sun, J.D Griffith, W.C. & Reddy, T.V. (1992a) Biological markers of exposure to benzene: *S*-Phenyl cysteine in albumin. *Carcinogenesis* 13, 1217–1220.

Bechtold, W.E., Sun, J.D., Birnbaum, L.S., Yin, S., Li, G.L., Kasicki, S., Lucier, G. & Henderson, R.F. (1992b) *S*-Phenylcysteine formation in hemoglobin as a biological exposure index to benzene. *Arch. Toxicol.* 66, 303–309.

Bleasdale, C., Golding, B.T., Kennedy, G., MacGregor, J.O. & Watson, W.P. (1993) Reactions of muconaldehyde isomers with nucleophiles including tri-O-acetylguanosine; formation of 1,2-disubstituted pyrroles from reactions of the (*Z,Z*)-isomer with primary amines. *Chem. Res. Toxicol.* 6, 407–412

Bleasdale, C., Kennedy, G., MacGregor, J.O., Nieschalk, J., Pearce, K., Watson, W.P. & Golding, B.T. (1996) Chemistry of muconaldehydes of possible relevance to the toxicology of benzene. *Environ. Health Perspectives* Suppl. 104, 1201–1209.

Bodell, W.J., Levat, G., Pongracz. & Pathak, D.N. (1994) DNA adducts formed by peroxidase activation of benzene metabolites. *Polycyclic Aromatic Comp.* 6, 1–8.

Bodell, W.J., Pathak, D.N., Levay, G., Ye, Q. & Pongracz, K. (1996) Investigation of the DNA adducts formed in B6C3F1 mice treated with benzene: Implications for molecular dosimetry. *Environ. Health Perspectives* Suppl. 104, 1189–1193.

Browning, E., ed. (1937) *Toxicity of Industrial Organic Solvents* (Medical Research Council Industrial Health Research Board Report No. 80). London, HMSO, Ch. I, pp. 7–63.

Chen, H. & Eastmond, D.A. (1995) Topoisomerase inhibition by phenolic metabolites; a potential mechanism for benzene's clastogenic effects. *Carcinogenesis* 16, 2301–2307.

Chung, F.L., Young, R. & Hecht, S.S. (1989) Detection of cyclic 1,N propanodeoxyguanosine adducts in DNA of rats treated with N-nitrosopyrrolidine and mice treated with crotonaldehyde. *Carcinogenesis* 10, 1291–1297.

CONCAWE (1995) Scientific basis for an air quality standard on benzene (CONCAWE Report No. 95/58R), Brussels.

CONCAWE (1996) Scientific basis for an air quality standard on benzene (CONCAWE Report No. 96/63), Brussels.

Creek, M.R., Mani, C., Vogel, J.S. & Turteltaub, K.W. (1997) Tissue distribution and macromolecular binding of extremely low doses of [^{14}C]-benzene in B6C3F1 mice. *Carcinogenesis* 18, 2421–2427

Crump, K.S. (1994) Risk of benzene-induced leukemia: A sensitivity analysis of the Pliofilm cohort with additional follow up and new exposure estimates. *J. Toxicol. Environ. Health* 43, 219–242.

Crump, K.S. (1996) Risk of benzene-induced leukemia predicted from the Pliofilm cohort. *Environ. Health Perspectives* Suppl. 104, 1437–1441.

Davies, S.G. & Whitham, G.H. (1977) Benzene oxide-oxepin. Oxidation to muconaldehyde. *J. Chem. Soc. Perkin I* 1346–1347.

Deloré, P & Bergomano, C. (1928) [Acute leukaemia during benzene poisoning; on the toxic origin of certain acute leukaemia and their relationship with severe anaemia.] *J. Méd. Lyon* 9, 227 (in French).

Frantz, C.E., Chen, H. & Eastmond, D.A. (1996) Inhibition of human topoisomerse II in vitro by bioactive benzene metabolites. *Environ. Health Perspectives* Suppl. 104, 1319–1323.

Fuchs, D. & von Soos, A. (1916/17) Über die Verbrennung des Benzols im Organismus des Menschen. *Hoppe-Seyler's Z. Physiol. Chem.* 98, 11–13.

Glatt, H., Padykula, R., Berchtold, G.A., Ludewig, G., Platt, K.L., Klein, J. & Oesch, F. (1989) Multiple activation pathways of benzene leading to products with varying genotoxic characteristics. *Environ. Health Perspectives* 82, 81–89.

Guengerich, F.P., Kim, D.-H. & Iwasaki, M. (1991) Role of human cytochrome P450 IIE1 in oxidation of low molecular weight suspects. *Chem. Res. Toxicol.* 4, 168–179.

Gut, I., Nedelcheva, V., Soucek, P., Stopka, P., Vodicka, P., Gelboin, H.V. & Ingelman-Sundberg (1996) The role of CYP2E1 and 2B1 in metabolic activation of benzene derivative. *Arch. Toxicol.* 71, 45–56.

Henderson, R.F. (1996) Species differences in the metabolism of benzene. *Environ. Health Perspectives* Suppl. 104, 1173–1175.

Huff, J.E., Haseman, J.K., DeMarini, D.M., Eustis, S., Maronpot, R.R., Peters, A.C. Pershing, R.L. Chrisp, C.C. & Jacobs, A.C. (1989) Multiple site carcinogenicity of benzene in Fischer 344 rats and B6C3F1 mice. *Environ. Health Perspectives* 82, 124–163.

IARC (1987) *IARC Monographs on the Evaluation of the Carcinogenic Risk to Humans*, Suppl. 7, *Overall Evaluations of Carcinogenicity: An Updating of* IARC Monographs *Volumes 1 to 42*, Lyon, IARC, p. 120.

Irons, R.D. & Stillmann, W.S. (1996) The process of leukemogenesis. *Environ. Health Perspectives* Suppl. 104, 1239–1246.

Jaffé, M. (1909) Über die Aufspaltung des Benzolrings im Organismus. *Hoppe-Seyler's Z. Physiol. Chem.* 62, 58–67.

Kolachana, P., Subrahmanyam, V.V., Meyer, K.B., Zhang, L.P. & Smith, M.T. (1993) Benzene and its phenolic metabolites produce oxidative DNA damage in HL60 cells in vitro and in the bone marrow in vivo. *Cancer Res.*, 53, 1023–1026.

von Koranyi, A. (1912) Der Beeinflussung der Leukämie durch Benzol. *Klin. Wschr.*, **49**, 1357.

Latriano, L., Goldstein, B.D. & Witz, G. (1986) Formation of muconaldehyde, an open-ring metabolite of benzene in mouse liver microsomes: An additional pathway for toxic metabolites. *Proc. Natl Acad. Sci. USA* 83, 8356–8360.

Le Noir, M.M. & Claude, H. (1897) [A case of purpura attributed to benzene poisoning.] *Bull. Soc. Méd. Hôp. Paris* 14, 1251–1260 (in French).

Levay, G., Pongracz, K. & Bodell, W.J. (1991) Detection of DNA adducts in HL-60 cells treated with hydroquinone and p-benzoquinone by ^{32}P-postlabelling. *Carcinogenesis* 12, 1181–1186.

Levay, G., Ross, D. & Bodell, W.J. (1993) Peroxidase activation of hydroquinone results in the formation of DNA adducts in HL-60 cells, mouse bone marrow macrophages and human bone marrow. *Carcinogenesis* 14, 2329–2334.

Levay, G., Pathak, D.N. & Bodell, W.J. (1996) Detection of DNA adducts in the white blood cells of B6C3F1 mice treated with benzene. *Carcinogenesis* 17, 151–153

Li, G., Wang, C., Xin, W. & Yin, S. (1996) Tissue distribution of DNA adducts and their persistence in blood of mice exposed to benzene. *Environ. Health Perspectives* Suppl. 104, 1337–1338.

Lindstrom, A.B., Yeowell-O'Connell, K., Waidyanatha, S., Golding, B.T., Tornero-Velez, R. & Rappaport, S.M. (1997) Measurement of benzene oxide in the blood of rats following administration of benzene. *Carcinogenesis* 18, 1637–1641.

Lindstrom, A.B., Yeowell-O'Connell, K., Waidyanatha, S., McDonald, T.A., Golding, B.T. & Rappaport, S.M. (1998) Formation of hemoglobin and albumin adducts of benzene oxide in mouse, rat and human blood. *Chem. Res. Toxicol.* 11, 301-310.

Lovern, M.R., Turner, M.J., Meyer, E.R, Kedderis, G.L., Bechtold, W.E. & Schlosser, P.M. (1997) Identification of benzene oxide as a product of benzene metabolism by mouse, rat and human liver microsomes. *Carcinogenesis* 18, 1695–1700.

Lutz, W.K. & Schlatter, C.H. (1979) Mechanism of the carcinogenic action of benzene: Irreversible binding to rat liver DNA. *Chem-Biol. Interactions* 18, 241–245.

Maltoni, C., Ciliberti, A., Cotti, G., Conti, B. & Belpoggi, F. (1989) Benzene, an experimental multipotential carcinogen: Results of the long term bioassays performed at the Bologna Institute of Oncology. *Environ. Health Perspectives* 82, 109–124.

McDonald, T.A.,Waidyanatha, S. & Rappaport, S.M. (1993a) Production of benzoquinone adducts with hemoglobin and bone marrow proteins following administration of [$^{13}C_6$]benzene to rats. *Carcinogenesis* 14, 1921–1925.

McDonald, T.A.,Waidyanatha, S. & Rappaport, S.M. (1993b) Measurement of adducts of benzoquinone with hemoglobin and albumin. *Carcinogenesis* 14, 1927–1932.

Norpoth, K.H., Mueller, G., Schell, C. & Jorg, E. (1996) Phenylguanine found in urine after benzene exposure. *Environ. Health Perspectives* Suppl. 104, 1159–1163.

Parke, D.V. (1996) Personal reflections on 50 years of study of benzene toxicology. *Environ. Health Perspectives* Suppl. 104, 1123–1128.

Pathak, D.N., Levay, G. & Bodell, W.J. (1995) DNA adduct formation in the bone marrow of B6C3F1 mice treated with benzene. *Carcinogenesis* 16, 1803–1808

Pongracz, K. & Bodell, W.J. (1996) Synthesis of N^2-(4-hydroxyphenyl)-2′-deoxyguanosine 3′-phosphate: Comparison by ^{32}P-postlabelling with the DNA adduct formed in HL-60 cells treated with hydroquinone. *Chem. Res. Toxicol.* 9, 593–598.

Rappaport, S.M, McDonald, T.A. & Yeowell-O'Connell, K. (1996) The use of protein adducts to investigate the disposition of reactive metabolites of benzene. *Environ. Health Perspectives* Suppl. 104, 1235–1237.

Reddy, M.V. & Randerath, K., (1986) Nuclease P1-mediated enhancement of sensitivity of ^{32}P-postlabelling test for structurally diverse adducts. *Carcinogenesis* 7, 1543–1547.

Reddy, M.V., Schultz, S.C., Blackburn, G. R. & Mackerer, C.R. (1994) Lack of DNA adduct formation in mice treated with benzene. *Mutat. Res.* 325, 149–155.

Ross, D. (1996) Metabolic basis of benzene toxicity. *Eur. J. Haemotol.* 57 (Suppl.), 111–118.

Rothman, N., Bechtold, W.E., Yin, S.-N., Dosemeci, M., Li, G., Wang, Y.-Z., Griffith, W.C., Smith, M.T. & Hayes, R.B. (1998) Urinary excretion of phenol, catechol, hydroquinone, and muconic acid by workers occupationally exposed to benzene. *Occup. Environ. Med.* 55, 705–711.

Schnatter, A.R., Armstrong, T.W., Thompson, L.S., Nicholich, M.J., Katz, A.M., Huebner, W.W. & Pearlman, E.D. (1996) The relationship between low level benzene exposure and leukemia in Canadian petroleum distribution workers. *Environ. Health Perspectives* Suppl. 104, 1375–1379.

Schultzen, O. & Naunyn, B. (1867) Uber das Verhalten der Kohlenwasserstoffe im Organismus. *Arch. Anat. Physiol.* 349–357.

Smith, M.T. (1996) The mechanism of benzene induced leukemia: A hypothesis and speculations on the causes of leukemia. *Environ. Health Perspectives* Suppl. 104, 1219–1225.

Snyder, R. & Hedli, C.C. (1996) An overview of benzene metabolism. *Environ. Health Perspectives* Suppl. 104, 1165–1171.

Snyder, R. & Kalf, G.F. (1994) A perspective on benzene leukemogenesis. *Crit. Rev. Toxicol.* 24, 177–209.

Snyder, R. & Kocsis, J.J. (1975) Current concepts of chronic benzene toxicity. *Crit. Rev. Toxicol.* 3, 265–288.

Snyder, R., Lee, E.W. & Kocsis, J.J. (1978) Binding of labeled benzene metabolites in mouse liver and bone marrow. *Res. Commun. Chem. Pathol. Pharmacol.* 20, 191–194.

Stark, A.A. & Rastetter, W.H. (1996) Structure–activity relationships in the mutagenicity and cytotoxicity of putative metabolites and related analogs of benzene derived from the valence tautomers benzene oxide and oxepin. *Environ. Mol. Mutag.* 28, 284–293.

Storck, W.J., Layman, P.L., Reisch, M., Thayer, A.M., Kirschner, E.M., Peaff, G.P. & Tremblay, J.-F. (1998) Facts and figures for the chemical industry. *Chem. Eng. News* 23 June, 38–79 (especially p. 41).

Taningher, M., Perrotta, A., Malacarne, D., Parodi, S., Zedda, A.I., Colacci, A. & Grilli, S. (1995) Lack of significant promoting activity by benzene in the rat liver model of carcinogenesis. *J. Toxicol. Environ. Health* 45, 481–488.

Törnqvist, M. & Ehrenberg, L. (1994) On cancer risk estimation of urban air pollution. *Environ. Health Perspectives* 102 (Suppl. 4), 173–182.

US National Toxicology Program (1986) *Toxicology and Carcinogenesis Studies of Benzene (CAS No. 71-43-2) in F344/N Rats and B6CF1 Mice (Gavage Studies)*

(NTP-TR 289; NIH Publication 86-2545), Research Triangle Park, North Carolina, US Department of Health and Human Services.

Vogel, E. & Günther, H. (1967) Benzene oxide-oxepin valence tautomerism. *Angew. Chem. Int. Ed.* 6, 385–476.

Weil, E.P. (1932) [Post-benzene leukaemia.] *Bull. Soc. Méd. Hôp. Paris* 46, 750 (in French).

Witz, G., Zhang, Z. & Goldstein, B.D. (1996) Reactive ring-opened aldehyde metabolites in benzene hematotoxicity. *Environ. Health Perspectives* **104**, 1195–1199.

Yin, S.-N., Li, Q., Liu, Y., Tian, F. & Jin, C. (1987a) Occupational exposure to benzene in China. *Br. J. Ind. Med.* **44**, 192–195.

Yin, S.-N., Li, G.-L., Tain, F.-D., Fu, Z.-I., Jin, C., Chen, Y.-J., Luo, S.-J., Ye, P.-Z., Zhang, J.-Z., Wang, G-C., Zhang, X.-C., Wu, H.-N. & Zhong,Q.-C. (1987b) Leukemia in benzene workers: A retrospective cohort study. *Br. J. Ind. Med.* **44**, 124–128.

Zbarsky, S.H. & Young L. (1943) Mercapturic acids. III. The conversion of benzene to phenylmercapturic acid in the rat. *J. Biol. Chem.* **151**, 487–492.

Exocyclic DNA Adducts in Mutagenesis and Carcinogenesis
B. Singer & H. Bartsch
IARC Scientific Publications No. 150
International Agency for Research on Cancer, Lyon, 1999

Synthesis of *para*-benzoquinone and 1,3-bis(2-chloroethyl)nitrosourea adducts and their incorporation into oligonucleotides

A. Chenna, H. Maruenda & B. Singer

Benzene is a widely known carcinogen and a cause of bone-marrow toxicity and leukaemia in humans. *para*-Benzoquinone is a stable metabolite of benzene. Its reaction with deoxycytidine, deoxyadenosine and deoxyguanosine produces the major stable exocyclic compounds (3-hydroxy)-1,N^4-benzetheno-2′-deoxycytidine, (9-hydroxy)-1,N^6-benzetheno-2′-deoxyadenosine and (7-hydroxy)-1,N^2-benzetheno-2′-deoxyguanosine, respectively, on a large scale and at high yield. The desired products were identified by fast atom bombardment–mass spectrometry, proton nuclear magnetic resonance and UV spectroscopy. These adducts were converted to the fully protected phosphoramidites and incorporated site-specifically into a series of oligonucleotides. 1,N^6-Ethano-2′-deoxyadenosine is one of the exocyclic adducts formed during DNA reaction with the antitumour agent, 1,3-bis(2-chloroethyl)nitrosourea. This compound was synthesized on a large scale with a high yield (62%) and then was converted to the phosphoramidite and incorporated site-specifically into oligonucleotides. The coupling efficiency of the incorporation of all these adducts was high (≥ 93%). After de-protection and purification of these oligomers, enzymatic hydrolysis and analysis by high-performance liquid chromatography confirmed the presence of the adduct in the oligomers. These oligomers are being used to investigate the biochemical and physical properties of these adducts.

Introduction

Benzene is a common environmental pollutant found all over the world. Human beings are exposed to benzene from many sources, including petroleum and petroleum products, automobile exhaust, cigarette smoke and in industries such as rubber and chemical manufacture. Benzene causes acute leukaemia in humans (IARC, 1987) and is toxic to the bone marrow (Goldstein, 1977). *para*-Benzoquinone is a stable metabolite of benzene (Snyder *et al.*, 1987; Huff *et al.*, 1989; Guengerich & MacDonald, 1990; Guengerich *et al.*, 1991) and a number of drugs and other chemicals (Pascoe *et al.*, 1988; Zheng & Hanzlik, 1992; McDonald *et al.*, 1993). Benzene must be metabolized to exert its toxic and leukaemogenic effects (Andrews *et al.*, 1977; Sammett *et al.*, 1979; Kalf, 1987)

The initial step in one metabolic pathway is oxidation by cytochrome P450 to form benzene oxide, which further yields phenol, then hydroquinone, which in turn is oxidized to *para*-benzoquinone. The latter accumulates in the bone marrow (Rickert *et al.*, 1979; Greenlee *et al.*, 1981), which is the site of the observed myelotoxic and leukaemogenic effects (Eastmond *et al.*, 1986; Kalf, 1987). Several adducts have been identified from the reaction of *para*-benzoquinone with nucleosides, nucleotides and DNA *in vitro* (Jowa *et al.*, 1986, 1990; Pongracz *et al.*, 1990; Pongracz & Bodell, 1991, 1996). The modified bases formed represent fewer than $1/10^6$ of the bases in DNA from human bone marrow or HL-60 cells when treated with *para*-benzoquinone *in vitro*. With such low levels of reaction in cells, ^{32}P-postlabelling was the only effective method of detection (Pongracz *et al.*, 1990; Pongracz & Bodell, 1991, 1996).

1,3-Bis(2-chloroethyl)nitrosourea (BCNU) is one of a family of therapeutic nitrosourea compounds used in cancer treatment. Reaction with DNA leads to several adducts, including saturated exocyclic adducts of adenine, cytosine and guanine: 1,N^6-

ethano-A (EA), 3,N^4-ethano-C (EC) and N^2,3-ethano-G (EG) (Ludlum, 1986, 1990), which resemble the exocyclic etheno adducts formed from the reaction of the chemical carcinogen, vinyl chloride, with DNA (Singer *et al.*, 1986). The first synthesis of an ethano deoxynucleoside was described by Zhang *et al.* (1995), who synthesized and incorporated 3,N^4-ethanodeoxycytidine into a site-specific oligonucleotide and investigated how this type of adduct affects replication.

In order to study the biochemical effects of *para*-benzoquinone and BCNU adducts, it was necessary to develop syntheses of the *para*-benzoquinone adducts with 2′-deoxyadenosine (dA), 2′-deoxycytidine (dC) and 2′-deoxyguanosine (dG) and 1,N^6-ethano-dA (EdA) and their phosphoramidites (Chenna & Singer, 1995, 1997; Maruenda *et al.*, 1998). In this work, we have summarized our previous findings describing the syntheses of *para*-benzoquinone-dA, *para*-benzoquinone-dC, *para*-benzoquinone-dG and EdA on a large scale, their conversion to the fully protected phosphoramidites and their site-specific incorporation into DNA oligonucleotides in high yield. These oligonucleotides are being studied for their effects on replication, mutation, repair *in vitro* and *in vivo* and their physical properties.

Results and Discussion

Synthesis of the phosphoramidites of para-*benzoquinone and 1,3-bis(2-chloroethyl)nitrosourea adducts and their incorporation into oligonucleotides*

1. Synthesis of the phosphoramidite of *para*-benzoquinone-deoxycytidine and its site-specific incorporation into oligonucleotides

2′-Deoxycytidine was reacted with *para*-benzoquinone at two pHs, 4.5 and 7.4, at 37 °C in a sodium acetate buffer. The reaction at pH 4.5 gave a much better yield of the desired product than that at pH 7.4, which, after purification, gave 61% of product **1** (Figure 1). When this product was analysed, the results were consistent with the addition of one molecule of *para*-benzoquinone to the base moiety followed by dehydration (–H_2O) (Chenna & Singer, 1995). The UV absorbance at various pHs was similar to that published by Pongracz *et al.* (1990) in Bodell's laboratory.

A major obstacle to preparing oligonucleotides containing *para*-benzoquinone-dC (**1**) and *para*-benzoquinone-dA (**5**) (Figures 1 and 2) is that these adducts have an extra reactive hydroxyl group on the base as well as the usual two hydroxyl groups (3′ and 5′-OH) on the sugar moiety. This extra hydroxyl on the exocyclic base definitely interferes in the synthesis of the oligonucleotides by branching or giving undesired by-products. Therefore, it must be protected with a suitable group which can be removed easily during the final de-protection of the oligonucleotides.

para-Benzoquinone-dC (**1**) was converted to its 5′-O-dimethoxytrityl (DMT) derivatives in a 68% yield by established methods (Schaller *et al.*, 1963; Chaudhary & Hernandez, 1979). The 5′-O-DMT-*para*-benzoquinone-dC (**2**) was successfully phosphitylated (Bodepudi *et al.*, 1992) by controlling the amount of phosphitylating reagent and the temperature to give product (**3**); however, a minor product (about 5%) was detected by thin-layer chromatography which was phosphitylated on both hydroxy groups (3′-OH and 3-OH) and was less polar than the major product (**3**). Derivative (**3**) was acylated by acetic anhydride (Schmidt *et al.*, 1956) to give product (**4**) (81%). The fully protected product (**4**) was used to make two oligonucleotides (**1a** and **1b**), 25 bases long, by standard phosphoramidite chemistry. The coupling rate was greater than 98%. When ammonia solution (28%) was used at 55 °C for 16 h to deprotect oligonucleotides (**1a**) and (**1b**), there was substantial loss of the desired oligomer. High-performance liquid chromatography (HPLC) of the oligomers after treatment with ammonia (Figure 4A) showed that only about 50% of the oligonucleotides contained the adduct (Figure 4A, peak 1), and another oligomer with a longer retention time (Figure 4A, peak 2) did not contain the adduct. Also, we observed that the UV of peak 1 absorbs from 340 to 220 nm; however, peak 2 absorbs from 310 to 220 nm, like unmodified DNA (Figure 4C). This indicated that peak 1 is the oligomer that contains the *para*-benzoquinone-dC (**1**), but peak 2 is a degradation product of peak 1. By deprotecting these oligomers (**1a** and **1b**) with 10% 1,3-diazabicyclo[5.4.0]undec-7-ene (DBU) in methanol at room temperature for three days (Xu & Swann, 1990; Seto *et al.*, 1991), however, HPLC showed the formation of only one oligomer (Figure 4B, peak 1), which contained *para*-benzoquinone-dC, and no degradation was

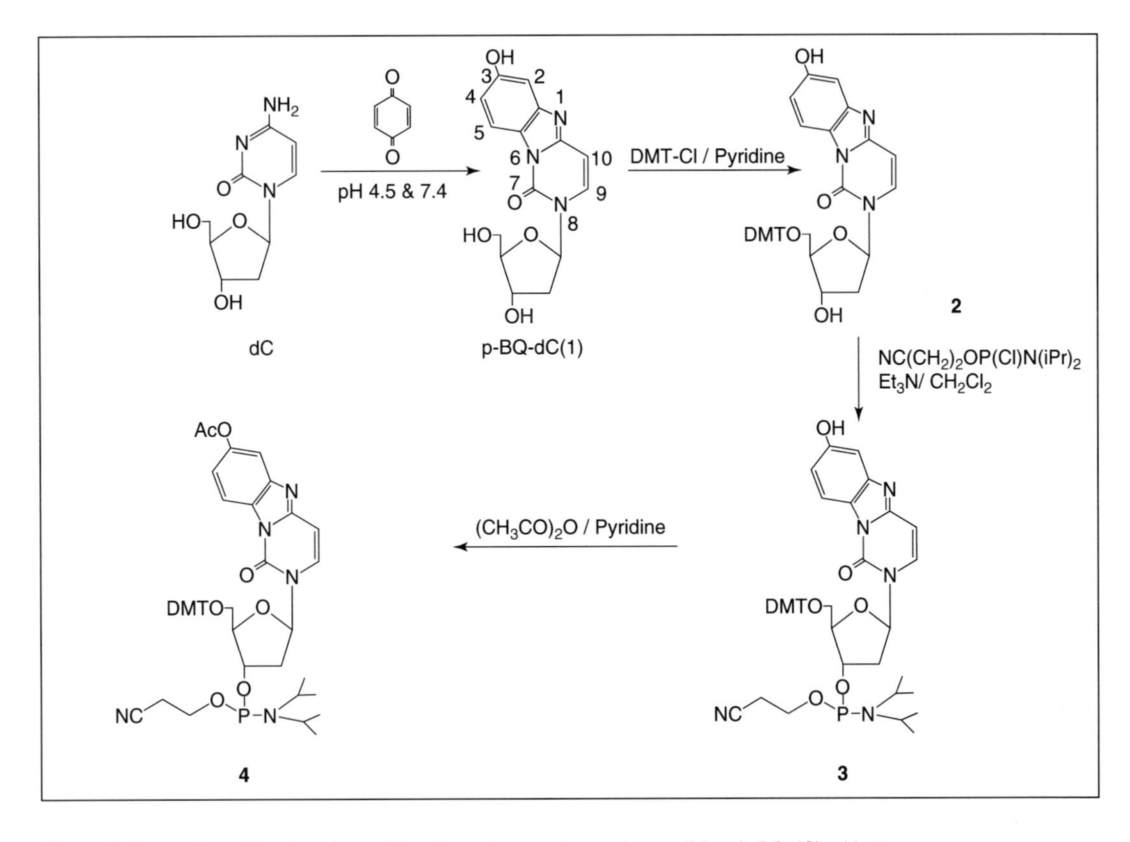

Figure 1. Preparation of the phosphoramidite of *para*-benzoquinone–deoxycytidine (p-BQ-dC) adduct

DMT-Cl, 4,4′-dimethoxytritylchloride

detected. The nucleoside composition of the oligomers was checked by enzyme digestion to nucleosides and subsequent analysis by HPLC. Enzyme hydrolysis of peak 1 showed that the oligomer contained the modified base *para*-benzoquinone-dC (Figure 5), indicating that the modified base survived the conditions used in the DNA synthesis and the de-protection procedure; however, the enzyme hydrolysis of peak 2 did not contain the modified base *para*-benzoquinone-dC (data not shown).

1a. 5′-CCGCTAG-**X**-GGGTACCGAGCTCGAAT-3′
1b. 5′-CCGCTAGCGGGTA-**X**-CGAGCTCGAAT-3′
X = ***para*-benzoquinone-dC**

2. Synthesis of the phosphoramidite of *para*-benzoquinone-deoxyadenosine (5) and its site-specific incorporation into oligonucleotides

2′-Deoxyadenosine reacted with *para*-benzoquinone in sodium acetate buffer at pH 4.5 at 37 °C gave *para*-benzoquinone-dA (**5**) (Figure 2) in good yield (Chenna & Singer, 1995). The UV absorbance of product (**5**) at various pHs was found to be similar to that reported (Pongracz & Bodell, 1991). When this reaction was conducted at pH 7.4 at 37 °C, HPLC of the reaction mixture showed formation of the same product but at a lower yield than at pH 4.5.

First, the 5′-hydroxy of *para*-benzoquinone-dA was protected by the DMT group, as for *para*-benzoquinone-dC, which gave about 71% of product (**6**) after purification. Phosphitylation of 5′-O-DMT-*para*-benzoquinone-dA (**6**) under anhydrous condition gave the phosphoramidite (**7**) as a diastereoisomeric pair, in 85% yield; however, protection of the 3′-OH with a phosphitylating reagent must be carefully controlled

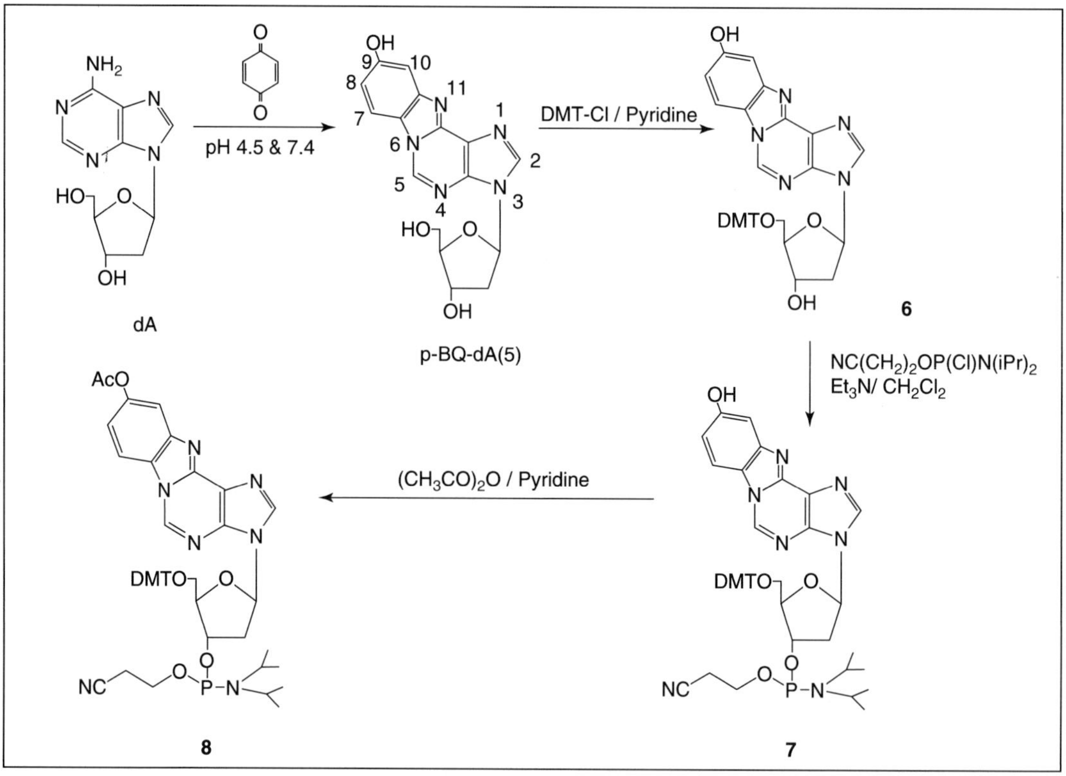

Figure 2. Preparation of the phosphoramidite of *para*-benzoquinone–deoxyadenosine (p-BQ-dA) adduct

DMT-Cl, 4,4´-dimethoxytritylchloride

(1–1.2 equivalents of the phosphitylating reagent was added at –40 °C) in order to avoid a secondary reaction with the phenolic exocyclic hydroxyl group on the base. The 9-hydroxy of compound (7) was protected by acylation with acetic anhydride, which gave product (8) in about 50% yield after purification (Chenna & Singer, 1995). This material was used for synthesizing two oligonucleotides by the standard phosphoramidite chemistry, which gave a high coupling efficiency.

2a. 5′-CCGCT-**Y**-GCGGGTACCGAGCTCGAAT-3′
2b. 5′-CCGCTAGCGGGT-**Y**-CCGAGCTCGAAT-3′
Y = ***para*-benzoquinone-dA**

These oligonucleotides were deprotected with DBU in methanol, instead of ammonia, at room temperature for three days in order to avoid likely degradation of the modified base, as was found previously for oligomers (**1a** and **1b**) containing *para* benzoquinone dC when treated with ammonia. The oligomers were purified by HPLC, which showed the formation of only one product. The purity of the oligomers and the presence of the modified *para*-benzoquinone-dA was confirmed by HPLC of the enzyme digest of the oligomers, which indicated survival of adduct (**5**).

Electrospray mass spectrometry (MS) was used for further identification of the oligonucleotides containing *para*-benzoquinone-dC or *para*-benzoquinone-dA. The 25-mer oligomer containing the adducts has a calculated molecular mass of 7795. Its electrospray mass spectrum showed that the molecular mass of oligomer **1a**, which contains *para*-benzoquinone-dC, was m/z 7793.88 ± 6.08 and that of oligomer **2a**, which contains *para*-benzo-

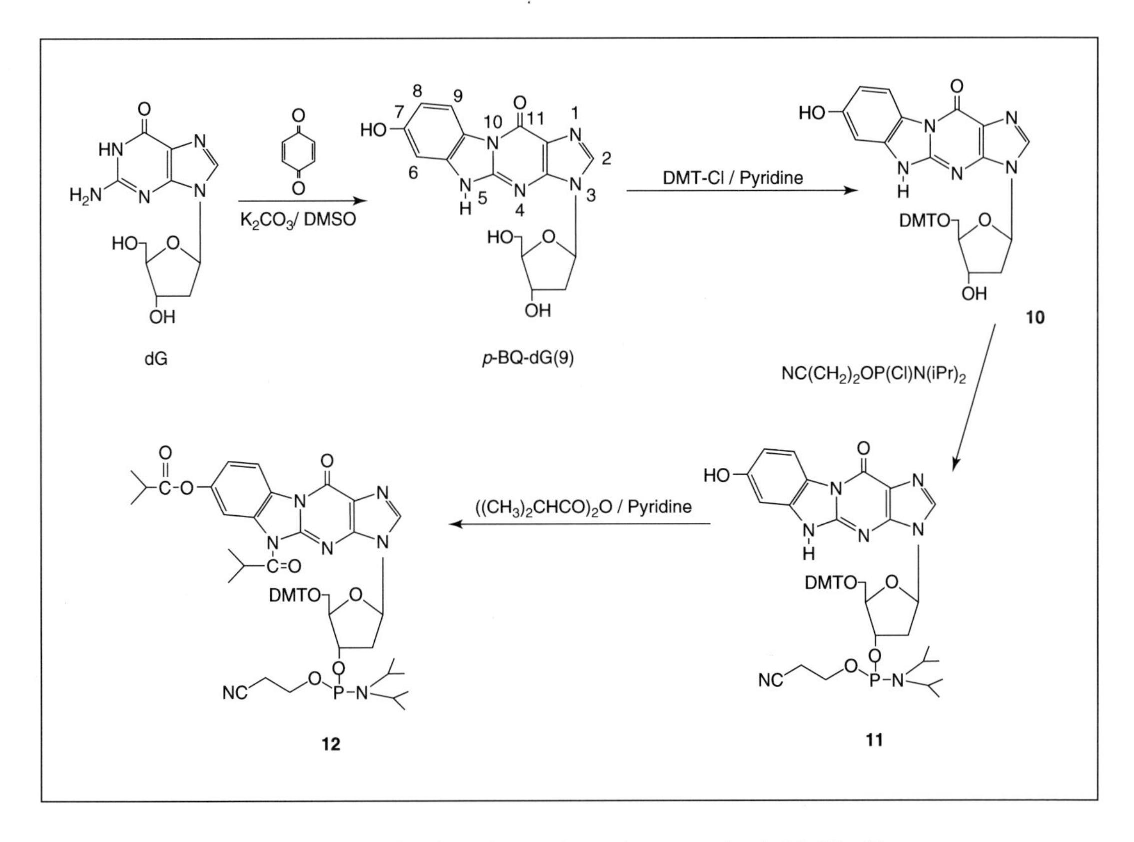

Figure 3. Preparation of the phosphoramidite of *para*-benzoquinone–deoxyguanosine (p-BQ-dG) adduct

DMT-Cl, 4,4′-dimethoxytritylchloride

quinone-dA, was *m/z* 7796.38 ± 21.83. These results are in agreement with the calculated molecular mass.

3. Synthesis of the phosphoramidite of *para*-benzoquinone-deoxyguanosine (9) and its site-specific incorporation into oligonucleotides

Reaction of 2′-deoxyguanosine with *para*-benzoquinone in aqueous solution at either pH 7.4 and 9.3 produced the same product as reported previously (Jowa *et al.*, 1986) and traces of other products, which were not studied. No reaction was detected at pH 4.5. The major product had the same UV spectrum as reported by Jowa *et al.* (1986), but the yield of the compound at both pH 7.4 and pH 9.3 was poor (< 2%). Therefore, the reaction was repeated in non-aqueous conditions, in dry dimethyl sulfoxide or *N,N*-dimethylformamide in the presence of K_2CO_3 or NaOH pellets. Under these conditions, substantial amounts of product (about 50%) were obtained, as shown by HPLC.

The reaction of dG with *para*-benzoquinone in dimethyl sulfoxide and K_2CO_3 first produced an unstable product, which was slowly converted to the final stable product (*para*-benzoquinone-dG). This reaction when monitored by HPLC or thin-layer chromatography over time clearly showed the conversion of the intermediate, which had a longer retention time (32–33 min), to the final product (Figure 6) and was eluted at 31 min. The UV spectrum of the unstable intermediate does not resemble that of the final product, as the final product has a significant trough at about 250 nm and a peak at about 270 nm, whereas the intermediates have a broad shoulder at 260–275 nm, in addition to red-shifted maxima (Figure 6).

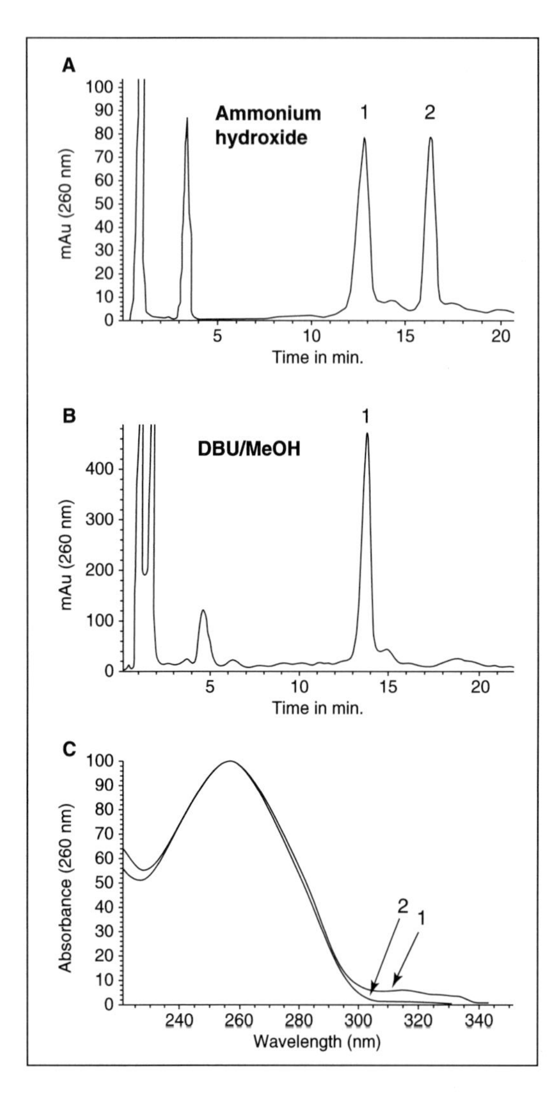

Figure 4. High-performance liquid chromatography (HPLC) of oligonucleotide containing *para*-benzoquinone (p-BQ)–deoxycytidine (dC) treated with ammonium hydroxide and 1,3-diazabicyclo[5.4.0]undec-7-ene (DBU)

A: Profile obtained after deprotection of oligonucleotide (**1a**) with 28% ammonia for 16 h. Peak 1 is the oligomer (**1a**) which contains the adduct *p*-BQ-dC; peak 2 is the oligomer which does not contain the adduct. B: Profile obtained after deprotection of oligomer (**1a**) with 10% DBU in methanol, which produced only one product: peak 1 contains the adduct *p*-BQ-dC. C: UV spectra of peaks 1 and 2. Spectrum 1 is that of the oligomer (**1a**) which contains *p*-BQ-dC; spectrum 2 for the degraded oligomer does not contain the adduct.

This intermediate showed two peaks very close to each other by HPLC (Figure 6). When either peak was collected and immediately reanalysed under the same HPLC conditions, there were again two peaks with the same retention time as each of the single peaks. This suggests that there is a single product in equilibrium in two forms, which are presumably isomers. The intermediate product was isolated by HPLC and analysed immediately by electrospray MS . The results suggest that the molecule carries two molecules of *para*-benzoquinone on the dG base, with removal of a molecule of water (dG + 2*para*-benzoquinone – H_2O) with a molecular mass of 463. A comparison of the UV spectra of the intermediate and stable adducts gives an indication of substitution on the N1 and N^2 of dG by *para*-benzoquinone, but we do not know where the second *para*-benzoquinone reacts with dG. Owing to the unstable nature of this adduct, it was not possible to use proton nuclear magnetic resonance (^{1}H-NMR) to obtain further information about the structure or the mechanism of the interconversion.

The reaction was scaled up to react 500 mg of dG with *para*-benzoquinone; after purification, this yielded 28% of pure product (**9**; Figure 3; Chenna & Singer, 1997). The UV spectra of this product (**9**) at various pHs were similar to that reported (Jowa *et al.*, 1986). *para*-Benzoquinone-dG (**9**) was converted to its 5′-O-DMT product in 78% yield and then to the phosphoramidite (**11**) in 69% yield. By controlling the amount of the phosphitylating reagent and the temperature (–70 °C), only the 3′-OH reacted. Derivative (**11**) was acylated with isobutyric anhydride (Schmidt *et al.*, 1956) to give product (**12**) in 81% yield. This fully protected compound (**12**) was used to synthesize two oligonucleotides: a 19-mer and a 25-mer. The phenoxyacetyl phosphoramidites were used to synthesize these DNA oligomers, and then standard phosphoramidite chemistry was followed. The coupling efficiency of *para*-benzoquinone-dG phosphoramidite was 96%. The oligonucleotides were deprotected under non-aqueous conditions with 10% DBU in methanol at room temperature for 24 h and then purified by 20% polyacrylamide gel electrophoresis followed by HPLC. DBU was used to de-protect these oligomers instead of ammonia, since *para*-benzoquinone-dG was not stable in 28% ammonia. In addition, the use of the phenoxyacetyl phosphoramidite protecting group allowed

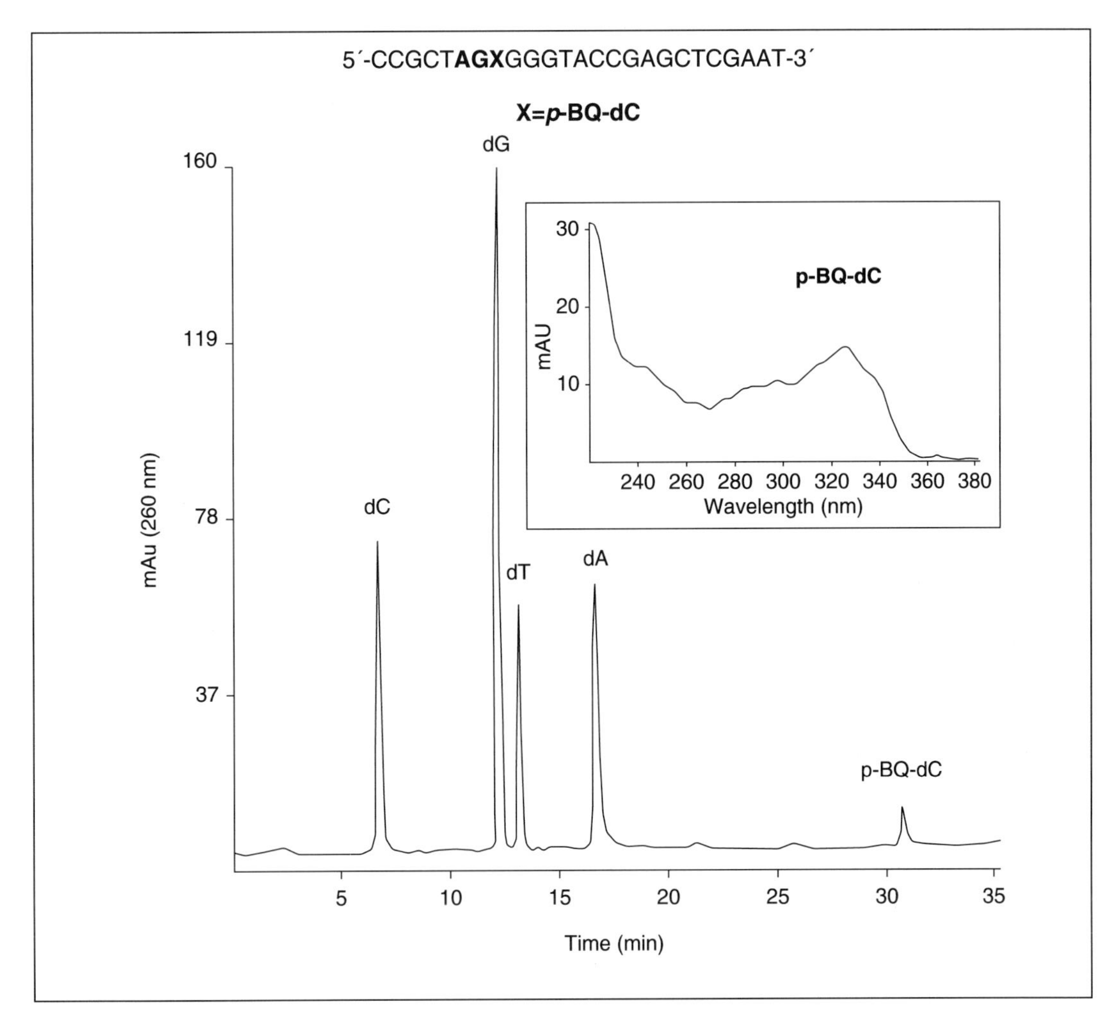

Figure 5. High-performance liquid chromatography profile obtained from the enzymatic digestion of the 25-mer oligonucleotide (**1a**) containing *para*-benzoquinone–deoxycytidine (p-BQ-dC) adduct with snake venom phosphodiesterase and alkaline phosphatase

mAu, milli-absorption units
2´-Deoxynucleosides obtained as a result of enzymatic digestion of the oligonucleotide (**1a**), which contains p-BQ-dC

deprotection of DNA oligomers at room temperature and in a shorter time (24 h) than the standard phosphoramidite, which requires three days.

3a. 5´- C C G C T A **X** C G G G T A C C G A G C T C G A A T - 3´
3b. 5´- X C G G G T A C C G A G C T C G A A T -3´
X = ***para*-benzoquinone-dG**

The composition of the oligomers was confirmed after enzymatic digestion of the DNA oligomers to nucleosides, followed by HPLC analysis.

The effect of *para*-benzoquinone adducts of DNA bases on the thermal stability of a 25-nucleotide oligodeoxyribonucleotide double helix is reported in a separate paper by Janos Sági and B. Singer in this volume (see also the Appendix for molecular models).

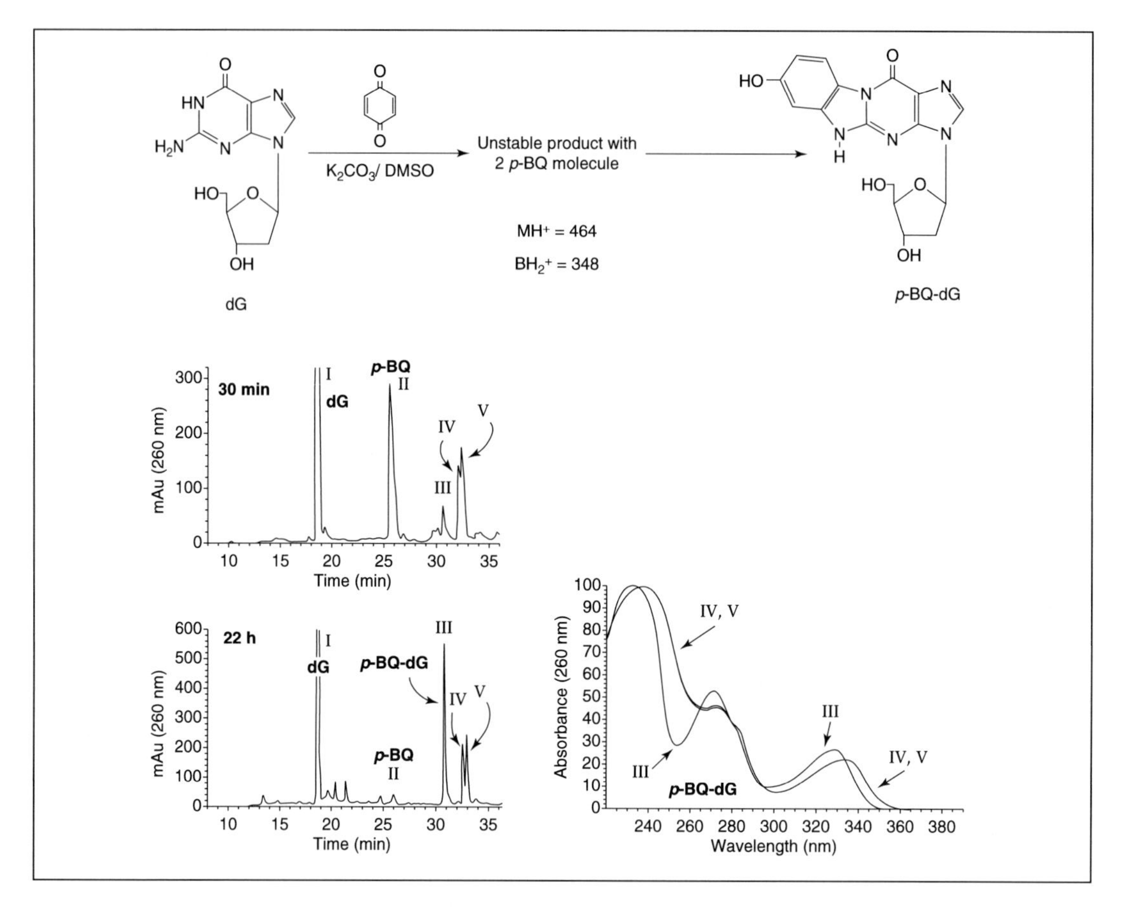

Figure 6. High-performance liquid chromatography (HPLC) of products of reactions of deoxyguanosine (dG) with *para*-benzoquinone (p-BQ) at 30 min and 22 h

The reaction was performed in *N,N*-dimethylformamide and K_2CO_3 at room temperature. (Top) 30-min reaction: peak I (~19 min) corresponds to the starting material, dG; peak II (~26 min) is *p*-BQ; peak III (~31 min) is the stable final adduct; peaks IV and V between 32 and 33 min are the two unstable intermediates. (Bottom) 22-h reaction: peak I is dG; only traces of *p*-BQ remain (peak II); the peak at ~31 min (peak III) is the stable *p*-BQ-dG product. The doublet at 32–33 min consists of two interconvertable forms of the unstable intermediates (peaks IV, V). When either of the doublets is collected and immediately reanalysed in the same HPLC system, the same two peaks (32–33 min) result (data not shown). The UV spectra of *p*-BQ-dG (III) and the intermediate products (IV, V) are shown; the arrows indicate the spectrum of each. Note that the three maxima for *p*-BQ-dG are shifted ~ 5 nm.

4. Synthesis of 1,N^6-ethano-2′-deoxyadenosine (17), its phosphoramidite and its incorporation into a DNA oligonucleotide

The reaction of BCNU with nucleosides and DNA produced several adducts, including 1,N^6-ethano-dA (**17**; Figure 7) in low yields (Ludlum, 1987). In order to study the biological and biophysical effects of this adduct, a total synthetic method of preparing it in high yield had to be developed. A suitable route was successfully developed (Maruenda *et al*, 1998). The 6-chloropurine-2′-deoxyriboside (**13**), was prepared (Robins & Basom, 1973) then derivatized to give compound (**14**) in 86% yield with *tert*-butyldimethylsilyl chloride, which was reacted with 2-hydroxyethylamine to produce compound (**15**) in 64% yield. Iodination

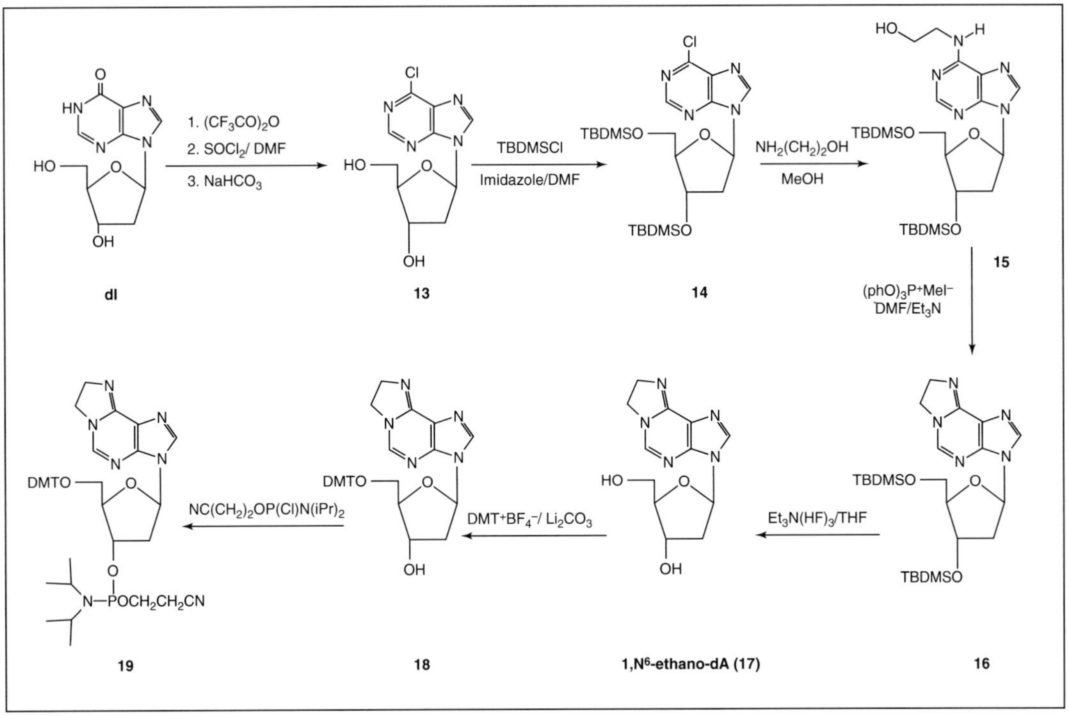

Figure 7. Synthesis of the phosphoramidite of 1,*N*⁶-ethano-deoxyadenosine adduct

dI, deoxyinosine; DMF, dimethylformamide; TBDMSCl, tert-butyldimethylsilyl chloride; THF, tetrahydrofuran; DMT, 4,4′-dimethyoxytrityl

of the β-hydroxyl group of (**15**) with methyltriphenoxyphosphonium iodide (Saito *et al.*, 1993) led to spontaneous cyclization of the resulting N^6-(2-iodoethyl) derivative, producing the 1,N^6-ethano-derivative (**16**) in 67% yield. Desilylation of the latter was accomplished by treatment with triethylamine trihydrofluoride (Pirrung *et al.*, 1994), which gave a quantitative yield (91%) of 1,N^6-ethano-dA (**17**). This adduct was analysed by thin-layer chromatography, HPLC, ^{1}H-NMR, ^{13}C-NMR, UV, electrospray MS, fast atom bombardment–MS and high-resolution MS. The UV spectrum of this compound (**17**) showed the same characteristics as reported previously (Tong & Ludlum, 1979; Ludlum, 1986). All of the data collected were in agreement with the structure of 1,N^6-ethano-dA (**17**) (Maruenda *et al.*, 1998).

The first attempt to tritylate compound (**17**) according to standard methods with 4,4′-dimethoxytritylchloride (Schaller *et al.*, 1963; Chaudhary & Hernandez, 1979) failed; however, treatment of compound (**17**) with $DMT^+BF^-_4$ (Bleasdale *et al.*, 1990) gave the 5′-O-DMT product (**18**) in 62% yield. Phosphitylation of 5′-O-DMT-1,N^6-ethano-dA (**18**) (Bodepudi *et al.*, 1992) yielded a 1:1 mixture of the expected diastereoisomers (**19**) in 71% yield. The phenoxyacetyl phosphoramidite nucleotides were used in the synthesis of DNA oligomer containing 1,N^6-ethano-dA by standard phosphoramidite chemistry. The overall yield was 87%, and the coupling efficiency of 1,N^6-ethano-dA phosphoramidite was 93%. The oligonucleotide was then de-protected under aqueous conditions with 28–30% ammonia for 1.5 h at 65 °C and purified by HPLC.

4. 5′- C C G C T **X** G C G G G T A C C G A G C T C G A A T - 3′
X = 1,N^6-etheno-dA

The composition of the DNA oligomer was confirmed after enzymatic digestion to nucleosides, followed by HPLC analysis, which showed that the modified base survived the conditions used in the DNA synthesis and deprotection procedure (Figure 8).

These new oligonucleotides containing *para*-benzoquinone-dC, *para*-benzoquinone-dA, *para*-benzoquinone-dG and 1,N^6-ethano-dA adducts are being used to obtain information about the repair, mutagenesis and structural properties of these adducts.

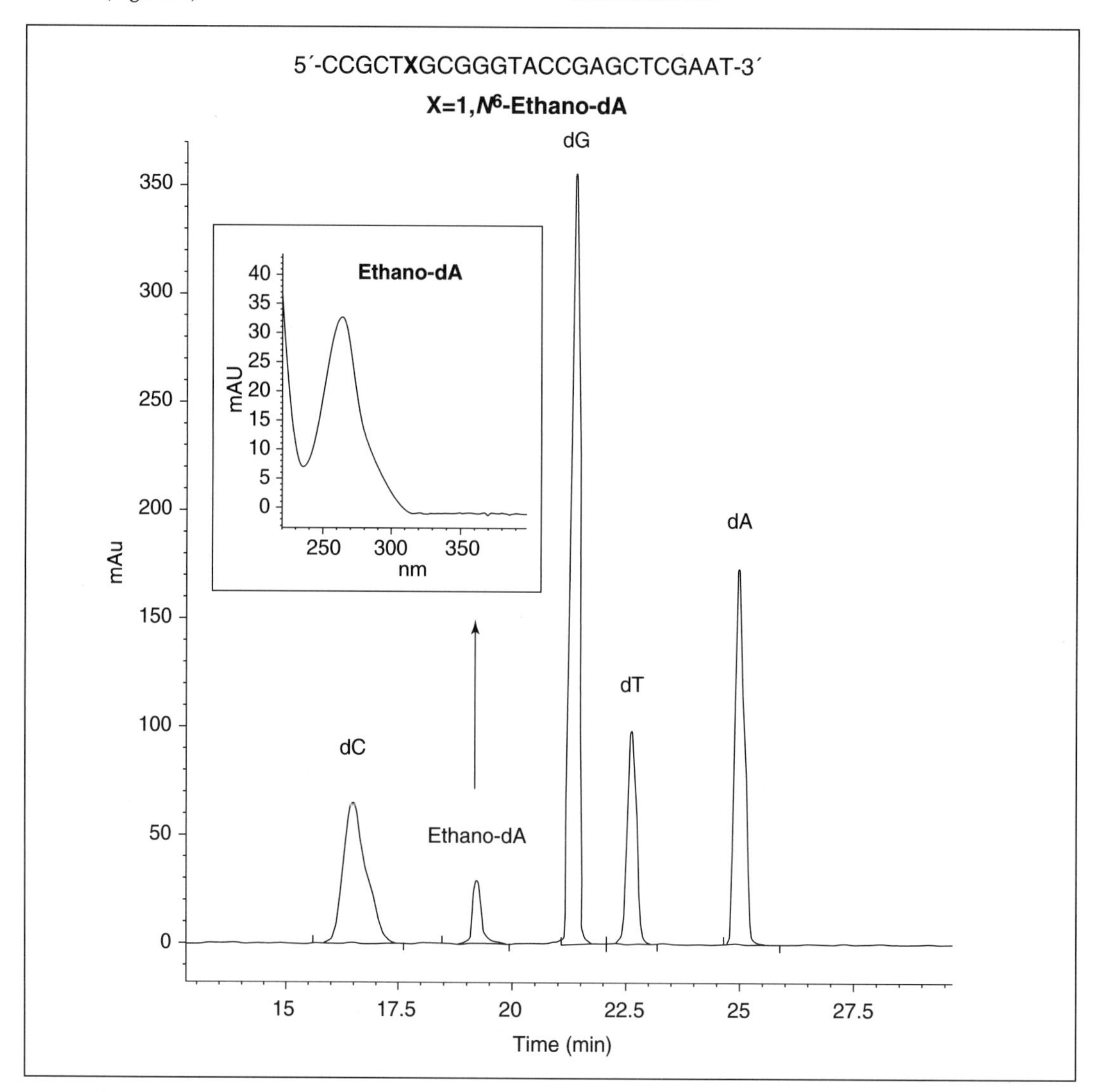

Figure 8. High-performance liquid chromatography profile obtained from the enzymatic digestion of 25-mer oligonucleotide containing 1, N^6-ethano-deoxyguanosine adduct

dA, deoxyadenosine; dG, deoxyguanosine; dC, deoxycytidine; dT, deoxythymidine. The retention times of the deoxynucleosides are dC, 16.5 min; 1,N^6-ethano-dA, 19.3 min; dG, 21.3 min; dT, 22.6 min and dA, 24.9 min. The UV spectrum of 1,N^6-ethano-dA at pH 4.5 is shown in the insert

Acknowledgements

This work was supported by grants number CA47723 and ES07368 from the National Institutes of Health to B.S. and was administered by the Lawrence Berkeley National Laboratory under Department of Energy Contract # DE-AC03-76SF00098.

References

Andrews, L.S., Lee, E.W., Witmer, C.M., Kocsis, J.J. & Snyder, R. (1977) Effects of toluene on the metabolism, disposition and hemopoietic toxicity of [^{3}H]benzene. *Biochem. Pharmacol.* 26, 293–300.

Bleasdale, C., Ellswood, S.B. & Golding, B. T. (1990) 4,4'-Dimethoxytrityl and 4-monomethoxytrityl tetrafluoroborate: Convenient reagents for the protection of primary alcohols including sugars. *J. Chem. Soc. Perkins Trans. I* 803–805.

Bodepudi, V., Shibutani, S. & Johnson, F. (1992) Synthesis of 2´-deoxy-7,8-dihydro-8-oxoguanosine and 2´-deoxy-7,8-dihydro-8-oxoadenosine and their incorporation into oligomeric DNA. *Chem. Res. Toxicol.* 5, 608–617.

Chenna, A. & Singer, B. (1995) Large scale synthesis of *p*-benzoquinone-2'-deoxycytidine and *p*-benzoquinone-2'-deoxyadenosine adducts and their site-specific incorporation into DNA oligonucleotides. *Chem. Res. Toxicol.* 8, 865–874.

Chenna, A. & Singer, B. (1997) Synthesis of a benzene metabolite adduct, 3''-hydroxy-1,N^2-benzetheno-2'-deoxyguanosine, and its site-specific incorporation into DNA oligonucleotides. *Chem. Res. Toxicol.*, 10, 165–171.

Chaudhary, S.K. & Hernandez, O. (1979) 4-Dimethylaminopyridine: An efficient and selective catalyst for the silylation of alcohols. *Tetrahedron Lett.* 2, 99–102.

Eastmond, D.A., Smith, M.T., Ruzo, L.O. & Ross, D. (1986) Metabolic activation of phenol by human myeloperoxidase and horseradish peroxidase. *Mol. Pharmacol.* 30, 674–679.

Goldstein, B.D. (1977) Benzene toxicity: a critical evaluation: Hematotoxicity in humans. *J. Toxicol. Environ. Health* Suppl. 2, 69–105.

Greenlee, W.F., Gross, E.A. & Irons, R.D. (1981) Relationship between benzene toxicity and the disposition of ^{14}C-labelled benzene metabolites in the rat. *Chem. Biol. Interactions* 74, 55–62.

Guengerich, F.P. & MacDonald, T.L. (1990) Mechanisms of cytochrome P-450 catalysis. *FASEB J.* 4, 2453–2459.

Guengerich, F.P., Kim, D.H. & Iwasaki, M. (1991) Role of human cytochrome P-450 IIE1 in the oxidation of many low molecular weight cancer suspects. *Chem. Res. Toxicol.* 4, 168–179.

Huff, J.E., Haseman, J.K., DeMarini, D.M., Eustis, S., Maronpot, R.R., Peters, A.C., Persing, R.L., Chrisp, C.E. & Jacobs, A.C. (1989) Multiple-site carcinogenicity of benzene in Fischer 344 rats and B6C3F1 mice. *Environ. Health Perspectives* 82, 125–163.

IARC (1987) *IARC Monographs on the Evaluation of the Carcinogenic Risks to Humans*, Suppl. 7, *Overall Evaluations of Carcinogenicity: An Updating of* IARC Monographs *Volumes 1–42*, Lyon.

Jowa, L., Winkle, S., Kalf, G., Witz, G. & Snyder, R. (1986) Deoxyguanosine adducts formed from benzoquinone and hydroquinone. *Adv. Exp. Med. Biol.* 197, 825–832.

Jowa, L., Witz, G., Snyder, R., Winkle, S. & Kalf, G.F. (1990) Synthesis and characterization of deoxyguanosine–benzoquinone adducts. *J. Appl. Toxicol.* 10, 47–54.

Kalf, G.F. (1987) Recent advances in the metabolism and toxicity of benzene. *Crit. Rev. Toxicol.* 18, 141–159.

Ludlum, D.B. (1986) Formation of cyclic adducts in nucleic acids by the haloethylnitrosoureas. In: *The Role of Cyclic Nucleic Acid Adducts in Carcinogenesis and Mutagenesis* (Singer, B. & Bartsch, H., eds), IARC Scientific Publications No. 70. Lyon, International Agency for Research on Cancer, pp. 137–146.

Ludlum, D.B. (1987) High performance liquid chromatographic separation of DNA adducts induced by cancer chemotherapeutic agents. *Pharmacol. Ther.* 34, 145–153.

Ludlum, D.B. (1990) DNA alkylation by the haloethylnitrosoureas: Nature of modifications produced and their enzymatic repair or removal. *Mutat. Res.* 233, 117–126.

Maruenda, H., Chenna, A., Liem, L.K. & Singer, B. (1998) Synthesis of 1,N-6-ethano-2′-deoxyadenosine, a metabolic product of 1,3-bis(2-chloroethyl)-nitrosourea, and its incorporation into oligomeric DNA. *J. Org. Chem.* 63, 4385–4389.

McDonald, T.A., Waidyanatha, S. & Rappaport, S.M. (1993) Measurement of adducts of benzoquinone with hemoglobin and albumin. *Carcinogenesis* 14, 1927–1932.

Pascoe, G.A., Calleman, C.J. & Baille, T.A. (1988) Identification of S-(2,5-dihydroxyphenyl)-cysteine and S-(2,5-dihydroxyphenyl)-N-acetyl-cysteine as urinary metabolites of acetaminophen in the mouse. Evidence for p-benzoquinone as a reactive intermediate in acetaminophen metabolism. *Chem.-Biol. Interactions* 68, 85–98.

Pirrung, M.C., Shuey, S.W., Lever, D.C. & Fallon, L.A. (1994) A convenient procedure for the deprotection of silylated nucleosides and nucleotides using triethylamine trihydrofluoride. *Bioorg. Med. Chem. Lett.* 4, 1345–1346.

Pongracz, K. & Bodell, W.J. (1991) Detection of 3′-hydroxy-1,N^6-benzetheno-2′-deoxyadenosine 3′-phosphate by ^{32}P-postlabeling of DNA reacted with *p*-benzoquinone. *Chem. Res. Toxicol.* 4, 199–202.

Pongracz, K. & Bodell, W.J. (1996) Synthesis of N^2-(4-hydroxyphenyl)-2′-deoxyguanosine 3′-phosphate: Comparison by ^{32}P-postlabeling with the DNA adduct formed in HL-60 cells treated with hydroquinone. *Chem. Res. Toxicol.* 9, 593–598.

Pongracz, K., Kaur, S., Burlingame, A.L. & Bodell, W.J. (1990) Detection of (3′-hydroxy)-3,N^4-benzetheno-2′-deoxycytidine-3′-phosphate by ^{32}P-postlabeling of DNA reacted with *p*-benzoquinone. *Carcinogenesis* 11, 1469–1472.

Rickert, D.E., Baker, T.S., Bus, J.S., Barrow, C.S. & Irons, R.D. (1979) Benzene disposition in the rat after exposure by inhalation. *Toxicol. Appl. Pharmacol.* 49, 417–423.

Robins, M.J. & Basom, G.L. (1973) Nucleic acid related compounds. 8. Direct conversion of 2′-deoxyinosine to 6-chloropurine 2′-deoxyriboside and selected 6-substituted deoxynucleosides and their evaluation as substrates of adenosine deaminase. *Can. J. Chem.* 51, 3161–3169.

Saito, T., Murakami, M., Inada, T., Hayashibara, H. & Fujii, T. (1993) Purines. 54. Intramolecular cyclization of 9-ethyl-1-(2-hydroxyethyl)adenine caused by nucleophiles formation of N-6, 1-ethanoadenine derivatives. *Chem. Pharm. Bull.* 41, 453–457.

Sammett, D., Lee, E.W., Kocsis, J.J. & Snyder, R. (1979) Partial hepatectomy reduces both metabolism and toxicity of benzene. *J. Toxicol. Environ. Health* 5, 785–792.

Schaller, H., Weimann, G., Lerch, B. & Khorana, H.G. (1963) Studies on polynucleotides. XXIV. The stepwise synthesis of specific deoxyribopolynucleotides. Protected derivatives of deoxyribonucleosides and new syntheses of deoxyribonucleoside-3′ phosphates. *J. Am. Chem. Soc.* 85, 3821–7.

Schmidt, O.T., Voigt, H., Puff, W. & Köster, R. (1956) Benzyläther der Ellagsäure und Hexaoxy-diphensäure. *Ann. Chem.* 586, 165–79

Seto, H., Seto, T., Ohkubo, T. & Saitoh, I. (1991) Reaction of malonaldehyde with nucleic acid. R. formation of pyrimido[1,2-*a*]purin-10(3H)-one nucleoside by thermal decomposition of diastereomers containing oxadiazabicyclononene residues linked to guanosine. *Chem. Pharm. Bull.* 39, 515–517.

Singer, B., Holbrook, S.R, Fraenkel-Conrat, H & Kusmierek, J.T. (1986) Neutral reactions of haloacetaldehydes with polynucleotides: Mechanisms, monomer and polymer products. In: *The Role of Cyclic Nucleic Acid Adducts in Carcinogenesis and Mutagenesis* (Singer, B. & Bartsch, H., eds), IARC Scientific Publications No. 70. Lyon, International Agency for Research on Cancer, pp.45–58.

Snyder, R., Jowa, L., Witz, G., Kalf, G. & Rushmore, T. (1987) Formation of reactive metabolites from benzene. *Arch. Toxicol.* 60, 61–64.

Tong, W.P. & Ludlum, D.B. (1979) Mechanism of action of the nitrosoureas. III. Reactions of bis-chloroethyl nitrosourea and bis-fluoroethyl nitrosourea with adenosine. *Biochem. Pharmacol.* 28, 1175–1179.

Xu, Y.Z. & Swann, P.F. (1990) A simple method for the solid phase synthesis of oligodeoxynucleotides containing O^4-alkylthymine. *Nucleic Acids Res.* 18, 4061–4065.

Zhang, W., Rieger, R., Iden, C. & Johnson, F. (1995) Synthesis of 3,N-4-etheno, 3,N-4-ethano, and 3-(2-hydroxyethyl) derivatives of 2′-deoxycytidine and their incorporation into oligomeric DNA. *Chem. Res. Toxicol.* 8, 148–156.

Zheng, J. & Hanzlik, R.P. (1992) Dihydroxylated mercapturic acid metabolites of bromobenzene. *Chem. Res. Toxicol.* 5, 561–567.

Exocyclic DNA Adducts in Mutagenesis and Carcinogenesis
B. Singer & H. Bartsch
IARC Scientific Publications No. 150
International Agency for Research on Cancer, Lyon, 1999

Purine DNA adducts of 4,5-dioxovaleric acid and 2,4-decadienal

J. Cadet, V.M. Carvalho, J. Onuki, T. Douki, M.H.G. Medeiros & P. Di Mascio

The present overview describes recent findings on the formation of cyclic adducts of purine DNA bases after reaction with two aldehyde compounds, 4,5-dioxovaleric acid (DOVA) and 2,4-decadienal (DDE), which are involved in 5-aminolaevulinic acid (ALA) accumulation and lipid peroxidation, respectively. ALA accumulates under pathological conditions and is associated with an increased incidence of liver cancer. The final oxidation product of ALA, DOVA, is an efficient alkylating agent of the guanine moieties in both nucleoside and isolated DNA. Adducts were produced through the formation of a Schiff base involving the *N*²-amino group of 2′-deoxyguanosine and the ketone function of DOVA, respectively. DDE is an important breakdown product of lipid peroxidation. It is cytotoxic to mammalian cells and is known to be implicated in DNA damage. It can bind to 2′-deoxyadenosine, yielding highly fluorescent products, including 1,*N*⁶-etheno-2′-deoxyadenosine and two other, related adducts. The reaction mechanism for the formation of DDE–2′-deoxyadenosine adducts involves epoxidation of DDE and subsequent addition of the resulting reactive intermediates to the *N*⁶ amino group of 2′-deoxyadenosine, followed by cyclization at the N1 site. Formation of endogenous DNA adducts may contribute to the genotoxic potential of ALA and DDE.

Introduction

A series of chemicals can alkylate DNA bases, some by addition to the N7-position of guanine residues or by involvement of the exocyclic amino groups of cytosine, adenine and guanine. Malondialdehyde (Basu *et al.*, 1988), glyoxal (Shapiro & Hachmann, 1966; Shapiro *et al.*, 1968) and α,β-unsaturated aldehydes (Chung *et al.*, 1984; Winter *et al.*, 1986; Eder & Hoffman, 1992, 1993; Nath & Chung, 1994; Douki & Ames, 1994) can be involved in the formation of modified bases with an additional ring. The latter structure can be either five-membered, as in the malondialdehyde–guanine adduct, or six-membered as in 1,*N*²-propano adducts of α,β-unsaturated aldehydes. In addition, etheno bases can be generated after reaction of DNA bases with vinyl chloride, ethyl carbamate, α-halogenated aldehydes, α-haloketones (Bartsch *et al.*, 1994) and epoxidized α,β-unsaturated aldehydes (Sodum & Chung, 1988, 1991). Most of these lesions have been characterized at the level of nucleosides and isolated DNA. Some of them have also been detected in cellular DNA, including that of rodent and human liver (Vaca *et al.*, 1992; Chaudhary *et al.*, 1994; Nath & Chung, 1994; Nair *et al.*, 1995). Emphasis is placed in this chapter on the identification of the main DNA bases produced by two aldehydic compounds, 4,5-dioxovaleric acid (DOVA; Douki *et al.*, 1998a) and 2,4-decadienal (DDE; Carvalho *et al.*, 1998). In both cases, the initial steps in adduct formation are similar to those identified for other reactive aldehydes (glyoxal and 2,3-epoxy-4-hydroxynonanal, respectively); however, additional rearrangements lead to the formation of unusual substituted rings.

DNA alkylation by 4,5-dioxovaleric acid

5-Aminolaevulinic acid (ALA), a precursor of porphyrin IX in the biosynthesis of haem, can accumulate in liver, brain and other organs under pathological conditions such as acute intermittent porphyria, tyrosinosis (Kappas *et al.*, 1983; Hindmarsh, 1986) and lead poisoning (Chisolm, 1971). A high incidence of primary liver cancer has been reported in patients with acute intermittent porphyria and has been associated with the frequency of acute attacks when the ALA plasma concentration rises about 100-

fold (Lithner & Wetterberg, 1984; Gubler *et al.*, 1990; Thunnissen *et al.*, 1991). This observation is an indication of possible mutagenic properties of ALA. Moreover, it has recently been observed that ALA can induce chromosomal aberrations in rat hepatocytes (Fiedler *et al.*, 1996). The determination of the carcinogenic potential of ALA has received additional attention as the result of the increasing use of ALA in cancer phototherapy (Kennedy & Pottier, 1992; Ortel *et al.*, 1993; Peng *et al.*, 1997). Nevertheless, little is known about the possible side-effects of ALA and the chemical mechanisms associated with its DNA damaging properties. The involvement of reactive oxygen species produced during the metal-catalysed oxidation of ALA by oxygen has been demonstrated by Bechara and coworkers (1993). In addition, ALA may catalyse its own oxidation, since it can release iron from ferritin (Oteiza *et al.*, 1995). Moreover, iron mobilization in the brain and liver of rat is triggered by long-term treatment with ALA (Demasi *et al.*, 1996). Induction of DNA damage by ALA is also indicated by the observation that DNA strand breaks are formed in plasmids in the presence of ferrous ions (Onuki *et al.*, 1994), copper ions (Hiraku & Kawanishi, 1996) or ferritin (Di Mascio *et al.*, unpublished data). In addition, increases in the levels of several radical-induced modified bases have been observed in isolated DNA after treatment with ALA (Fraga *et al.*, 1994; Hiraku & Kawanishi, 1996; Douki *et al.*, 1998b). Interestingly, the relative yield of the latter lesions was similar to those obtained by gamma irradiation of isolated DNA in aerated aqueous solution (Douki *et al.*, 1998b), which strongly suggests the involvement of hydroxyl radicals in ALA-mediated DNA damage. Formation of oxidized bases, including 8-oxo-7,8-dihydroguanine and 5-hydroxycytosine, has also been found in liver DNA of rats treated with ALA (Fraga *et al.*, 1994; Douki *et al.*, 1998b). Taken together, these data provide evidence for the biological relevance of ALA metal-catalysed oxidation. In that respect, the final oxidation product of ALA, DOVA, is an interesting compound, and its dicarbonyl structure makes it a good candidate for alkylating the exocyclic amino group of DNA bases, as already reported for acetaldehyde (Vaca *et al.*, 1995) and glyoxal derivatives (Shapiro & Hachmann, 1966; Shapiro *et al.*, 1968). Therefore, we investigated the reactivity of DOVA towards 2′-deoxyguanosine (dG) and identified two diastereoisomeric adducts. Their formation involves addition of the ketone function of DOVA to the N^2-amino group of dG, leading to formation of a Schiff base. The reaction was also shown to take place within isolated double-stranded DNA.

Addition of 4,5-dioxovaleric acid to 2′-deoxyguanosine and reduction of the reaction products

The reaction between DOVA and dG led to the formation of two major products (Figure 1), which were separated by reversed-phase high-performance liquid chromatography (HPLC). The UV absorption features of **1** and **2** were very similar to those of dG, except for a slight red shift of approximately 5 nm (Figure 2). Attempts to isolate compounds **1** and **2** were unsuccessful because the adducts rapidly reverted to dG with a half-life of 25 min in neutral aqueous solution at room temperature. Another interesting feature of the reaction between dG and DOVA is the dependence of the efficiency of adduct formation on the concentration of the reagents. The yield of products **1** and **2** was near 90% at high concentration of reagents (13 mmol/L dG, 130 mmol/L DOVA), but limited to 40% under more dilute conditions (5 mmol/L dG, 20 mmol/L DOVA). Altogether, these results strongly suggest that the formation of DOVA–dG adducts was reversible in water. Adducts **1** and **2** are likely to be Schiff bases resulting from the initial addition of a carbonyl group of DOVA to the exocyclic amino group of dG. The similarity of the chromatographic behaviour and UV spectra of adducts **1** and **2** suggests that they are diastereoisomers. The reaction mixture of dG with DOVA was treated with $NaBH_4$ in order to convert the labile Schiff base linkage of adducts **1** and **2** into a stable secondary amine function, and two new compounds, **3** and **4**, were isolated. These two products were clearly generated at the expense of adducts **1** and **2**. Compounds **3** and **4** exhibited UV absorption spectra similar to those of adducts **1** and **2**, but a major difference between the compounds and their reduction products was the high stability of adducts **3** and **4**, with no detectable degradation after more than three days in aqueous solution at room temperature.

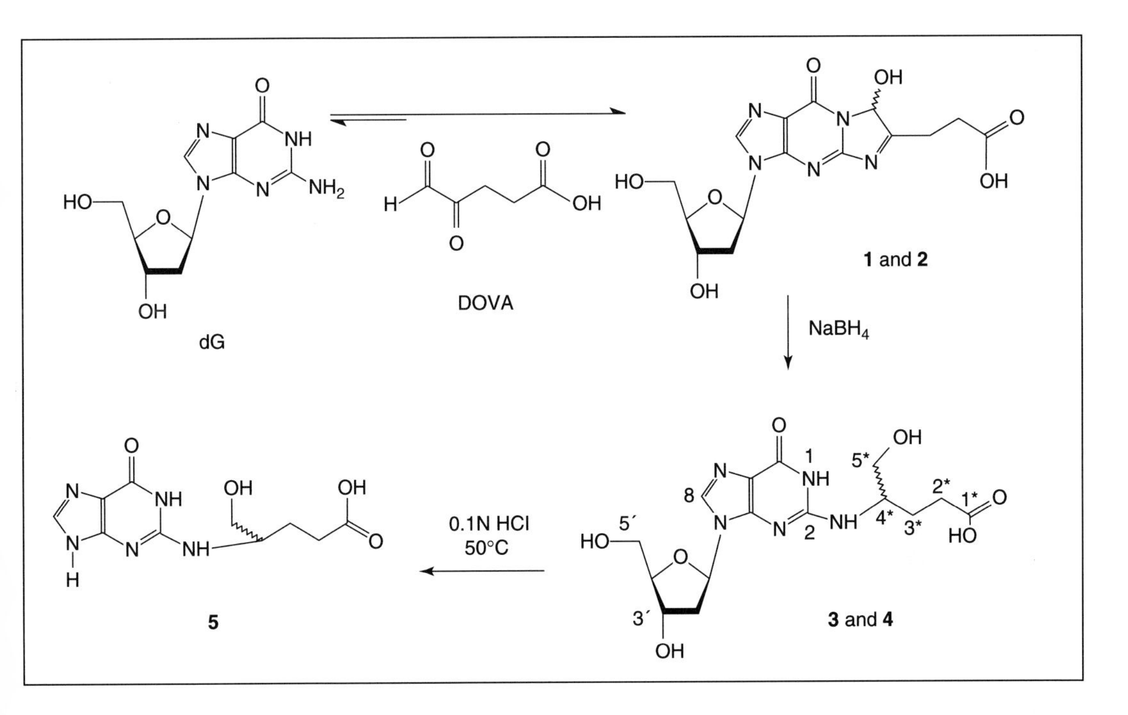

Figure 1. Structure and formation of 4,5-dioxovaleric acid (DOVA)–2′-deoxyguanosine (dG) adducts and their reduction products

The nomenclature used for the labelling of the atom is shown by the bold numbers on the structures of 3 and 4.

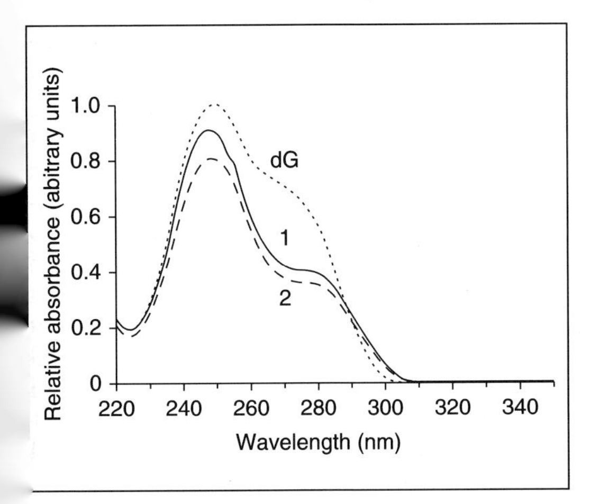

Figure 2. UV absorption spectra of 2′-deoxyguanosine (dG) and 4,5-dioxovaleric acid (DOVA)–dG adducts 1 and 2

The UV spectra were obtained by high-performance liquid chromatography with a diode array detector. The mobile phase was a pH 7, 25-mmol/L ammonium formate solution containing 5% methanol.

Characterization of the reduced 4,5-dioxovaleric acid–2′-deoxyguanosine adducts

Nucleosides **3** and **4** were characterized as the two diastereoisomers of the reduction products of the Schiff base generated by addition of DOVA to the N^2-amino group of the guanine moiety. The molecular mass of **3** and **4** was shown to be 383, as inferred from electrospray ionization mass spectrometry. This can be rationalized in terms of a [1:1] DOVA adduct to dG that had a reduced Schiff base and a reduced carbonyl group. Indeed, the data from infrared spectroscopy unambiguously showed that a carboxylate group (ν = 1604 cm^{-1}) was still present in the molecule. The nucleosidic structure of **3** and **4** was confirmed by the observation in the electrospray ionization–mass spectrometry spectrum of a [MH^+–116] peak which arose from loss of the 2′-deoxyribose ring (Figure 3). The presence of an unmodified 2-deoxyribose ring was also shown by ^1H- and ^{13}C-nuclear magnetic resonance (NMR). All of the signals of the sugar moiety of **3** and **4** were similar to those of unmodified dG. In addition, the multiplicity of the signals was

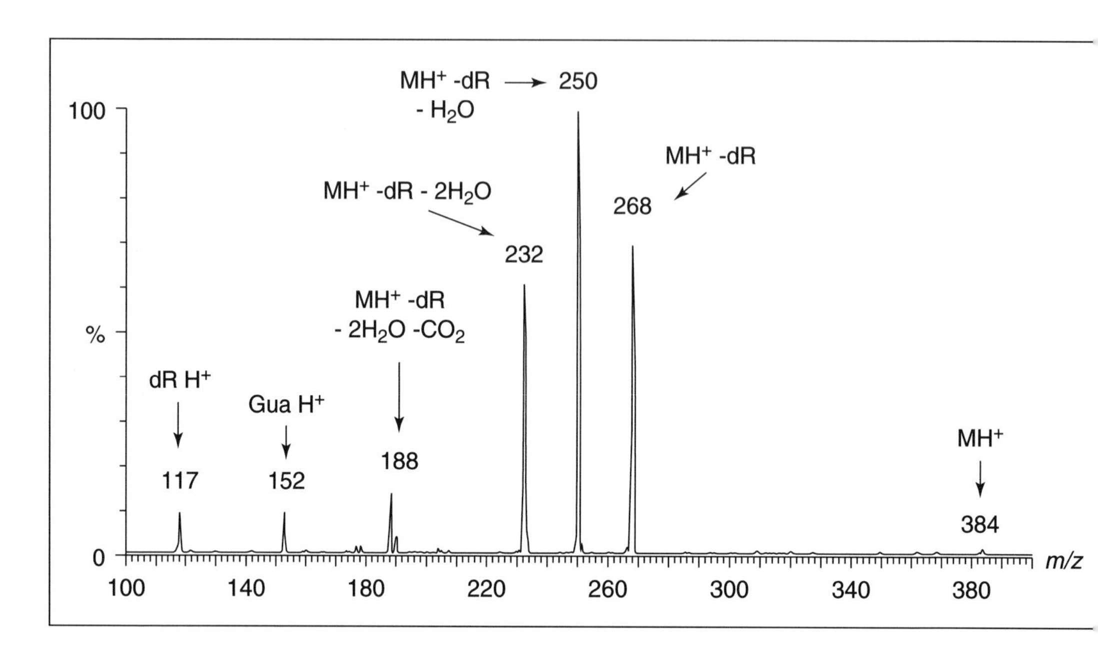

Figure 3. Electrospray ionization tandem mass spectrum in the positive mode of an equimolar mixture of the reduced 4,5-diox-ovaleric acid–2′-deoxyguanosine adducts 3 and 4

dR, 2-deoxyribose; Gua, guanine. The spectrum represents the daughter ions of the parent ion at *m/z* = 384 ([MH+]).

identical for both adducts. The chemical shifts of H8 and the carbon atoms of the guanine moiety of **3** and **4** were similar to those of dG. The latter observations, and the strong similarity between the UV spectra of **3** and **4** on the one hand and dG on the other hand, suggest that the purine ring was not modified by the addition reaction.

Evidence for alkylation of the exocyclic amino group of the guanine moiety was provided by the observation of the loss of the corresponding signal (6.47 ppm for dG) in the ^{1}H-NMR spectrum recorded in dimethyl sulfoxide-*d6*. Interestingly, an exchangeable proton which resonated at $\delta = 6.85$ ppm was assigned to a NH group. This amino proton underwent scalar coupling with the vicinal H-C4* methinic proton of the DOVA moiety. It can thus be concluded that the initial addition of the amino group of guanine occurred on the ketone function of DOVA. The NH would have been linked to a methylene group and should resonate as a triplet in the case of an addition to the aldehyde function. The chemical shift of the H4* proton (4.21 ppm) is also in agreement with the presence of a vicinal secondary amine (atoms labelled with an asterisk represents those of the aldehyde moiety numbered as in the free aldehyde). In adducts **3** and **4**, the aldehyde function was converted into hydroxymethyl group after reduction by $NaBH_4$. This is in agreement with the observation in the ^{1}H NMR spectrum of two signals around 3.8 ppm which were assigned to the two H5* atoms. The ^{1}H NMR signals observed in the low field region (2.5 > δ > 1.9 ppm) were assigned to the C2* and C3 methylene groups of the DOVA moiety. The ^{13}C NMR chemical shifts of C2* and C3* in the high field region are in agreement with the latter assignments. The higher values of the chemical shifts of C5* and C4* can be accounted for by the attachment of a heteroatom to both carbons. The presence of the C1* signal in the very low field region (177.1 ppm) of the spectrum confirmed that the $NaBH_4$ reduction step had not modified the carboxylic acid function of the DOVA moiety.

Detection of 4,5-dioxovaleric acid–2′-deoxyguanosine adducts in DNA

Reduced DOVA–dG adducts **3** and **4** were found to be hydrolysed quantitatively into their corre sponding base after treatment with 0.1 N HCl at 50 °C for 2 h. The UV absorption spectrum of the hydrolysis product **5** was similar to those of the modified nucleosides **3** and **4**. In addition, the ^{1}H-NMR features of **5** were very similar to that of the base moiety of **3** and **4**. Interestingly, hydrolysis of reduced adduct **3** or **4** led to the formation of the same compound. This provides an additional indication that the nucleosides were diastereoisomers. The hydrolysed reduced DOVA–dG adduct **5** was detected by taking advantage of its electroactivity with the detection potential of the HPLC–electron capture system set at 850 mV with respect to a AgO/AgCl reference electrode. The detection limit was close to 10 pmol, and the assay could be used to search for **5** in isolated DNA after incubation with DOVA and subsequent reduction by $NaBH_4$. DNA was first hydrolysed into nucleosides by successive incubation with nuclease P1 and alkaline phosphatase. The resulting mixture was injected onto a first HPLC system, and the fractions corresponding to the retention time of the reduced DOVA–dG adducts were collected. The purified fraction containing the mixture of adducts **3** and **4** was then hydrolysed under acidic conditions and analysed by HPLC with electrochemical and UV detection. Both HPLC elution profiles exhibited a peak corresponding to hydrolysed reduced adduct **5**. The level of modification was determined to be 1300 adducts per 10^6 normal nucleotides, although this value might be an underestimate since no evidence could be obtained for completion of the $NaBH_4$ reduction step.

Mechanism of formation of 4,5-dioxovaleric acid–2′-deoxyguanosine adducts and biological significance

Identification of reduction products **3** and **4** provided evidence for the formation of a covalent link between the exocyclic N^2 atom of guanine and the C4* atom of the DOVA moiety. This showed that the Schiff base expected to be present in adducts **1** and **2** involves the ketone function of DOVA and the exocyclic amino group of dG. It is likely that a subsequent attack on the N1 imine group of the guanine moiety occurs at the aldehyde function of DOVA (Figure 1). This would lead to the formation of a five-membered ring carrying an asymmetric carbon, namely the former C5* atom of DOVA. Unfortunately, the instability of adducts **1** and **2** prevented the extensive spectroscopic studies that would be required for more complete characterization. It was also not possible to determine whether dehydration of the -NH-CHOH- link leading to the formation of a Schiff base occurred before or after cyclization of the adduct. In the latter case, a compound with a structure similar to that of the products obtained by addition of either glyoxal or its derivatives to 2′-deoxyguanosine (Shapiro & Hachmann, 1966; Shapiro *et al.*, 1968) could be an intermediate in the formation of **1** and **2**.

The structure of the DOVA–dG adducts is interesting since, in contrast to glyoxal adducts which exhibit a diol function, they carry a Schiff base and a hydroxyl group on vicinal carbons. The detection of the DOVA–dG adducts in isolated DNA showed that the involvement of the N^2 amino group of guanine in the hydrogen bonding with cytosine does not decrease its ability to react with DOVA. It is likely that the double-helical structure of DNA increases the stability of the adducts over that of isolated dG, as observed for other hydrolytic processes such as deamination. Other aldehydes that can induce formation of Schiff bases in DNA are known to be mutagenic (Marnett *et al.*, 1985; Dellarco, 1993). Interestingly, glyoxal and other α-dicarbonyl compounds are both mutagens (Bjeldanes & Chew, 1979) and efficient alkylating agents of the N^2 exocyclic amino group of guanine (Shapiro & Hachmann, 1966; Shapiro *et al.*, 1968). Acetaldehyde adducts to guanine with a Schiff base structure have recently been detected in human leukocytes (Fang & Vaca, 1997). Altogether, these data suggest that the formation of adducts between DOVA and guanine moieties of DNA might be biologically relevant in the mutagenicity of ALA.

Etheno-2′-deoxyadenosine adducts with 2,4-decadienal

DDE has been reported to be one of the most cytotoxic of the α,β-unsaturated aldehydes arising from the breakdown of lipid peroxides. This was inferred from the lethal effects exerted by DDE on human diploid fibroblasts and umbilical vein endothelial cells (Kaneko *et al.*, 1987, 1988). DDE is also highly cytotoxic to human erythroleukaemia cells (Nappez *et al.*, 1996): it inhibits cell growth, affects cell viability, changes the cellular glutathione concentration and is involved in DNA fragmentation. Furthermore, similar amounts of DDE and 4-hydroxynonenal were formed after oxidation of low-density lipoproteins (Thomas *et al.*, 1994). As it is a substituted α,β-unsaturated aldehyde, DDE was expected to react with DNA bases. Therefore, attempts were made to characterize the products of the reaction of DDE with 2′-deoxyadenosine (dA), which was used as a model for DNA. Identification of the nucleoside adducts produced by such reactions is of fundamental importance in order to unveil the structural features that govern the ability of α,β-unsaturated aldehydes to alkylate DNA.

Isolation and characterization of the main adducts between 2,4-decadienal and 2′-deoxyadenosine and isolated DNA

Reversed-phase HPLC analysis of the reaction mixture of DDE and dA showed the presence of two major well-retained adducts, **7** and **8** (Figure 4), with other minor products, including 1,N^6-etheno-2′-deoxyadenosine (εdA) (Barrio *et al* 1972) on the basis of a comparison of its UV an HPLC features with those of an authentic star dard. Adducts **7** and **8** had UV absorption fea tures similar to those of εdA, strongly suggestin an etheno structure similar to that of th adducts reported by Sodum and Chung (1991 The bathochromic shift (Δ = 3 nm) is consister with the presence of an alkyl side-chain on th etheno ring. Moreover, **7** and **8** show strong flu orescence, with maxima of excitation and emis sion at 320 and 410 nm, respectively. The latte feature is also characteristic of εdA derivative (Kochetkov *et al.*, 1971; Secrist *et al.*, 1972).

The electrospray ionization–mass spectrum c adduct **7** in the positive mode shows a pseudc molecular ion [MH^+] at m/z = 436. This may b due to a [1:1] adduct between dA and DDE wit a gain of two oxygen atoms. The H8 and H2 res

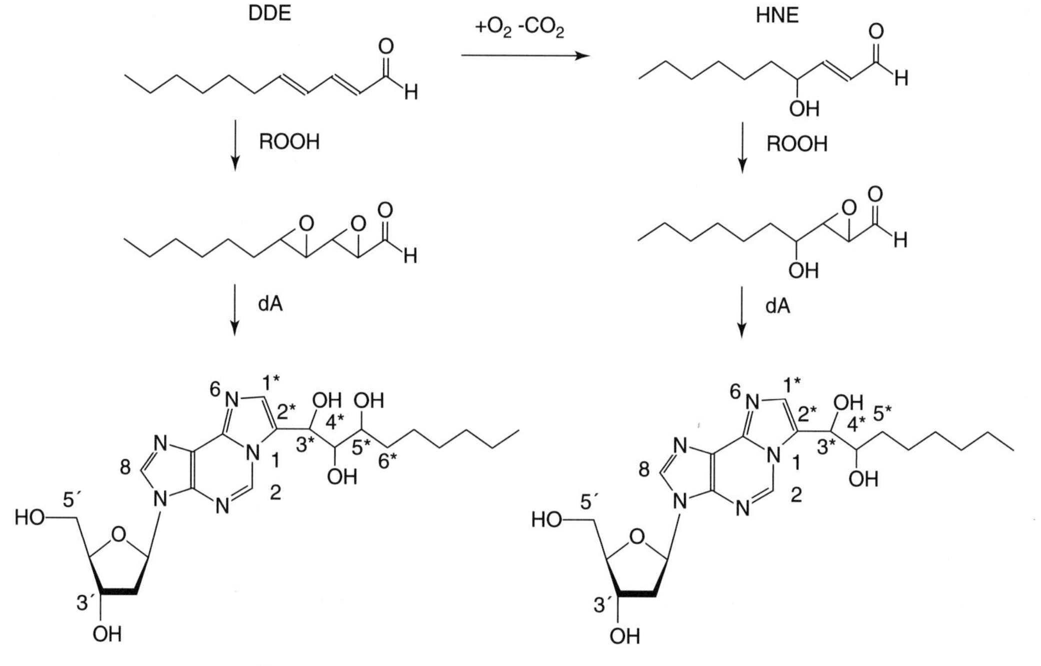

Figure 4. Proposed mechanism for the formation of the 2,4-decadienal (DDE)–2′-deoxyadenosine (dA) adduct

HNE, *trans*-4-hydroxy-2-nonenal. The nomenclature used for the labelling of the atom is shown by the bold numbers on the structures of **7** and **8**.

onances of the adenine moiety of adduct 7 were observed at δ = 8.27 and 9.22 ppm, respectively (Figure 5). The high values for the chemical shift suggest that the aromaticity of the purine ring was retained during adduct formation. The five characteristic signals of the DDE moiety, comprising four vinyl and one aldehyde protons, were not observed in the ^{1}H-NMR spectrum of adduct 7. This ruled out either a direct reaction of the aldehyde carbonyl group with the exocyclic NH_2 of the adenine ring or a Michael addition mechanism for the formation of adduct 7.

The presence of an alkyl chain in adduct 7 was inferred from the observation of signals in the high field region of the ^{1}H-NMR spectrum ($0.64 < \delta < 1.50$ ppm). Integration of the related signals indicated that these 11 protons involve one methyl and four methylene groups. Three other signals that overlapped with the bulk of the sugar resonances over the range 3.30–5.32 ppm were assigned to H3*, H4* and H5* protons. A last signal which resonated as a singlet was observed at δ = 7.44 ppm. This is typical of an etheno proton (H1*) which does not exhibit scalar couplings due to lack of vicinal hydrogen atoms. Altogether, the above data are consistent with an etheno structure exhibiting an octyl substituent at C2*. The chemical shifts of H3* (δ = 5.32), H4* (δ = 3.86) and H5* (δ = 3.30) of the side-chain are consistent with those of carbinol groups. This is in agreement with the molecular mass inferred from electrospray ionization–mass spectrometry only if the oxygen of the aldehyde function of DDE was lost during the addition to dA. Therefore, it is likely that the C1*–C2* double bond arises from a dehydration reaction involving the former C1 position of DDE, subsequent to its covalent binding to the adenine exocyclic amino group. This is in agreement with the mechanism proposed for the formation of etheno bases.

The electrospray ionization–mass spectrometry spectrum in the positive mode of adduct **8** showed a significant pseudomolecular ion [MH$^+$] at *m/z* 406 and a predominant fragment at *m/z* 290 [MH$^+$ - 2-deoxy-D-*erythro*-pentose]. These two pieces of information are indicative of the loss of 30 amu in the molecular mass of adduct **8** with respect to that of 7. This may be explained in terms of a difference of one carbinol group in the structures of adducts 7 and **8**. This conclusion received confirmation

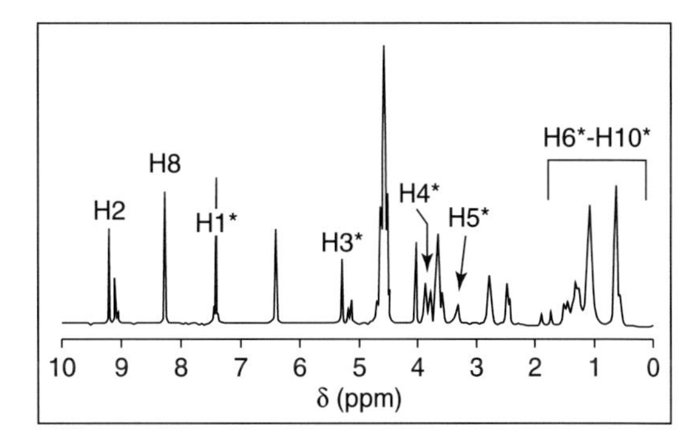

Figure 5. 400-MHz ^{1}H-nuclear magnetic resonance spectrum of 2,4-decadienal–2′-deoxyadenosine adduct 7 in D_2O

Only the protons of the purine ring and the 2,4-decadienal moiety are labelled.

from careful consideration of the ^{1}H-NMR features. The signals of the H2 and H8 protons of adenine were observed at δ = 9.29 and δ = 8.39 ppm, with an etheno proton (δ = 7.52 ppm) in the low field region of the spectrum (Figure 5). The presence of a heptyl chain was inferred from the observation of pentane signals ($0.74 < \delta < 1.44$ ppm) and two carbinol protons assigned as H3* and H4* at δ = 5.09 and δ = 4.13 ppm, respectively. These data indicate that adduct **8** is identical to the substituted etheno base adduct generated by incubation of dA with 2,3-epoxy-4-hydroxynonanal (Sodum & Chung, 1991).

Isolated calf thymus DNA was allowed to react with DDE under conditions similar to those described for dA. After enzymatic hydrolysis, the resulting mixture of nucleosides was analysed by reversed-phase-HPLC with fluorescence detection (Figure 6). Three main altered nucleosides were formed, which had retention times identical to those of εA and adducts 7 and **8**, as inferred from co-elution with authentic standards.

Mechanistic considerations

The reaction of adenine nucleosides with 2,3-epoxy-4-hydroxynonanal yields adducts with an extended 1,N^6-etheno ring (Sodum & Chung, 1991). Interestingly, DDE could react with 2′-deoxyadenosine to yield cyclic adducts by a similar mechanism. Preliminary experiments on the reaction of calf thymus DNA with DDE showed generation of εdA, adduct 7, adduct **8** and other

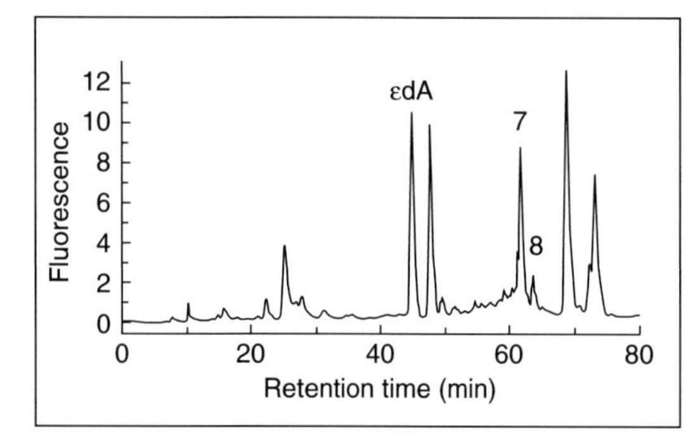

Figure 6. High-performance liquid chromatography elution profile of enzymatically digested 2,4-decadienal-treated DNA obtained by fluorescence detection

εdA, etheno-2′-deoxyadenosine
The excitation and emission wavelengths of the spectrofluorimeter were set at 310 and 420 nm, respectively. The sample was injected onto a C18 reverse column and eluted by a gradient of an aqueous 25-mmol/L ammonium formate solution and acetonitrile.

fluorescent products that are not yet characterized. The observation that adducts **7** and **8** were produced only in the presence of peroxides suggests the following mechanism. The formation of adduct **7** is likely to involve the initial epoxidation of two ethylenic bonds of DDE. Then, subsequent hydrolysis of the C4–C5 epoxide would give rise to a diol. The formation of adduct **8** may require initial generation of 4-hydroxynonenal from DDE, as reported by Grein *et al.* (1993). This reaction is expected to be followed by epoxidation of the C2–C3 double bond. In a subsequent step, the formation of adducts **7** and **8** requires nucleophilic addition of the N^6 exocyclic amino group of dA to the C2 position of 2,3-epoxidized DDE and *trans*-4-hydroxy-2-nonenal, respectively. Then, the N1 amino proton of the purine ring adds to the aldehyde function. Ring closure followed by dehydration of the C1*–C2* bond gives rise to adducts **7** and **8**, respectively. Interestingly, both reactions were found to be dependent on the kind of peroxide present in the reaction mixture, and the reactions of DDE with dA and calf thymus DNA were carried out with peroxidized terahydrofuran, H_2O_2 or *tert*-butylhydroperoxide. Of these peroxides, H_2O_2 gave the highest yield of adduct formation.

Biological relevance

There is evidence in the literature for the formation of fluorescent products from the reaction between adenine nucleosides and oxidative degradation products of fatty acids (Frankel *et al.*, 1983; Fujimoto *et al.*, 1984; Frankel *et al.*, 1987; Grein *et al.*, 1993). DDE and 2-octenal were found to be the most active products in the formation of fluorescent products with dA (Hasegawa *et al.*, 1988, 1989); however, the adducts obtained were not successfully characterized. The present results provide further insight. The formation of substituted etheno bases after reaction of DDE with DNA bases further emphasizes the importance of this class of cyclic DNA adduct. Several carcinogenic and mutagenic compounds, including vinyl monomers, α-halogenated aldehydes and α-haloketones were shown to form etheno adducts with DNA (Bartsch *et al.*, 1994). Etheno nucleotides appear to accumulate and persist in the livers of rodents exposed to vinyl chloride and ethyl carbamate. The detection of low levels of etheno bases in tissues of untreated animals and humans (Nair *et al.*, 1995) suggests the existence of endogenous sources of reactive aldehyde intermediates, possibly from lipid peroxidation processes. Formation of 1,N^6-etheno adducts from the reaction of purine nucleosides with 2,3-epoxy-4-hydroxynonanal was reported by Sodum and Chung (1991). α,β-Unsaturated aldehydes are converted *in vitro* into the corresponding epoxides by either microsomal enzymes or reaction with endogenously generated peroxides (Chen & Chung, 1996). The latter process was considered to be a possible endogenous route for the formation of DNA etheno adducts *in vivo*. Furthermore, εdA has been reported to be highly mutagenic in mammalian cells (Pandya & Moriya, 1996).

Acknowledgments

This work was supported by the Fundaçào de Amparo à Pesquisa do Estado de Sào Paulo, Brazil, the Conselho Nacional para o Desenvolvimento Científico e Tecnológico, Brazil, the Programa de Apoio aos Núcleos de Excelência, Brazil, and the Universidade de Sào Paulo, Comité Français d'Evaluation de la Coopération Universitaire avec le Brésil. V.M.C. and J.O. are recipients of fellowships from the Fundaçào de Amparo à Pesquisa do Estado de Sào Paulo.

References

Barrio, J.R., Secrist, J.A. & Leonard, N.J. (1972) Fluorescent adenosine and cytidine derivatives. *Biochem. Biophys. Res. Commun.* 46, 597–604.

Bartsch, H., Barbin, A., Marion, M.-J, Nair, J. & Guichard, Y. (1994) Formation, detection, and role in carcinogenesis of ethenobases in DNA. *Drug Metab. Rev.* 26, 349–371.

Basu, A.K., O'Hara, S.M., Valladir, P., Stone, K., Mols, O. & Marnett, L.J. (1988) Identification of adducts formed by reaction of guanine nucleosides with malondialdehyde and structurally related aldehydes. *Chem. Res. Toxicol.* 1, 53–59.

Bechara, E.J.H., Medeiros, M.H.G., Monteiro, H.P., Hermes-Lima, M., Pereira, B., Demasi, M., Costa, C.A., Abdalla, D.S.P., Onuki, J., Wendel, C.M.A. & Di Mascio, P. (1993) A free radical hypothesis of lead poisoning and inborn porphyrias associated with 5-aminolevulinic acid overload. *Quim. Nova* 16, 385–392.

Bjeldanes, L.F. & Chew, H. (1979) Mutagenicity of 1,2-dicarbonyl compounds: Maltol, kojic acid, diacetyl and related substances. *Mutat. Res.* 67, 367–371.

Carvalho, V.M., Di Mascio, P., de Arruda Campos, I.P., Douki, T., Cadet, J. & Medeiros, M.H.G. (1998) Formation of 1,N^6-etheno-2'-deoxyadenosine adducts by *trans,trans*-2,-decadienal. *Chem. Res. Toxicol.* 11, 1042–1047.

Chaudhary, A.K., Nokubo, N., Ramachandra Reddy, G., Yeola, S.N., Morrow, J.D., Blair, I.A. & Marnett, L.J. (1994) Detection of endogenous malondialdehyde–deoxyguanosine adducts in human liver. *Science* 265, 1580–1582.

Chen, H.-J.C. & Chung, F.-L. (1996) Epoxidation of *trans*-4-hydroxy-2-nonenal by fatty acid hydroperoxides and hydrogen peroxide. *Chem. Res. Toxicol.* 9, 306–312.

Chisolm, J.J., Jr (1971) Lead poisoning. *Sci. Am.* 224, 15–23.

Chung, F.-L., Young, R. & Hecht, S.S. (1984) Formation of cyclic 1,N^2-propanodeoxyguanosine adducts in DNA upon reaction with acrolein or crotonaldehyde. *Cancer Res.* 44, 990–995.

Dellarco, V.L. (1993) A mutagenicity assessment of acetaldehyde. *Mutat. Res.* 195, 1–20.

Demasi, M., Penatti, C.A.A., DeLucia, R. & Bechara, E.J.H. (1996) The prooxidant effect of 5-aminolevulinic in the brain tissue of rats: Implications in neuropsychiatric manifestations in porphyrias. *Free Radicals Biol. Med.* 20, 291–299.

Douki, T. & Ames, B.N. (1994) An HPLC–EC assay for 1,N^2-propano adducts of 2'-deoxyguanosine with 4-hydroxynonenal and other α,β-unsaturated aldehydes. *Chem Res. Toxicol.* 7, 511–518.

Douki, T., Onuki, J., Medeiros, M.H.G., Bechara, E.J. H., Cadet, J. & Di Mascio, P. (1998a) DNA alkylation by 4,5-dioxovaleric acid, the final oxidation product of 5-aminolevulinic acid. *Chem. Res. Toxicol.* 11, 150–157.

Douki, T., Onuki, J., Medeiros, M.H.G., Bechara, E.J.H., Cadet, J. & Di Mascio, P (1998b) Hydroxyl radicals are involved in the oxidation of isolated and cellular DNA bases by 5-aminolevulinic acid. *FEBS Lett.* 428, 93–96.

Eder, E. & Hoffman, C. (1992) Identification and characterization of deoxyguanosine–crotonaldehyde adducts. Formation of 7,8-cyclic adducts and 1,N^2, 7,8 *bis*-cyclic adducts. *Chem. Res. Toxicol.* 5, 802–808.

Eder, E. & Hoffman, C. (1993) Identification and characterization of deoxyguanosine adducts of mutagenic β-alkyl-substituted acrolein congeners. *Chem. Res. Toxicol.* 6, 486–494.

Fang, F.L. & Vaca, C.E. (1997) Detection of DNA adducts of acetaldehyde in peripheral white blood cells of alcohol abusers. *Carcinogenesis* 18, 627–632.

Fiedler, D.M., Eckl, P.M. & Krammer, B. (1996) Does δ-aminolaevulinic acid induce genotoxic effects? *J. Photochem. Photobiol. B: Photobiol.* 33, 39–44.

Fraga, C.G., Onuki, J., Lucesoli, F., Bechara E.J.H. & Di Mascio, P. (1994) 5-Aminolevulinic acid mediates the *in vivo* and *in vitro* formation of 8-hydroxy-2'-deoxyguanosine in DNA. *Carcinogenesis* 15, 2241–2244.

Frankel, E.N., Neff, W.E. & Selke, E. (1983) Analysis of autoxidized fats by gas chromatography–mass spectrometry: VIII. Volatile thermal decomposition products of hydroperoxy cyclic peroxides. *Lipids* 18, 353–357.

Frankel, E.N., Neff, W.E., Brooks, D.D. & Fujimoto, K. (1987) Fluorescence formation from the interaction of DNA with lipid oxidation degradation products. *Biochim. Biophys. Acta* 919, 239–244.

Fujimoto, K., Neff, W.E. & Frankel, E.N. (1984) The reaction of DNA with lipid oxidation products, metals and reducing agents. *Biochim. Biophys. Acta* 795, 100–107.

Grein, B., Huffer, M., Scheller, G. & Schreier, P. (1993) 4-Hydroxy-2-nonenal and other products formed by water-mediated oxidative decomposition of α,β-unsaturated aldehydes. *J. Agric. Food Chem.* 41, 2385–2390.

Gubler, J.G., Bergetzi, M.J. & Meyer, U.A. (1990) Primary liver carcinoma in two sisters with acute intermittent porphyria. *Am. J. Med.* 89, 540–541.

Hasegawa, K., Fujimoto, K., Kaneda, T. & Frankel, E.N. (1988) Characterization of fluorescent products from reaction of methyl linoleate hydroperoxides with adenine in the presence of Fe^{2+} and ascorbic acid. *Biochim. Biophys. Acta* 962, 371–376.

Hasegawa, K., Fujimoto, K., Kaneda, T., Neff, W.E. & Frankel, E.N. (1989) Formation of fluorescent products in the reaction of adenine and other bases with secondary oxidation products of methyl linoleate hydroperoxides. *Agric. Biol. Chem.* 53, 1575–1581.

Hindmarsh, J. T. (1986) The porphyrias: Recent advances. *Clin. Chem.* 32, 1255–1263.

Hiraku, Y. & Kawanishi, S. (1996) Mechanism of oxidative DNA damage induced by δ-aminolevulinic acid in the presence of copper ion. *Cancer Res.* 56, 1786–1793.

Kaneko, T., Honda, S., Nakano, S. & Matsuo, M. (1987) Lethal effects of a linoleic acid hydroperoxide and its autoxidation products, unsaturated aliphatic aldehydes, on human diploid fibroblasts. *Chem.–Biol. Interactions* 63, 127–137.

Kaneko, T., Kaji, K. & Matsuo, M. (1988) Cytotoxicities of a linoleic acid hydroperoxide and its related aliphatic aldehydes toward cultured human umbilical vein endothelial cells. *Chem.–Biol. Interactions* 67, 295–304.

Kappas, A., Sassa, S. & Anderson, K.E. (1983) The porphyrias. In: *The Metabolic Basis of Inherited Diseases* (Stanbury, J.B., Wyngaarden, J.B., Fredrickson, D.S., Goldstein, J.L. & Brown, M.S., eds). New York, McGraw-Hill, pp. 1301–1384.

Kennedy, J.C. & Pottier, R.H. (1992) Endogenous protoporphyrin IX, a clinically useful photosensitizer for photodynamic therapy. *J. Photochem. Photobiol. B: Biol.* 14, 275–292.

Kochetkov, N.K., Shibaev, V.N., Kost, A.A. & Zelinsky, N.D. (1971) New reaction of adenine and cytosine derivatives, potentially useful for nucleic acids modification. *Tetrahedron Lett.* 22, 1993–1996.

Lithner, F. & Wetterberg, L. (1984) Hepatocellular carcinoma in patients with acute intermittent porphyria. *Acta Med. Scand.* 215, 271–274.

Marnett, L.J., Hurd, H.K., Hollstein, M.C., Levin, D.E., Esterbauer, H. & Ames, B.N. (1985) Naturally occurring carbonyl compounds are mutagens in *Salmonella* tester strain TA104. *Mutat. Res.* 148, 25–34.

Nair, J., Barbin, A. & Bartsch, H. (1995) 1,N^6-Ethenodeoxyadenosine and 3,N^4-ethenodeoxycytidine in liver DNA from humans and untreated rodents detected by immunoaffinity/^{32}P-postlabelling. *Carcinogenesis* 16, 613–617.

Nappez, C., Battu, S. & Beneytout, J.L. (1996) *trans, trans*-2,4-Decadienal: Cytotoxicity and effect on glutathione level in human erythroleukemia (HEL) cells. *Cancer Lett.* 99, 115–119.

Nath, R.G. & Chung, F.-L. (1994) Detection of exocyclic 1,N^2-propanodeoxyguanosine adducts as common DNA lesions in rodents and humans. *Proc. Natl Acad. Sci. USA* 91, 7491–7495.

Onuki, J., Medeiros, M.H.G., Bechara, E J.H. & Di Mascio, P. (1994) 5-Aminolevulinic acid induces single strand breaks in plasmid DNA in the presence of Fe^{2+} ions. *Biochim. Biophys. Acta* 1225, 259–263.

Ortel, B., Tanew, A. & Hπnigsmann, H. (1993) Lethal photosensitization by endogenous porphyrins of PAM cells modification by desferrioxamine. *J. Photochem. Photobiol. B: Photobiol.* 17, 273–278.

Oteiza, P.I., Kleinman, C.G., Demasi, M. & Bechara, E.J.H. (1995) 5-Aminolevulinic acid induces iron release from ferritin. *Arch. Biochem. Biophys.* 316, 607–611.

Pandya, G.A. & Moriya, M. (1996) 1,*N*6-Ethenodeoxyadenosine, a DNA adduct highly mutagenic in mammalian cells. *Biochemistry* 35, 11487–11492

Peng, Q., Berg, K., Moan, J., Kongshaug, M., Giercksky, K.E. & Nesland, J.M. (1997) 5-Aminolevulinic acid-based photodynamic therapy: Principles and experimental research. *Photochem. Photobiol.* 65, 235–251.

Secrist, J.A., Barrio, J.R., Leonard, N.J. & Weber, G. (1972) Fluorescent modification of adenosine-containing coenzymes. Biological activities and spectroscopic properties. *Biochemistry* 11, 3499–3506.

Shapiro, R. & Hachmann, J. (1966) The reaction of guanine derivatives with 1,2-dicarbonyl compounds. *Biochemistry* 5, 2799–2807.

Shapiro, R., Cohen, B.I., Shiuey, S.-J. & Mauer, H. (1968) On the reaction of guanine with glyoxal, pyruvaldehyde and kethoxal and the structure of the acylguanines. A new synthesis of N^2-alkylguanines. *Biochemistry* 8, 238–245.

Sodum, R.M. & Chung, F.-L. (1988) 1, N^2-Ethenodeoxyguanosine as a potential marker for DNA adduct formation by *trans*-4-hydroxy-2-nonenal. *Cancer Res.* 48, 320–323.

Sodum, R.M. & Chung, F.-L (1991) Stereoselective formation of *in vitro* nucleic acid adducts by 2,3-epoxy-4-hydroxynonanal. *Cancer Res.* 51, 137–143.

Thomas, C.E., Jackson, R.L., Ohlweiler, D.L. & Ku, G. (1994) Multiple lipid oxidation products in low-density lipoproteins induce interleukin-1 beta release from human blood mononuclear cell. *J. Lipid Res.* 35, 853–856.

Thunnissen, P.L., Meyer, J. & de Koning, R.W. (1991) Acute intermittent porphyria and primary liver cancer. *Neth. J. Med.* 38, 171–174.

Vaca, C.E., Vodicka, P. & Hemminki, K. (1992) Determination of malondialdehyde modified 2-deoxyguanosine 3′-monophosphate and DNA by ^{32}P-postlabelling. *Carcinogenesis* 13, 593–599.

Vaca, C.E., Fang, J.-L. & Schweda, E.K.H. (1995) Studies of the reaction of acetaldehyde with deoxynucleosides. *Chem.-Biol. Interactions* 98, 51–67.

Winter, C.K., Segall, H.J. & Haddon, W.F. (1986) Formation of cyclic adducts of deoxyguanosine with the aldehydes *trans*-4-hydroxy-2-hexenal and *trans*-4-hydroxy-2-nonenal *in vitro*. *Cancer Res.* 46, 5682–5686.

Exocyclic DNA Adducts in Mutagenesis and Carcinogenesis
B. Singer & H. Bartsch
IARC Scientific Publications No. 150
International Agency for Research on Cancer, Lyon, 1999

Adducts of chlorohydroxyfuranones and of aldehyde conjugates

L. Kronberg, T. Munter, J. Mäki, F. Le Curieux, D. Pluskota & R. Sjöholm

The work reported here concerned the structural determination of adducts formed in reactions of nucleosides with chlorohydroxyfuranones and with conjugates of malonaldehyde and acetaldehyde. The chlorohydroxyfuranones are bacterial mutagens produced during chlorine disinfection of drinking-water, malonaldehyde is the major end-product of lipid peroxidation, and acetaldehyde is the major metabolite of ethanol. Exogenous sources of acetaldehyde are, for example, tobacco smoke and automobile emissions. The chlorohydroxyfuranones were found to form substituted and unsubstituted cyclic etheno adducts with adenosine, cytidine and guanosine and substituted cyclic propeno adducts with adenosine. The malonaldehyde–acetaldehyde conjugates reacted with adenosine to produce a strongly fluorescent dihydropyridine purine derivative and a substituted propenoformyl adenosine derivative. Most of the adducts identified in this study are new, and their mutagenic properties are as yet unknown.

Introduction

Numerous studies have shown that chlorine disinfection of drinking-water processed from surface water produces genotoxic compounds (Meier, 1988). In most drinking-water, about one-third of the mutagenicity can be attributed to the presence of the directly acting mutagen 3-chloro-4-(dichloromethyl)-5-hydroxy-2(5*H*)-furanone (MX, see Scheme 1) (Smeds *et al.*, 1997); however, other structurally analogous chlorohydroxyfuranones are also produced during chlorine disinfection, such as 3-chloro-4-(chloromethyl)-5-hydroxy-2(5*H*)-furanone (CMCF), 3-chloro-4-methyl-5-hydroxy-2(5*H*)-furanone (MCF) and 3,4-dichloro-5-hydroxy-2(5*H*)-furanone (MCA) (Kronberg & Franzén, 1993). MX generates about 6000 revertants per nmol in *Salmonella typhimurium* strain TA100 (Tikkanen & Kronberg, 1990) and has been found to be mutagenic in mammalian cells *in vitro* and *in vivo* (Brunborg *et al.*, 1991; Fekadu *et al.*, 1994); it has also been reported to be a multisite carcinogen in rodents (Komulainen *et al.*, 1997). CMCF, MCF and MCA generate about 1000, 10 and 3 net revertants per nmol, respectively, in TA100 (Kronberg & Franzén, 1993; LaLonde *et al.*, 1997). As most genotoxic chemicals form adducts with base units of DNA, the purpose of our work was to determine whether chlorohydroxyfuranones react with the DNA bases.

R	Compound
Cl	3,4-Dichloro-5-hydroxy-2(5*H*)-furanone
CH_3	3-Chloro-4-methyl-5-hydroxy-2(5*H*)-furanone
CH_2Cl	3-Chloro-4-(chloromethyl)-5-hydroxy-2(5*H*)-furanone
$CHCl_2$	3-Chloro-4-(dichloromethyl)-5-hydroxy-2(5*H*)-furanone

Scheme 1. Structures of the chlorohydroxyfuranones studied

During these studies, we found that MX forms an adenosine adduct that had previously been shown to be produced by malonaldehyde (Le Curieux *et al.*, 1997). In order to obtain the reference compound, we reacted malonaldehyde with adenosine and found an unknown adduct, besides the known malonaldehyde adducts. This adduct was found to be formed from a malonaldehyde–acetaldehyde condensation product. Acetaldehyde has been shown to be mutagenic in several biological assays and is considered to be carcinogenic in experimental animals (IARC,

1985). In studies of the reactions of acetaldehyde with nucleosides, acetaldehyde forms unstable Schiff bases, mainly with deoxyguanosine (Hemminki & Susi, 1984; Vaca *et al.*, 1995). The finding that the adduct of the malonaldehyde–acetaldehyde conjugate is rather stable prompted us to explore adducts of these conjugates further.

Adducts of chlorohydroxyfuranones

We have studied the reactions of chlorohydroxyfuranones with nucleosides and found that they undergo relatively complex reactions, forming many adducts. Chlorohydroxyfuranones exist mainly in a ring form (hydroxyfuranone) at low pH and in an open-chain form under neutral conditions (Scheme 1). The open forms are α,β-unsaturated carbonyl compounds; however, none of the adducts hitherto identified is produced by reaction of the α,β-unsaturated carbonyl arrangement shown in Scheme 1. This is due either to initial degradation of the compounds (MCA) or to internal proton transfer leading to a change in the conjugation (MCF, CMCF and MX).

We reacted MCA with adenosine, guanosine and cytidine, identified the adducts and proposed likely mechanisms for the formation of these adducts. The proposed mechanisms are supported by the observation of the location of the ^{13}C-labelled atom in the adducts. It is known that the chlorine at the β-carbon of MCA is easily displaced by a hydroxyl group and mucoxychloric acid is formed (Scheme 2) (Wasserman & Precopio, 1952). The aldehyde of mucoxychloric acid is then attacked by the exocyclic amino group of the nucleosides. In the resulting carbinolamine **A** (Scheme 2), the halogen is displaced by a conjugate attack *via* the endocyclic nitrogen in the respective nucleoside, and a dihydroimidazole ring is formed. After dehydration, the etheno-oxalo derivatives are formed. We have found that etheno-oxalo derivatives of adenosine and of cytidine undergo decarboxylation and the etheno-formyl derivatives are produced (Kronberg *et al.*, 1993). The non-cyclic chloropropenal derivative may be formed from the carbinolamine intermediate (**A**) through dehydration, decarboxylation and proton transfer (Kronberg *et al.*, 1996)

The etheno-oxalo adenosine derivative that has an additional adenosine unit bound to the etheno bridge (εoA,A) is most likely formed through attack of an additional molecule of adenosine on an imine derived from the carbinolamine (**A**), resulting in an *N,N*-acetal. The chlorine in the acetal is displaced by a conjugate attack via the endocyclic nitrogen (N1) of one of the adenosine units, and a saturated imidazoline ring is obtained. Finally, εoA,A is obtained by oxidation (dehydrogenation) of the imidazoline ring unit to an imidazole ring. It is likely that the oxidizing agent is MCA. The corresponding etheno-formyl derivative (εcA,A) is obtained by loss of carbon dioxide from the oxalo group (Asplund *et al.*, 1995).

The etheno derivatives may be produced from chloroacetaldehyde formed by further break-down of mucoxychloric acid (Kronberg *et al.*, 1992).

While MCA yields cyclic etheno derivatives and non-cyclic propeno derivatives, the compounds MCF , CMCF and MX, which have a carbon atom attached at C4 (the ring form), will form cyclic propeno adducts (CMCF, MX) or non-cyclic buteno derivatives (MCF). We have proposed that in these three compounds the reaction is initiated by a proton transfer from the methyl group to the labelled carbon (Scheme 3). Next, the exocyclic amino group of adenosine attacks the former methyl carbons, and enol intermediates are formed. The enol intermediates differ structurally only in the number of chlorine atoms. Upon tautomerization to the keto form, the enol intermediates lose HCl. The intermediate produced from MCF has only one chlorine atom, and when this chlorine has been lost no further substitution reactions are possible. After a proton transfer, the formyl butenoic acid adduct is obtained (Munter *et al.*, 1996). The enol intermediate produced by CMCF has two chlorine atoms and that produced by MX has three chlorine atoms, and, after tautomerization to the keto form, the chlorine at the carbon next to N^6 of adenosine will be lost. The chlorine at the saturated carbon is displaced by a conjugate attack *via* the endocyclic nitrogen (N1) of adenine, and cyclic β-ketoacids are obtained. The acids undergo decarboxylation and the final stable adducts are the cyclic propenoformyl adducts (Munter *et al.*, 1998).

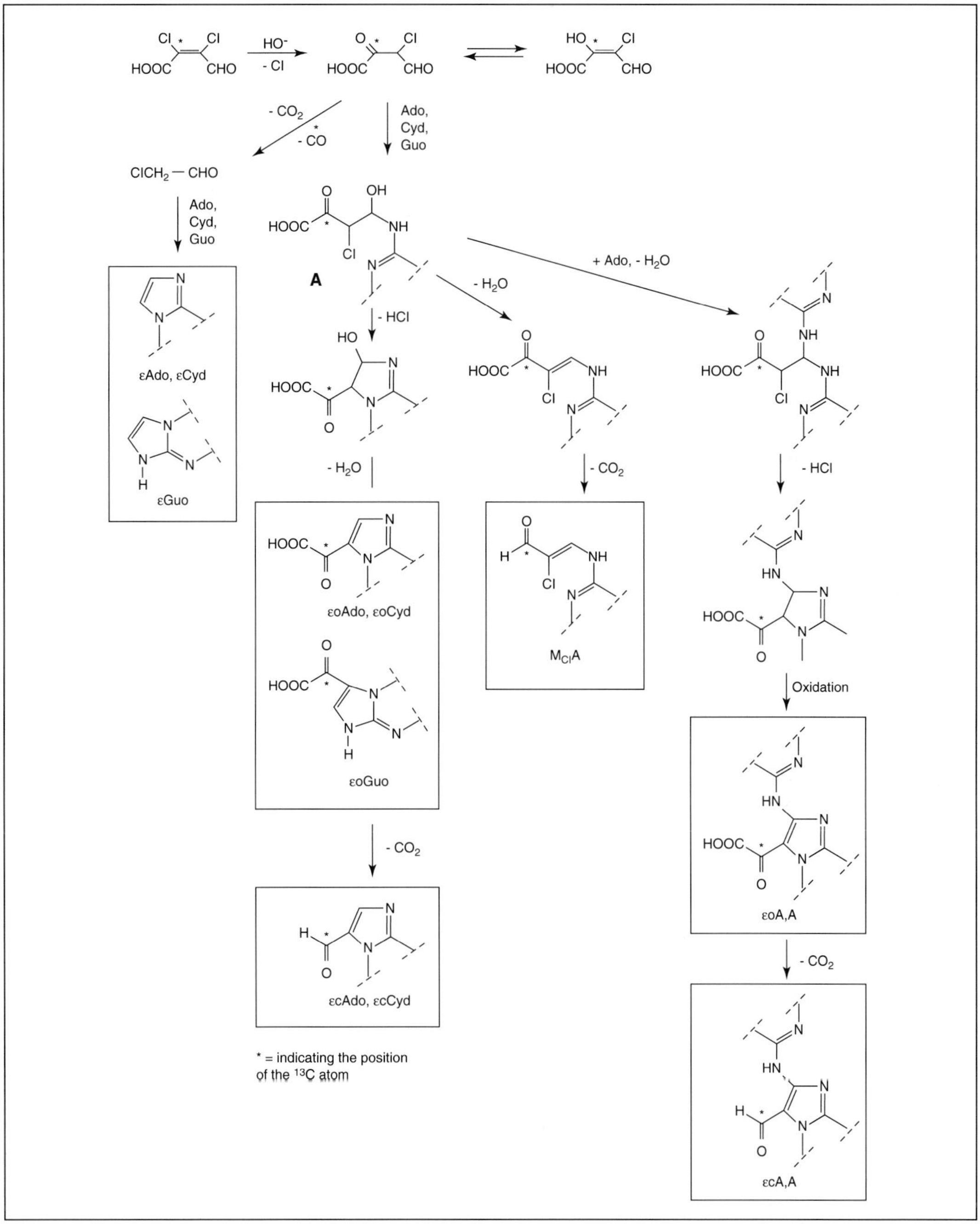

* = indicating the position of the ^{13}C atom

Scheme 2. Mechanisms for the formation of adducts of 3,4-dichloro-5-hydroxy-2(*5H*)-furanone (MCA) with adenosine (Ado), guanosine (Guo) and cytidine (Cyd)

ε, etheno, εo, etheno-oxalo; εc, etheno-formyl

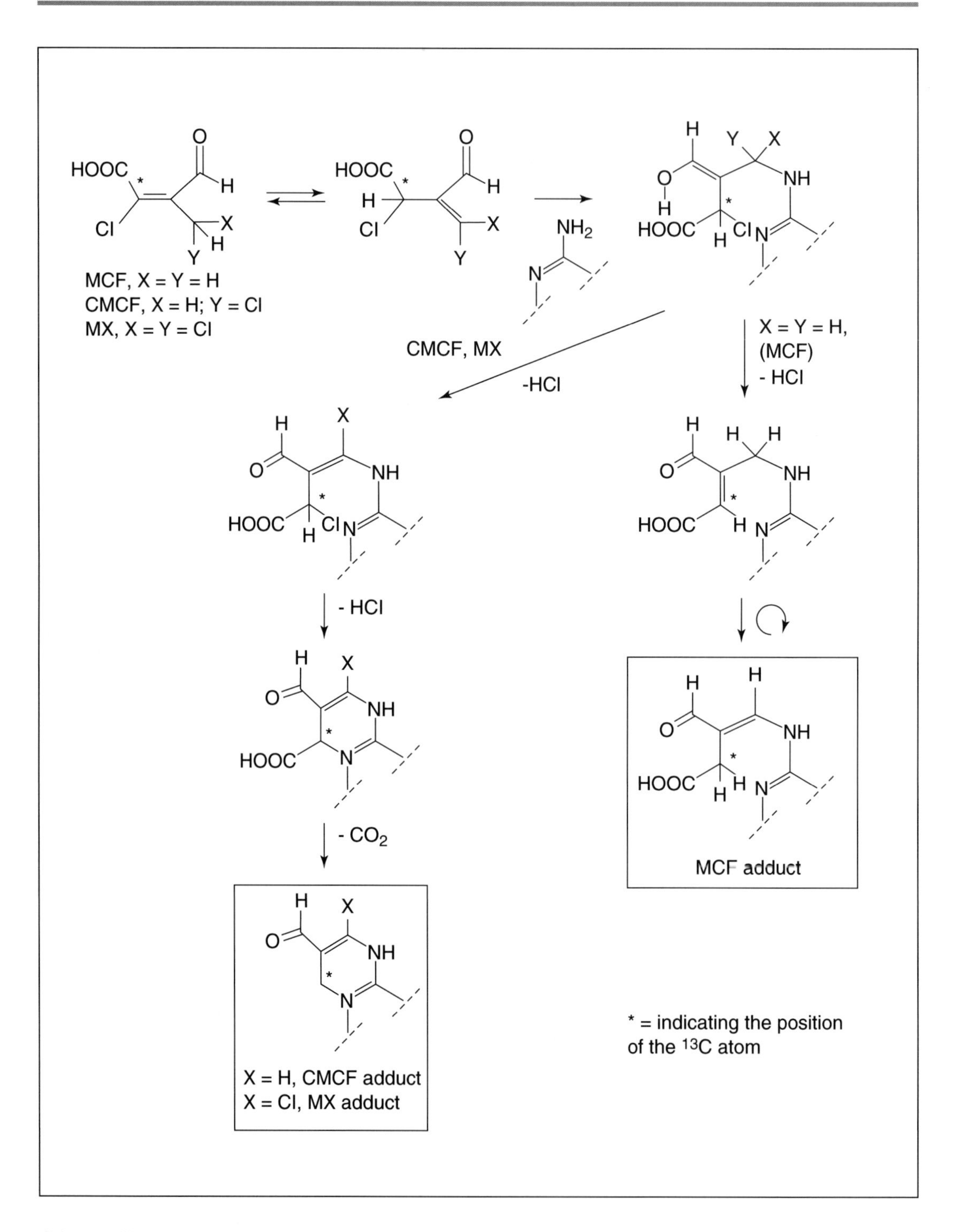

Scheme 3. Mechanisms for the formation of adducts of 3-chloro-4-(dichloromethyl)-5-hydroxy-2(*5H*)-furanone (MX), 3-chloro-4-(chloromethyl)-5-hydroxy-2(*5H*)-furanone (CMCF) and 3-chloro-4-methyl-5-hydroxy-2(*5H*)-furanone (MCF)

The mutational spectrum of MX in *S. typhimurium* strain TA100 indicates that MX causes primarily GC→TA transversions at the second position of the target sequence CCC in the *hisG46* allele (Knasmüller *et al.*, 1996). These mutational events suggest a mechanism in which MX forms an adduct with guanine or *via* depurination of the guanine base. Marsteinstredet *et al.* (1997) used DNA sequence analysis of a MX-treated DNA fragment and demonstrated that MX reacts preferentially with the guanine base *in vitro* to produce chemically stable DNA adducts. Recently, we reacted MX with guanosine and identified an adduct in which an etheno-formyl bridge is incorporated between the N1 and N^2 positions of guanine (Scheme 4) (Munter *et al.*, 1999). This adduct might be involved in the genotoxicity of MX; however, it could not be detected by high-performance liquid chromatography of the hydrolysate of calf thymus DNA reacted with MX.

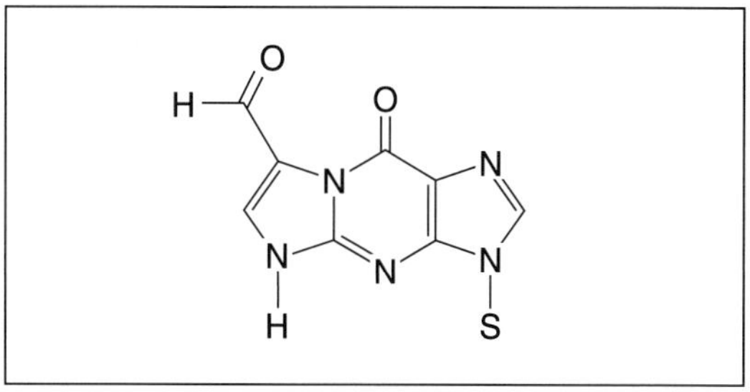

Scheme 4. 3-Chloro-4-(dichloromethyl)-5-hydroxy-2(*5H*)-furanone–guanosine adduct

Adducts of malonaldehyde–acetaldehyde conjugates

One of the adducts formed by MX was identified as 3-(N^6-deoxyadenosinyl)propenal (M_1dA) (Le Curieux *et al.*, 1997). In order to be sure of the correct structural assignment, we prepared a reference compound by reacting malonaldehyde with deoxyadenosine (dA). In the chromatogram of the reaction mixture, we observed peaks representing M_1dA and N^6-oxazocinyl-dA (M_3dA) (Stone *et al.*, 1990) and a peak due to an unknown, strongly fluorescent compound. The compound was isolated and identified as a dihydropyridyl-purine derivative (M_2AA-dA in Scheme 5) (Le Curieux *et al.*, 1998). The formation of the adduct can be explained by hydroly-

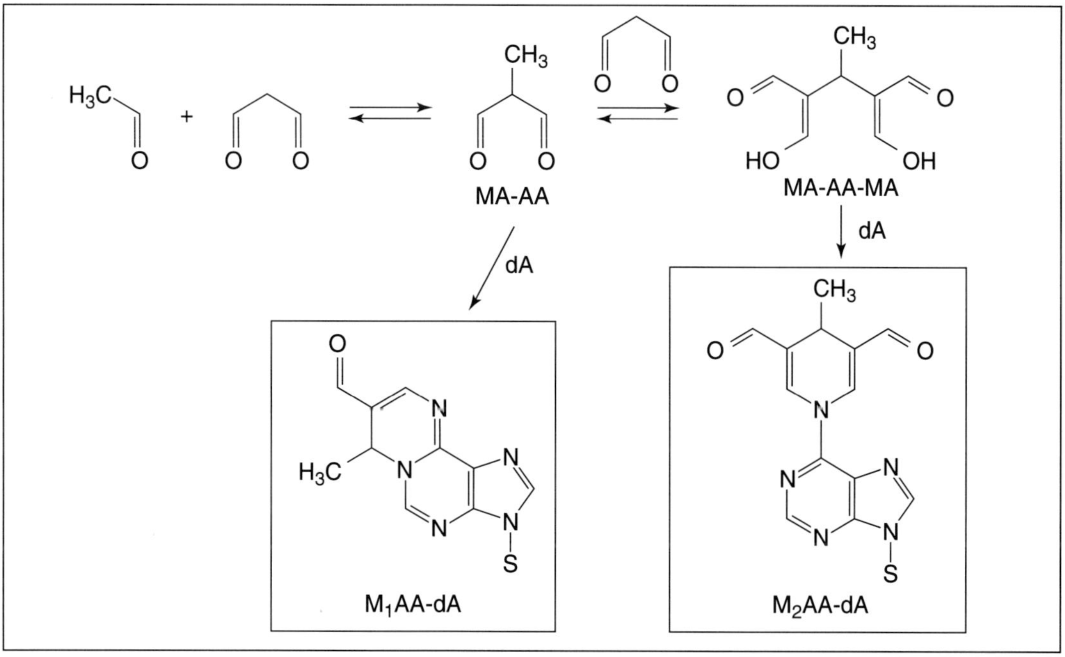

Scheme 5. Adducts of malonaldehyde–acetaldehyde conjugates

sis of malonaldehyde to acetaldehyde and subsequent condensation of one molecule of acetaldehyde with two molecules of malonaldehyde (Scheme 5). A 10-fold increase in adduct yield could be achieved by reacting adenosine with malonaldehyde in the presence of acetaldehyde.

Acetaldehyde is formed endogenously through enzymatic oxidation of ethanol in the liver. It is known that ethanol consumption leads to an increase in the cellular concentration of malonaldehyde and thus an increased possibility for malonaldehyde conjugation with acetaldehyde. The findings of Tuma *et al.* (1996) and Xu *et al.* (1997) show that acetaldehyde–malonaldehyde condensation takes place in physiological fluids, and they reported that the dihydropyridyl unit could be detected in proteins in the liver of rats fed ethanol. Our work shows that the conjugate may also react with the exocyclic amino group of adenosine. Therefore, dihydropyridine DNA adducts may also be formed as a consequence of ethanol metabolism or of exogenous sources of acetaldehyde.

Recently, we found that the condensation product derived from one molecule of malonaldehyde and one molecule of acetaldehyde (MA-AA in Scheme 5) produces another fluorescent adduct by reaction with adenosine. The adduct was found to consist of a methyl-substituted propeno-formyl bridge incorporated between N1 and N^6 of adenosine (M_1AA-dA in Scheme 5).

These findings stress the importance of the presence of reactive aldehyde condensation products in physiological fluids. The condensation products may form more stable and possibly more harmful adducts than those formed by a specific aldehyde.

References

Asplund, D., Kronberg, L., Sjöholm. R. & Munter, T. (1995) Reactions of mucochloric acid with adenosine: Formation of 8-(N^6-adenosinyl)ethenoadenosine derivatives. *Chem. Res. Toxicol.* 8, 841–846

Brunborg, G., Home, J.A., Søderlund, E.J., Hongslo, J.K., Vartiainen, T., Lötjönen, S. & Becher, G. (1991) Genotoxic effects of the drinking water mutagen 3-chloro-4-(dichloromethyl)-5-hydroxy-2(5*H*)-furanone (MX) in mammalian cells *in vitro* and *in vivo*. *Mutat. Res.* 260, 55–64

Fekadu, K., Parzefall, W., Kronberg, L., Franzén, R., Schulte-Hermann, R. & Knasmüller, S. (1994) Induction of genotoxic effects in *E. coli* K-12 cells recovered from various organs of mice by chlorohydroxyfuranones, byproducts of water disinfection. *Environ. Mol. Mutag.* 24, 317–324

Hemminki, K. & Susi, R. (1984) Sites of reaction of glutaraldehyde and acetaldehyde with nucleosides. *Arch. Toxicol.* 55, 186–190

IARC (1985) *IARC Monographs on the Evaluation of the Carcinogenic Risk of Chemicals to Humans: Allyl Compounds, Aldehydes, Epoxides and Peroxides*, Vol 36, Lyon, 101–132

Knasmüller, S., Zöhrer, E., Kronberg, L., Kundi, M., Franzén, R. & Schulte-Hermann, R. (1996) Mutational spectra of *Salmonella typhimurium* revertants induced by chlorohydroxyfuranones, byproducts of chlorine disinfection of drinking water. *Chem. Res. Toxicol.* 9, 374–381

Komulainen, H., Kosma, V-M., Vaittinen, S-L., Vartiainen, T., Kaliste-Korhonen, E., Lötjönen, S., Tuominen, R. K. & Tuomisto, J. (1997) Carcinogenicity of the drinking water mutagen 3-chloro-4-(dichloromethyl)-5-hydroxy-2(5*H*)-furanone in the rat. *J. Natl Cancer Inst.* 89, 848–856

Kronberg, L. & Franzén, R. (1993) Determination of chlorinated furanones, hydroxyfuranones, and butenedioic acids in chlorine-treated water and pulp bleaching liquor. *Environ. Sci. Technol.* 27, 1811–1818

Kronberg, L., Sjöholm, R. & Karlsson, S. (1992) Formation of 3,N^4-ethenocytidine, 1,N^6-ethenoadenosine and 1,N^2-ethenoguanosine in reactions of mucochloric acid with nucleosides. *Chem. Res. Toxicol.* 5, 852–855

Kronberg, L., Karlsson, S. & Sjöholm, R. (1993) Formation of ethenocarbaldehyde derivatives of adenosine and cytidine in reactions with mucochloric acid. *Chem. Res. Toxicol.* 6, 495–499

Kronberg, L., Asplund, D., Mäki, J. & Sjöholm, R. (1996) Reactions of mucochloric and mucobromic acid with adenosine and cytidine: Formation of chloro- and bromopropenal derivatives. *Chem. Res. Toxicol.* 9, 1257–1263

LaLonde, R.T., Bu, L., Henwood, A., Fiumano, J. & Zhang, L. (1997) Bromine-, chlorine-, and mixed halogen-substituted 4-methyl-2(5*H*)-furanones: Synthesis and mutagenic effects of halogen and hydroxyl group replacements. *Chem. Res. Toxicol.* 10, 1427–1436

Le Curieux, F., Munter, T. & Kronberg, L. (1997) Identification of adenine adducts formed in reaction of calf thymus DNA with mutagenic chlorohydroxy-furanones found in drinking water. *Chem. Res. Toxicol.* 10, 1180–1185

Le Curieux, F., Pluskota, D., Munter, T., Sjöholm, R. & Kronberg, L. (1998) Formation of a fluorescent adduct in the reaction of 2′-deoxyadenosine with a malonaldehyde-acetaldehyde condensation product. *Chem. Res. Toxicol.* 11, 989–994

Marsteinstredet, U., Brunborg, G., Bjørås, M., Søderlund, E., Seeberg, E., Kronberg, L. & Holme, J. A. (1997) DNA damage induced by 3-chloro-4-(dichloromethyl)-5-hydroxy-2(5*H*)-furanone (MX) in HL-60 cells and purified DNA in vitro. *Mutat. Res.* 390, 171–178

Meier, J.R. (1988) Genotoxic activity of organic chemicals in drinking water. *Mutat. Res.* 196, 211–245

Munter, T., Kronberg, L.& Sjöholm, R. (1996) Identification of the adducts formed in reactions of adenosine with 3-chloro-4-methyl-5-hydroxy-2(5*H*)-furanone—A bacterial mutagen present in chlorine disinfected drinking water. *Chem. Res. Toxicol.* 9, 703–708

Munter, T., Le Curieux, F., Sjöholm, R. & Kronberg, L. (1998) Reactions of the potent bacterial mutagen 3-chloro-4-(dichloromethyl)-5-hydroxy-2(5*H*)-furanone (MX) with 2′-deoxyadenosine and calf thymus DNA: Identification of fluorescent propenoformyl derivatives. *Chem. Res. Toxicol.* 11, 226–233

Munter, T., Le Curieux, F., Sjöholm, R. & Kronberg, L. (1999) Identification of an ethenoformyl adduct formed in the reaction of the potent bacterial mutagen 3-chloro-4-(dichloromethyl)-5-hydroxy-2(5*H*)-furanone with guanosine. *Chem. Res. Toxicol.* 12, 46–52

Smeds, A., Vartiainen, T., Mäki-Paakkanen, J. & Kronberg, L. (1997) Concentrations of Ames mutagenic chlorohydroxyfuranones and related compounds in drinking water. *Environ. Sci. Technol.* 31, 1033–1039

Stone, K., Ksebati, M.B. & Marnett, L. J. (1990) Investigation of the adducts formed by reaction of malonaldehyde with adenosine. *Chem. Res. Toxicol.* 3, 33–38

Tikkanen, L, & Kronberg, L. (1990) Genotoxic effects of various chlorinated butenoic acids identified in chlorinated drinking water. *Mutat. Res.* 240, 109–116

Tuma, D., Thiele, G., Xu, D., Klassen, L. & Sorrell, M.F. (1996) Acetaldehyde and malondialdehyde react together to generate distinct protein adducts in the liver during long-term ethanol administration. *Hepatology* 23, 872–880

Vaca, C.E., Fang, J.-L. & Schweda, E.K.H. (1995) Studies of the reaction of acetaldehyde with deoxynucleosides. *Chem-Biol-Interact.* **98**, 51–67

Wasserman, H.H. & Precopio, F.M. (1952) Studies of mucohalic acids. I. The structure of mucoxychloric acid. *J. Am. Chem. Soc.* 74, 326–328

Xu, D., Thiele, G.M., Kearley, M.L., Haugen, M.D., Klassen, L.W., Sorrell, M.F. & Tuma, D.J. (1997) Epitope characterization of malondialdehyde–acetaldehyde adducts using an enzyme-linked immunosorbent assay. *Chem. Res. Toxicol.* 10, 978–986

Exocyclic DNA Adducts in Mutagenesis and Carcinogenesis
B. Singer & H. Bartsch
IARC Scientific Publications No. 150
International Agency for Research on Cancer, Lyon, 1999

Cyclic adducts and intermediates induced by simple epoxides

J.J. Solomon

Simple epoxides such as ethylene oxide, propylene oxide, epichlorohydrin and glycidol are mutagenic and carcinogenic compounds that are important industrial chemicals. Mutagenic and carcinogenic epoxides can also be formed metabolically from industrially important compounds such as alkenes (ethylene, butadiene, propylene and styrene), vinyl halides (vinyl chloride and vinyl bromide) and other vinyl monomers (acrylonitrile and acrylamide). Simple epoxides react with nucleosides and DNA predominantly by the S_N2 mechanism at the most nucleophilic sites (ring nitrogens) in DNA to form 2-hydroxy-2-alkyl adducts. The major hydroxyalkyl adducts that form at N7 of deoxyguanosine and N3 of deoxyadenosine are chemically unstable owing to the presence of a charged quaternary nitrogen at the site of alkylation, and they depurinate spontaneously to remove the charge, forming potentially mutagenic abasic sites. Hydroxyalkylation at N1 of deoxyadenosine and N3 of deoxycytidine also results in the production of charged, unstable species because the pKa increases dramatically after alkylation. The charge can be lost from these adducts by the formation of cyclic adducts, which occurs when there is a good leaving group on the hydroxyalkyl side-chain. Most simple epoxides remove the charge on hydroxyalkyl adducts at N1 of deoxyadenosine and N3 of deoxyxytidine by competitive rearrangements, such as hydrolytic deamination, to form 1-hydroxyalkyl-deoxyinosine and 3-hydroxyalkyl-deoxyuridine adducts and Dimroth rearrangement to form N^6-hydroxyalkyl-deoxyadenosine adducts. These rearrangements are facilitated intramolecularly by the formation of cyclic intermediates, with the participation of the hydroxyl group of the hydroxyalkyl side-chain. These adducts are uncharged, stable and potentially mutagenic and are likely to contribute to the biological activity of simple epoxides.

Introduction

Many simple mono-substituted epoxides are important industrial chemicals which pose occupational health problems because of their mutagenic and carcinogenic properties. The volatile epoxides ethylene oxide and propylene oxide are among the 50 chemicals produced in greatest volume in the United States (Kirschner, 1996). Mutagenic and carcinogenic epoxides can also be formed metabolically during the detoxification of parent alkenes (ethylene, butadiene, propylene and styrene), vinyl halides (vinyl chloride and vinyl bromide) and vinyl monomers (acrylonitrile and acrylamide). Many of these precursors to epoxides are also important industrial chemicals, acrylonitrile, butadiene, ethylene, propylene, styrene and vinyl chloride being among the 50 with the highest production volumes in the United States (Kirschner, 1996). The genotoxicity of the parent compounds and their corresponding epoxides has been reviewed (Ehrenberg & Hussain, 1981; IARC, 1987, 1992, 1994). This review focuses on the epoxide-induced formation of cyclic DNA adducts and the role of cyclic intermediates in the formation of other potentially mutagenic epoxide-induced DNA adducts. Figure 1 shows the structures of some mono-substituted epoxides that are covered.

Reactions of epoxides with nucleic acid components

The chemical reactivity of mono-substituted epoxides and their DNA adduct-forming ability has been reviewed (Ehrenberg & Hussain, 1981; Hemminki, 1983; Segerbäck, 1994). Under physiological conditions, the mono-substituted epoxides react primarily at the unsubstituted β-carbon with the strong nucleophilic centres in DNA (ring

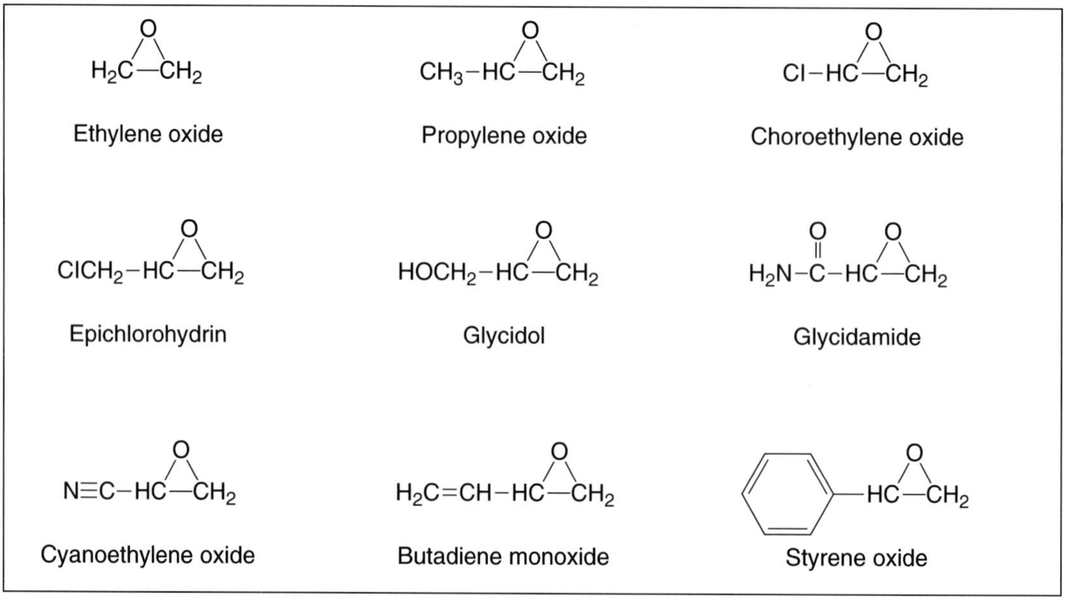

H_2C—CH_2 (epoxide O)	CH_3–HC—CH_2 (epoxide O)	Cl–HC—CH_2 (epoxide O)
Ethylene oxide	Propylene oxide	Choroethylene oxide
$ClCH_2$–HC—CH_2 (epoxide O)	$HOCH_2$–HC—CH_2 (epoxide O)	H_2N–C(=O)–HC—CH_2 (epoxide O)
Epichlorohydrin	Glycidol	Glycidamide
N≡C–HC—CH_2 (epoxide O)	H_2C=CH–HC—CH_2 (epoxide O)	Phenyl–HC—CH_2 (epoxide O)
Cyanoethylene oxide	Butadiene monoxide	Styrene oxide

Figure 1. Structures of some mono-substituted epoxides

nitrogens) *via* the S_N2 mechanism. This results in the formation of diastereomeric 2-hydroxy-2-alkyl (HA) adducts in DNA (Figure 2) because of the chiral centre at the α-carbon. The substituted α-carbon is less reactive because of steric hindrance and the electron-withdrawing nature of many of the substituents (R). Thus, most simple epoxides form diastereomeric HA adducts predominantly at ring nitrogens: N7 Gua, N3 and N1 Ade, N3 Cyt and N3 Thy (Hemminki, 1983; Solomon *et al.*, 1988; Segal *et al.*, 1990; Li *et al.*, 1992; Solomon *et al.*, 1993; Segerback, 1994; Singh *et al.*, 1996). When an electron-releasing group or one that can stabilize the α-carbon through resonance, such as R = phenyl in styrene oxide or R = vinyl in butadiene monoxide, is directly attached to the α-carbon, the reactivity of the α-carbon is increased, permitting the formation of both S_N2 HA and S_N1 2-hydroxy-1-alkyl regioisomers (Figure 2). These products have been carefully characterized in the cases of butadiene monoxide (Selzer & Elfarra 1996a,b, 1997a,b) and styrene oxide (Latif *et al.*, 1988; Qian & Dipple, 1995; Barlow *et al.*, 1997, 1998) with nucleic acid components. Because of the S_N1 character of styrene oxide and butadiene monoxide, hydroxyalkylation at exocyclic sites such as O^6-Gua (Latif *et al.*, 1988) and N^2-Gua (Latif *et al.*, 1988; Selzer & Elfarra, 1996b) is seen with these epoxides. With the typical S_N2 epoxides, exocyclic adducts are rarely found (Hemminki, 1983; Segerback, 1994). Most studies with simple epoxides were conducted *in vitro* with nucleic acid components and DNA. Studies *in vivo* have primarily shown the major alkylation product at N7 of Gua (Swenberg *et al.*, 1992; Walker *et al.*, 1992; Segerback *et al.*, 1995; Koivisto *et al.*, 1996, 1997, 1998; Plna & Segerback, 1997; Rios-Blanco *et al.*, 1997; Tretyakova *et al.*, 1998).

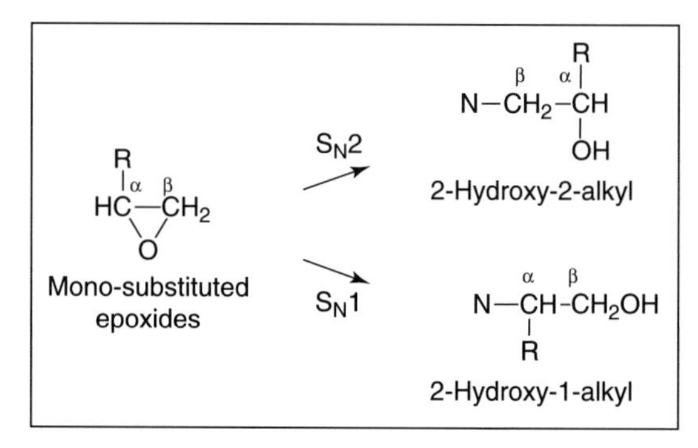

Figure 2. *N*-Hydroxyalkylation of DNA by mono-substituted epoxides

Cyclic adducts induced by simple epoxides

Vinyl monomers such as vinyl chloride, vinyl bromide and vinyl carbamate are carcinogens that require metabolic activation to mono-substituted epoxides to form adducts with DNA. In addition to the major DNA adduct, 7-(2-oxoethyl)guanine (7-OE-Gua), which forms with chloroethylene oxide, the epoxide of vinyl chloride, cyanoethylene oxide, the epoxide of acrylonitrile, and the epoxide of vinyl carbamate, minor cyclic etheno (ε) adducts are formed. Chloroethylene oxide reacts with nucleic acid components to form all four cyclic ε adducts 1,N^6-εdA, 3,N^4-εdC, N^2,3-εdG and 1,N^2-εdG (see top of Figure 3) (Guengerich, 1992; Guengerich *et al.*, 1993; Muller *et al.*, 1997). Guengerich and Raney (1992), using ^{13}C-labelling, elucidated the mechanism of formation of 1,N^6-εdA and 3,N^4-εdC from 1-halooxiranes such as chloroethylene oxide. Initial electrophilic attack occurs at the most basic endocyclic nitrogens (N1 of Ade and N3 of Cyt) by the β-methylene carbon of the epoxide. This is followed by epoxide ring opening and loss of a good leaving group (lost as HCl for chloroethylene oxide) to form an aldehyde (1-OE-dA and 3-OE-dC) which cyclizes with an exocyclic nitrogen (N^6 of dA and N^4 of dC) in a Schiff base reaction to form the new ethano ring, which becomes an ε adduct after dehydration. This is probably the mechanism of formation of 1,N^6-εdA which forms after reaction of cyanoethylene oxide with dA and DNA *in vitro* (Solomon *et al.*, 1993). In this case, the cyano group is lost as hydrogen cyanide. 1,N^6-εAdo also forms in the reaction of the epoxide of vinyl carbamate with Ado (Guengerich & Kim, 1991), where the leaving group is lost as carbamic acid. The formation, repair and biological consequences of ε adducts are discussed elsewhere in this volume.

Other mono-substituted epoxides have also been shown to form cyclic adducts (Figure 3, bottom) in reaction with 2′-deoxynucleosides, when there is a good leaving group. These include the

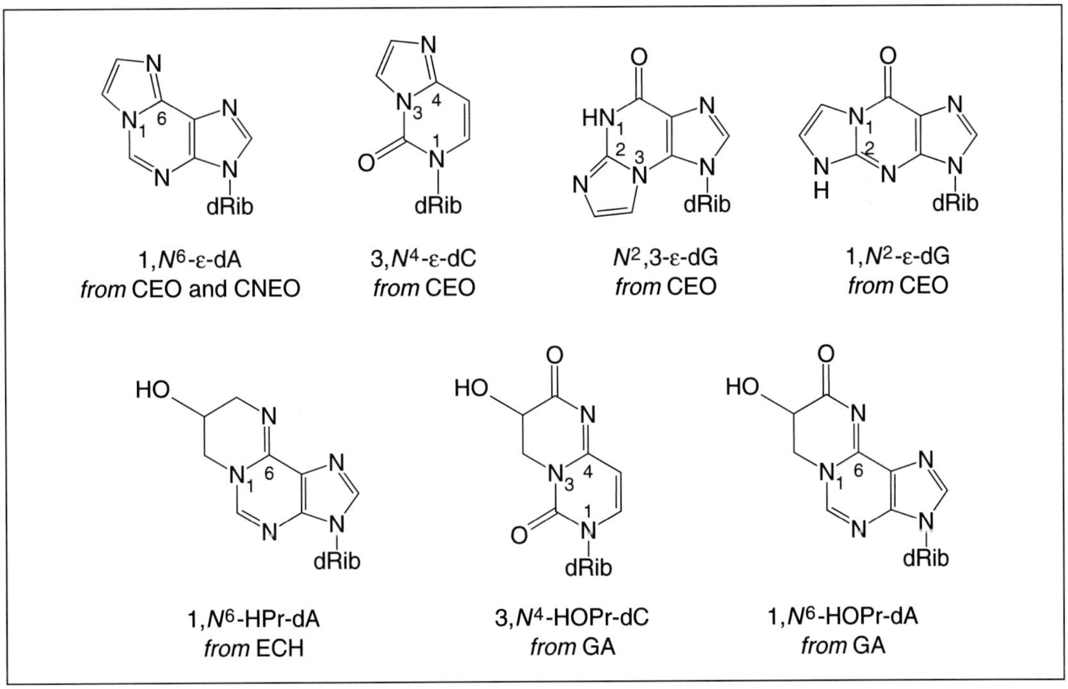

1,N^6-ε-dA *from* CEO and CNEO

3,N^4-ε-dC *from* CEO

N^2,3-ε-dG *from* CEO

1,N^2-ε-dG *from* CEO

1,N^6-HPr-dA *from* ECH

3,N^4-HOPr-dC *from* GA

1,N^6-HOPr-dA *from* GA

Figure 3. Cyclic adducts of mono-substituted epoxides

ε, etheno; CEO, chloroethylene oxide; CNEO, cyanoethylene oxide; HPr, 2-hydroxypropano; ECH, epichlorohydrin; HOPr, 2-hydroxy-3-oxopropano; GA, glycidamide

formation of 1,N^6-(2-hydroxypropano)-dA from epichlorohydrin, where HCl is lost (Singh *et al.*, 1996), and the cyclic adducts 1,N^6-(2-hydroxy-3-oxopropano)-dA and 3,N^4-(2-hydroxy-3-oxopropano)-dC (B. Zhu., W. Winnik & J.J. Solomon, unpublished data) induced by glycidamide, the mutagenic epoxide metabolite of acrylamide (IARC, 1994), where NH_3 is lost. These cyclic glycidamide adducts have been tentatively characterized by ultraviolet and mass spectroscopy, and details of their chromatographic and spectroscopic properties and chemical stability will be published elsewhere. Glycidaldehyde, another monosubstituted mutagenic and carcinogenic bifunctional epoxide, in which the substituted group R is CHO (formyl), has been shown to form ε adducts (1,N^2-εdG) and hydroxymethyl-substituted etheno adducts (Goldschmidt *et al.*, 1968; Golding *et al.*, 1990; Steiner *et al.*, 1992); however, these adducts are formed in DNA by a completely different mechanism from that for the typical S_N2 or S_N1 epoxides discussed above, because the epoxide ring of glycidaldehyde is not involved in the initial reaction with DNA. In this case, the carbonyl of glycidaldehyde reacts like other aldehydes, such as formaldehyde, at exocyclic nitrogens in DNA, followed by cyclization *via* reaction of an endocyclic nitrogen with the α- or β-carbon of the epoxide ring (Golding *et al.*, 1990).

Cyclic intermediates involved in adduct formation

As discussed above, the primary sites of hydroxyalkylation for most epoxides is at the most nucleophilic centres in DNA, at N7 of Gua and N3 of Ade, with very little reaction at exocyclic sites. These 7-HA-dG and 3-HA-dA adducts are chemically unstable and spontaneously depurinate to form abasic sites in DNA in order to get rid of the charge on the quaternary nitrogen which forms after hydroxyalkylation. At elevated pH, hydrolysis competes with depurination for the 7-alkyl-dG adducts, where hydroxide ion attack at C8 results in the formation of stable, uncharged, imidazole ring-opened products (Shuker & Farmer, 1992).

Hydroxyalkylation at N3 of dC and N1 of dA also results in charged species. Methylation at N3 of dC causes the pKa of dC to increase from 4.3 (Dunn & Hall, 1975) to 8.8 (Sowers *et al.*, 1989). More than 95% of 3-methyl-dC adducts are therefore protonated at physiological pH. Similarly, hydroxyethylation by ethylene oxide at N1 of Ado raises the pKa of Ado from 3.5 (Dunn & Hall, 1975) to 8.3 (Windmueller & Kaplan, 1961). Whereas alkylation of N7 of Gua and N3 of Ade results in charged species which spontaneously depurinate to remove the charge, the epoxide-induced hydroxyalkyl adducts at N1 of dA and N3 of dC do not get rid of their charge *via* cleavage of the glycosidic bond. One way that these adducts can remove their charge is by cyclization, which occurs when there is a good leaving group on the hydroxyalkyl side-chain, as discussed above. The cyclic adducts that form are not protonated at physiological pH (Krzyzosiak *et al.*, 1986), as indicated by their pKa values, which are less than 4 (Singer & Gruenberger, 1983; Kusmierek *et al.*, 1987; Kuśmierek & Singer, 1992). Competitive reactions to cyclization are hydrolytic deamination of 3-HA-dC and 1-HA-dA to form 3-HA-dU and 1-HA-dI adducts, respectively, and the Dimroth rearrangement of 1-HA-dA to form N^6-HA-dA adducts, all of which are stable and unprotonated in DNA. These rearrangements are probably facilitated by cyclic intermediates as discussed below.

A cyclic intermediate is involved in the hydrolytic deamination of 3-(2-hydroxy-2-alkyl)deoxycytidine

The half-life for deamination of dC (pH 7.4, 37 °C) is 189 years (Sowers *et al.*, 1989). When dC is methylated at N3, the half-life for deamination to 3-methyl-dU under physiological conditions decreases by a factor of 4000 to 406 h (Sowers *et al.*, 1989). This increase in deamination rate has been attributed to the increased fraction of protonated molecules in 3-methyl-dC, since alkylation dramatically increases the pKa, as discussed above. Solomon and co-workers have demonstrated that 3-HA-dC adducts induced by several monosubstituted S_N2 epoxides, including ethylene oxide, propylene oxide, cyanoethylene oxide, glycidol and epichlorohydrin, deaminate under physiological conditions even more rapidly, with half-lives of 10 h or less, to form 3-HA-dU adducts (Solomon *et al.*, 1988; Segal *et al.*, 1990; Li *et al.*, 1992; Solomon *et al.*, 1993; Singh *et al.*, 1996).

A mechanism for the rapid hydrolytic deamination of epoxide-induced 3-HA-dC to 3-HA-dU has been postulated (Li *et al.*, 1992; Solomon, 1994) in which the side-chain hydroxyl group intramolecularly assists in the deamination through the formation of a cyclic intermediate (Figure 4). The role of the side-chain hydroxyl group was established by comparing the rate of deamination of ethylene oxide-induced 3-(2-hydroxyethyl)-dC with 3-ethyl-dC. The presence of the side-chain hydroxyl group increased the rate of deamination 60-fold to 10 h at pH 7.4 and 37 °C (Li *et al.*, 1992). The attack of the hydroxide ion was demonstrated to occur at the C4 position of the pyrimidine ring of the cyclic intermediate by hydrolysis with ^{18}O-water and analysis of the reaction product by mass spectrometry (C. Lai & J.J. Solomon, unpublished data). These studies were necessary because attack of the hydroxide ion at the substituted carbon in the newly formed oxazolinium ring system of the cyclic intermediate (Figure 4) could have resulted in the same 3-HA-dU final product. To establish conclusively that a cyclic intermediate was involved in the hydrolytic deamination, the cyclic intermediate in the deamination of 3-HA-Cyt formed from ethylene oxide and propylene oxide was isolated and characterized by ultraviolet, mass and nuclear magnetic resonance spectrometry at the base level (C. Lai & J.J. Solomon, unpublished data). The cyclic intermediate can be detected at the base level because of the presence of the labile N1 hydrogen, which can be lost after cyclization of protonated 3-HA-Cyt (Figure 5). Kinetic analysis of the formation of 3-HA-Ura from 3-HA-Cyt has revealed that the formation of the cyclic intermediate is the slower step in the deamination at the base level (C. Lai & J.J. Solomon, unpublished data).

Figure 6 shows the two competing pathways, deamination and cyclization, that remove the charge that forms after epoxide-induced hydroxyalkylation at N3 of dC. These pathways produce chemically stable, potentially mutagenic adducts.

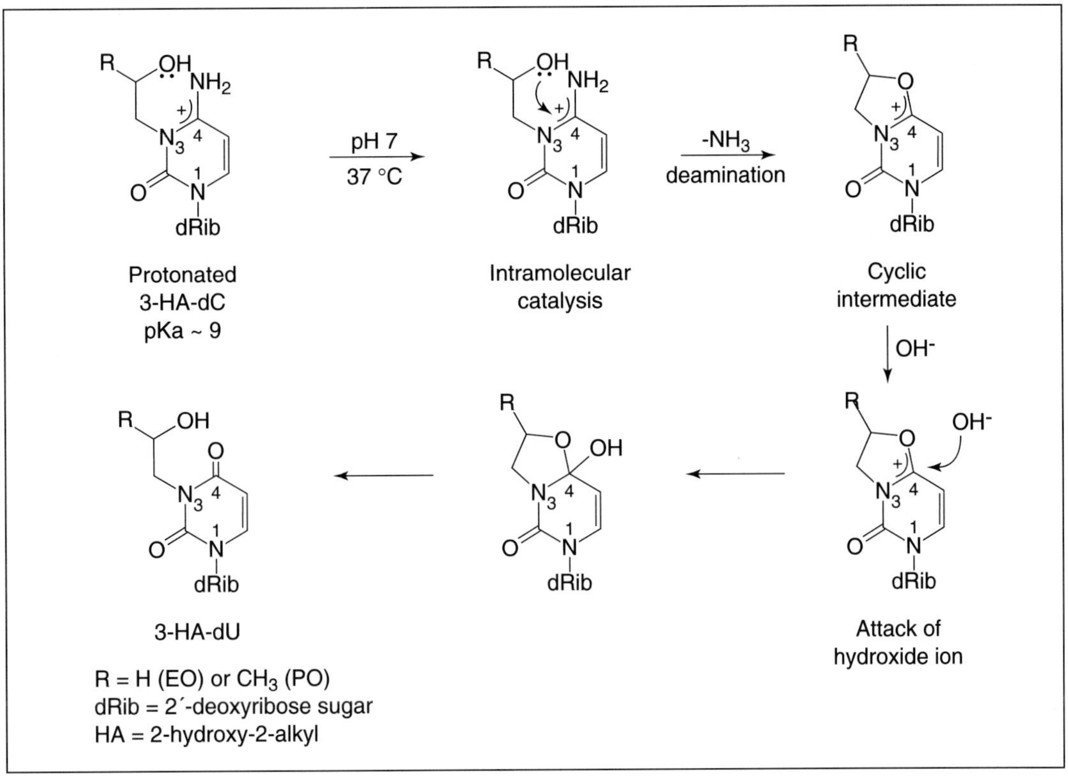

Protonated
3-HA-dC
pKa ~ 9

pH 7
37 °C

Intramolecular
catalysis

$-NH_3$
deamination

Cyclic
intermediate

OH^-

Attack of
hydroxide ion

3-HA-dU

R = H (EO) or CH_3 (PO)
dRib = 2´-deoxyribose sugar
HA = 2-hydroxy-2-alkyl

Figure 4. Postulated mechanism of hydrolytic deamination of 3-(2-hydroxy-2-alkyl)-deoxycytidine (3-HA-dC) to 3-(2-hydroxy-2-alkyl)-deoxyuridine (3-HA-dU)

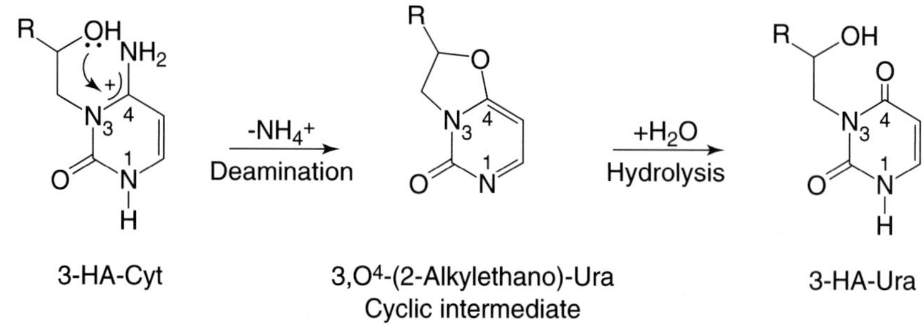

$-NH_4^+$ Deamination

$+H_2O$ Hydrolysis

3-HA-Cyt

$3,O^4$-(2-Alkylethano)-Ura
Cyclic intermediate

3-HA-Ura

Figure 5. Cyclic intermediate detected during deamination of 3-(2-hydroxy-2-alkyl)-cytosine

In addition to the epoxides mentioned above, recent studies with glycidamide (B. Zhu, W. Winnik & J.J. Solomon, unpublished data), allyl glycidyl ether (Plna *et al.*, 1996) and butadiene monoxide (Selzer & Elfarra, 1997b) have also detected 3-HA-dU in reactions with nucleosides

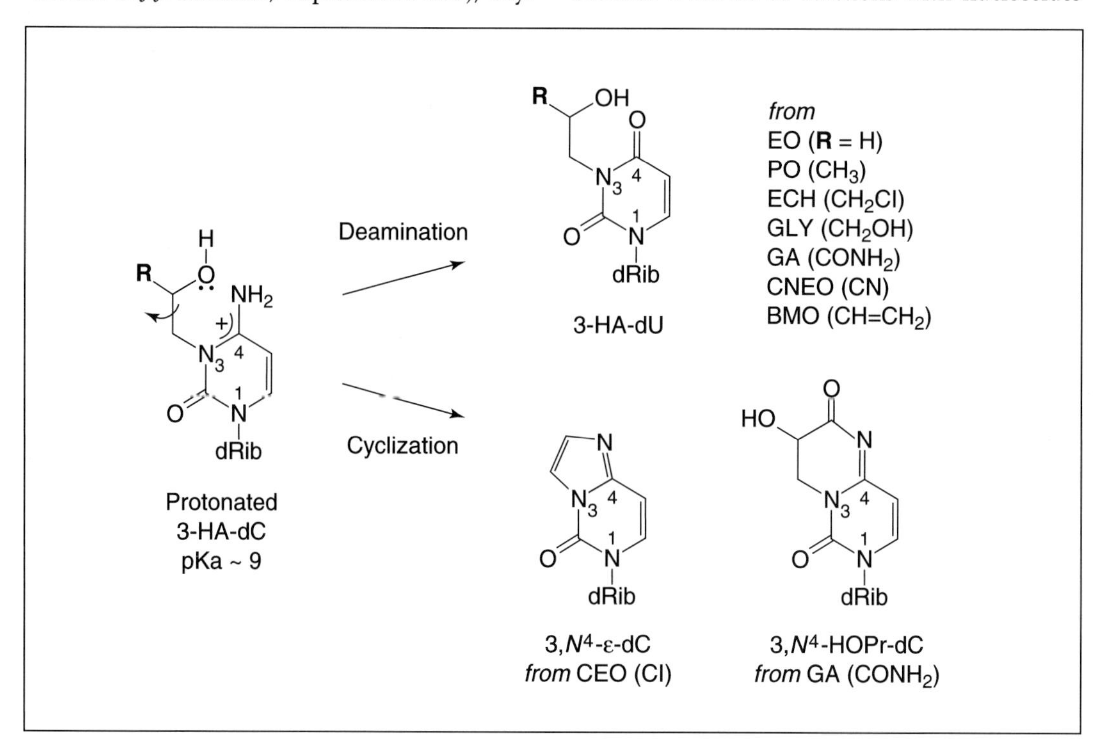

Figure 6. Rearrangements after epoxide-induced hydroxyalkylation at N3 of deoxycytidine

HA, 2-hydroxy-2-alkyl; EO, ethylene oxide; PO, propylene oxide; ECH, epichlorohydrin; GLY, glycidol; GA, glycidamide; CNEO, cyanoethylene oxide; BMO, butadiene monoxide; CEO, chloroethylene oxide

and/or DNA. Allyl glycidyl ether is a mutagenic and carcinogenic epoxide (IARC, 1989) in which the hydrogen of the hydroxyl group of glycidol is replaced by an allyl group. The adduct pattern found for allyl glycidyl ether in reaction with nucleic acid components and DNA is typical of those of other monosubstituted S_N2 epoxides (Plna *et al.*, 1996). Because butadiene monoxide has an S_N1 character, the diastereomers of both α-substituted and β-substituted regioisomers, α-3-(2-hydroxy-1-vinyl)-dU and β-3-(2-hydroxy-2-vinyl)-dU, were detected. The half-lives for deamination (pH 7.4, 37 °C) of the β-3-(2-hydroxy-2-vinyl)-dC diastereomers were 2.3 and 2.5 h. The inability to detect α-3-(2-hydroxy-1-vinyl)-dC diastereomers was attributed to their faster rates of deamination (Selzer & Elfarra, 1997b). The faster deamination of the α-3-substituted-dC adducts was accounted for by (i) steric hindrance from the vinyl group when it is attached to the same carbon as the hydroxyl involved in the formation of the five-membered cyclic intermediate in β-3-substituted-dC adducts or (ii) the greater nucleophilicity of the primary hydroxyl in α-3-substituted-dC adducts.

Hydrolytic deamination of 1-(2-hydroxy-2-alkyl)-deoxyadenosine

An analogous reaction scheme to the deamination of 3-HA-dC (Figure 4) involving a five-membered cyclic intermediate can be envisioned for the hydrolytic deamination of 1-HA-dA to 1-HA-dI, because hydroxyalkylation at N1 of dA results in a protonated species at N^6 at physiological pH. Inosine adducts formed from the hydrolytic deamination of 1-HA-(d)A have been detected only with epoxides that show S_N1 character, i.e. butadiene monoxide (Selzer & Elfarra, 1996a) and styrene oxide (Qian & Dipple, 1995; Barlow *et al.*, 1998). A cyclic intermediate for this deamination has been proposed (Fujii *et al.*, 1986) but has not been isolated. The site of attack of the hydroxide ion in this facile deamination reaction has been established to be at C6 in styrene oxide-induced α-1-(2-hydroxy-1-phenylethyl)-Ado by hydrolysis with ^{18}O-water (Barlow *et al.*, 1997). Here again, the α-1-(2-hydroxy-1-phenylethyl)-Ado adducts deaminate much more rapidly (35-fold) than the β-1-(2-hydroxy-2-phenylethyl)-Ado adducts (Qian & Dipple, 1995).

Dimroth rearrangement of 1-(2-hydroxy-2-alkyl)-deoxyadenosine to N^6-(2-hydroxy-2-alkyl)-deoxyadenosine

In addition to cyclization and deamination, 1-HA-dA adducts induced by simple epoxides can remove the charge that forms after hydroxyalkylation at N1 of dA (pKa ~ 8) by Dimroth rearrangement to form stable N^6-HA-dA adducts (Figure 7). The Dimroth rearranged products, N^6-HA-dA adducts, are uncharged, with a pKa of about 4.0 (Dunn & Hall, 1975). Dimroth-rearranged products have also been found in the reaction of nucleic acid components and DNA with most epoxides, including allyl glycidyl ether (Plna *et al.*, 1996), butadiene monoxide (Selzer & Elfarra, 1996a), cyanoethylene oxide (Singh *et al.*, 1996), epichlorohydrin (Hemminki *et al.*, 1980), ethylene oxide (Li *et al.*, 1992), glycidamide (B. Zhu & J.J. Solomon, unpublished data), glycidol (Hemminki *et al.*, 1980), propylene oxide (Lawley & Jarman, 1972; Solomon *et al.*, 1988) and styrene oxide (Savela *et al.*, 1986; Qian & Dipple, 1995; Barlow *et al.*, 1998). The base-catalysed Dimroth rearrangement of 1-substituted-(d)Ado is well known, but 1-alkyl-(d)Ado requires a basic pH to rearrange, with a half-life of hours instead of days (Macon & Wolfenden, 1968). The introduction of a hydroxyl group into the side-chain of 1-alkyl-adenine derivatives was found to speed up the Dimroth rearrangement at neutral pH (Fujii *et al.*, 1986). The rate of rearrangement of 1-substituted-adenine derivatives correlated with their pKa values: the lower the pKa, the faster the rearrangement. The increase in the rate of Dimroth rearrangement when there is an electron-withdrawing hydroxyl group on the side-chain was explained as an inductive effect, whereby the hydroxyl group promotes attack of hydroxide ion on the positively charged purine ring (Fujii *et al.*, 1986; Barlow *et al.*, 1998).

An alternative hypothesis to explain the rapid Dimroth rearrangement that occurs under physiological conditions with many epoxide-induced 1-HA-dA adducts and which is not seen in analogous adducts that do not contain a hydroxyl group at the 2-position of the hydroxyalkyl side-chain (Chen *et al.*, 1981; Segal *et al.*, 1979; Solomon *et al.*, 1988; Li *et al.*, 1992; Solomon *et al.*, 1993; Solomon 1994) is presented in Figure 8. In this postulated mechanism,

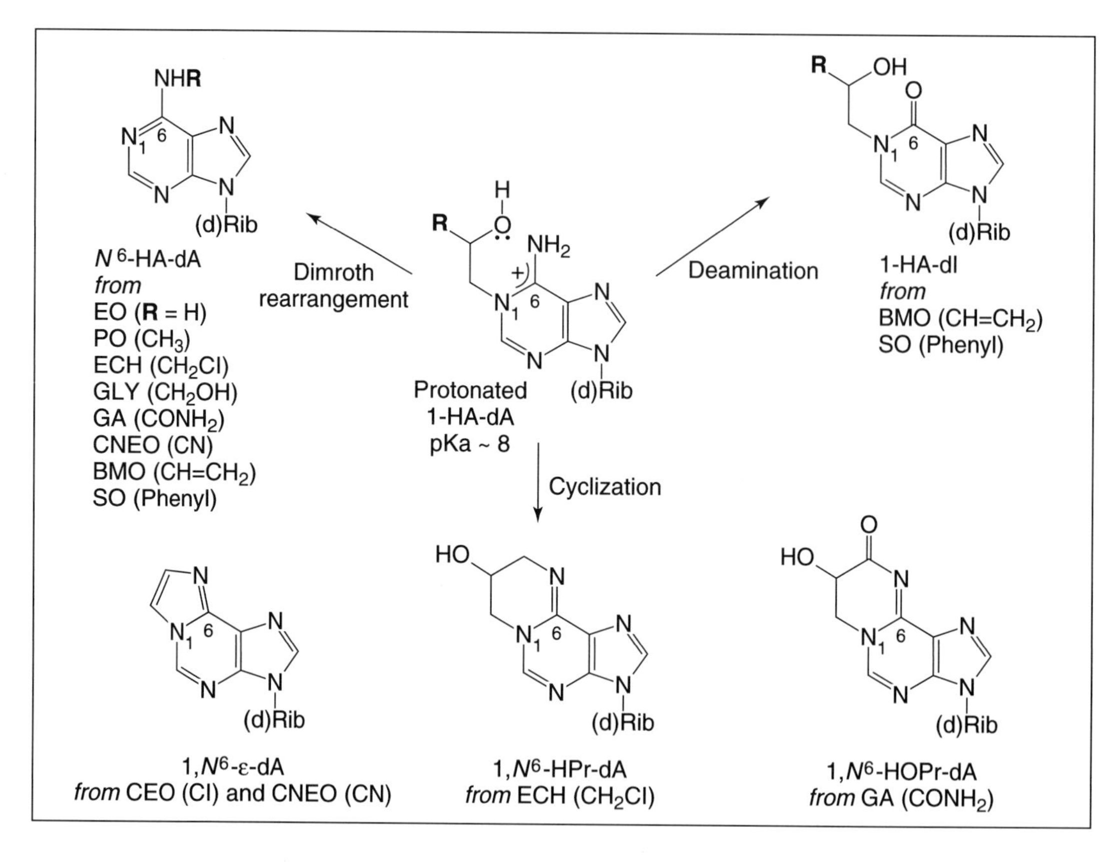

Figure 7. Rearrangements after epoxide-induced hydroxyalkylation at N1 of deoxyadenosine

HA, 2-hydroxy-2-alkyl; EO, ethylene oxide; PO, propylene oxide; ECH, epichlorohydrin; GLY, glycidol; GA, glycidamide; CNEO, cyanoethylene oxide; BMO, butadiene monoxide; SO, styrene oxide; CEO, chloroethylene oxide; ε, etheno; HPr, 2-hydroxypropano; HOPr, 2-hydroxy-3-oxopropano

after epoxide-induced hydroxyalkylation, the charge is delocalized within the purine ring. The side-chain hydroxyl group then attacks the C2 carbon in the purine ring, which causes the N1–C2 bond to break and a nine-membered cyclic intermediate to form, which still contains the charge. The exocyclic nitrogen at N^6 then attacks the positively charged carbon at C2 since there is free rotation around the C5–C6 bond and the newly formed ring is large enough not to interfere. The new bond that forms between N^6 and C2 causes the recently formed O–C2 bond to break, which produces the stable, uncharged N^6-HA-dA adduct. This intramolecular assistance of the side-chain hydroxyl group through the formation of a cyclic intermediate may explain the rapid Dimroth rearrangement observed under physiological conditions with many epoxides after hydroxyalkylation at N1 of deoxyadenosine.

Biological consequences of epoxide-induced adducts

The major adducts of the simple mono-substituted epoxides at 7-HA-dG and 3-HA-dA are not promutagenic. Their chemical instability (depurination) and/or glycosylase-mediated repair results in potentially mutagenic apurinic sites (Loeb & Preston, 1986). 7-Alkyl-dG can also undergo imidazole ring-opening to chemically stable, potentially mutagenic lesions. Adducts at O^6 of Gua are promutagenic (Loechler *et al.*,

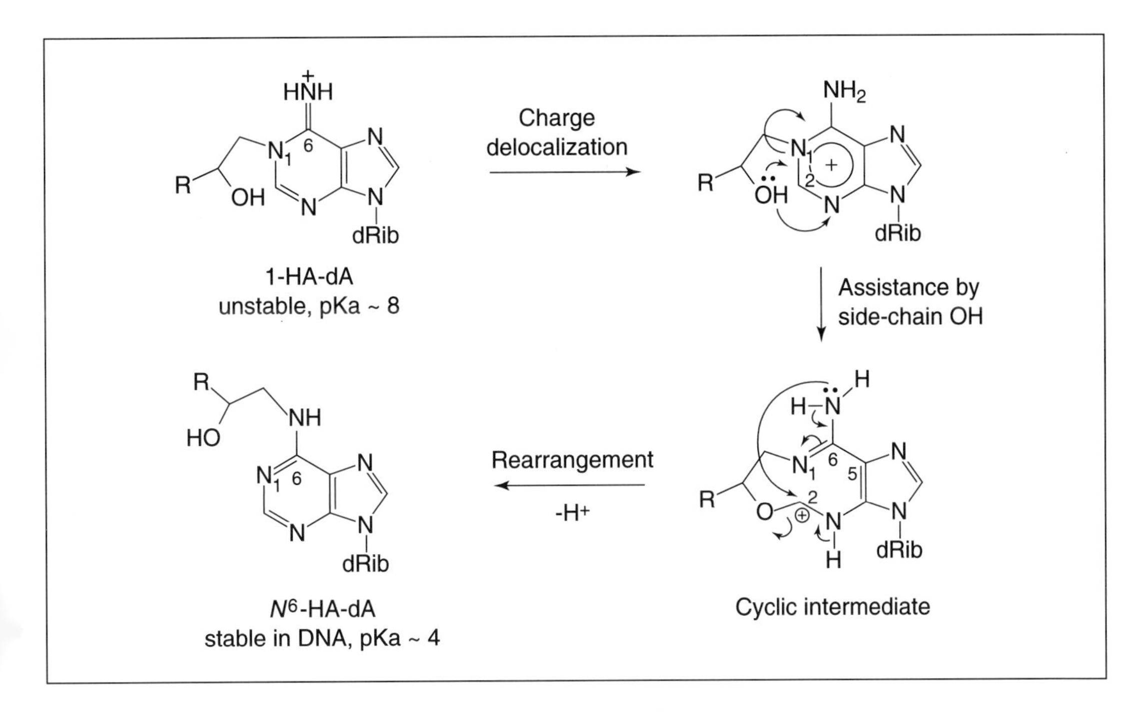

Figure 8. Dimroth rearrangement assisted at neutral pH by side-chain hydroxyl group

HA, 2-hydroxy-2-alkyl

1984), but are probably important only for epoxides with an S_N1 character, such as styrene oxide (Latif *et al.*, 1988).

Hydroxyalkylation by simple epoxides at N1 of dA and/or N3 of dC results in charged, unstable species which can form cyclic adducts when there is a good leaving group on the hydroxyalkyl side-chain (Guengerich & Kim, 1991; Guengerich, 1992; Solomon *et al.*, 1993; Singh *et al.*, 1996). The persistence and mutagenic potential of cyclic adducts that tie up Watson-Crick sites in DNA are discussed elsewhere in this volume. As discussed above, the unstable 1-HA-dA adducts in DNA undergo rapid deamination and/or Dimroth rearrangement under physiological conditions to form chemically stable, potentially mutagenic 1-HA-dI and N^6-HA-dA lesions. The rate of Dimroth rearrangement in styrene-oxide-induced 1-substituted-dA adducts was found to be slower than that for 1-substituted-Ado adducts, resulting in increased yields of the deaminated 1-substituted-dI adducts in deoxynucleoside reactions (Barlow *et al.*, 1998). In DNA, the yields of β-1-substituted-dI products were increased over those found in reactions with nucleosides (Barlow *et al.*, 1998). These results suggest that 1-HA-dI adducts in DNA may be important. The mutagenic potential of 1-HA-dI adducts has not been studied, but both Watson-Crick sites of dA are disrupted by the HA group at N1 and by replacement of the N^6-amino group with a carbonyl oxygen, which can participate in erroneous hydrogen-bonding. The Dimroth rearranged product, N^6-HA-dA, may not be mutagenic because this lesion still has a hydrogen atom at N^6 which can correctly hydrogen-bond by rotating the HA group out of the way. Nuclear magnetic resonance and by-pass studies with site-modified oligomers that could address the potential mutagenicity of epoxide-induced N^6-HA-dA have not been done.

With many of the epoxides discussed above, the chemically unstable 3-HA-dC lesions in DNA rapidly deaminate to stable, potentially mutagenic 3-HA-dU lesions. This lesion blocks a central Watson-Crick hydrogen-bonding site and replaces the N^4-amino of dC with a carbonyl oxygen. Mutagenic by-pass was found for site-modified oligomers containing ethylene oxide-induced 3-(2-hydroxyethyl)-dU (Bhanot *et al.*, 1994; Zhang *et al.*, 1995) and 3-HA-dU lesions induced by epichlorohydrin, glycidol and propylene oxide (O.S. Bhanot & J.J. Solomon, unpublished data). Propylene oxide was found to induce mutations at template cytosine residues, probably as a result of the 3-(2-hydroxypropyl)-dU adduct (Snow *et al.*, 1994). There is no known repair of 3-HA-dU lesions. Studies with site-modified oligomers containing 3-HA-dU lesions have established that these lesions cannot be repaired by uracil DNA glycosylase from human placenta and bacteria (O.S. Bhanot & J.J. Solomon, unpublished data). Thus, an understanding of the mutagenic potential of cyclic adducts and adducts that form from cyclic intermediates, such as 3-HA-dU and 1-HA-dI, may play a key role in revealing the mechanisms of mutagenesis and carcinogenesis of simple mono-substitutes epoxides.

Acknowledgments

This work was supported by a Research Grant ES05694 from the National Institute of Environmental Health Sciences and by Center Grants ES00260 and CA16087 from the National Institutes of Health. The author wishes to thank Dr Witold Winnik and Dr Udai Singh for helpful discussions and Dr Opinder Bhanot and graduate students Cong Lai and Bing Zhu for sharing their unpublished results.

References

Barlow, T., Ding, J., Vouros, P. & Dipple, A. (1997) Investigation of hydrolytic deamination of 1-(2-hydroxy-1-phenylethyl)adenosine. *Chem. Res. Toxicol.* 10, 1247–1249

Barlow, T., Takeshita, J. & Dipple A. (1998) Deamination and Dimroth rearrangement of deoxyadenosine–styrene oxide adducts in DNA. *Chem. Res. Toxicol.* 11, 838–845

Bhanot, O.S., Singh, U.S. & Solomon, J.J. (1994) The role of 3-hydroxylethyldeoxyuridine in mutagenesis by ethylene oxide. *J. Biol. Chem.* 269, 30056–30064

Chen, R., Mieyal, J.J. & Golthwaite, D.A. (1981) The reaction of β-propiolactone with derivatives of adenine and DNA. *Carcinogenesis* 2, 73–80

Dunn, D.B. & Hall, R.H. (1975) In: *Handbook of Biochemistry and Molecular Biology Nucleic Acids* (Fasman, G.D., ed.). Cleveland, CRC Press, 3rd Ed., pp. 65–215

Ehrenberg, L. & Hussain, S. (1981) Genetic toxicity of some important epoxides. *Mutat. Res.* 86, 1–113

Fujii, T., Saito, T. & Terahara, N. (1986) Purines. XXVII. Hydrolytic deamination *versus* Dimroth rearrangement in the 9-substituted adenine ring: Effect of an ω-hydroxyalkyl group at the 1-position. *Chem. Pharm. Bull.* 34, 1094–1107

Golding, B.T., Kennedy, G. & Watson, W.P. (1990) Structure determination of adducts from the reaction of (R)-glycidaldehyde and guanosine. *Carcinogenesis* 11, 865–868

Goldschmidt, B.M., Blazej, T.P. & Van Duuren, B.L. (1968) The reaction of guanosine and deoxyguanosine with glycidaldehyde. *Tetrehedron Lett.* 13, 1583–1586

Guengerich, F.P. (1992) Roles of the vinyl chloride oxidation products 1-chlorooxirane and 2-chloroacetaldehyde in the *in vitro* formation of etheno adducts of nucleic acid bases. *Chem. Res. Toxicol.* 5, 2–5

Guengerich, F.P. & Kim, D.H. (1991) Enzymatic oxidation of ethyl carbamate to vinyl carbamate and its role as an intermediate in the formation of 1,N^6-ethenoadenosine. *Chem. Res. Toxicol.* 4, 413–421

Guengerich, F.P. & Raney, V.M. (1992) Formation of etheno adducts of adenosine and cytidine from 1-halooxiranes. Evidence for a mechanism involving initial reaction with the endocyclic nitrogen atoms. *J. Am. Chem. Soc.* 114, 1074–1080

Guengerich, F.P., Persmark, M. & Humphreys, W.G. (1993) Formation of 1,N^2- and N^2,3-ethenoguanine from 2-halooxiranes: Isotopic labeling studies and isolation of a hemiaminal derivative of N^2-(2-oxoethyl)guanine. *Chem. Res. Toxicol.* 6, 435–468

Hemminki, K. (1983) Nucleic acid adducts of chemical carcinogens and mutagens. *Arch. Toxicol.* 52, 249–285

Hemminki, K., Paasivirta, J., Kurkirinnr, T. & Virkki, L. (1980) Alkylation products of DNA bases by simple epoxides. *Chem.-Biol. Interactions* 30, 259–270

IARC (1987) *IARC Monographs on the Evaluation of the Carcinogenic Risk of Chemicals to Humans,* Supplement 7. Lyon

IARC (1989) *IARC Monographs on the Evaluation of the Carcinogenic Risk of Chemicals to Humans,* Vol. 47, Lyon, IARC, pp. 237–261

IARC (1992) *IARC Monographs on the Evaluation of the Carcinogenic Risk of Chemicals to Humans,* Vol. 54, Lyon, IARC, pp. 237–285

IARC (1994) *IARC Monographs on the Evaluation of the Carcinogenic Risk of Chemicals to Humans,* Vol. 60, Lyon, IARC, pp. 45–435

Kirschner, E.M. (1996) Growth of top 50 chemicals slowed in 1995 from very high 1994 rate. *Chem. Eng. News* 8 April, 16–17

Koivisto, P., Adler, I.D., Sorsa, M. & Peltonen, K. (1996) Inhalation exposure of rats and mice to 1,3-butadiene induces N^6-adenine adducts of epoxybutene detected by ^{32}P-postlabeling and HPLC. *Environ. Health Perspectives* 104 (Suppl. 3), 655–657

Koivisto, P., Sorsa, M., Pacchierotti. F. & Peltonen, K. (1997) ^{32}P-Postlabelling/HPLC assay reveals an enantioselective adduct formation in N7 guanine residues *in vivo* after 1,3-butadiene inhalation exposure. *Carcinogenesis* 18, 439–443

Koivisto, P., Adler, I.D., Pacchierotti, F. & Peltonen, K. (1998) DNA adducts in mouse testis and lung after inhalation exposure to 1,3-butadiene. *Mutat. Res.* 397, 3–10

Krzyzosiak, W.J., Wiewiorowski, M., & Jaskolski, M. (1986) Chemical modification of adenine and cytosine residues with chloroacetaldehyde at the nucleoside and the tRNA level: The structural effect of chloroacetaldehyde modification. In: *The Role of Cyclic Nucleic Acid Adducts in Carcinogenesis and Mutagenesis* (Singer, B. & Bartsch, H., eds), IARC Scientific Publications, 70. Lyon, IARC, pp. 75–81

Kusmierek, J.T. & Singer, B. (1992) 1,N^2-Ethenodeoxyguanosine: Properties and formation in chloroacetaldehyde-treated polynucleotides and DNA. *Chem. Res. Toxicol.* 5, 634–638

Kusmierek, J.T., Jensen, D.E., Spengler, S.J., Stolarski, R. & Singer, B. (1987) Synthesis and properties of N^2,3-ethenoguanosine and N^2,3-ethenoguanosine 5′-diphosphate. *J. Org. Chem.* 52, 2374–2378

Latif, F., Moschel, R.C., Hemminki, K. & Dipple, A. (1988) Styrene oxide as a stereochemical probe for the mechanism of aralkylation at different sites on guanosine. *Chem. Res. Toxicol.* 1, 364–369

Lawley, P.D. & Jarman, M. (1972) Alkylation by propylene oxide of deoxyribonucleic acid, adenine, guanosine and deoxyguanylic acid. *Biochem. J.* 126, 893–900

Li, F., Segal, A. & Solomon, J.J. (1992) *In vitro* reaction of ethylene oxide with DNA and characterization of DNA adducts. *Chem.-Biol. Interactions* 83, 35–54

Loeb, L.A., & Preston, B.D. (1986) Mutagenesis by apurinic/apyrimidinic sites. *Ann. Rev. Genet.* 20, 201–230

Loechler, E.L., Green, C.L. & Essigmann, J.M. (1984) *In vivo* mutagenesis by O^6-methylguanine built into a unique site in a viral genome. *Proc. Natl Acad. Sci. USA* 81, 13–17

Macon, J.B. & Wolfenden, R. (1968) 1-Methyladenosine. Dimroth rearrangement and reversible reduction. *Biochemistry* 7, 3453–3458

Muller, M., Belas, F.J., Blair, I.A. & Guengerich, F.P. (1997) Analysis of 1,N^2-ethenoguanine and 5,6,7,9-tetrahydro-7-hydroxy-9-oxoimidazo[1,2-a]purine in DNA treated with 2-chlorooxirane by high performance liquid chromatography/electrospray mass spectrometry and comparison of amounts to other DNA adducts. *Chem. Res. Toxicol.* 10, 242–247

Plna, K. & Segerback, D. (1997) ^{32}P-Postlabelling of DNA adducts formed by allyl glycidyl ether *in vitro* and *in vivo*. *Carcinogenesis* 18, 1457–1462

Plna, K., Segerback, D. & Schweda, E.K.H. (1996) DNA adduct formation by allyl glycidyl ether. *Carcinogenesis* 17, 1465–1471

Qian, C. & Dipple, A. (1995) Different mechanisms of alkylation of adenosine at the 1- and N^6-positions. *Chem. Res. Toxicol.* 8, 389–395

Rios-Blanco M.N., Plna, K., Faller, T., Kessler, W., Hakansson, K., Kreuzer, P.E., Ranasinghe, A., Filser J.G., Segerback, D. & Swenberg, J.A. (1997) Propylene oxide: Mutagenesis, carcinogenesis and molecular dose. *Mutat. Res.* 380, 179–197

Savela, K., Hesso, A. & Hemminki, K. (1986) Characterization of reaction products between styrene oxide and deoxynucleosides and DNA. *Chem.-Biol. Interactions* 60, 235–246

Segal, A., Mate, U. & Solomon, J.J. (1979) *In vitro* Dimroth rearrangement of 1-(2-carboxyethyl)adenine to N^6-(2-carboxyethyl)-adenine in single-stranded calf thymus DNA. *Chem.-Biol. Interactions* 28, 333–344

Segal, A., Solomon, J.J. & Mukai, F. (1990) *In vitro* reactions of glycidol with pyrimidine bases in calf thymus DNA. *Cancer Biochem. Biophys.* 11, 59–67

Segerback, D. (1994) DNA alkylation by ethylene oxide and some mono-substituted epoxides. In: *DNA Adducts: Identification and Biological Significance* (Hemminki, K., Dipple, A., Shuker, D.E.G., Kadlubar, F.F., Segerback, D. & Bartsch, H., eds) IARC Scientific Publications No. 125. Lyon, IARC, pp. 37–47

Segerback, D., Calleman, C.J., Schroeder, J.L., Costa, L.G. & Faustman, E.M. (1995) Formation of N-7-(2-carbamoyl-2-hydroxyethyl)guanine in DNA of the mouse and the rat following intraperitoneal administration of [14C]acrylamide. *Carcinogenesis* 16, 1161–1165

Selzer, R.R., & Elfarra, A.A. (1996a) Characterization of N^1- and N^6-adenosine adducts and N^1-inosine adducts formed by the reaction of butadiene monoxide with adenosine: Evidence for the N^1-adenosine adducts as major initial products. *Chem. Res. Toxicol.* 9, 875–881

Selzer, R.R., & Elfarra, A.A. (1996b) Synthesis and biochemical characterization of N^1-, N^2-, and N^7-guanosine adducts of butadiene monoxide. *Chem. Res. Toxicol.* 9, 126–132

Selzer, R.R., & Elfarra, A.A (1997a) Characterization of four N-3-thymidine adducts formed *in vitro* by the reaction of thymidine and butadiene monoxide. *Carcinogenesis* 18, 1993–1998

Selzer, R.R., & Elfarra, A.A (1997b) Chemical modification of deoxycytidine at different sites yields adducts of different stabilities: Characterization of N3- and O^2-deoxycytidine and N3-deoxyuridine adducts of butadiene monoxide. *Arch. Biochem. Biophys.* 343, 63–72

Shuker, D.E. & Farmer, P.B. (1992) Relevance of urinary DNA adducts as markers of carcinogen exposure. *Chem. Res. Toxicol.* 5, 450–460

Singer, B. & Gruenberger, D. (1983) *Molecular Biology of Mutagens and Carcinogens*, New York, Plenum Press

Singh, U.S., Decker-Samuelian, K. & Solomon, J.J. (1996) Reaction of epichlorohydrin with 2′-deoxynucleosides: Characterization of adducts *Chem.-Biol. Interactions* 99, 109–128

Snow, E.T., Singh, J., Koenig, K.L. & Solomon, J.J. (1994) Propylene oxide mutagenesis at template cytosine residues. *Environ. Mol. Mutag.* 23, 274–280

Solomon, J.J. (1994) DNA adducts of lactones, sultones, acylating agents and acrylic compounds. In: *DNA Adducts: Identification and Biological Significance* (Hemminki, K., Dipple, A., Shuker, D.E.G., Kadlubar, F.F., Segerback, D. & Bartsch, H., eds), IARC Scientific Publications No. 125. Lyon, IARC, pp. 179–198

Solomon, J.J., Mukai, F., Fedyk, J. & Segal, A. (1988) Reactions of propylene oxide with 2′-deoxynucleosides and *in vitro* with calf thymus DNA. *Chem.-Biol. Interactions* 67, 275–294

Solomon, J.J., Singh, U.S. & Segal, A. (1993) *In vitro* reactions of 2-cyanoethylene oxide with calf thymus DNA. *Chem.-Biol. Interactions* 88, 115–135

Sowers, L.C., Sedwick, W.D. & Shaw, B.R. (1989) Hydrolysis of N^3–methyl-2′-deoxycytidine: Model compound for reactivity of protonated cytosine residues in DNA. *Mutat. Res.* 215, 131–138

Steiner, S., Crane, A.E. & Watson, W.P. (1992) Molecular dosimetry of DNA adducts in C3H mice treated with glycidaldehyde. *Carcinogenesis* 13, 119–124

Swenberg, J.A., Fedtke, N., Ciroussel, F., Barbin, A. & Bartsch, H. (1992) Etheno adducts formed in DNA of vinyl chloride-exposed rats are highly persistent in liver. *Carcinogenesis* 13, 727–729

Tretyakova, N.Yu., Chiang, S.Y., Walker, V.E. & Swenberg, J.A. (1998) Quantitative analysis of 1,3-butadiene-induced DNA adducts *in vivo* and *in vitro* using liquid chromatography electrospray ionization tandem mass spectrometry. *J. Mass Spectrom.* 33, 363–376

Walker, V.E., Fennel, T.R., Upton, P.B., Skopek, T.R., Prevost, V., Shuker, D.E. & Swenberg, J.A. (1992) Molecular dosimetry of ethylene oxide: Formation and persistence of 7-(2-hydroxyethyl)guanine in DNA following repeated exposures of rats and mice. *Cancer Res.* 52, 4328–4334

Windmueller, H.G. & Kaplan, N.O. (1961) The preparation and properties of N-hydroxyethyl derivatives of adenosine, adenosine triphosphate, and nicotinamide adenine dinucleotide. *J. Biol. Chem.* 236, 2716–2726

Zhang, W., Johnson, F., Grollman, A.P. & Shibutani, S. (1995) Miscoding by the exocyclic and related DNA adducts 3,N^4-etheno-2′-deoxycytidine, 3,N^4-ethano-2′-deoxycytidine and 3-(2-hydroxyethyl)-2′-deoxyuridine. *Chem. Res. Toxicol.* 8, 157–163

Exocyclic DNA Adducts in Mutagenesis and Carcinogenesis
B. Singer & H. Bartsch
IARC Scientific Publications No. 150
International Agency for Research on Cancer, Lyon, 1999

Chapter III. Formation from exogenenous and endogenous sources

Formation of etheno adducts and their effects on DNA polymerases

F.P. Guengerich, S. Langouët, A.N. Mican, S. Akasaka, M. Müller & M. Persmark

Etheno (ε) and related DNA adducts are formed from the reaction of certain bifunctional electrophiles with DNA. Our interest has been focused on oxiranes substituted with leaving groups, e.g. 2-chlorooxirane, the epoxide derived from the carcinogen vinyl chloride. The chemical mechanisms of the formation of the major etheno products derived from adenine, cytosine and guanine have been elucidated by nuclear magnetic resonance analysis and ^{13}C-labelled precursors. The amounts of all major etheno adducts have been quantified in DNA treated with 2-chlorooxirane by coupled high-performance liquid chromatography of nucleoside and base products. 1,*N*^{2}-ε-Gua, its formally hydrated but stable hemiaminal HO-ethanoGua (5,6,7,9-tetrahydro-7-hydroxy-9-oxoimidazo[1,2-*a*]purine) and 1,*N*^{2}-ethanoGua have all been inserted at a single site in oligonucleotides. All three of these bases block polymerases, cause misincorporations and produce some mutations in bacteria. The patterns of blockage and substitution vary among polymerases. In nucleotide excision repair-deficient *Escherichia coli*, 1,*N*^{2}-ε-Gua yielded a calculated 16% mutation frequency (base-pair substitutions) when the results were corrected for strand usage. 1,*N*^{2}-ε-Gua was also examined in Chinese hamster ovary cells with a stable integration system; the mutants are more complex than observed in bacteria and include rearrangements, deletions and base-pair substitutions other than at the adduct site.

Introduction

Our studies with etheno (ε) and related exocyclic DNA adducts began with vinyl chloride (Guengerich *et al.*, 1979, 1981a), a chemical that attracted considerable interest owing to the epidemiological evidence for human carcinogenicity obtained in the 1970s (Apfeldorf & Infante, 1981). Since then, we have studied the epoxides generated from vinyl chloride (Guengerich *et al.*, 1979), vinyl bromide (Guengerich *et al.*, 1981a), acrylonitrile (Guengerich *et al*, 1981b), vinyl carbamate (Guengerich & Kim, 1991) and *N*-nitroso-*N*-methylvinylamine (Okazaki *et al.*, 1993), all of which yield the typical array of etheno products in DNA. Most of these vinyl monomers are oxidized to the corresponding epoxides by cytochrome P450 2E1 (Guengerich *et al.*, 1991). The halogenated epoxides (i.e. 2-halooxiranes) rapidly rearrange to 2-haloacetaldehydes, which can also react with DNA to yield some of the same etheno products; however, the haloacetaldehyde reactions are considerably slower than those with the substituted oxiranes, and several lines of evidence indicate that the oxiranes are more important in DNA modification under relatively physiological settings (Guengerich *et al.*, 1981a, 1992).

The etheno adducts are also of interest because they can be derived from products of lipid peroxidation (4-hydroxynonenal) and components of chlorinated drinking-water (Sodum & Chung, 1988; Kronberg *et al.*, 1996). The chemistry of these reactions is less well understood than those of the simpler oxiranes; however, the burden of etheno DNA adducts from these sources may be of greater concern to the general population, because some of the etheno adducts appear to be relatively resistant to repair and are found at levels of $1/10^7$ DNA bases in human liver (J.A. Swenberg and F.P. Guengerich, unpublished data). Nevertheless, vinyl halides are reognized human carcinogens and can produce the same unusual liver tumours (haemangiosarcomas) in animal models; therefore, they continue to serve as excellent models for etheno adduct-related cancer.

Since the 1984 meeting on exocyclic adducts (Singer & Bartsch, 1986), considerable progress has been made in studying the chemistry of etheno adducts and in the development of analytical methods. What remains unclear is exactly how these adducts cause mutations and, possibly, cancer. Some of them have been incorporated into site-specific vectors and used in misincorporation experiments, e.g. 1,N^6-εAde (derived from adenine) and 3,N^4-εCyt (derived from cytosine) (Basu *et al.*, 1993). The results have shown considerable variation in different systems, and relatively little has yet been done with the guanine derivatives.

Chemistry of formation of etheno adducts

Intuitively, the establishment of the chemical mechanisms of etheno adduct formation would appear to be a relatively simple task, since the reacting oxiranes have only two carbons and the etheno products are relatively simple structures. Methods were developed for the analysis of all of the major known etheno and related adducts in DNA treated with 2-chlorooxirane (Müller *et al.*, 1997). The procedures involve chemical or enzymatic digestion and high-performance liquid chromatography (HPLC) coupled with sensitive detection methods. 1,N^6-εdAdo (derivative of adenosine) is highly fluorescent; some of the other adducts show weak fluorescence but can be analysed at the required level of sensitivity needed for these assays. Mass spectrometry was required to achieve sufficient sensitivity in some cases. The problem in analysis is clearly with the guanine reactions (Table 1). The dominant chemical reaction of 2-chlorooxirane is at the N7 atom, and the etheno products are very minor. Therefore, these adducts may be formed *via* what might normally be considered unfavoured pathways. The experiments presented in Table 1 were all done under conditions in which DNA was modified for only ~10 epoxide half-lives, in order to minimize reactions involving the rearrangement product 2-chloroacetaldehyde (Guengerich, 1992).

Other problems are the symmetrical nature of the 2-carbon unit incorporated into the etheno products and the lack of vicinal proton coupling for use in nuclear magnetic resonance assignments. A further problem is that synthesis of a substituted halooxirane with a single radioactive or stable atom label (e.g. ^{13}C) could be confounded by the rapid rearrangement to the 2-haloacetaldehyde, which would also slowly react (Guengerich, 1992).

Table 1. Adducts formed by treatment of DNA with 2-chlorooxirane

Adduct	pmol formed/μmol DNA base[a]	Method
N^7-(2-Oxoethyl)Gua	10 000	HPLC/fluorescence
1,N^6-ε-dAdo	500	HPLC/fluorescence
HO-ethanoGua (dGuo)	24–29	HPLC/MS/MS
N^2,3-ε-Gua	16	HPLC/fluorescence
3,N^4-ε-dCyd	7	HPLC/fluorescence
1,N^2-ε-Gua	1–2.5	HPLC/fluorescence, HPLC/MS/MS

From Müller *et al.* (1997); HPLC, high-performance liquid chromatography; MS, mass spectrometry; MS/MS, tandem electrospray

[a] DNA (calf thymus, 5 mg/ml) treated with 10 mmol/L 2-chlorooxirane

We addressed these issues with an approach suggested by W.G. Humphreys, who worked in this laboratory at the time (Guengerich *et al.*, 1993). 2,2-Dibromoethanol was synthesized with a single ^{13}C label under carefully controlled conditions by a Hell-Volhard-Zelinsky bromination. At pH 9.2, the alcohol was slowly converted to 2-bromooxirane, which rapidly reacted with nucleosides and did not give side-reactions associated with 2-bromoacetaldehyde (Guengerich & Raney, 1992). The conclusions reached in these studies (Guengerich & Raney, 1992; Guengerich *et al.*, 1993) are presented in Figure 1. With Ado and Cyd, the initial attack is by the ring nitrogen (Ado N1, Cyd N^4) on the unsubstituted methylene of the 2-halooxirane, followed by cyclization of the resulting aldehyde. The reactions leading to the formation of N^2,3- and 1,N^2-εGuo are very minor (Table 1) and reflect disfavoured reactions of ring Gua nitrogens on the halide-substituted methylene of the 2-halooxirane, followed by cyclization of the aldehyde with the N^2 atom to yield the cyclic product. It should be emphasized that these reactions cannot be explained by the rearrangement of 2-bromooxirane to 2-bromoacetaldehyde because we carried out such labelling experiments and found a different ^{13}C labelling pattern (Guengerich & Persmark, 1994).

In the course of studies on the mechanism of formation of 1,N^2-εGua, we identified HO-ethanoGua as a new reaction product, a stable hemiaminal that does not readily dehydrate to 1,N^2-εGua. This compound is believed to be formed by attack of the Gua N^2 atom on the unsubstituted methylene of the 2-halooxirane (Figure 1).

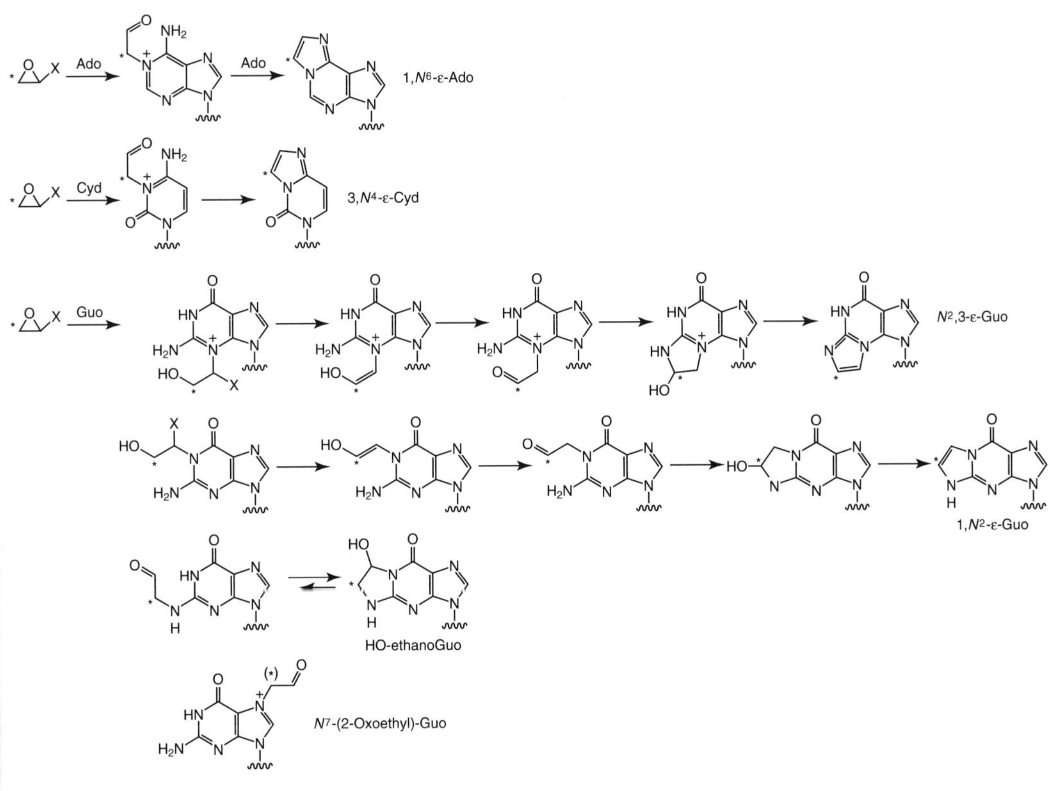

Figure 1. Mechanisms of formation of etheno and related adducts from 2-substituted oxiranes (Guengerich & Raney, 1992; Guengerich *et al.*, 1993)

One of the unexpected results we observed in studies with deuterium was the exchange of the H5 proton of N^2,3-εGuo in acid or at neutrality and the loss of H7 of 1,N^2-εdGuo at neutral pH (Figure 2). Viable mechanisms have been presented (Guengerich *et al.*, 1993). Recently, Marnett and his associates (Niedernhofer *et al.*, 1997) reported the pH-dependent ring opening of the six-membered ring homologue of 1,N^2-εGua, pyrimido[1,2-*a*]purin-10-(3*H*)-one (M_1Gua) (Figure 2). The base-catalysed exchange of the H7 proton of 1,N^2-εGua could also be explained by such an equilibrium, which is actually a stepwise variation of Scheme IIIB of Guengerich *et al.* (1993). HO-ethanodGuo (Figure 1) was shown to exchange the 7-OH oxygen atom with $H_2{}^{18}O$, indicating reversible ring opening (Langouët *et al.*, 1997).

Interactions of 1,*N*²-ethenoguanine, HO-ethanoguanine and ethanoguanine with polymerases

Methods were developed for the synthesis of phosphoramidite derivatives of 1,N^2-εGua, HO-ethanoGua and ethanoGua (the saturated derivative of 1,N^2-εGua) and their incorporation into oligonucleotides at defined positions (Langouët *et al.*, 1997, 1998). Purity and identity were rig orously established by capillary gel electrophore sis, HPLC and mass spectrometry. All three of th adducts strongly blocked all of the polymerase tested, including *Escherichia coli* polymerase (pol) exo⁻ (Klenow fragment, Kf⁻), *E. coli* pol I exo⁻, bacteriophage T7 exo⁻, human immunode ficiency virus-1 reverse transcriptase and rat po β. Kf⁻ was more inclined to extend the primer b one base than the other polymerases (Figure 3) EthanoGua was a particularly strong block, an HO-ethanoGua tended to be more blocking tha 1,N^2-εGua (Langouët *et al.*, 1997).

Sequence analysis of products of the prime extension incubations with Kf⁻ indicated exten sive incorporation of G into both the one-bas extended and full-length products (Langouët *e al.*, 1997, 1998). Some T and A, as well as C, wer incorporated into the full-length product with Kf and pol II⁻. The incorporation patterns differed as least qualitatively, depending on the poly merase used; e.g. pol II⁻ and T7⁻ yielde 'frameshift' products (one- and two-base dele tions) (Langouët *et al.*, 1997).

The misincorporation of deoxyribonucleo sides-5′-triphosphates (dNTPs) opposite th

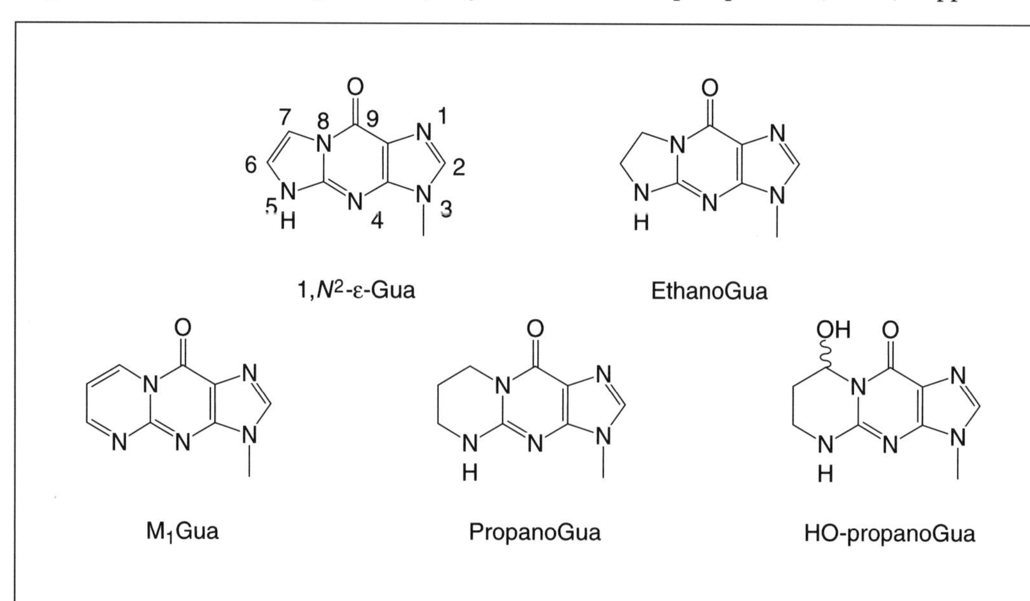

Figure 2. Structures of modified guanine bases used in site-specific misincorporation and mutagenesis assays (see Figure 1 for other structures used)

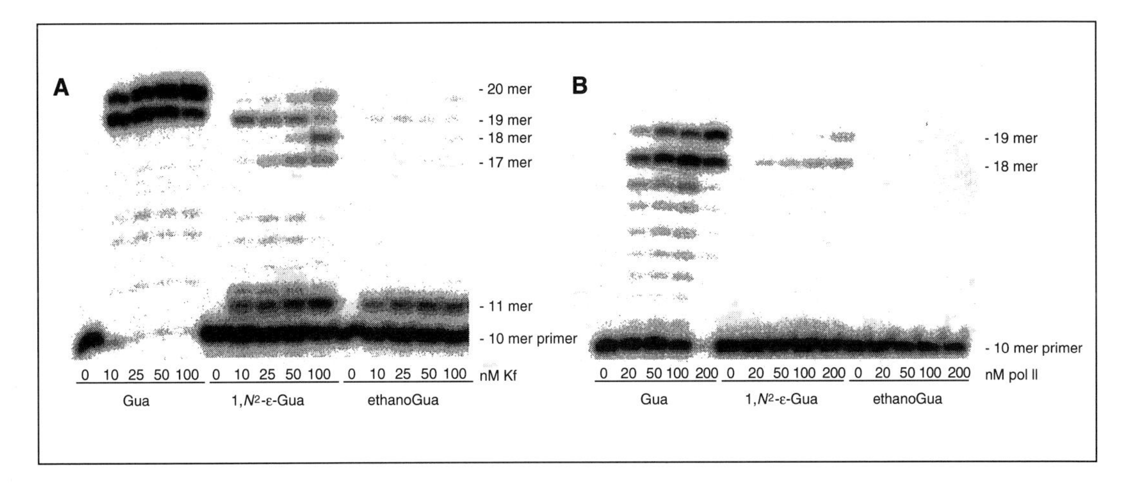

Figure 3. Blockage of polymerases by 1,N^2-ε-Gua and Gua (Langouët *et al.*, 1997)

The 5′ ^{32}P-end-labelled primer 5′ TCAATTCGGA (10 mer) was annealed with the 19-mer 3′ AGTTAAGCCTG*TGGGTGAC, where G* = Gua, 1,1,N^2-ε-Gua or ethanoGua, as indicated (Langouët *et al.*, 1997). Incubations were done in the presence of mixtures of dATP, dCTP, dGTP and dTTP for 30 min with the indicated current ratios of (A) *E. coli* polymerase I (Kf$^-$) or (B) *E. coli* polymerase II$^-$ (both exonuclease minus enzymes).

adducts in the template was analysed in more detail by steady-state kinetics and the general approach of Goodman (Boosalis *et al.*, 1987) with Kf$^-$, pol II$^-$ and T7$^-$ (Langouët *et al.*, 1997, 1998). The results with Kf$^-$ are summarized in Table 2. The kinetics of incorporation was greatly retarded by the substitution of any of the adducts for Gua, as suggested by the results of extension studies in the presence of all four dNTPs (Figure 3). The rate of incorporation of dCTP was retarded by several orders of magnitude when Gua was replaced by any of the three derivatives. With Kf$^-$, insertion of dGTP became the most favoured across all adducts examined. Even though the k_{cat}:K_m ratio is less favourable for each of these misincorporations of dGTP than at Gua, the decreased efficiency with dCTP has the effect of increasing the misincorporation ratio.

As the misincorporation ratios varied with the polymerases examined (Langouët *et al.*, 1997,

Table 2 .Steady-state parameters for misincorporation by Kf$^-$ (Langouët *et al.*, 1997, 1998)

	dNTP							
	k_{cat}/K_m, min^{-1} $mmol/L^{-1}$				Misincorporation ratio [a]			
Template base	A	C	G	T	A	C	G	T
Gua	ND	364	0.010	0.006	ND	—	3×10^{-5}	2×10^{-5}
1,N^2-ε-Gua	1.6×10^{-3}	8×10^{-3}	9×10^{-3}	ND	0.2	—	1.1	ND
HO-ethanoGua	0.024	0.056	8.5×10^{-3}	ND	0.45	—	0.15	ND
EthanoGua	ND	1.0×10^{-6}	5.6×10^{-5}	1.2×10^{-5}	ND	—	53	11

ND, no detectable incorporation (see Langouët *et al.*, 1997)
[a]Calculated as (k_{cat}/K_m)dNTP/(k_{cat}/K_m)dCTP, where dNTP ≠ dCTP (Boosalis *et al.*, 1987)

1998), misincorporation must be considered a kinetic phenomenon, which cannot be understood only in terms of thermodynamically favourable mispairing. The deletions observed in the full-length extension studies can be rationalized in terms of a 'slipped intermediate' model, at least for the two-base deletions (Langouët *et al.*, 1997), in which the retardation of incorporation allows time for the primer–template pair to re-orientate in the polymerase (Figure 4). It should be noted that all of the results are from steady-state experiments, and the meaning of the parameters k_{cat} and K_m in these systems is not well understood (Langouët *et al.*, 1997). In such experiments, release of the oligonucleotide and binding of the next can dominate kinetics (Johnson, 1995). We suggest that k_{cat} probably reflects the rate of formation of the phosphodiester bond in the cases considered here, simply because it is so slow. Whether the chemical reaction itself or a step involving conformational change limits the reaction is unclear. In other work, we have developed pre-steady-state kinetic approaches (Lowe & Guengerich, 1996), but these might not be useful here if the first-cycle burst were obliterated.

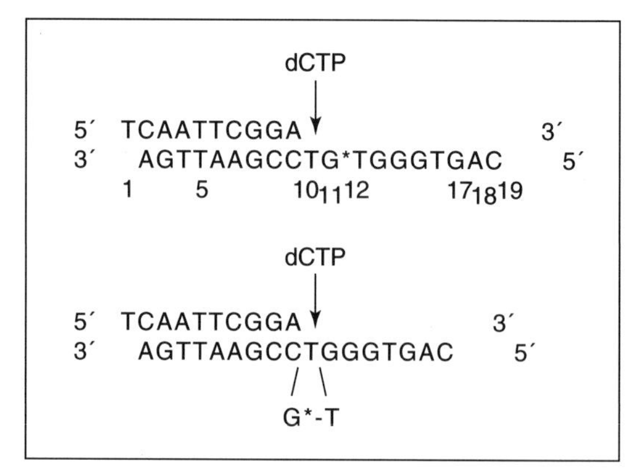

Figure 4. Postulated template slippage in Kf⁻ (Langouët *et al.*, 1997)

Analysis of mutations produced by etheno adducts

Two systems have been used for the analysis of mutants generated in cells. The first involves a system identical to that developed by Marnett and used in the analysis of propanoGua and M_1Gua mutants (Fink *et al.*, 1997), allowing direct comparisons to be made without regard to factors such as sequence context. The mutants seen opposite 1,N^2-εGua and ethanoGua were highly dependent on the expression of UvrA, an essential protein in nucleotide excision repair, and also appeared to require functional SOS response. HO-ethanoGua was not very mutagenic. In the absence of UvrA, both 1,N^2-ε-Gua and ethanoGua produced G→A and G→T substitutions, 1,N^2-εGua being more mutagenic after correction for strand usage (Table 3). The mutation frequency with 1,N^2-εGua is similar to that reported for the six-membering ring homologue M_1Gua (Fink *et al.*, 1997). It should also be noted that these two mutagenic events correspond to insertion of T and A opposite the modified Gua adducts, events that were seen more often with pol II⁻ than with Kf⁻ *in vitro* (Langouët *et al.*, 1997, 1998).

The results of bacterial mutagenicity experiments do not always reflect events seen in mammalian cells, and major differences have been noted with 3,N^4-εCyt (Moriya *et al.*, 1994). We examined events related to replication of a sequence containing 1,N^2-εGua in Chinese hamster ovary (CHO) cells in culture (Akasaka & Guengerich, 1999). A system developed by Essigmann and his associates was used, in which a plasmid with 1,N^2-εGua incorporated at a single site is inserted into the chromosome (Ellison *et al.*, 1989). This approach, although technically more

Table 3 .Mutation frequencies in *Escherichia coli* LM103 (uvrA⁻) (Langouët *et al.*, 1998)

Base	G→ A (%)	G→ T(%)	G→ C (%)	Total corrected for strand usage (%)
1,N^2-ε-Gua	2.0	0.74	0.09	16
HO-ethanoGua	0.11	0.19	0.07	0.8
EthanoGua	0.71	0.71	< 0.2	2.2

demanding than the use of shuttle vectors, has the advantage of chromosomal integration and avoids any ambiguity in replication differences. Both wild-type and UV-5 cells (deficient in repair) were used. The mutant frequency could be estimated by recovery of the DNA, amplification of the integrated section by the polymerase chain reaction and quantification of the fraction resistant to *Nhe*I, which cuts at the unmodified site. The mutant frequency was increased by 10-fold in the wild-type cells when 1,N^2-εGua was inserted in place of Gua. The mutant frequency was even higher in the UV-5 cells, but the background is considerably higher.

The cells with the pCNheIA vector integrated into the chromosome were selected by resistance to G418. DNA was isolated from the integrated cells, and appropriate 288-base-pair fragments (which included the 1,N^2-εGua-inserted site) were amplified by polymerase chain reaction. Mutants were selected by resistance to the restriction enzyme *Nhe*I. Sequence analysis of the mutants isolated from the cells transfected with the control oligonucleotide showed only G→A transitions at positions 3 and 7 in the sequence (Table 4). The mutants isolated from cells transfected with the vector containing the 1,N^2-εGua adduct yielded a variety of mutations, including deletions, rearrangements and base-pair mutations at sites removed from 1,N^2-εGua.

The mutant frequency in the AA8 cells replicating 1,N^2-εGua was considerably higher than for Gua in these cells, and the mutant sequences presented in Table 2 can be considered representative. With the UV-5 cells, the background mutant frequency is higher, but, as with the AA8 cells, the mutant sequences were quite varied and differed considerably from the simple base-pair substitutions obtained with the unmodified vectors, even though the number of mutants analysed was small.

The results of analysis of the CHO mutants suggests that 1,N^2-εGua produces complex mutations in mammalian cell systems. These may be related to the strong blocking effects on polymerases (Langouët *et al.*, 1997, 1998). The possibility of ring-opening (see mention regarding M_1Gua adduct above) and the involvement of N^1-(2-oxoethyl)Gua (ring-opened form (Langouët *et al.*, 1998)) cannot be ruled out. The possibility exists that complex mutations may also be produced in the bacterial assays (Langouët *et al.*, 1998) but are not observed because of the screening system used.

Table 4. Mutations induced by 1,N^2-ε-Gua-containing pCNheIA inserted into Chinese hamster ovary cell genome

Plasmid G* =	UV background	Mutation rate, *Nhe* I	Sequence at target site	Change
G			ATGCTA**G** CAT	
Gua	AA8 (Wild type)	0.36%	ATGCTA**A**CAT	G_7'A
1,N^2-ε-Gua	AA8	4.6%	**T**TG**G**TAGCAT	C_4'G, A1'T
			ATGCTA**A**CAT	G_7'A
			ATGCTAGC---	Rearrangement
			ATG-------	Rearrangement
			AT**A**CTAGCAT	G_3'A
Gua	UV-5 (UV-sensitive)	4.5%	AT**A**CTAGCAT (×3)	G_3'A
1,N^2-ε-Gua	UV-5	7.8%	ATGCTAG**A**AT	C_8'A
			ATGC**AA**----	T_5'A, 4-base deletion
			ATGCTAGC--	Rearrangement
			ATGCTAG---	Rearrangement
			ATG----	Rearrangement
			ATGCTAGCA-	Rearrangement
			ATGCTA**T**CAT	G_7'T

From Akasaka & Guengerich (1999)

Conclusions

The chemistry of the formation of etheno DNA adducts from vinyl monomers is now known in detail, as is much of the enzymatic formation of the epoxide intermediates, which is dominated by cytochrome P450 2E1. How such DNA adducts are derived from other chemical sources is still not as clear. All DNA adducts in this group that have been considered to date can show effects on polymerases, either in the isolated form or in cells. The rank order of the biological properties seems to vary depending on the particular system, and these activities must be considered in the light of the relative abundance of the adducts in DNA. Exactly how these adducts relate to cancer is not yet clear, although the high levels of some adducts found in humans and their demonstrated biological effects suggest a number of possibilities. The high fraction of normal incorporation past these lesions, in which most of the normal base-pairing region is blocked, is an intriguing observation and suggests that other forces are involved in recognition.

References

Akasaka, S. & Guengerich, F.P. (1999) Mutagenicity of site-specifically located 1,N^2-ethenoguanine in Chinese hamster ovary cell chromosomal DNA. *Chem. Res. Toxicol.* 12, 501–507

Apfeldorf, R. & Infante, P.F. (1981) Review of epidemiologic study results of vinyl chloride-related compounds. *Environ. Health Perspectives* 41, 221–226

Basu, A.K., Wood, M.L., Niedernhofer, L.J., Ramos, L.A. & Essigmann, J. M. (1993) Mutagenic and genotoxic effects of three vinyl chloride-induced DNA lesions: 1,N^6-ethenoadenine, 3,N^4-ethenocytosine, and 4-amino-5-(imidazol-2-yl)imidazole. *Biochemistry* 32, 12793–12801

Boosalis, M.S., Petruska, J. & Goodman, M.F. (1987) DNA polymerase insertion fidelity: Gel assay for site-specific kinetics. *J. Biol. Chem.* 262, 14689–14696

Ellison, K.S., Dogliotti, E., Connors, T.D., Basu, A.K. & Essigmann, J.M. (1989) Site-specific mutagenesis by O^6-alkylguanines located in the chromosomes of mammalian cells: Influence of the mammalian O^6-alkylguanine-DNA alkyltransferase. *Proc. Natl Acad. Sci. USA* 86, 8620–8624

Fink, S.P., Reddy, G.R. & Marnett, L.J. (1997) Mutagenicity in *Escherichia coli* of the major DNA adduct derived from the endogenous mutagen, malondialdehyde. *Proc. Natl. Acad. Sci. USA* 94, 8652–8657

Guengerich, F.P. (1992) Roles of the vinyl chloride oxidation products 1-chlorooxirane and 2-chloroacetaldehyde in the *in vitro* formation of etheno adducts of nucleic acid bases. *Chem. Res. Toxicol.* 5, 2–5

Guengerich, F.P. & Kim, D.-H. (1991) Enzymatic oxidation of ethyl carbamate to vinyl carbamate and its role as an intermediate in the formation of 1,N^6-ethenoadenosine. *Chem. Res. Toxicol.* 4, 413–421

Guengerich, F.P. & Persmark, M. (1994) Mechanism of formation of ethenoguanine adducts from 2-haloacetaldehydes: ^{13}C labeling patterns of 2-bromoacetaldehyde. *Chem. Res. Toxicol.* 7, 205–208

Guengerich, F.P. & Raney, V.M. (1992) Formation of etheno adducts of adenosine and cytidine from 1-halooxiranes. Evidence for a mechanism involving initial reaction with the endocyclic nitrogens. *J. Am. Chem. Soc.* 114, 1074–1080

Guengerich, F.P., Crawford, W.M., Jr & Watanabe, P.G. (1979) Activation of vinyl chloride to covalently bound metabolites: Roles of 2-chloroethylene oxide and 2-chloroacetaldehyde. *Biochemistry* 18, 5177–5182

Guengerich, F.P., Mason, P.S., Stott, W.T., Fox, T.R. & Watanabe, P.G. (1981a) Roles of 2-haloethylene oxides and 2-haloacetaldehydes derived from vinyl bromide and vinyl chloride in irreversible binding to protein and DNA. *Cancer Res.* 41, 4391–4398

Guengerich, F.P., Geiger, L.E., Hogy, L.L. & Wright, P.L. (1981b) *In vitro* metabolism of acrylonitrile to 2-cyanoethylene oxide, reaction with glutathione, and irreversible binding to proteins and nucleic acids. *Cancer Res.* 41, 4925–4933

Guengerich, F.P., Kim, D.-H. & Iwasaki, M. (1991) Role of human cytochrome P-450 IIE1 in the oxidation of many low molecular weight cancer suspects. *Chem. Res. Toxicol.* 4, 168–179

Guengerich, F.P., Persmark, M. & Humphreys, W.G. (1993) Formation of 1,N^2- and N^2,3-ethenoguanine derivatives from 2-halooxiranes: isotopic labeling studies and formation of a hemiaminal derivative of N^2-(2-oxoethyl)guanine. *Chem. Res. Toxicol.* 6, 635–648

Johnson, K.A. (1995) Rapid quench kinetic analysis of polymerases, adenosinetriphosphates, and enzyme intermediates. *Meth. Enzymol.* 249, 38–61

Kronberg, L., Asplund, D., Mäki, J. & Sjöholm, R. (1996) Reaction of mucochloric and mucobromic acids with adenosine and cytidine: Formation of chloro- and bromopropenal derivatives. *Chem. Res. Toxicol.* 9, 1257–1263

Langouët, S., Mican, A.N., Müller, M., Fink, S.P., Marnett, L.J., Muhle, S.A. & Guengerich, F.P. (1998) Misincorporation of nucleotides opposite three 5-membered exocyclic ring guanine derivatives: 1,N^2-ethenoguanine, 5,6,7,9-tetrahydro-9-oxoimidazo[1,2-*a*]purine, and 5,6,7,9-tetrahydro-7-hydroxy-9-oxoimidazo[1,2-*a*]purine. *Biochemistry* 37, 5184–5193

Langouët, S., Müller, M. & Guengerich, F.P. (1997) Misincorporation of dNTPs opposite 1,N^2-ethenoguanine and 5,6,7,9-tetrahydro-7-hydroxy-9-oxoimidazo[1,2-*a*]purine in oligonucleotides by *Escherichia coli* polymerases I exo$^-$ and II exo$^-$, T7 polymerase exo$^-$, human immunodeficiency virus-1 reverse transcriptase, and rat polymerase β. *Biochemistry* 36, 6069–6079

Lowe, L.G. & Guengerich, F.P. (1996) Steady-state and pre-steady-state kinetic analysis of dNTP insertion opposite 8-oxo-7,8-dihydroguanine by *Escherichia coli* polymerases I exo$^-$ and II exo$^-$. *Biochemistry* 35, 9840–9849

Moriya, M., Zhang, W., Johnson, F. & Grollman, A.P. (1994) Mutagenic potency of exocyclic DNA adducts: Marked differences between *Escherichia coli* and simian kidney cells. *Proc. Natl Acad. Sci. USA* 91, 11899–11903

Müller, M., Belas, F.J., Blair, I.A. & Guengerich, F.P. (1997) Analysis of 1,N^2-ethenoguanine and 5,6,7,9-tetrahydro-7-hydroxy-9-oxoimidazo[1,2-*a*]purine in DNA treated with 2-chlorooxirane by high performance liquid chromatography/mass spectrometry and comparison of amounts with other adducts. *Chem. Res. Toxicol.* 10, 242–247

Niedernhofer, L.J., Riley, M., Schnetz-Boutaud, N., Sanduwaran, G., Chaudhary, A.K., Reddy, G.R. & Marnett, L.J. (1997) Temperature-dependent formation of a conjugate between tris(hydroxymethyl)aminomethane buffer and the malondialdehyde–DNA adduct pyrimidopurinone. *Chem. Res. Toxicol.* 10, 556–561

Okazaki, O., Persmark, M. & Guengerich, F.P. (1993) *N*-Nitroso-*N*-methylvinylamine: Reaction of the epoxide with guanyl and adenyl moieties to yield adducts derived from both parts of the molecule. *Chem. Res. Toxicol.* 6, 168–173

Singer, B. & Bartsch, H., eds (1986) *The Role of Cyclic Nucleic Acid Adducts in Carcinogenesis and Mutagenesis*, IARC Scientific Publications No. 70. Lyon, IARC

Sodum, R.S. & Chung, F.L. (1988) 1,N^2-Ethenodeoxyguanosine as a potential marker for DNA adducts formation by *trans*-4-hydroxy-2-nonenal. *Cancer Res.* 48, 320–323

Exocyclic DNA Adducts in Mutagenesis and Carcinogenesis
B. Singer & H. Bartsch
IARC Scientific Publications No. 150
International Agency for Research on Cancer, Lyon, 1999

Reactions of α-acetoxy-*N*-nitroso-pyrrolidine and crotonaldehyde with DNA

S.S. Hecht, P. Upadhyaya & M. Wang

α-Acetoxy-*N*-nitrosopyrrolidine (α-acetoxyNPYR) is a stable precursor to α-hydroxyNPYR, the initial product of metabolism and proposed proximate carcinogen of NPYR. Crotonaldehyde (2-butenal) is a metabolite of NPYR and also a mutagen and carcinogen. Both α-acetoxyNPYR and crotonaldehyde are known to form DNA adducts, but these reactions have not been completely characterized. In previous studies, we detected substantial amounts of unidentified radioactivity in hydrolysates of DNA that had been reacted with radiolabelled α-acetoxyNPYR. We have now characterized these products as 2-hydroxytetrahydrofuran, the cyclic form of 4-hydroxybutanal, and paraldol, the dimer of 3-hydroxybutanal. They were characterized by comparison with standards and by comparison of their derived 2,4-dinitrophenylhydrazones with standards. [^{3}H]H_2O was also identified. 2-Hydroxytetrahydrofuran is the major product in neutral thermal hydrolysates of α-acetoxyNPYR-treated DNA and is derived predominantly from N^2-(tetrahydro-furan-2-yl)deoxyguanosine 8. Paraldol is present to a lesser extent than 2-hydroxytetrahydrofuran in these reactions and is formed from paraldol-releasing adducts, which in turn are produced by the reaction of crotonaldehyde or paraldol, solvolysis products of α-acetoxyNPYR, with DNA. Paraldol is a major product in hydrolysates of crotonaldehyde-treated DNA, being present in amounts 100 times greater than those of previously identified adducts. These results provide a more complete picture of the reactions of α-acetoxyNPYR with DNA and yield some new insights on possible endogenous DNA adducts formed from crotonaldehyde.

Introduction

N-Nitrosopyrrolidine (NPYR) is a well-established hepatocarcinogen in rats and a respiratory carcinogen in Syrian golden hamsters (Preussmann & Stewart, 1984). It occurs in the human diet and in tobacco smoke and is a probable product of nitric oxide-mediated endogenous nitrosation in humans (Tricker *et al.*, 1991a; Hoffmann & Hoffmann, 1997; Wang & Hecht, 1997). Indeed, NPYR has been detected in human urine (Tricker *et al.*, 1991b; Mostafa *et al.*, 1994; van Maanen *et al.*, 1996). NPYR is representative of the class of cyclic nitrosamines, which includes a number of other important carcinogens such as *N*-nitrosopiperidine, *N*-nitrosomorpholine and *N*′-nitrosonornicotine. In this paper, we describe our research on DNA adduct formation from α-acetoxyNPYR and crotonaldehyde; both compounds react with DNA to yield cyclic adducts.

One goal of our work is to better understand the role of specific DNA adducts in organ-specific carcinogenesis by cyclic nitrosamines. There are some striking structure–activity relationships among these compounds. For example, NPYR induces only liver tumours in rats; oesophageal tumours have never been observed (Preussmann & Stewart, 1984; Gray *et al.*, 1991). In contrast, *N*-nitrosopiperidine induces mainly oesophageal tumours under the same conditions (Gray *et al.*, 1991; Lijinsky & Reuber, 1981). It is likely that these differences in carcinogenicity relate to differences in metabolic activation and DNA adduct formation between these structurally related nitrosamines. Our second major goal is to develop a DNA-based biomarker as an indicator of NPYR formation and metabolic activation in humans. As NPYR is extensively metabolized *in vivo* (Hecht *et al.*, 1981), detection of the parent compound is not a practical or useful biomarker. Moreover, its major urinary metabolites undergo extensive further metabolism. A DNA adduct or hydrolysis product could provide a collective measure of NPYR formation and metabolic activation.

The central role of α-hydroxyNPYR in the metabolic activation of NPYR was first demonstrated in our laboratory 20 years ago (Hecht *et al.*, 1978). We have used α-acetoxyNPYR as a stable precursor to this important intermediate and have carried out studies of its reactions with DNA. α-HydroxyNPYR is unstable and spontaneously ring-opens to 4-oxobutanediazohydroxide (**2**, Figure 1). The two electrophilic centres in this intermediate lead to complex adduct chemistry, including the formation of cyclic DNA adducts. While adduct formation from simple acyclic nitrosamines such as *N*-nitrosdimethylamine and *N*-nitrosodiethylamine is well characterized and results from predictable chemistry of alkanediazonium ion reactions with DNA bases, adduct formation from cyclic nitrosamines such as NPYR is more complex.

Figure 1 summarizes our current understanding of DNA adduct formation from α-acetoxyNPYR.

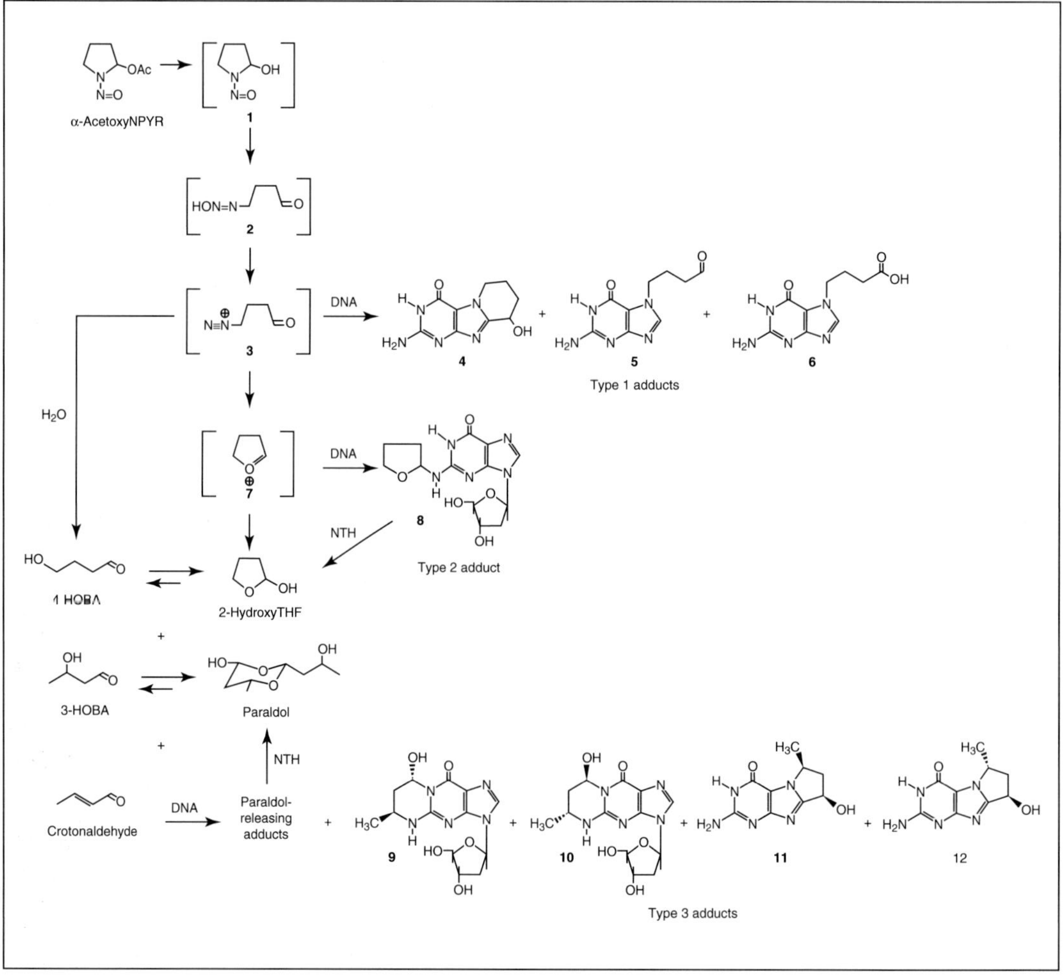

Figure 1. Products formed in the solvolysis of α-acetoxy-*N*-nitrosopyrrolidine (NPYR) and in its reactions with DNA

4-HOBA, 4-hydroxybutanal; THF, tetrahydrofuran; NTH, neutral thermal hydrolysis.

We have somewhat arbitrarily classified adducts as types 1, 2 or 3. Type 1 adducts result from the reaction of intermediate **2** or **3** with the N7 position of guanine, in much the same way that simple alkylating diazonium ions react. Adducts **4–6** are formed in this way; adduct **4** is the major product among these (Chung *et al.*, 1989a; Wang *et al.*, 1989; Wang & Hecht, 1997). Its cyclic structure is a clear result of the doubly electrophilic intermediate **3**. Shank and colleagues first identified a fluorescent adduct of NPYR in rat liver, but the structure was unknown (Hunt & Shank, 1982). We identified this adduct as **4**, and Shank and co-workers carried out studies on its formation and persistence in NPYR-treated rats (Hunt & Shank, 1991). Adduct **5** has also been detected in rat liver after NPYR treatment (Wang *et al.*, 1992). Type 2 adducts result from reactions with DNA of the cyclic oxonium ion **7** and related intermediates. The major adduct formed in this reaction is N^2-(tetrahydrofuran-2-yl)deoxyguanosine (**8**) (Young-Sciame *et al.*, 1995; Wang *et al.*, 1995). Type 3 adducts are formed from stable products of α-acetoxyNPYR hydrolysis, such as crotonaldehyde (Wang *et al.*, 1988). The cyclic 1,N^2-propanodeoxyguanosine adducts **9** and **10** were first characterized in our laboratory, along with the *N*7,C8 cyclic adducts **11** and **12** (Chung & Hecht, 1983; Chung *et al.*, 1984, 1989a). Others have made similar observations and have isolated bis-adducts of crotonaldehyde (Eder & Hoffman, 1992). Analysis of rat liver DNA after treatment with NPYR demonstrates the presence of small amounts of adducts **9** and **10** (Chung *et al.*, 1989b). An important series of studies by Chung and Nath showed that these adducts are present in human and laboratory animal DNA (Nath & Chung, 1994; Chung *et al.*, 1996; Nath *et al.*, 1996, 1998). They are apparently formed *via* generation of crotonaldehyde in lipid peroxidation, although the role of NPYR in this process has not been excluded.

In spite of what appears to be a relatively complete characterization of DNA adduct formation from α-acetoxyNPYR, chromatograms of neutral thermal hydrolysates of DNA that had been reacted with radiolabelled α-acetoxyNPYR showed substantial amounts of unidentified material. One of these chromatograms is illustrated in Figure 2; P9 and P12 were uncharacterized. A goal of our recent studies has therefore been to characterize these unknown peaks.

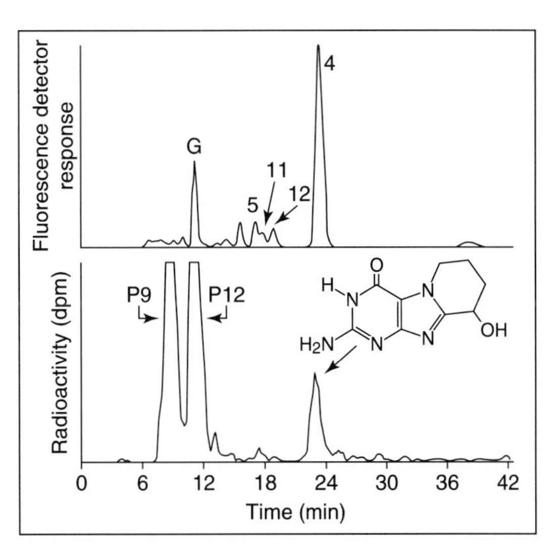

Figure 2. High-performance liquid chromatography trace obtained upon analysis of a neutral thermal hydrolysate of DNA treated with [^{3}H]α-acetoxy-*N*-nitrosopyrrolidine

Upper panel, fluorescence detection of standards; lower panel, radioflow detection of products; G, guanine. Numbers in upper panel refer to structures in Figure 1.

Results

Preliminary data indicated that P9 and P12 shown in Figure 2 had the same retention times and formed the same 2,4-dinitrophenylhydrazones (2,4-DNPs) as two products of the solvolysis of α-acetoxyNPYR—paraldol, the dimer of 3-hydroxybutanal (3-HOBA), and 2-hydroxytetrahydrofuran (2-hydroxyTHF), the cyclic form of 4-HOBA, respectively (Figure 1). DNA that had been reacted with α-acetoxyNPYR was precipitated, washed once with 70% ethanol and twice with ethanol, then dried and subjected to neutral thermal hydrolysis. The hydrolysates were analysed by strong cation exchange high-performance liquid chromatography (HPLC). Material that eluted at the retention times of paraldol and 2-hydroxyTHF was collected, derivatized with 2,4-DNP reagent and analysed by HPLC. The results showed that the 2,4-DNPs were identical to standard 3-HOBA-2,4-DNP and crotonaldehyde-2,4-DNP (from paraldol) and 4-HOBA-2,4-DNP (from 2-hydroxyTHF).

The same DNA hydrolysates were analysed by reversed-phase HPLC, and material that eluted at the retention times of paraldol and 2-hydroxyTHF was collected and analysed by gas chromatography–mass spectrometry (MS). The MS patterns of standard paraldol and 2-hydroxyTHF and the corresponding isolated products were essentially identical (Figure 3A,B). These results demonstrate that paraldol and 2-hydroxyTHF are products of the neutral thermal hydrolysis of α-acetoxyNPYR-treated DNA.

DNA reacted with α-acetoxyNPYR was also hydrolysed enzymatically, and these hydrolysates were treated with 2,4-DNP reagent. The results of this analysis demonstrated the presence of 3-HOBA-2,4-DNP, crotonaldehyde-2,4-DNP and 4-HOBA-2,4-DNP, as in the neutral thermal hydrolysates.

It seemed likely that the source of the paraldol-releasing adducts in α-acetoxyNPYR-treated DNA was crotonaldehyde, which is known to be a product of the solvolysis of α-acetoxyNPYR and is known to react with DNA. Therefore, we investigated the reaction of crotonaldehyde with DNA. This DNA was extensively purified, subjected to either neutral thermal or enzyme hydrolysis and derivatized with 2,4-DNP reagent. The results of this analysis demonstrated the presence of 3-HOBA-2,4-DNP and crotonaldehyde-2,4-DNP, as in the studies described above.

Products of the neutral thermal or enzymatic hydrolysis of DNA reacted with either α-acetoxyNPYR or crotonaldehyde are shown in Table 1. 2-HydroxyTHF is the major product of both neutral thermal and enzyme hydrolysis of DNA reacted with α-acetoxyNPYR. The sum of the amounts of 2-hydroxyTHF and the N^2-tetrahydrofuranyl adduct **8** in the enzyme hydrolysates is similar to the amount of 2-hydroxyTHF in the neutral thermal hydrolysate, consistent with the known conversion of **8** to 2-hydroxyTHF under neutral thermal hydrolysis conditions. The levels of paraldol were significantly lower than those of 2-hydroxyTHF in hydrolysates of α-acetoxyNPYR-treated DNA.

In hydrolysates of DNA reacted with crotonaldehyde, the amount of paraldol is more than 100 times greater than that of adducts **9–12**, previously characterized in these reactions (Table 1). The levels of paraldol are greater in neutral ther-

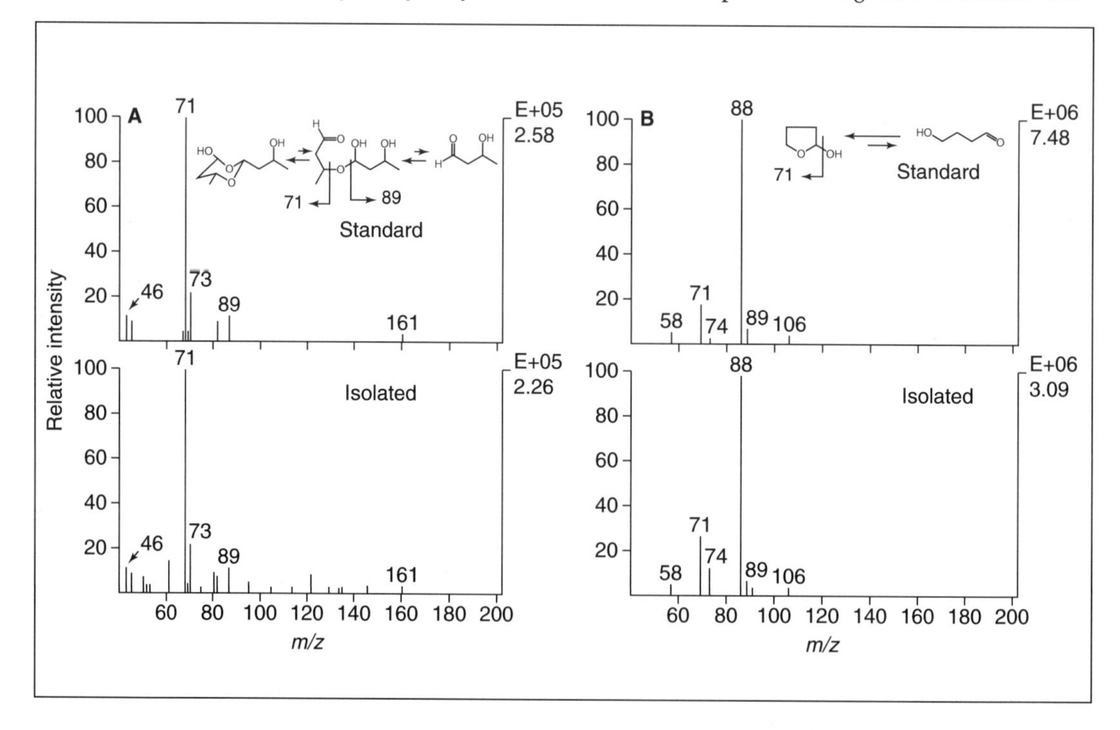

Figure 3. Positive chemical ionization–mass spectrometry of standard and isolated paraldol (A) and 2-hydroxytetrahydrofuran (B)

Table 1. Quantification of products formed in the reactions of α-acetoxy-*N*-nitrosopyrrolidine (α-acetoxyNPYR) or crotonaldehyde with DNA

Reactant	Hydrolysis conditions	Products (mmol/mol G or dG)[a]						
		Paraldol[b]	2-HydroxyTHF[b]	4[c]	8	9 + 10[d]	11	12
α-AcetoxyNPYR	Neutral thermal or acid[e]	9.9 ± 2.1	54 ± 3.5	24.3 ± 4.4	ND	1.8 ± 0.2	0.07 ± 0.01	0.08 ± 0.01
	Enzyme	4.1 ± 0.8	33 ± 0.6	NQ	27 ± 2.3	NQ	NQ	NQ
Crotonaldehyde	Neutral thermal or acid[e]	9.6 ± 3.0	ND	ND	ND	0.07 ± 0.04	0.033 ± 0.006	0.043 ± 0.006
	Enzyme	4.3 ± 0.5	ND	ND	ND	NQ	NQ	NQ

α-AcetoxyNPYR (38 mmol) was allowed to react with calf thymus DNA (5.0 g) for 20 h, 37 ºC, in 0.1 mol/L phosphate buffer (pH 7.0). Crotonaldehyde (0.76 mmol) was similarly reacted with DNA (100 mg). ND, not detected; NQ, not quantified

[a] Neutral thermal and acid hydrolysis products expressed per mol G, enzyme hydrolysis, per mol dG. Values are mean ± SD ($n = 3$).

[b] Quantified as their 2,4-dinitrophenylhydrazones (DNPs): 3-hydroxybutanal-2,4-DNP and crotonaldehyde-2,4-DNP (ratio, 6:1) from paraldol and 4-hydroxybutanal-2,4-DNP from 2-hydroxytetrahydrofuran

[c] Numbers refer to structures in Figure 1. Adducts 5 and 6 were not quantified.

[d] Quantified as guanine base

[e] Acid hydrolysis for adducts 9 and 10 only

mal hydrolysates than in enzyme hydrolysates of both types of DNA, suggesting the presence of a thermally unstable paraldol-releasing adduct.

Figure 2 shows a chromatogram of a neutral thermal hydrolysate of [^{3}H]α-acetoxyNPYR-treated DNA from an earlier study (Wang *et al.*, 1995). P9 and P12 in this chromatogram are about the same size, but the data in Table 1 demonstrate that the levels of paraldol are less than those of 2-hydroxyTHF in neutral thermal hydrolysates of α-acetoxyNPYR treated DNA. Therefore, paraldol is not the only constituent of P9 in Figure 1. Our data indicate that the other major component is [^{3}H]H_2O.

Discussion

In previous work, we noted the presence of substantial amounts of unidentified, early-eluting radiolabel in neutral thermal hydrolysates of [^{3}H]α-acetoxyNPYR-treated DNA (Figure 2, Wang *et al.*, 1995). This finding was in contrast to those obtained by fluorescence detection, in which adduct **4** predominated (Chung *et al.*, 1989a). We have now identified this unknown radiolabel. The main early-eluting products derived from α-acetoxyNPYR-treated DNA are paraldol, the dimer of 3-HOBA, and 2-hydroxyTHF, the cyclic and predominant form of 4-HOBA. [^{3}H]H_2O was also identified. Thus, P9 in Figure 2 is comprised of paraldol and [^{3}H]H_2O, while P12 is 2-hydroxyTHF.

The N^2 tetrahydrofuranyl adduct **8** is the source of most of the 2-hydroxyTHF observed in these experiments. We showed previously that this adduct is unstable at the nucleoside level, decomposing to 2-hydroxyTHF and deoxyguanosine with a half-life of 26 min at pH 7, 37 °C (Wang & Hecht, 1997; Young-Sciame *et al.*, 1995). Thus, the finding that 2-hydroxyTHF is formed after neutral thermal and enzyme hydrolysis of α-acetoxyNPYR-treated DNA is completely consistent with previous data.

Comparison of the amounts of 2-hydroxyTHF present in neutral thermal and enzyme hydrolysates indicates that adduct **8** is the main source of 2-hydroxyTHF. There may also be some minor 2-hydroxyTHF-releasing adducts in this DNA, since in experiments in which we analysed each portion of the HPLC chromatogram of the enzyme hydrolysate for material that gives 4-HOBA-2,4-DNP after derivatization we obtained evidence of an additional 2-hydroxyTHF-releasing adduct.

The most likely sources of the paraldol released by hydrolysis of α-acetoxyNPYR-treated DNA is an adduct or adducts formed by the reaction of crotonaldehyde or paraldol with DNA. Crotonaldehyde is known to be a solvolysis product of α-acetoxyNPYR (Wang *et al.*, 1988). Our results suggest that DNA damage by crotonaldehyde may be far more extensive than previously realized, since the levels of paraldol-releasing adducts are about 100 times greater than those of the 1,N^2-propanodeoxyguanosine adducts **9–12**. It will be important to determine whether similar ratios of paraldol-releasing adducts to 1,N^2-propanodeoxyguanosine adducts are found in human DNA. Chung and Nath have shown that the levels of adducts **9** and **10** range from approximately 0.003 to 1 μmol/mol guanine in human DNA. If the ratio of paraldol-releasing adducts to **9** and **10** is the same as that observed here, one would expect the levels of paraldol-releasing adducts in human DNA to range from 0.3 to 100 μmol/mol guanine. For comparison, the levels in human lung DNA of 8-oxodeoxyguanosine, a common indicator of oxidative DNA damage, are approximately 2–12 μmol/mol guanine (Asami *et al.*, 1997).

We are currently exploring the nature of the paraldol-releasing DNA adducts formed from crotonaldehyde. Enzymatic hydrolysis of crotonaldehyde-treated DNA gives at least five peaks which, after treatment with 2,4-DNP reagent, yield 3-HOBA-2,4-DNP and crotonaldehyde-2,4-DNP. MS and tandem MS data for these peaks suggest that some of them result from cross-linking of DNA bases by crotonaldehyde, while others are formed by reaction of paraldol with DNA.

In summary, 2-hydroxyTHF and paraldol have been identified as substantial products in neutral thermal and enzyme hydrolysates of DNA reacted with α-acetoxyNPYR. They are formed from the N^2-tetrahydrofuranyl adduct **8** and paraldol-releasing adducts, respectively. Paraldol-releasing adducts appear to be major products in the reaction of crotonaldehyde with DNA. These results define more completely the reactions of α-acetoxy-NPYR and crotonaldehyde with DNA. Moreover, 2-hydroxyTHF and paraldol may be useful as biomarkers of DNA modification by NPYR and crotonaldehyde.

Acknowledgements

This study was supported by Grant CA-44377 from the National Cancer Institute. We thank Alex Alvarado, Carol Powers, Yongli Shi, Guang Cheng and Hong Liang for their technical assistance.

References

Asami, S., Manabe, H., Miyake, J., Tsurudome, Y., Hirano, T., Yamaguchi, R., Itoh, H. & Kasai, H. (1997) Cigarette smoking induces an increase in oxidative DNA damage, 8-hydroxydeoxyguanosine, in a central site of the human lung. *Carcinogenesis* 18, 1763–1766

Chung, F.-L. & Hecht, S.S. (1983) Formation of cyclic 1,N^2-adducts by reaction of deoxyguanosine with α-acetoxy-*N*-nitrosopyrrolidine, 4-(carbethoxynitrosamino)butanal, or crotonaldehyde. *Cancer Res.* 43, 1230–1235

Chung, F.-L., Young, R. & Hecht, S.S. (1984) Formation of cyclic 1,N^2-propanodeoxyguanosine adducts in DNA upon reaction with acrolein or crotonaldehyde. *Cancer Res.* 44, 990–995

Chung, F.-L., Wang, M. & Hecht, S.S. (1989a) Detection of exocyclic guanine adducts in hydrolysates of hepatic DNA of rats treated with *N*-nitrosopyrrolidine and in calf thymus DNA reacted with α-acetoxy-*N*-nitrosopyrrolidine. *Cancer Res.* 49, 2034–2041

Chung, F.-L., Young, R. & Hecht, S.S. (1989b) Detection of cyclic 1,N^2-propanodeoxy-guanosine adducts in DNA of rats treated with *N*-nitrosopyrrolidine and mice treated with crotonaldehyde. *Carcinogenesis* 10, 1291–1297

Chung, F.-L., Chen, H.-J.C. & Nath, R.G. (1996) Lipid peroxidation as a potential source for the formation of exocyclic DNA adducts. *Carcinogenesis* 17, 2105–2111

Eder, E. & Hoffman, C. (1992) Identification and characterization of deoxyguanosine–crotonaldehyde adducts. Formation of cyclic adducts and 1,*N*,7,8 *bis*-cyclic adducts. *Chem. Res. Toxicol.* 5, 802–808

Gray, R., Peto, R., Branton, P. & Grasso, P. (1991) Chronic nitrosamine ingestion in 1040 rodents: The effect of the choice of nitrosamine, the species studied, and the age of starting exposure. *Cancer Res.* 51, 6470–6491

Hecht, S.S., Chen, C.B. & Hoffmann, D. (1978) Evidence for metabolic α-hydroxylation of *N*-nitrosopyrrolidine. *Cancer Res.* 38, 215–218

Hecht, S.S., McCoy, G.D., Chen, C.B. & Hoffmann, D. (1981) The metabolism of cyclic nitrosamines. In: *N-Nitroso Compounds* (Scanlan, R.A. & Tannenbaum, S.R., eds). Washington DC, American Chemical Society, pp. 49–75

Hoffmann, D. & Hoffmann, I. (1997) The changing cigarette, 1950–1995. *J. Toxicol. Environ. Health* 50, 307–364

Hunt, E.J. & Shank, R.C. (1982) Evidence for DNA adducts in rat liver after administration of *N*-nitrosopyrrolidine. *Biochem. Biophys. Res. Commun.* 104, 1343–1348

Hunt, E.J. & Shank, R.C. (1991) Formation and persistence of a DNA adduct in rodents treated with *N*-nitrosopyrrolidine. *Carcinogenesis* 12, 571–575

Lijinsky, W & Reuber, M.D. (1981) Carcinogenic effect of nitrosopyrrolidine, nitrosopiperidine and nitrosohexamethyleneimine in Fischer rats. *Cancer Lett.* 12, 99–103

van Maanen, J.M.S., Welle, I.J., Hageman, G., Dallinga, J.W., Mertens, P.L.J.M. & Kleinjans, J.C.S. (1996) Nitrate contamination of drinking water: Relationship with HPRT variant frequency in lymphocyte DNA and urinary excretion of *N*-nitrosamines. *Environ. Health Perspectives* 104, 522–528

Mostafa, M.H., Helmi, S., Badawi, A.F., Tricker, A.R., Spiegelhalder, B. & Preussmann, R. (1994) Nitrate, nitrite and volatile *N*-nitroso compounds in the urine of *Schistosoma haematobium* and *Schistosoma mansoni* infected patients. *Carcinogenesis* 15, 619–625

Nath, R.G. & Chung, F.-L. (1994) Detection of exocyclic 1,N^2-propanodeoxyguanosine adducts as common DNA lesions in rodents and humans. *Proc. Natl Acad. Sci. USA* 91, 7491–7495

Nath, R.G., Ocando, J.E. & Chung, F.-L. (1996) Detection of 1,N^2-propanodeoxyguanosine adducts as potential endogenous DNA lesions in rodent and human tissues. *Cancer Res.* 56, 452–456

Nath, R.G., Ocando, J.E., Guttenplan, J.B. & Chung, F.-L. (1998) 1,N^2-Propanodeoxyguanosine adducts: Potential new biomarkers of smoking-induced DNA damage in human oral tissue. *Cancer Res.* 58, 581–584

Preussmann, R. & Stewart, B.W. (1984) *N*-Nitroso carcinogens. In: *Chemical Carcinogenesis* (Searle, C.E., ed.). Washington DC, American Chemical Society, pp. 643–828

Tricker, A.R., Pfundstein, B., Theobald, E., Preussmann, R. & Spiegelhalder, B. (1991a) Mean daily intake of volatile *N*-nitrosamines from foods and beverages in West Germany in 1989–1990. *Food Chem. Toxicol.* 29, 729–732

Tricker, A.R., Stickler, D.J., Chawla, J.C. & Preussmann, R. (1991b) Increased urinary nitrosamine excretion in paraplegic patients. *Carcinogenesis* 12, 943–946

Wang, M. & Hecht, S.S. (1997) A cyclic N7, C-8 guanine adduct of N-nitrosopyrrolidine (NPYR): Formation in nucleic acids and excretion in the urine of NPYR-treated rats. *Chem. Res. Toxicol.* 10, 772–778

Wang, M., Chung, F.-L. & Hecht, S.S. (1988) Identification of crotonaldehyde as a hepatic microsomal metabolite formed by α-hydroxylation of the carcinogen *N*-nitrosopyrrolidine. *Chem. Res. Toxicol.* 1, 28–31

Wang, M., Chung, F.-L. & Hecht, S.S. (1989) Formation of acyclic and cyclic guanine adducts in DNA reacted with α-acetoxy-*N*-nitrosopyrrolidine. *Chem. Res. Toxicol.* 2, 423–428

Wang, M., Chung, F.-L. & Hecht, S.S. (1992) Formation of 7-(4-oxobutyl)guanine in hepatic DNA of rats treated with *N*-nitrosopyrrolidine. *Carcinogenesis* 13, 1909–1911

Wang, M., Young-Sciame, R., Chung, F.-L. & Hecht, S. S. (1995) Formation of *N*-tetrahydrofuranyl and *N*-tetrahydropyranyl adducts in the reactions of α-acetoxy-*N*-nitrosopyrrolidine and α-acetoxy-*N*-nitrosopiperidine with DNA. *Chem. Res. Toxicol.* 8, 617–624

Young-Sciame, R., Wang, M., Chung, F.-L. & Hecht, S.S. (1995) Reactions of α-acetoxy-*N*-nitrosopyrrolidine and α-acetoxy-*N*-nitrosopiperidine with deoxyguanosine: Formation of *N*-tetrahydrofuranyl or *N*-tetrahydropyranyl adducts. *Chem. Res. Toxicol.* 8, 607–616

Exocyclic DNA Adducts in Mutagenesis and Carcinogenesis
B. Singer & H. Bartsch
IARC Scientific Publications No. 150
International Agency for Research on Cancer, Lyon, 1999

Glyoxal–guanine DNA adducts: Detection, stability and formation *in vivo* from nitrosamines

R.N. Loeppky, W. Cui, P. Goelzer, M. Park & Q. Ye

The glyoxal–deoxyguanosine adduct (gG) is formed from α-nitrosamino aldehydes and dG *in vitro* and *in vivo* from nitrosamines carrying the 2-hydroxyethyl side-chain as well as from *N*-nitrosomorpholine. The structures of all of the diastereomeric forms of both the *cis* and *trans* isomers of the adduct have been investigated by *ab initio* calculations and with nuclear magnetic resonance spectroscopy at 500 MHz. The preferred orientation of the OH groups is *trans*, but at equilibrium a small amount of the *cis* isomer was observed. The pH-independent equilibrium constant for the hydrolysis of the gG adduct is $K = 1.36 \times 10^{-4}$ mol/L, and its rate of formation at pH 7.3 is $k = 5.3$ $min^{-1}mol^{-1}$. In acid (pH 2), the hydrolysis of the nucleosidic linkage is nearly twice as rapid as the hydrolysis of gG to glyoxal and dG. We used a gG analogue to explore a number of reductive methods for derivatization of the adduct, but all of the processes either gave low yields or product mixtures which rendered them impractical for derivatizing the adduct in DNA. A ^{32}P-postlabelling method for detection of the pH-sensitive gG adduct has been developed, which permitted detection of the adduct in the liver DNA of male Wistar rats after administration of selected nitrosamines. The levels of adducts found were: *N*-nitrosodiethanolamine > 2-hydroxyethylmethynitrosamine > *N*-nitrosomorpholine > 2-hydroxyethyethylnitrosamine. In separate experiments, *N*-nitrosodiethanolamine gave greater adduct levels than its metabolite 2-hydroxy-*N*-nitrosomorpholine. Mechanistic pathways for the generation of gG adducts *in vitro* and *in vivo* are discussed.

Introduction

The glyoxal–guanine (gG) adduct **1** has been known for many years. Interest in this adduct first arose as a result of the viral mutagenic properties of glyoxal, which was shown to bind to both RNA and, subsequently, DNA (Frankel-Conrat, 1954; Staehlin, 1958, 1959). The structure, and many of the properties of the adduct were first elucidated by Shapiro and his colleagues (Shapiro & Hachmann, 1966; Shapiro *et al.*, 1969, 1986). Leonard's group provided proof of the definitive structure, showing that the glyoxal moiety was bound to N1 and the NH_2 rather than N3 of the purine ring (Czarnik & Leonard, 1980). Many of the aspects of the early history of this adduct have been reviewed (Shapiro *et al.*, 1986).

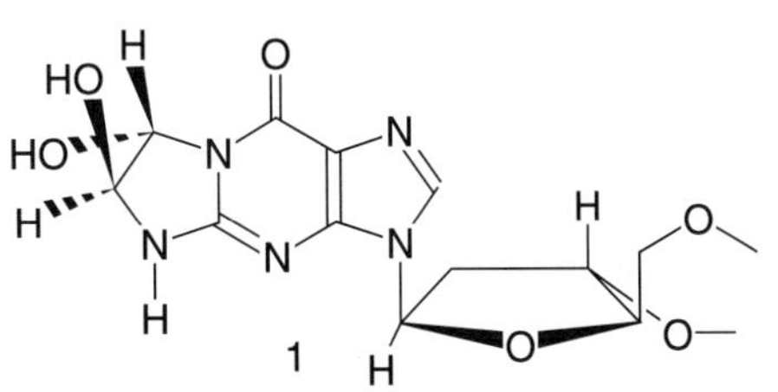

More recent interest in the gG adduct stems form its formation from α-nitrosamino aldehydes **4**, which yield glyoxal or glyoxal equivalents when reacted with water or amines. Chung and Hecht (1985; Chung *et al.*, 1986) demonstrated that *N*-nitroso-2-hydroxymorpholine **5**, the hemiacetal form of the α-nitrosamino aldehyde **3**, a metabolite of both *N*-nitrosodiethanolamine (NDELA) **2** and *N*-nitrosomorpholine, forms the gG adduct upon reaction with guanosine. Further studies by our group have shown that the gG adduct forms from NHMOR and all of the α-nitrosamino aldehydes examined when these substances are incubated in H_2O (or buffer) with deoxyguanosine

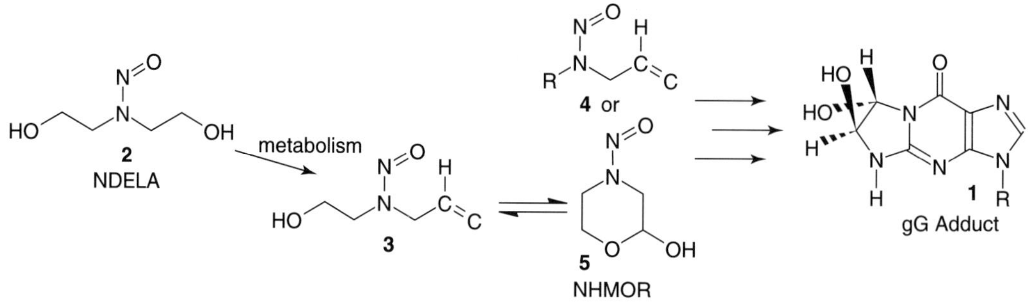

Scheme 1

(dG), oligonucleotides containing guanine or calf-thymus DNA (see Scheme 1) (Loeppky *et al.*, 1984, 1987a,b, 1993a,b). These *trans*formations occur 'spontaneously', without further activation; the probable mechanisms are discussed later. Additional data for a fuller understanding of the properties of the gG adduct and its mode of formation came from our finding that it is formed in rat liver DNA from a number of nitrosamines.

Structural properties of the glyoxal–guanine adduct

There are four possible diastereomeric forms of the glyoxal-deoxyguanosine (gdG) adduct. Two of these, one with the two OH groups *trans* and one in which they are *cis*, are depicted in Figure 1. Low-resolution nuclear magnetic resonance (NMR) spectroscopy was used previously to assign the stereochemistry of the adduct to the *trans* geometry. This assignment was based on an absence of coupling between the C–H protons of the glyoxal moiety, which indicates a dihedral angle of close to 90° between these hydrogen atoms. Because the adduct is formed *in vivo*, we have undertaken a more thorough investigation of its structure. *Ab initio* calculations were performed first at the PM3 level, proceeding to RHF/321-G* to generate minimized, gas-phase structures for all four gdG diastereomers. Only one *trans* and one *cis* form are discussed here. The HC–CH dihedral angle is 119.7° in the *trans* form and 21.3° in the *cis* form. The five-membered ring formed by glyoxal adduction is displaced from planarity by about 10° by the upward thrust of one of the carbons and is not significantly different in the *cis* and *trans* forms.

NMR (500 MHz) analysis of the gG adduct in D_6DMSO revealed no coupling between the CH hydrogens, but each of these protons is coupled to the adjacent OH proton. A similar spectrum is observed for the gdG adduct in D_6DMSO and permits the assignments shown below. The glyoxal-derived ring hydrogens can be distinguished by the coupling between the NH hydrogen and the hydrogen at δ 4.85. (Couplings were identified from the COSY spectrum and are noted by the double-headed arrows.) Addition of D_2O to the sample results in the disappearance of all couplings indicated by the arrows. These data are consistent with the assignment of the *trans* geometry to the adduct produced from the reaction of glyoxal with either guanine or deoxyguanosine and, by implication and comparison with data in the literature, guanosine as well. Use of the HC–CH dihedral angle from the *ab initio* calculations (≈120°) in the Karplus equation results in an estimate of J of 2.5 Hz. As can be seen from inspection of the partial spectrum shown in Figure 2, the two CH peaks (δ 5.74 and δ 5.18) from the glyoxal-derived protons are relatively broad but are not split. Assuming that J = 1 Hz, given the line width limits, use of the

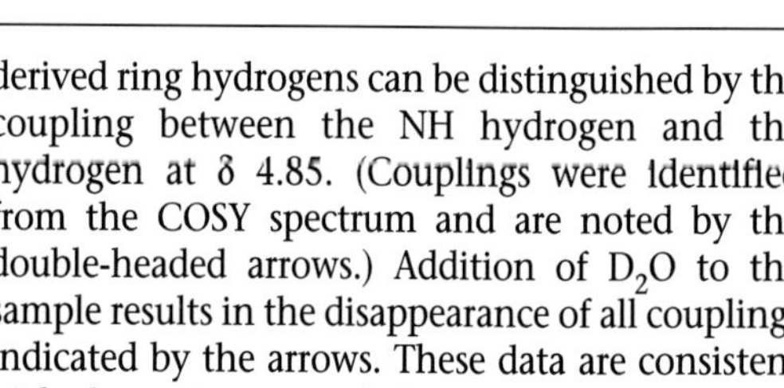

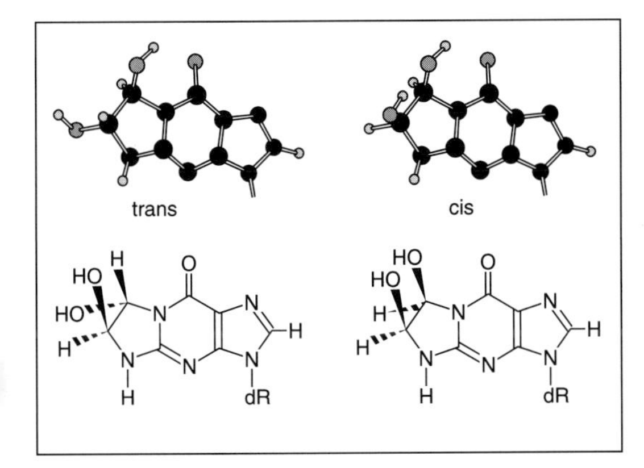

Figure 1. Structures produced from *ab initio* calculations of the glyoxal–deoxyguanosine *cis* and *trans* isomers (top) with the line structures shown below

The deoxyriboside portion of the structure has been eliminated from the top portion for brevity of presentation.

Karplus equation results in an estimated dihedral angle of 108°.

In the course of our studies on the hydrolysis rate of gdG to glyoxal and dG, we followed the transformation by high-resolution ^{1}H-NMR (500 MHz). The reaction reaches an equilibrium (see below) which favours the adduct. A portion of the ^{1}H-NMR spectrum of the equilibrium mixture is shown in Figure 2 and reveals two new sets of peaks in the same region as the protons of the *trans* isomer. We propose that these result from the diastereomeric *cis* adducts. This assignment is supported by a new set of peaks at ≈ δ 7.8 assigned to the C8 H atoms of the diastereomers, which integrate ≈ 1:1 with each set of *cis* proton peaks. Careful inspection reveals a J = 5.4 Hz for the major *cis* diastereomer. Use of the simple Karplus relationship results in an estimate of a dihedral angle of 42° between these hydrogen atoms, which is somewhat greater than that calculated *ab intio*. There are two possible reasons for the discrepancies between the angles estimated from the Karplus relationship and those calculated. The *ab initio* calculations are gas-phase determinations and show the presence of intramolecular hydrogen bonds, which certainly do not exist in aqueous solution; these may perturb the dihedral angle as the OH groups arrange themselves to maximize these interactions. Secondly, it is well known that electronegative substituents, such as the N and OH groups, significantly affect the magnitude of the coupling constant. Karplus relationships are available that take these substituents into

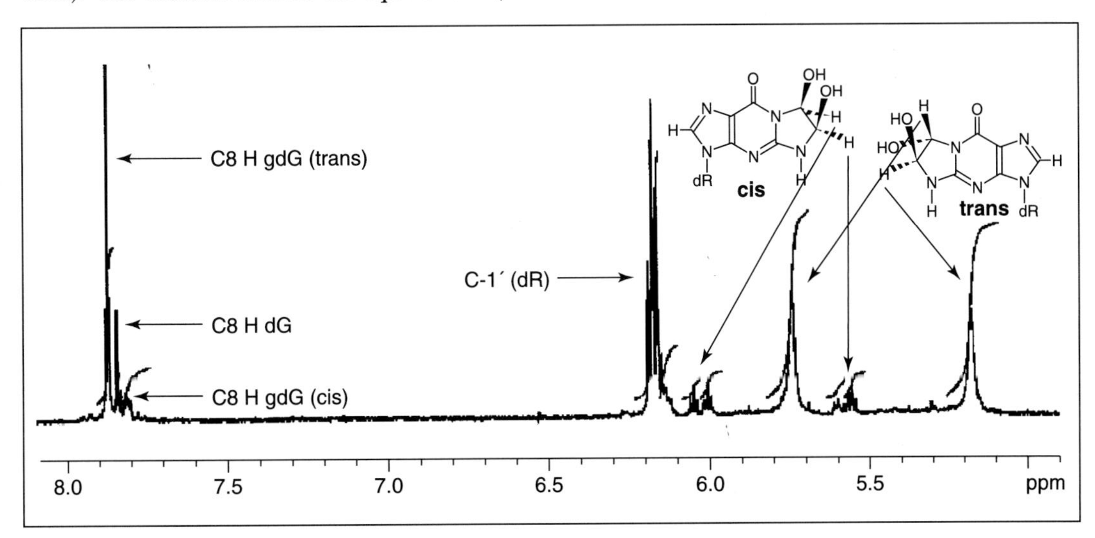

Figure 2. The partial 500 MHz ^{1}H-nuclear magnetic resonance spectrum of the glyoxal–deoxyguanosine (gdG) adduct after hydrolytic equilibrium has been reached at an initial [gdG] = 10 mmol/L at pH 7

Signals for both the *cis* and *trans* adducts and dG are visible as indicated.

account, but none was found that was applicable to our system. The usual effect of an electronegative substituent is to diminish the magnitude of the coupling constant from that calculated from the simple Karplus relationship.

Kinetics and equilibrium of hydrolysis

Because several nitrosamines have been shown to produce the gG adduct *in vivo*, it is important to understand the characteristics of its stability. From a structural point of view, the adduct is a bis-hemiaminal. Normally hemiaminals dehydrate to give imines (Schiff's bases). Dehydration at the position of attachment to N1 of the guanine is not possible for gG, however. The failure to observe dehydration at the other carbon has precedent in aldehyde chemistry. Aldehydes with electron withdrawing groups attached directly to the carbonyl, such as chloral and glyoxal, exist as hydrated gem-diols in aqueous solution. Hemiacetals, structurally close relatives of hemiaminals, are known to undergo facile acid- and base-catalysed hydrolysis to the corresponding aldehyde. This propensity for hydrolysis creates problems for the detection of traces of these adducts. To gain further insight into this property of the gG adduct, we determined the equilibrium constant for its hydrolysis/formation and began an investigation of the kinetics and mechanism of its hydrolysis.

Shapiro and his colleagues (1986) showed that the gG adduct is relatively stable to acid but hydrolysed in base. Using high-performance liquid chromatography (HPLC) as the primary analytical tool, we found that these generalizations were also applicable to the hydrolytic stability of the gdG adduct, depending on the pH. As is shown in Scheme 2, at low pH (2), hydrolysis of the nucleosidic linkage competes with cleavage of the glyoxal moiety from gdG. The former transformation is approximately twice as rapid as the latter under these conditions at 25 °C. Under basic conditions, only glyoxal is cleaved from gdG.

Several estimates of the equilibrium constant for gG-type adduct hydrolysis have been published. In their review of gG and similar adducts, Shapiro *et al.* (1986) reported that the *K* for formation of the 9-methylguanine–glyoxal adduct at pH 7 is about 12 000 per mol ($1/K = 8.3 \times 10^{-5}$ mol/L), but this determination is based on an unpublished thesis. Shapiro *et al.* noted that the formation equilibrium is adversely effected by the formation of 9-methylguanine anion in basic solution but is compensated for by the generation of a 9-MegG anion. A more recent determination by Montoya *et al.* (1995), who used both ultraviolet and electrochemical methods, gives a *K* (formation of the gG adduct) of 5.8×10^8 per mol at 25 °C and pH 7. In the latter determination, the sample of gG adduct was not pure, and the extinction coefficients for the adduct were estimated by adding successively larger amounts of glyoxal to guanosine or guanine. We have previously shown that more than a single molecule of glyoxal can adduct guanosine under such conditions (Loeppky *et al.*, 1987a).

Because of the discrepancies in these reports and our own concerns about our ability to clearly distinguish between dG and gdG through the use of ultraviolet spectroscopy, we used two

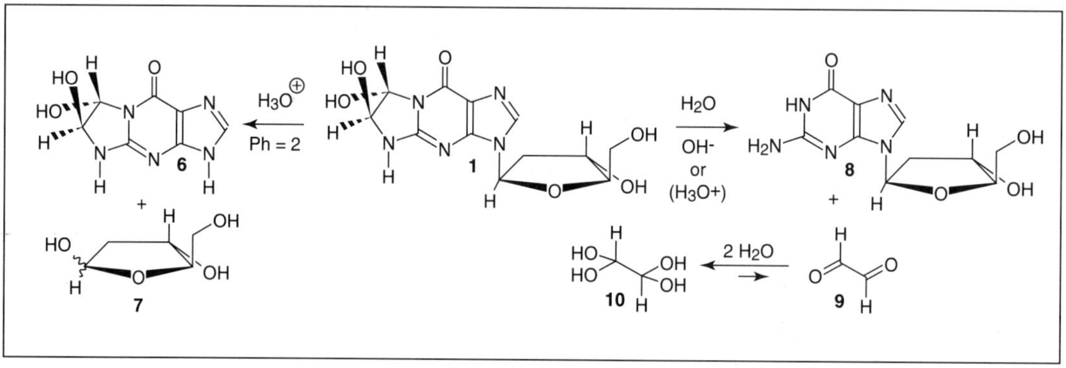

Scheme 2

methods, HPLC and NMR, to determine the equilibrium constant of this process and to measure the kinetics of gdG hydrolysis (Scheme 2). The most accurate concentrations of gdG and dG can be obtained by HPLC, but these data show no indication of the presence of the *cis* gdG observed in low concentration by NMR at neutral pH. At present, therefore, in our kinetic and equilibrium analyses we ignore this possible isomerization and consider only the total gdG concentration. Because of limits to the sensitivity and signal averaging requirements, the NMR studies were performed with $[gdG]_0$ = 10 mmol/L, whereas the HPLC determinations were done with 1.5 mmol/L < $[gdG]_0$ < 2 mmol/L. The values of K and the rate constants k and k_{-1} were slightly smaller in the NMR experiments than in the HPLC studies. The constants are reported in Table 1.

Inspection of Table 1 shows that the equilibrium constant (K_{obs}) for the dissociation of gdG to g and dG decreases with increasing pH: the deviation becomes more significant at higher pH. A plot of log k versus pH for the first three values of Table 1 is linear with a slope of 0.92, which is indicative of base-catalysed hydrolysis of the adduct which is first order in $[OH^-]$. Inspection of the placement of rate constants for the highest two pH values shows that the first-order dependence on $[OH^-]$ breaks down with these rate constants, not increasing as much with $[OH^-]$ as would be anticipated. While the exact nature of this deviation and the chemistry behind it must await further exploration, it suggests that gdG may react with OH^- reversibly to form a product that is not on the pathway to hydrolysis. This hypothesis is shown in equations 1 and 2, and leads to the definition of K_{obs} shown in equation 3. The hypothesis was tested by plotting $1/K_{obs}$ versus $[OH^-]$. If the hypothesis is true, then a straight line with an intercept of $1/K$ and a slope of K_p/K should result. Indeed, such a plot is linear (r = 0.96) and gives $K = 1.36 \times 10^{-4}$ mol/L and $K_p = 1.37 \times 10^5$. It should be noted that this differs from the interpretation of Shapiro *et al.* presented above. The equilibria they discuss can reasonably be expected to be on the coordinate of the reaction from gdG to g + dG and will be incorporated into K, which should not vary with base. While we can only speculate on the nature of the substance that we have designated as P^-, it could reasonably involve either OH^- addition to a carbon of the purine ring or an anion derived from dehydration of the adduct five-membered ring. It should also be noted that there is reasonable agreement between the equilibrium constant estimated for 9-Me-gG dissociation (Shapiro *et al.*, 1986) and that found by us for gdG.

The rate (k_{-1}) of the gG adduction is rapid under physiological conditions. The equilibrium significantly favours adduct formation. Since DNA represents a source of G for reaction with glyoxal, hydrolysis of one adduct is likely to be followed by formation of another. From this perspective, DNA can be considered to be a good trap for glyoxal, and, in the absence of specific enzymatic repair mechanisms, the adducts should persist and be mutagenic, although they may move from base to base. The latter hypotheses is being tested, although use of relatively high pH condi-

Table 1. Rate constants for the reversible hydrolysis of glyoxal–deoxyguanosine

pH	$k \times 10^{-4}$	k_{-1}	$K_{obs}{}^a \times 10^{-4}$
6.52	1.4	1	1.4
7.32	7.00	5.3	1.3
8.11	38	32.1	1.2
8.47	41	45.6	0.9
8.76	47	60.2	0.78

37°C; k_1 (min^{-1}), k_{-1} (min^{-1} mol^{-1}), K_{obs} (mol)

[a] See text for explanation

$$gdG \underset{k'_{-1}}{\overset{k'}{\rightleftharpoons}} g + dG \quad (1)$$

$$gdG + OH^- \overset{K_p}{\rightleftharpoons} P^- \quad (2)$$

$$K_{obs} = k/k_{-1} = \frac{K}{1 + K_p[OH^-]} \quad (3)$$

tions during DNA hydrolysis and gG isolation procedures will lead to loss of adducts, as we have demonstrated and discuss below.

Chemical transformations of the glyoxal–guanine adduct

In order to develop chemical methods for stabilizing the inherently unstable gG adduct, we investigated a number of transformations. The goal of this work was to find a method or methods for sensitive detection of the gG adduct in DNA. This chemistry is summarized in Scheme 3. Leonard's group (Stattsangi *et al.*, 1977; Czarnik & Leonard, 1980) showed previously that the gG adduct **6** can be converted into the etheno adduct **11** by reaction with hydroiodic acid (HI) at elevated temperatures for several days. Shapiro *et al.* (1969) showed that the gG adduct **6** can be cleaved by HIO_4 to give the formamide **12**. Neither of the transformations appeared to be useful for stabilizing or the subsequent detection of the adduct in DNA.

In investigating other methods, we used a guanine derivative that contains a 'nucleoside-like' benzyloxymethyl (BOM) linkage at N9 and which confers greater solubility in organic solvents. Our initial attempts were directed at reductive stabilization through conversion of one or both of the HOCH–N links into a CH_2–N linkage by using $NaCNBH_3$ and other reagents. We attempted several procedures for converting one of the hemiaminal linkages to a Schiff's base, which would be reduced to the amine *in situ*. Reaction of **13** with trifluoroacetic acid followed by direct addition of $NaCNBH_3$ had only modest success, producing a 25% yield of **14** and a 30% yield of the further reduction product **15**. A similar attempt with use of acetic anhydride to dehydrate the hemiaminal **13** gave only a 10% yield of **16** after reduction. Greater success was obtained when the ylide reagent ($CH_3(O_2C)N$-SO_2-$N(C_2H_5)_3$) of Atkins and Burgess (1968), a dehydrating agent for alcohols, was used in combination with $NaCNBH_3$. Under these conditions, an equimolar mixture of **14** and **17** was obtained in 80% yield, but this transformation required high temperatures, and generation of the interconverting isomeric mixture was impractical. Reaction of **13** with an excess of either ethanol or methanol resulted in the formation of the respective aminal **18** in high yield. Use of this transformation type for appropriate derivatization of the gG adduct is being explored further. Overall, this search for possible derivatization schemes was discouraging, and successful development of a ^{32}P-postlabelling method, described below, has decreased our enthusiasm for this line of research at present.

Detection of the glyoxal–guanine adduct by ^{32}P-postlabelling

As stated above, incubation of α-nitrosamino aldehydes with DNA *in vitro* results in the formation of gG adducts (Loeppky *et al.*, 1993b). Since NHMOR **5** is an intermediary metabolite of NDELA (Airoldi *et al.*, 1984), we sought to develop a sensitive method for detection of the gG adduct in DNA taken from animal tissue after administration of NDELA or related nitrosamines, in order to test various mechanistic hypotheses for activation of these environmentally prevalent compounds. Although we were concerned that the gG adducts might not survive the steps required in a typical ^{32}P-postlabelling assay, the relative lack of promise shown by our chemical derivatization experiments left us with little choice. Application and development of this method to our pH-sensitive adduct was greatly facilitated by the use of HPLC coupled to a radiochemical detector. While the details of this procedure are being published elsewhere, the method is summarized in Scheme 4. For the purposes of method development, standard gG adducts were prepared from guanine, deoxyguanosine, 3′- or 5′-nucleotides or the 3′,5′-diphosphonucleotide by reaction with glyoxal under conditions similar to those first described by Chung and Hecht (1985) for synthesis of the gdG adduct. Our initial experiments were directed at determining the stability of the adduct during various hydrolytic and enzymatic modifications, with HPLC–ultraviolet separation and detection of the modified nucleoside, which was freed from its phosphate esters by alkaline phosphatase digestion. These procedures were time-saving and obviated use of radioactivity. Calf thymus DNA modified by incubation with glyoxal was used in these experiments. Throughout the method development, standards were used to assess the stability of the adducts, and all of the steps were optimized sequentially for maximum adduct detection.

DNA modified *in vitro* by incubation with glyoxal, NHMOR or another α-nitrosamino-aldehyde or obtained from rat liver after nitrosamine administration was ^{32}P-postlabelled as shown in Scheme 4 (**19**). When DNA was isolated from liver tissue, care was taken to free it of protein and RNA by treatment with proteinase K and RNase I, respectively, before ^{32}P-postlabelling. The gG

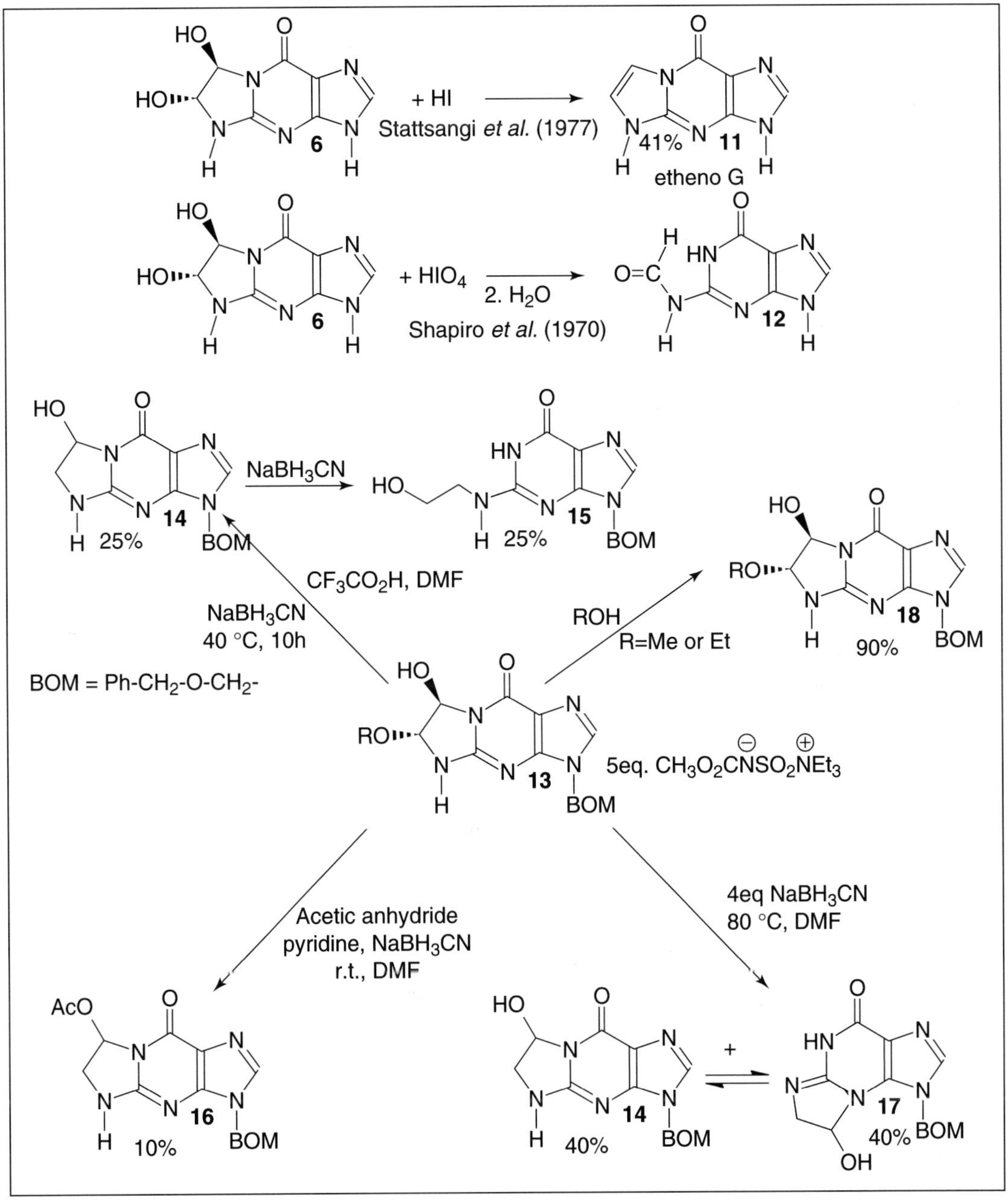

Scheme 3

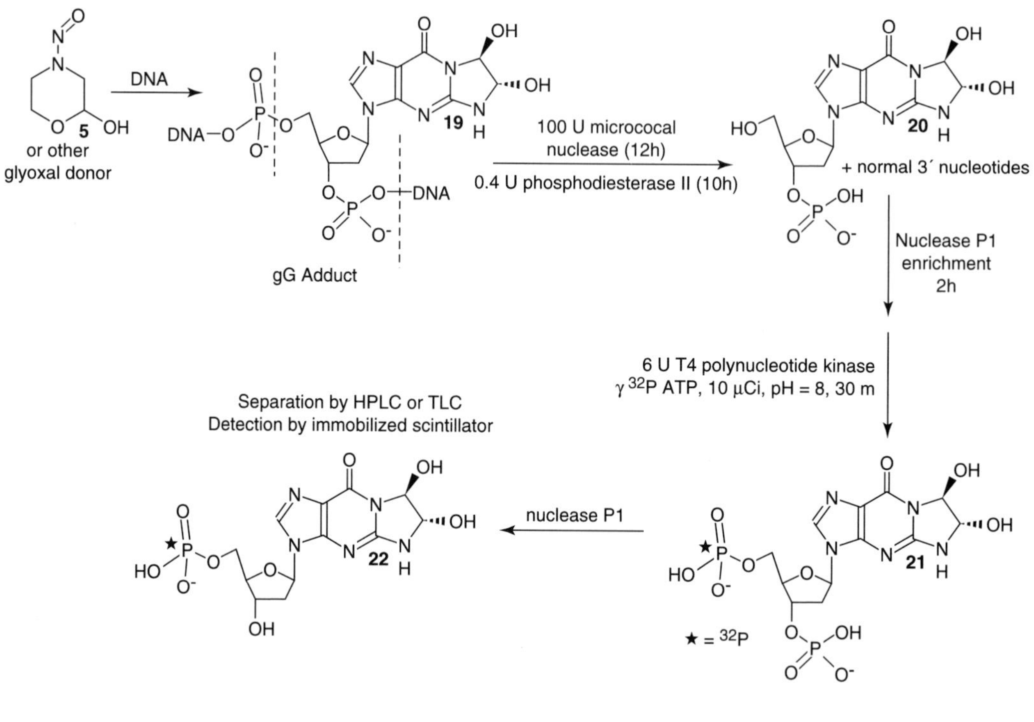

Scheme 4

adducts are well preserved under the conditions of enzymatic DNA hydrolysis used (Scheme 4). In separate experiments, we established that 3'P-gdG (**20**) was a relatively poor substrate for nuclease P1 (in comparison with non-adducted nucleotides), so that we could significantly enrich the nucleotide content by using it to cleave the unmodified phosphonucleotides and thereby reduce the background. The most problematic step from the perspective of preserving the unstable gG adduct is the actual ^{32}P-labelling, in which γ-^{32}P-ATP and T4 polynucleotide kinase are used at a pH of 9.5. Operation at this high pH results in a high rate of loss of adducts by hydrolysis; labelling at pH 7 was better but still inefficient because the reaction was slow. Optimization studies revealed that the maximum amount of adduct could be labelled at pH 8 (30 min).

In our first study, we detected the adduct as the bisphosphate (**21**) but found that it was slowly being converted into 5'P-gdG as the samples stood waiting for HPLC analysis. To avoid this, we incorporated a final nuclease P1 incubation, which permitted detection of the gG adduct as the 5'P-gdG (**22**). Using only the radiochemical detector and 10 μCi of γ-^{32}P-ATP, we could detect five adducts in 10^7 nucleotides. In every case, we also submitted our samples to two-dimensional thin-layer chromatography (TLC) on either polyethyleneimine-cellulose or cellulose followed by autoradiography and computer-based phosphor-image analysis. Sensitivity can be enhance approximately 10-fold by use of the TLC method, and it gives independent confirmation of adduct identity in comparison with standards. In all cases, quantification involved spiking samples (successive addition) with standards. Incubation of NHMOR with calf thymus DNA *in vitro* was found to produce gG adducts even when 10^{-12} mol of NHMOR were incubated with 100 mg of DNA. When 10^{-9} mol of NHMOR were incubated with DNA, the yield increased with time up to 24 h to 450 pmol of gG

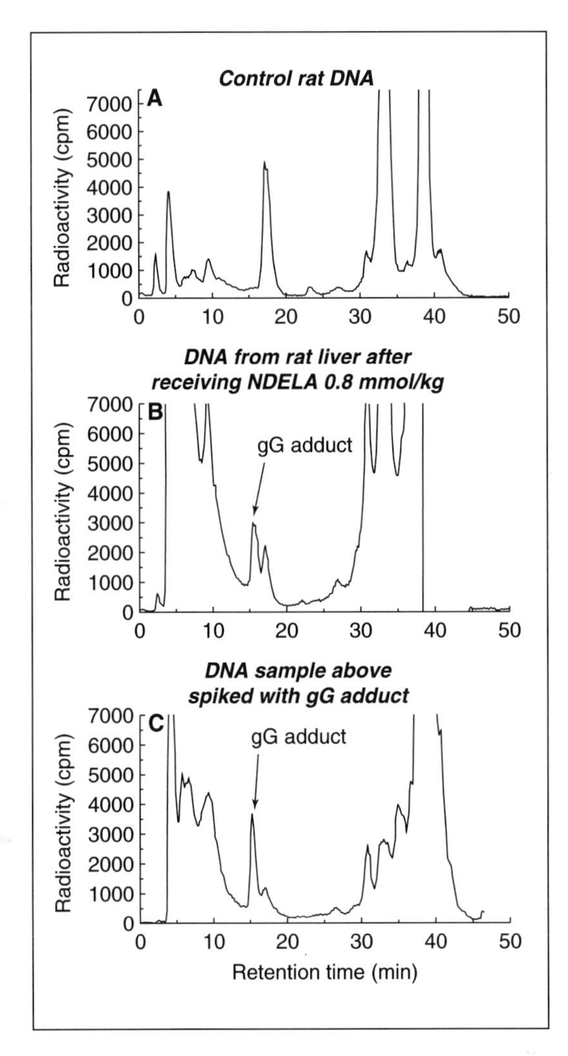

Figure 3. High-performance liquid radiochromatograms of (A) control rat liver DNA after ^{32}P-postlabelling; (B) ^{32}P-postlabelling of rat liver DNA after administration of 0.8 mmol *N*-nitrosodiethanolamine (NDELA) per kg body weight; and (C) the same sample spiked with authentic glyoxal–guanine (gG) adduct

adduct per 100 µg of DNA. Possible routes to the formation of gG adducts under these conditions are discussed below.

This method was used to determine whether gG adducts formed *in vivo* after administration of specific nitrosamines. In all cases, the dose of nitrosamine was 0.8 mmol/kg body weight, a dose ascertained as non-toxic in other experiments in which the extent of DNA single-strand breaks by these nitrosamines was determined (Loeppky *et al.*, 1998a). In the first experiments, NDELA and its intermediary metabolite NHMOR, were administered. Both of these nitrosamines produced gG adducts in male Wistar rat liver DNA, while no adducts were seen in the control. There were approximately twice as many adducts from NDELA as from NHMOR. In a second set of experiments with the four different nitrosamines, **2**, **23–25** (see diagram below and Table 2), gG adducts were produced from all and could be detected by HPLC and, of course, by the more sensitive two-dimensional TLC technique. The quantities given in Table 2 were determined by HPLC and are in good agreement with the results produced in the TLC determinations.

The observation that **2**, **23** and **24**, all ethanol nitrosamines, produce gG adducts is significant. NDELA is an environmentally prevalent nitrosamine that is found as an inadvertent cont-

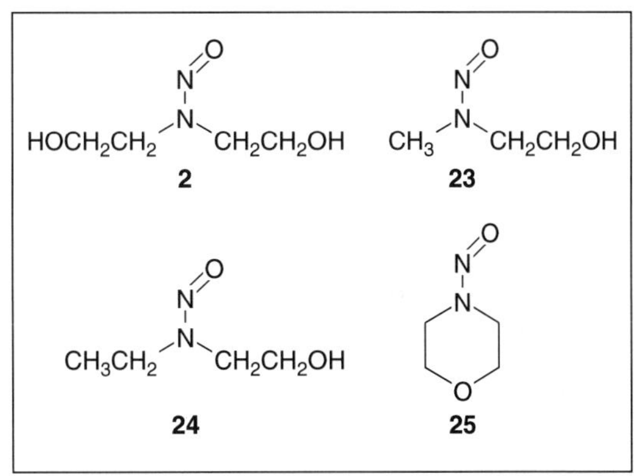

Table 2. DNA glyoxal–guanine adducts produced *in vivo* from nitrosamines[a]

No.	Substrate	Adducts[b]
2	*N*-Nitrosodiethanolamine	6.5
23	Methylethanolnitrosamine	4.8
25	*N*-Nitrosomorpholine	2.1
24	Ethylethanolnitrosamine	1.4

[a] Administered to 4-week-old male Wistar rats at 0.8 mmol/kg body weight in H_2O by gavage
[b] per 10^6 nucleotides

aminant in many products that contain di- or triethanolamine or their derivatives (Loeppky *et al.*, 1998a). These include cosmetics, personal-care items, metalworking fluids, pesticides and tobacco. NMOR is known to undergo microsomal conversion to NHMOR, but no adducts have been isolated after its administration *in vivo*. The lower levels of gG adducts from **23** and **24** are probably due to competitive oxidation of these nitrosamines at the methyl and ethyl groups, respectively, a process that is likely to form 2-hydroxyethyl adducts from the corresponding diazonium ion intermediate. Eisenbrand and we have shown independently (Scherer *et al.*, 1991; Loeppky *et al.*, 1998b) that NDELA also produces 2-hydroxyethyl-O^6-G adducts.

Possible mechanisms of glyoxal–guanine formation from α-nitrosamino-aldehydes *in vitro*

As stated in the introduction, it has been known for some time that NHMOR and other α-nitrosamino-aldehydes generate glyoxal equivalents in buffer and form gG adducts with nucleosides, but, because no evidence existed until now for the importance of this process *in vivo*, the mechanism of this interesting transformation has not been seriously investigated. Three possible pathways, shown in Schemes 5–7, have been considered. The first, Scheme 5, proposed by Chung and Hecht(1985), envisions oxidation of the carbon chain by elimination of NO^- from **27** after initial Schiff's base formation with the NH_2 group of G. While there is no question that the α-hydrogen atom is acidic in such a structure, the elimination of NO^-, a poor leaving group, from nitrosamines is relatively rare and is doubtful.

We have advanced another postulate, Scheme 6, which involves formation of a 1,2,3-oxadiazolinium ion **30** as an intermediate in the isomerization of **4** to **31a**. We have demonstrated the conversion of **4** (R = CH_3) to **30** (Loeppky *et al.*, 1988) and, on the basis of model chemistry (Loeppky & Srinivasan, 1995), propose that glyoxal **9** forms hydrolytically after the intriguing charge reversal ('Umpolung') of **31a** to **31b**. We do not know, however, whether the conditions for the formation of **30**, a process that we have seen only in acid, are likely to be present in DNA.

Recent findings in our laboratory have led us to consider the pathway shown in Scheme 7 as a more probable route to glyoxal equivalents in DNA from NHMOR and related α-nitrosamino-aldehydes. Two key pieces of evidence support this mechanistic hypothesis, which is still under active investigation. α-Nitrosaminoalkyl radicals, shown here as **33**, rapidly lose NO to form an imine (**34**). We have independently generated these radicals and shown that the NO loss is a clean, rapid process

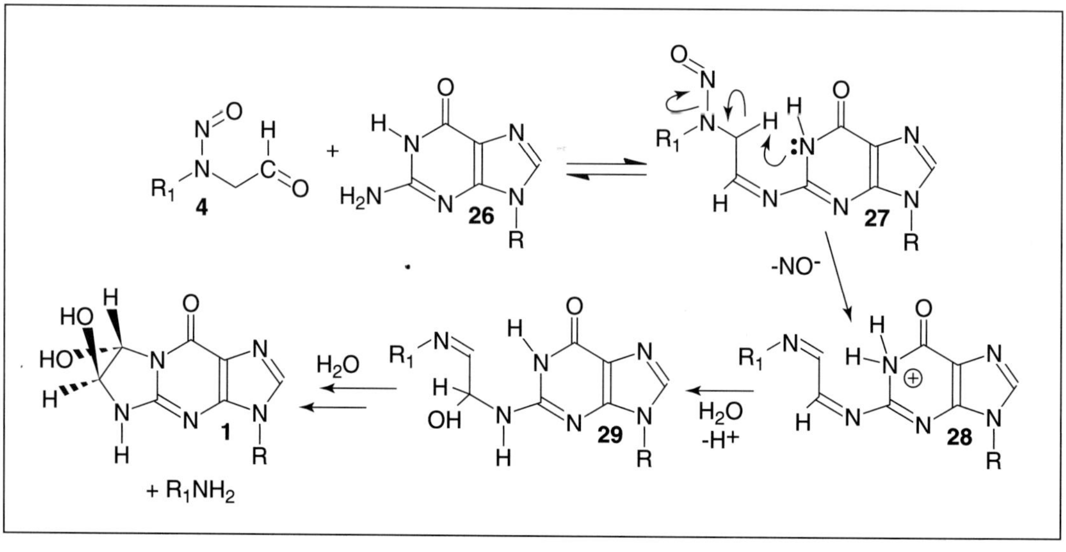

Scheme 5

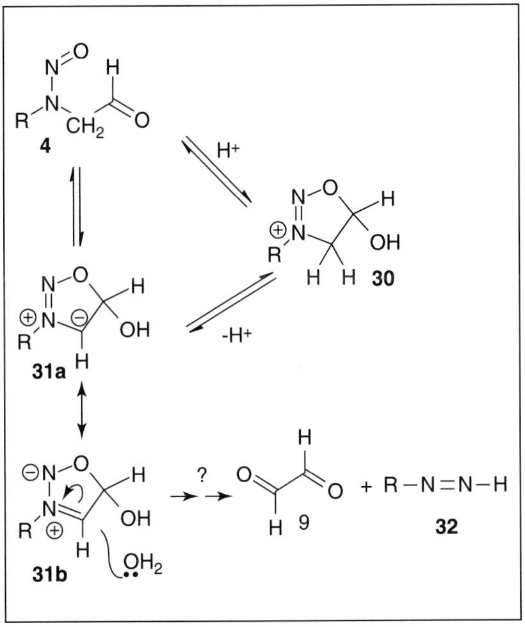

Scheme 6

(Loeppky & Li, 1991). In preliminary experiments, we have shown that bubbling O_2 through a solution of NHMOR results in the formation of glyoxal, which can be detected as its 2,4-dinitrophenylhydrazone. The pathway depicted in Scheme 7 shows generation of the radical by O_2-mediated H-atom abstraction from NHMOR, but this oxidation could also occur through the corresponding enolate ion and produce superoxide ion as the product. A related pathway could be involved in the formation of gG adducts *in vivo*, as discussed below.

Possible mechanisms of glyoxal–guanine adduct formation from *N*-nitrosodiethanolamine *in vivo*

Up to this point in the discussion, we have focused our attention on gG adduct formation from NHMOR. We have demonstrated that low levels of gG adducts are formed from NHMOR in rat liver *in vivo*, but Hecht *et al.* (1989) failed to observe significant carcinogenic activity with this NDELA metabolite in AJ mice or rats. By using specifically deuterated derivatives of NDELA and NHMOR, we have shown that DNA single-strand breaks in the livers of male Wistar rats are diminished significantly when the α-H of NDELA is replaced with D (Loeppky *et al.*, 1998a). This finding suggests that NDELA undergoes α-oxidation, as shown in Scheme 8. This process is expected to lead to 2-hydroxyethyl adducts, with the production of glycolaldehyde. We have shown that rat liver microsomes rich in P450 2E1 and pre-induced with isoniazid can mediate oxidation of NDELA at the α-position to produce glycolalde-

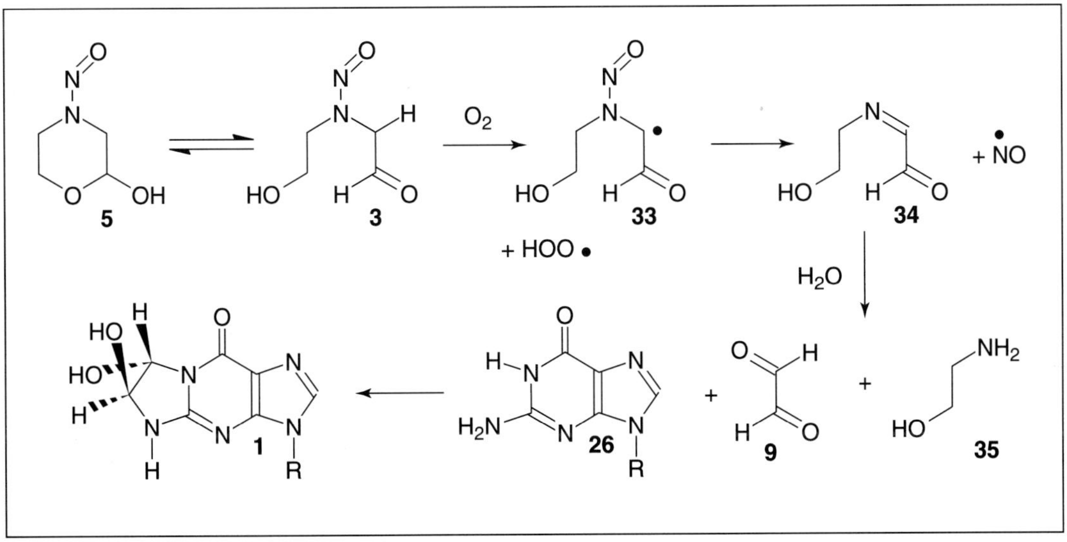

Scheme 7

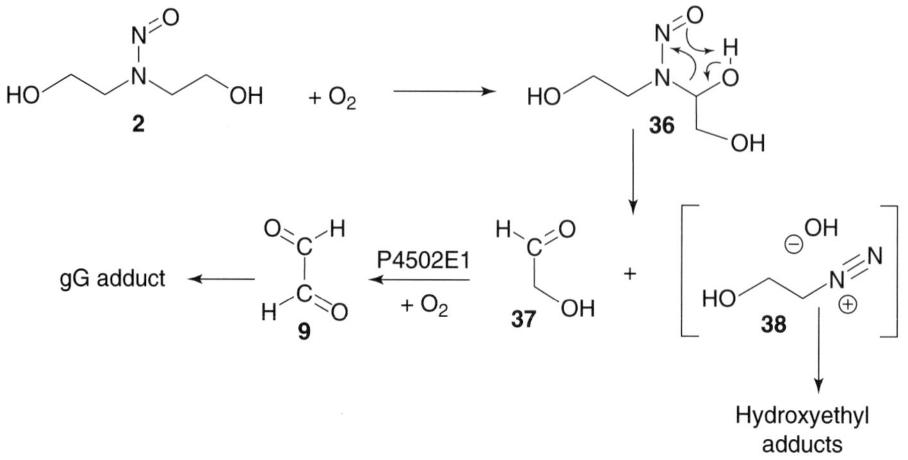

Scheme 8

hyde. We have further shown that these same microsomes are capable of mediating the oxidation of glycolaldehyde to glyoxal. If these processes occur *in vivo*, which is likely, they represent a pathway to the formation of gG adducts. It is important to note that this pathway could result in two DNA adducts from a single oxidation. The diazonium **38** will produce 2-hydroxyethyl adducts, accompanying the gG adducts formed from oxidation of glycolaldehyde to glyoxal. This process—two potential, simultaneous adducts from a single carcinogen molecule—also occurs from the P450-mediated epoxidation of methylvinylnitrosamine (Okazaki *et al.*, 1993) and may explain the carcinogenic potency of these compounds.

A significant observation made during our studies of the microsomal oxidation of NDELA, NHMOR and the other nitrosamines listed in Table 2 is that not only do they all produce glycol aldehyde and then glyoxal, but those compounds carrying the 2-hydroxyethyl chain are also oxidized at the OH-bearing carbon to produce the corresponding α-nitrosamino-aldehydes **4**. In the case of NDELA, the oxidation at the OH-bearing carbon β-position is twice as extensive as is α-oxidation and produces NHMOR. The microsome-mediated oxidation of NHMOR results in the formation of both glycolaldehyde and glyoxal, indicating how it could produce gG adducts. Oxidation of NHMOR at the positions adjacent to either nitrogen atom to give the corresponding α-hydroxynitrosamines will lead to glycolaldehyde and/or glyoxal. We have independently synthesized the α-hydroxynitrosamine that would be formed by α-oxidation on the side-chain of NHMOR with more oxidation and shown that it produces the gG adduct when generated in the presence of dG (Park *et al.*, 1998). Thus, both α- and β-oxidation appear to be important in the bioactivation of NDELA and possibly of structurally related nitrosamines.

Conclusions

Our investigation into the mechanism of activation of NDELA has led to the development of a sensitive ^{32}P-postlabelling assay for the gG adduct and shown that it is present in the liver DNA of rats fed NDELA and related nitrosamines. This adduct must now be considered seriously in terms of tumour etiology and could be used as a marker of human exposure if simpler, more sensitive methods of detection are developed. The ^{32}P-postlabelling assay for the gG adduct has the required sensitivity but, because of the lability of the adduct and the many steps involved, would be cumbersome to use in an epidemiological study. While additional work is required to elucidate the mechanisms of bioactivation of 2-hydroxyethylnitrosamine and NMOR, our work to date shows that the gG adduct could be formed by oxidation of the glycolaldehyde by-product of α-oxidation through the formation of glyoxal. The role that NHMOR may play in activation of NDELA remains unclear, but we have evidence for several pathways to the gG adduct. Although significant proportions of NDELA doses are excreted unchanged, it is a potent carcinogen. The fact that two adduct-producing moieties are generated in a single enzymatic oxidation of this compound could provide some explanation for this unusual phenomenon.

Acknowledgement

The support of this research by a grant (ES 03946) from the National Institutes of Environmental Health Sciences is gratefully acknowledged.

References

Airoldi, L., Bonfanti, M., Fanelli, R., Bove, B., Benfenati, E. & Gariboldi, P. (1984) Identification of a nitrosamino aldehyde and a nitrosamino acid resulting from β-oxidation of *N*-nitrosodiethanolamine. *Chem. Biol. Interactions* 51, 103–113

Atkins, G.M. & Burgess, E.M. (1968) The reactions of an *N*-sulfolylamine inner salt. *J. Am. Chem. Soc.* 90, 4744–4745

Chung, F.L. & Hecht, S.S. (1985) Formation of the cyclic 1,N^2-glyoxal-deoxyguanosine adduct upon reaction of *N*-nitroso-2-hydroxymorpholine with deoxyguanosine. *Carcinogenesis* 6, 1671–1673

Chung, F.L., Hecht, S.S. & Palladino, G. (1986) Formation of cyclic nucleic acid adducts from some simple α,β-unsaturated carbonyl compounds and cyclic nitrosamines. In: *The Role of Cyclic Nucleic Acid Adducts in Carcinogenesis and Mutagenesis* (B. Singer & H. Bartsch, eds), IARC Scientific Publications No. 70. Lyon, IARC, pp. 207–225

Czarnik, A.W. & Leonard, N.J. (1980) Unequivocal assignment of the skeletal structure of the guanine–glyoxal adduct. *J. Org. Chem.* 45, 3514–3517

Frankel-Conrat, H. (1954) Reactions of nucleic acid with formaldehyde. *Biochim. Biophys. Acta* 15, 307–309

Hecht, S.S., Lijinsky, W., Kovatch, R.M., Chung, F.L. & Saavedra, J.E. (1989) Comparative tumorigenicity of *N*-nitroso-2-hydroxymorpholine, *N*-nitrosodiethanolamine, and *N*-nitrosomorpholine in A/J mice and F344 rats. *Carcinogenesis* 10, 1475–1477

Loeppky, R.N. & Li, Y.E. (1991) Nitrosamine activation and detoxication through free radicals and their derived cations. In: *Relevance to Human Cancer of* N-*Nitroso Compounds, Tobacco Smoke and Mycotoxins* (O'Neill, I.K., Chen, J. & Bartsch, H., eds), IARC Scientific Publications No. 105. Lyon, IARC, pp. 375–382

Loeppky, R.N. & Srinivasan, A. (1995) Thiol oxidation by 1,2,3-oxadiazolinium ions, presumed carcinogens. *Chem. Res. Toxicol.* 8, 817–820

Loeppky, R.N., Tomasik, W., Kovacs, D.A., Outram, J.R. & Byington, K.H. (1984) Alternative bioactivation routes for β-hydroxynitrosamines. Biochemical and chemical model studies. In: N-*Nitroso Compounds: Occurrence, Biological Effects and Relevance to Human Cancer* (O'Neill, I.K., von Borstel, R.C., Miller, C.T., Long, J. & Bartsch, H., eds), IARC Scientific Publications No. 57. Lyon, IARC, pp. 429–436

Loeppky, R.N., Tomasik, W., Denkel, E. & Eisenbrand, G. (1987a) α-Nitrosaminoaldehydes: Highly reactive metabolites. In: *The Relevance of* N-*Nitroso Compounds to Human Cancer: Exposure and Mechanisms* (Bartsch, H., O'Neill, I.K. & Schulte-Hermann, R., eds), IARC Scientific Publications No. 84. Lyon, IARC, pp. 94–99

Loeppky, R.N., Tomasik, W. & Kerrick, B.E (1987b) Nitroso transfer from alpha-nitrosamino aldehydes: Implications for carcinogenesis. *Carcinogenesis* 8, 941–946

Loeppky, R.N., Fleischmann, E.D., Adams, J.E., Tomasik, W., Schlemper, E.O. & Wong, T.C. (1988) The 5-hydroxy-1,2,3-oxadiazolinium ion: Neighboring group interaction between *N*-nitroso and aldehyde carbonyl. *J. Am. Chem. Soc.* 110, 5946–5951

Loeppky, R.N., Erb, E. & Srinivasan, A. (1993a) The chemistry of putative intermediates in the bioactivation of β-oxidized nitrosamines. In: *The Chemistry and Biochemistry of Nitrosamines and Other* N-*Nitroso Compounds* (Loeppky, R.N. & Michejda, C.J., eds). Washington DC, American Chemical Society, pp. 334–336

Loeppky, R.N., Lee, M.P. & Mueller, S. (1993b) Modification of DNA by α-nitrosamino aldehydes. In: *DNA Adducts: Identification and Biological Significance* (Hemminki, K., Dipple, A., Shuker, D.E.G., Kadlubar, F.F., Segerbäck, D. & Bartsch, H., eds) IARC Scientific Publications No. 125. Lyon, IARC , pp. 429–432

Loeppky, R.N., Fuchs, A., Janzowski, C., Humberd, C., Goelzer, P., Schneider, H. & Eisenbrand, G. (1998a) Probing the mechanism of the carcinogenic activation of *N*-nitrosodiethanolamine with deuterium isotope effects: In vivo induction of DNA single strand breaks and related in vitro assays. *Chem. Res. Toxicol.* 11, 1556–1566

Loeppky, R.N., Goelzer, P. & Gu, F. (1998b) in preparation

Montoya, M.R., Miguel, J. & Mellado, R. (1995) Addition of α-dicarbonylcompounds to guanine and guanosine. *Gaz. Chim. Ital.* 125, 309–313

Okazaki, O., Persmark, M. & Guengerich, F. P. (1993) *N*-Nitroso-*N*-methylvinylamine: Reaction of the epoxide with guanyl and adenyl moieties to yield adducts derived from both parts of the molecule. *Chem. Res. Toxicol.* 6, 168–173

Park, M., Gu, F. & Loeppky, R.N. (1998) The synthesis of highly reactive multifunctional α,β-epoxy- and α-acetoxy-nitrosamines. *Tetrahedron Lett.* 39, 1287–1290

Shapiro, R. & Hachmann, J. (1966) The reaction of guanine derivatives with 1,2-dicarbonyl compounds. *Biochemistry* 5, 2799–2807

Shapiro, R., Cohen, B.I., Shiuey, S. & Maurer, H. (1969) On the reactions of guanine with glyoxal, pyruvaldehyde, and kethoxal, and the structure of the acylguanines. A new synthesis of N2-alkylguanines. *Biochemistry* 8, 238–245

Shapiro, R., Cohen, B.I. & Clagett, D.C. (1970) *J. Biol. Chem.*, 245, 2633–2639

Shapiro, R., Sodum, R.S., Everett, D.W. & Kundu, S.K. (1986) Reactions of nucleosides with glyoxal and acrolein. In: *The Role of Cyclic Nucleic Acid Adducts in Carcinogenesis and Mutagenesis* (B. Singer & H. Bartsch, eds), IARC Scientific Publications No. 70. Lyon, IARC, pp. 165–173

Staehlin, M. (1958) Reaction of tobacco mosaic virus nucleic acid with formaldehyde. *Biochem. Biophys. Acta* 29, 410–417

Staehlin, M. (1959) Inactivation of virus nucleic acid with glyoxal derivatives. *Biochim. Biophys. Acta* 31, 448–454

Stattsangi, P.D., Leonard, N.J. & Frihart, C.R. (1977) 1,N^2-Ethenoguanine and $N^{2,}$3-ethenoguanine. Synthesis and comparison of the electronic spectral properties of these linear and angular triheterocycles related to the Y bases. *J. Org. Chem.* 42, 3292–3296

Exocyclic DNA Adducts in Mutagenesis and Carcinogenesis
B. Singer & H. Bartsch
IARC Scientific Publications No. 150
International Agency for Research on Cancer, Lyon, 1999

Chapter IV. Physiochemical approaches to structural elucidation

Effects of 3,N^4-ethenodeoxycytidine on duplex stability and energetics

G.E. Plum, C.A. Gelfand & K.J. Breslauer

The exocyclic cytosine adduct 3,N^4-ethenocytosine is highly mutagenic in mammalian cells. We describe the impact of this adduct on DNA duplex stability. The adduct does not disrupt the overall B-form DNA structure; however, structural accommodation of the adduct is necessary at the lesion site. Despite the relatively small structural perturbation imparted by the adduct, there is a large adduct-induced destabilization of the DNA duplex. This destabilization is observed to be independent of the cross-strand partner base and neighbouring base pairs. The thermodynamic origins of the destabilization are, however, strongly dependent on the cross-strand partner base and neighbouring base pairs. Comparisons are made between the impact of the 3,N^4-ethenocytosine adduct and other lesions on DNA thermodynamics. The lesions are similar in that all result in destabilization of the DNA duplex. The magnitudes and the thermodynamic origins of that destabilization vary widely, the 3,N^4-ethenocytosine adduct being dramatically more destabilizing than other lesions. The impact of damaged sites on the stability of the DNA helix suggests that energetic differences between damaged and normal DNA may contribute to the recognition of damage by the cellular DNA repair machinery.

Introduction

One of the enduring mysteries in the repair of damage to DNA is the ability of the cellular repair system to identify specifically and repair efficiently a few hundred or thousand damaged sites in a sea of thousands of millions of base pairs of undamaged DNA. Over the past several years, we have been exploring an idea that we call 'energetic recognition' (Plum & Breslauer, 1994). Simply stated, the initial event in damage recognition may be the sensing by the scanning repair enzyme of a region of relative instability in the duplex. This instability need not be a disruption of base pairing, i.e. a bubble, or any other conformationally altered structure but may be manifest in a more subtle way. Detection of this region of instability would be followed by a more specific incursion by the repair protein that may or may not result in recognition of a substrate at the site.

The first step in evaluating the concept of energetic recognition is to determine whether DNA duplexes are destabilized by the presence of a lesion. In this paper, we describe recent studies of the highly mutagenic adduct 3,N^4-ethenocytosine (εC). The effects of the adduct on DNA duplex structure and stability will be described, with emphasis on the thermodynamic origins of that stability. Our complete study of the adduct is reported elsewhere (Gelfand *et al.*, 1998a); here, we highlight some of the more interesting results. The discussion includes reference to our previous work on several other lesions, including the abasic site (Gelfand *et al.*, 1996, 1998b) and the 1,N^2-propanoguanosine (Plum *et al.*, 1992) and 8-oxoguanosine (8-oxoGuo) adducts (Plum *et al.*, 1995).

Reaction of DNA with metabolites of the carcinogens vinyl chloride (Laib *et al.*, 1981; Kusmierek & Singer, 1982; Ciroussel *et al.*, 1990;

Matsuda *et al.*, 1995; Guichard *et al.*, 1996) and ethyl carbamate (Dahl *et al.*, 1978; Leithauser *et al.*, 1990; Park *et al.*, 1990), reaction with products of lipid peroxidation (El Ghissassi *et al.*, 1995; Nair *et al.*, 1997) or hepatic copper poisoning (Nair *et al.*, 1996) can result in formation of the εC adduct. The presence of the etheno moiety partially obstructs the Watson-Crick base-pairing face of the cytosine base (Figure 1). Therefore, it is reasonable to expect physiochemical, i.e. structural and thermodynamic, consequences of formation of the εC adduct in DNA.

The inability to form Watson-Crick base-pairing interactions is presumed to be the origin of the mutagenic potential of the εC adduct. Extensive miscoding due to εC is observed; each of the four nucleotide bases may appear opposite the adduct (Zhang *et al.*, 1995; Shibutani *et al.*, 1996). The εC adduct is highly mutagenic in mammalian cells but only weakly so in *Escherichia coli* (Moriya *et al.*, 1994). Enzyme activities that repair εC with good efficiency have been demonstrated to be due to double-stranded uracil-DNA glycosylase in *E. coli* and to mismatch-specific thymine DNA glycosylase in humans (Saparbaev & Laval, 1998).

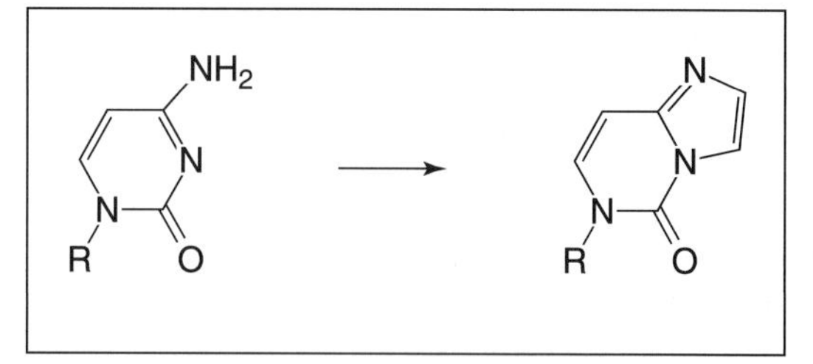

Figure 1. Formation of the 3,N^4-ethenocytosine adduct (right) obstructs the Watson-Crick base pairing face of cytosine (left).

CGCATG**N**GTACGC
GCGTAC**M**CATGCG

where N and M are
A, C, G, T or εC

Figure 2. The series of 12 oligodeoxyribonucleotides studied

N and M represent the four standard nucleotide bases, as well as 3,N^4-ethenocytosine (εC). In the text, we refer to the duplexes using the notation 'strand 1', 'strand 2'. Two families of lesion containing duplexes are cited: four in which the εC adduct is in strand 1 (εC.N) and four in which it is in strand 2 (N.εC).

The system studied and methods employed

Adduct-containing oligonucleotides

A series of 13-mer oligodeoxyribonucleotide duplexes was studied. The 12 duplexes were identical except for the central 'base pair' (Figure 2): four of the duplexes contained the canonical Watson-Crick base pairs in the central position; four contained the εC in the central position, with each of the four normal bases opposite. The remaining four duplexes also contained the adduct in the central position, but on the other strand with each of the four normal bases opposite. The presence of the adduct on each of the strands allows for preliminary examination of the effects of sequence context. Below, the duplexes are identified by the central base pair (N·M), as seen in Figure 2.

Spectroscopic and calorimetric methods

The duplexes were characterized structurally by circular dichroism spectropolarimetry and thermodynamically by a combination of absorbance-monitored thermal dissociation (ultraviolet melting) and differential scanning calorimetry (DSC). Melting temperature (T_m) values were derived from the ultraviolet melting experiments at 50 μmol/L oligonucleotide concentration. The values for the enthalpy change associated with duplex disruption, ΔH°, are derived from DSC thermograms. The values for the free energy change associated with duplex disruption, ΔG°, were derived by combining the DSC-determined ΔH° value and the T_m over a range of oligonucleotide concentrations (Plum *et al.*, 1995; Gelfand *et al.*, 1998b).

Equation 1 was used to calculate the free energy change

$$\Delta G^\circ = \Delta H^\circ \left(1 - \frac{T}{T_m}\right) - RT\,(n_{eff} - 1)\ln\left(\frac{C_t}{2n_{eff}}\right)$$

Eq.1

where C_t is the total DNA strand concentration and n_{eff} is the effective molecularity. We introduced the parameter n_{eff} to absorb deviations from the two-state, bimolecular behaviour that is assumed in the conventional model based on van't Hoff equation. In the van't Hoff model, the slope of a $1/T_m$ versus ln C_t plot (m) is dependent on the molecularity (n = 2 for duplexes) and the transition enthapy change (equation 2).

$$m = \frac{(n-1)R}{\Delta H^\circ}$$

Eq.2

Rearranging equation 2 and solving for n leads to equation 3.

$$n = \frac{m\Delta H^\circ}{R} + 1$$

Eq.3

Substitution of the calorimetrically determined ΔH° value and the value of m from the concentration dependence of T_m gives the value for the effective molecularity, n_{eff}, which may deviate significantly from the formal molecularity determined by counting strands in the complex.

Measurement of the transition enthalpy change

The preferred method of measurement of the transition enthalpy is calorimetry. DSC provides a direct, model-independent measurement of ΔH°. With this method, the difference in heat capacity between the DNA duplex-containing solution of interest and a reference solution is measured as a function of temperature. The area under the resultant curve of C_p versus T corresponds to the enthalpy change of the duplex dissociation transition.

In addition to calorimetric measurement of ΔH°, methods based on models of the transition process as in the van't Hoff equation may be used (Marky & Breslauer, 1987). The shape of the melting curve (monitored by either optical methods or calorimetry) contains information about the value of ΔH°. A sharp curve (the process goes to completion over a narrow range of temperature) indicates a large value for ΔH°. Conversely, a broad transition indicates a low ΔH° value. Such differences in the shapes of transitions are frequently confused with differences in the cooperativity of the processes. Differences in cooperativity can exist and will manifest themselves similarly; however, the underlying thermodynamic reasons for the shape of the transitions are different. The only way to evaluate cooperativity effects experimentally is by comparing the ΔH° value determined from a direct calorimetric measurement to a ΔH° value determined indirectly from the shape of the transition curve (Sturtevant, 1987; Breslauer *et al.*, 1992).

Definition of 'ΔΔ' parameters

Note that all of the thermodynamic parameters discussed in this paper refer to the duplex dissociation process. Differences in the dissociation thermodynamics due to the presence of the lesion are presented as 'ΔΔ' parameters. These parameters are computed by subtracting the appropriate value measured for the Watson-Crick control duplex from the value measured for the adduct-containing duplex. For all 'ΔΔ' parameters, a positive sign indicates stabilization of the duplex by the modification, whereas a negative sign indicates destabilization.

The lesion does not disrupt duplex global structure: Circular dichroism spectra indicate that the presence of the εC adduct has only modest structural effects (Gelfand *et al.* 1998a), regardless of the base opposite the adduct and the sequence context. In all cases, the global B-form structure of the duplex is retained. Detailed nuclear magnetic resonance (NMR)-derived structures from the laboratory of de los Santos (Cullinan *et al.*, 1996; Korobka *et al.*, 1996; Cullinan *et al.*, 1997) reinforce the conclusion that the B-form DNA structure is retained despite the presence of εC. These structures also give insight into the possibilities for structural accommodation of the adduct, either by rotation of the bases away from the central helix axis, by rotation about the glycosidic bond or by partial stacking of the opposing base with the adduct.

The lesion reduces duplex thermal and thermodynamic stability: Regardless of the base opposite or bases next to the εC adduct, the thermal stability of the duplex is reduced significantly. The value of ΔT_m does depend on the sequence context and ranges from –10.3 to –18.1 °C. The magnitudes of adduct-induced thermal destabilization observed for εC-containing duplexes are consistent with the effects of the other lesions we have studied (Plum *et al.*, 1992, 1995; Gelfand *et al.*, 1996, 1998b). Such comparisons are appropriate because oligonucleotides of identical sequence are compared under identical conditions of solution and DNA concentration. When the exocyclic adduct of guanine, 1,N^2-propanoguanine, is placed opposite A or C, a ΔT_m of approximately –12 is observed (Plum *et al.*, 1992). The introduction of an abasic site results in ΔT_m values ranging from –12.1 to –20.7 °C depending on the sequence context (Gelfand *et al.*, 1998b). The effects of the 8-oxoGuo adduct on T_m are more modest (Plum *et al.*, 1995), with ΔT_m values between –3.2 and –1.0 when C, G or T is opposite the adduct. When the 8-oxoGuo adduct is opposite A, thermal stability is increased relative to the duplex containing a GA mismatch (ΔT_m = 7.6 °C).

The thermodynamic destabilization of the duplex by the εC adduct is very large. The values of the free energy change, ΔG°, associated with the disruption of the normal Watson-Crick duplexes range from 19.6 to 20.5 kcal/mol. Therefore, the parent duplexes are of identical stability, within experimental error. Introduction of the εC adduct reduces the ΔG° to 5.1 to 7.6 kcal/mol, with an average value of 6.4 kcal/mol. The observed changes in ΔG° due to the lesion fall in a narrow range, –12 to –15.3 kcal/mol. This reduction in favourable free energy (ΔΔG° = –13.0 on average) corresponds to a factor of 10^{-10} in the association constant of lesion-containing duplexes. This adduct-induced destabilization is greater than that observed for the other adducts studied.

The impact of an abasic site on duplex stability varies widely (ΔΔG° = –3.0 to –11.3 kcal/mol), depending both on sequence context and the base opposite the lesion (Gelfand *et al.*, 1998b). The 8-oxoGuo adduct imparts relatively small, though thermodynamically significant, free energy effects (ΔΔG° = –2.0 to +3.4 kcal/mol) (Plum *et al.*. 1995). The 1,N^2-propanoguanine adduct destabilizes the duplex by about 4 kcal/mol (Plum *et al.*, 1992). Thus, the exocyclic etheno adduct of cytosine is dramatically more destabilizing than a 'complementary' propano adduct of guanine. Taken together, these data indicate that the mere disruption of Watson-Crick base pairing is not sufficient to determine the impact of adducts on duplex stability. How that base pairing is blocked is important in determining a damaged site's impact on duplex stability.

Changes in T_m are not necessarily good predictors of changes in thermodynamic stability: It is often assumed that observed changes in T_m can be used to evaluate the impact of adducts on the thermodynamic stability (ΔG°) of the DNA duplex. The relative ease with which T_m is measured makes the measurement popular but not rigorous. Frequently, the trends in T_m and in ΔG° are similar; however, there is in general no simple correspondence between the two parameters. As is clear from examination of Figure 3, there is not an exact correlation between ΔT_m and ΔΔG° values. This finding is consistent not only with observations of lesion-containing duplexes but with duplexes comprised exclusively of normal Watson-Crick base pairs.

The reason for the failure of ΔT_m to reflect reliably the lesion-induced changes in thermodynamic stability (ΔΔG°) is neglect of the temperature dependence of the duplex stability. T_m values reflect the behaviour of the duplex at high temperature. The free-energy changes are evaluated for a low-temperature standard state, arbitrarily defined but typically 25 or 37 °C. We use 25 °C as the standard temperature in our studies of duplex thermodynamics. These low-temperature standard states are not merely of academic interest to the physical chemist but correspond to the tem-

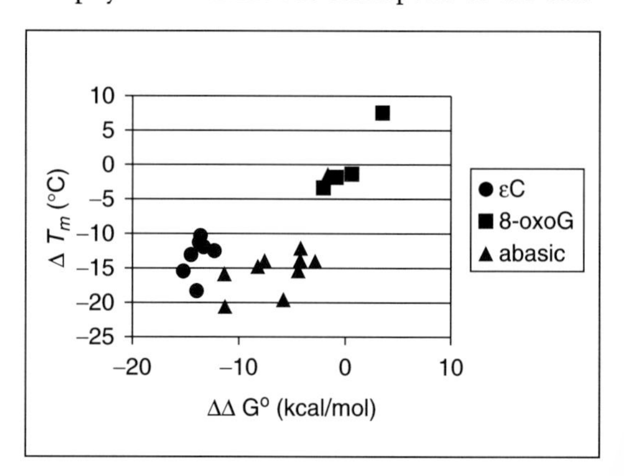

Figure 3. Plot of the change in duplex T_m and the change in ΔG° due to the 3,N^4-ethenocytosine (εC) (Gelfand *et al.* 1998b) and 8-oxoguanine (8-oxoG) (Plum *et al.*, 1995) adducts and the abasic site (Gelfand *et al.*, 1996, 1998a) indicates that the two parameters are not well correlated.

perature domain of biologically interesting processes. To relate data collected at high temperature to the low-temperature values that are desired requires knowledge of the temperature dependence of duplex stability. That dependence is embodied in the enthalpy change associated with the process. Measurement of ΔH^o forms the foundation for the thermodynamic characterization of the oligonucleotides described here.

The thermodynamic origins of duplex destabilization depend on the sequence context: The changes in stability ($\Delta\Delta G^o$) induced by the εC adduct reflect a combination of enthalpic ($\Delta\Delta H^o$) and entropic effects ($\Delta\Delta S^o$). The magnitudes of these effects vary with the sequence context and with the base opposite the adduct. The relative contributions of the adduct-induced changes in the enthalpy and entropy terms for disruption of the duplexes studied can be seen in Figure 4. Recalling that the impact of the εC adduct on duplex stability is independent of sequence context (within the limited set studied) and opposing base leads to the most remarkable observation about the thermodynamic impact of εC. The value for $\Delta\Delta H^o$ ranges from –4.6 to –32.8 kcal/mol, yet that for $\Delta\Delta G^o$ ranges from only –12.0 to –15.3 kcal/mol. Regardless of the magnitude of $\Delta\Delta H^o$, there is an almost exactly compensating change in the entropy term, resulting in a large impact on duplex stability that is invariant with sequence context and opposing base. In all cases examined, the impact of εC on the enthalpy term is destabilizing; however, the entropy term may be either stabilizing or destabilizing.

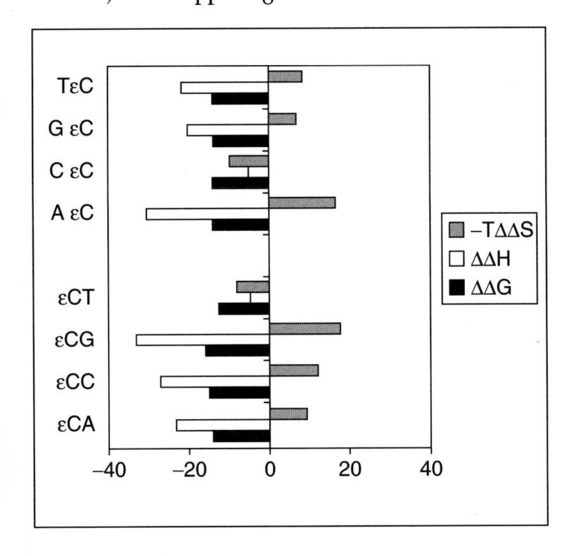

Figure 4. Impact of the 3,N^4-ethenocytosine (εC) adduct on duplex thermodynamics

The contributions of enthalpic and entropic effects to duplex stability (free energy change) are presented for each adduct-containing duplex (Gelfand *et al.*, 1998a). The units for each parameter ($\Delta\Delta G^o$, $\Delta\Delta H^o$ and $-T\Delta\Delta S^o$) plotted on the abscissa are kcal/mol and T = 25 °C.

The thermodynamic origins of duplex destabilization depend strongly on the nature of the lesion. The enthalpic cost of an abasic site is great and depends very strongly on the sequence context and only secondarily on the opposing base (Gelfand *et al.*, 1998b). When two guanines flank the abasic site, $\Delta\Delta H^o$ is on average –9.9 kcal/mol. With two flanking cytosines, $\Delta\Delta H^o$ is on average –38.6 kcal/mol. In both cases, substantial entropic compensation is observed. The compensation does not result in invariant stability with respect to sequence context and opposing base as is observed for εC. In contrast, the 8-oxoGuo adduct is enthalpically destabilizing only when placed opposite a C residue (Plum *et al.*, 1995). Regardless of the opposing base, the 8-oxoGuo adduct-induced change in the entropy term is unfavourable for duplex stability.

The presence of a lesion may alter the mechanism of duplex melting: Oligonucleotide duplex dissociation occurs by a combination of two processes. Figure 5 depicts the two equilibria that describe these processes. The melting process begins with breaking of a subpopulation of base pairs, typically strating at the ends. This process, which is usually called 'unzipping' or 'fraying', does not involve a change in the number of species and is, therefore, independent of concentration. The strand separation process is dependent on concentration because the equilibrium is between the duplex (one species) and the single strands (two species).

The relative contributions of the unzipping and strand separation processes depend on the length of the oligonucleotide duplex. The melting of short oligonucleotides is dominated by the strand separation process. This dominance of strand separation is the basis of the well-known two-state approximation. That is, there are no

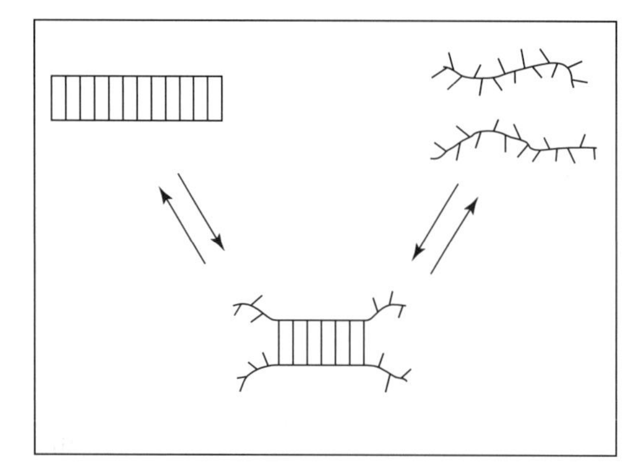

Figure 5. Mechanism of oligonucleotide dissociation

The equilibrium (depicted on the left) between the fully base-paired and partially base-paired states ('fraying' or 'unzipping') is coupled to the equilibrium (depicted on the right) between the partially base-paired states and the fully dissociated single strands.

thermodynamically significant partially base-paired intermediate states. The effective molecularity equals the formal molecularity of 2. In contrast to the behaviour of short oligonucleotides, polymer melting is independent of strand concentration. It is dominated by the expansion of melted regions and not by strand separation. The effective molecularity is 1. The melting of intermediate-length oligonucleotides may have significant contributions from partially base-paired intermediate states. The monomolecular fraying and the bimolecular strand separation processes both contribute. The effective molecularity assumes intermediate values, $1 < n_{eff} < 2$.

Examination of Table 1 reveals that the four normal Watson-Crick 13-mer duplexes studied melt with the expected molecularity of 2. Such behaviour is expected for oligonucleotide duplexes of this length. Interestingly, incorporation of the εC adduct results in significant deviations in the effective molecularity from the expected value of 2. We have observed deviation from $n_{eff} = 2$ for many lesions or mismatch containing duplexes (Plum *et al.*, 1995; Gelfand *et al.*, 1996, 1998a,b). Most of these duplexes have n_{eff} values between 1 and 2; however, a few have $n_{eff} > 2$. The normal and adduct-containing duplexes examined are identical in length (13 base pairs); therefore, the observed differences in n_{eff} are not trivial consequences of duplex length. While any explanation for this behaviour should be viewed as tentative, it seems likely that the relative contributions of unzipping and strand separation are altered for adduct-containing duplexes. The apparent tendency towards monomolecular behaviour may reflect a reduction in the free-energy cost of strand separation. Thus, the anomalous value of n_{eff} may indicate an adduct-induced predisposition of the duplex towards melting. Alternatively, the deviation from the expected value of n_{eff} for lesion-containing duplexes may reflect some other as yet unknown deviation from the two-state, bimolecular melting behaviour typically observed for short oligonucleotide duplexes.

Concluding remarks

The thermal and thermodynamic stabilities of DNA are reduced significantly by the presence of an εC adduct. The magnitude of the thermodynamic destabilization induced by εC is greater than that observed for other DNA lesions. Significantly, unlike most other lesions, the thermodynamic destabilization does not depend on the cross-strand partner base or the neighbouring

Table 1. Effective molecularity of normal and 3,N^4-ethenocytosine (ε)-containing duplexes

Normal duplexes				3,N^4-Ethenocytosine-containing duplexes							
A.T	C.G	G.C	T.A	ε C.A	ε C.C	ε C.G	ε C.T	A. ε C	C. ε C	G. ε C	T. ε C
2.01	2.09	2.04	2.06	1.87	1.85	1.67	2.01	1.73	2.20	1.87	1.88

From Gelfand *et al.* (1998a)

base pairs. Regardless of the sequence context, the enthalpic effect of the εC adduct is destabilizing. The entropic effects of the adduct depend on the sequence context and can be either stabilizing or destabilizing, resulting in a common net destabilization. Differences in the effective molecularity observed for adduct-containing oligonucleotide duplexes can be interpreted in terms of an altered mechanism of duplex dissociation, although alternative explanations cannot currently be eliminated. Clearly, the εC adduct exerts significant thermodynamic perturbations on DNA in the absence of large-scale structural perturbations.

To date, several single-base damaged DNA lesions have been subjected to extensive thermodynamic characterization. In all cases, the introduction of a damaged site is thermally and thermodynamically destabilizing relative to the normal Watson-Crick parent DNA. The structural perturbations associated with the damaged sites are limited to the immediate vicinity of the damage—the base opposite and perhaps the nearest neighbours. The details of the thermodynamic origins, however, differ markedly among the various lesions. The magnitudes of the thermodynamic effects of the lesions suggest that the energetic consequences of damage to DNA may propagate beyond the structurally perturbed sites. Taken together, the structural and thermodynamic data are consistent with the idea of energetic recognition. The initial recognition event of the lesion repair process may be dependent on the thermodynamic anomalies imparted by damage and not on conformationally altered states of the DNA that promote formation of specific interactions with the repair proteins.

Acknowledgements

The authors thank their collaborators Professor Arthur P. Grollman and Professor Francis Johnson of the State University of New York at Stony Brook. They also thank Robert Rieger for synthesizing the lesion containing oligonucleotides. This research was supported by NIH grants GM-23509 and GM-34469 to K.J.B. and CA-47795 to Arthur P. Grollman.

References

Breslauer, K.J., Friere, E. & Straume, M.S. (1992) Calorimetry: A tool for DNA and ligand–DNA studies. *Meth. Enzymol.* 211, 533–567

Ciroussel, F., Barbin, A., Eberle, G. & Bartsch, H. (1990) Investigations on the relationship between DNA ethenobase adduct levels in several organs of vinyl chloride-exposed rats and cancer susceptibility. *Biochem. Pharmacol.* 39, 1109–1113

Cullinan, D., Johnson, F., Grollman, A.P., Eisenberg, M. & de los Santos, C. (1997) Solution structure of a DNA duplex containing the exocyclic lesion 3,N^4-etheno-2′-deoxycytidine opposite 2′-deoxyguanosine. *Biochemistry* 36, 11933–11943

Cullinan, D., Korobka, A., Grollman, A.P., Patel, D.J., Eisenberg, M. & de los Santos, C. (1996) NMR solution structure of an oligodeoxynucleotide duplex containing the exocyclic lesion 3,N^4-etheno-2′-deoxycytidine opposite thymidine: Comparison with the duplex containing deoxyadenosine opposite the adduct. *Biochemistry* 35, 13319–13327

Dahl, G.A., Miller, J.A. & Miller, E.C. (1978) Vinyl carbamate as a promutagen and a more carcinogenic analog of ethyl carbamate. *Cancer Res.* 38, 3793–3804

El Ghissassi, F., Barbin, A., Nair, J. & Bartsch, H. (1995) Formation of 1,N^6-ethenoadenine and 3,N^4-ethenocytosine by lipid peroxidation products and nucleic acid bases. *Chem. Res. Toxicol.* 8, 278–283

Gelfand, C.A., Plum, G.E., Grollman, A.P., Johnson, F. & Breslauer, K.J. (1996) Impact of a bistrand abasic lesion on DNA duplex properties. *Biopolymers* 38, 439–445

Gelfand, C.A., Plum, G.E., Grollman, A.P., Johnson, F. & Breslauer, K.J. (1998a) The impact of an exocyclic cytosine adduct on DNA duplex properties: Significant thermodynamic consequences despite modest lesion-induced structural alterations. *Biochemistry* 37, 12507–12512

Gelfand, C.A., Plum, G.E., Grollman, A.P., Johnson, F. & Breslauer, K.J. (1998b) The thermodynamic consequences of an abasic lesion in duplex DNA are strongly dependent on base sequence. *Biochemistry* 37, 7321–7327

Guichard, Y., El Ghissassi, F., Nair, J., Bartsch, H. & Barbin, A. (1996) Formation and accumulation of

DNA ethenobases in adult Sprague-Dawley rats exposed to vinyl chloride. *Carcinogenesis* 17, 1553–1559

Korobka, A., Cullinan, D., Cosman, M., Grollman, A.P., Patel, D.J., Eisenberg, M. & de los Santos, C. (1996) Solution structure of an oligodeoxynucleotide duplex containing the exocyclic lesion 3,N^4-etheno-2′-deoxycytidine opposite 2′-deoxyadenosine, determined by NMR spectroscopy and restrained molecular dynamics. *Biochemistry* 35, 13310–13318

Kusmierek, J.T. & Singer, B. (1982) Chloroacetaldehyde-treated ribo- and deoxyribopolynucleotides. 2. Errors in transcription by different polymerases resulting from ethenocytosine and its hydrated intermediate. *Biochemistry* 21, 5723–5728

Laib, R.J., Gwinner, L.M. & Bolt, H.M. (1981) DNA alkylation by vinyl chloride metabolites: Etheno derivatives or 7-alkylation of guanine? *Chem.-Biol. Interactions* 37, 219–231

Leithauser, M.T., Liem, A., Steward, B.C., Miller, E.C. & Miller, J.A. (1990) 1,N^6-Ethenoadenosine formation, mutagenicity and murine tumor induction as indicators of the generation of an electrophilic epoxide metabolite of the closely related carcinogens ethyl carbamate (urethane) and vinyl carbamate. *Carcinogenesis* 11, 463–473

Marky, L.A. & Breslauer, K.J. (1987) Calculating thermodynamic data for transitions of any molecularity from equilibrium melting curves. *Biopolymers* 26, 1601–1620

Matsuda, T., Yagi, T., Kawanishi, M., Matsui, S. & Takebe, H. (1995) Molecular analysis of mutations induced by 2-chloroacetaldehyde, the ultimate carcinogenic form of vinyl chloride, in human cells using shuttle vectors. *Carcinogenesis* 16, 2389–2394

Moriya, M., Zhang, W., Johnson, F. & Grollman, A.P. (1994) Mutagenic potency of exocyclic DNA adducts: Marked differences between *Escherichia coli* and simian kidney cells. *Proc. Natl Acad. Sci. USA* 91, 11899–11903

Nair, J., Sone, H., Nagao, M., Barbin, A. & Bartsch, H. (1996) Copper-dependent formation of miscoding etheno-DNA adducts in the liver of Long Evans cinnamon (LEC) rats developing hereditary hepatitis and hepatocellular carcinoma. *Cancer Res.* 56, 1267–1271

Nair, J., Vaca, C.E., Velic, I., Mutanen, M., Valsta, L.M. & Bartsch, H. (1997) High dietary omega-6 polyunsaturated fatty acids drastically increase the formation of etheno-DNA base adducts in white blood cells of female subjects. *Cancer Epidemiol. Biomarkers Prev.* 6, 597–601

Park, K.K., Surh, Y.J., Steward, B.C. & Miller, J.A. (1990) Synthesis and properties of vinyl carbamate epoxide, a possible ultimate electrophilic and carcinogenic metabolite of vinyl carbamate and ethyl carbamate. *Biochem. Biophys. Res. Commun.* 169, 1094–1098

Plum, G.E. & Breslauer, K.J. (1994) DNA lesions: A thermodynamic perspective. *Ann. NY Acad. Sci.* 726, 45–56

Plum, G.E., Grollman, A.P., Johnson, F. & Breslauer, K.J. (1992) Influence of an exocyclic guanine adduct on the thermal stability, conformation, and melting behavior of a DNA duplex. *Biochemistry* 31, 12096–12102

Plum, G.E., Grollman, A.P., Johnson, F. & Breslauer, K.J. (1995) Influence of the oxidatively damaged adduct 8-oxodeoxyguanosine on the conformation, energetics, and thermodynamic stability of a DNA duplex. *Biochemistry* 34, 16148–16160

Saparbaev, M. & Laval, J. (1998) 3,N^4-Ethenocytosine, a highly mutagenic adduct, is a primary substrate for *Escherichia coli* double-stranded uracil-DNA glycosylase and human mismatch-specific thymine-DNA glycosylase. *Proc. Natl Acad. Sci. USA* 95, 8508–8513

Shibutani, S., Suzuki, N., Matsumoto, Y. & Grollman, A.P. (1996) Miscoding properties of 3,N^4-etheno-2′-deoxycytidine in reactions catalyzed by mammalian DNA polymerases. *Biochemistry* 35, 14992–8

Sturtevant, J.M. (1987) Biochemical applications of differential scanning calorimetry. *Annu. Rev. Phys. Chem.* 38, 463–488

Zhang, W., Johnson, F., Grollman, A.P. & Shibutani, S. (1995) Miscoding by the exocyclic and related DNA adducts 3,N^4-etheno-2′-deoxycytidine, 3,N^4-ethano-2′-deoxycytidine, and 3-(2-hydroxyethyl)-2′-deoxyuridine. *Chem. Res. Toxicol.* 8, 157–163

Exocyclic DNA Adducts in Mutagenesis and Carcinogenesis
B. Singer & H. Bartsch
IARC Scientific Publications No. 150
International Agency for Research on Cancer, Lyon, 1999

Solution structures of DNA duplexes containing the exocyclic lesion 3,N^4-etheno-2′-deoxycytidine

D. Cullinan, M. Eisenberg & C. de los Santos

Introduction

Since the mid-1970s, it has been known that exposure to vinyl chloride, used in the manufacture of synthetic polymers, increases the incidence of human liver sarcoma (Creech & Johnson, 1974; Maltoni *et al.*, 1974). Since then, it has been shown that vinyl chloride and some related compounds are metabolized by the hepatic P450 system to produce chloroacetaldehyde (Guengerich *et al*, 1979; Barbin *et al.*, 1985; Scherer *et al.*, 1986). This alkylating agent can then react with cellular DNA to produce several exocyclic lesions, including 3,N^4-etheno-2′-deoxycytidine (εdC) (Barbin & Bartsch, 1986; Kusmierik & Singer, 1992). εdC has also been shown to be formed endogenously in human and rat liver after lipid peroxidation (Nair *et al.*, 1995).

There have been a number of investigations of the mutagenic properties of εdC, both *in vitro* and *in vivo*. With the Klenow fragment of DNA polymerase I, replication of a template containing εdC showed that primarily dA and T were incorporated opposite the εdC (Singer & Spengler, 1986; Simha *et al.*, 1991; Zhang *et al.*, 1995). Tranformation of *Escherichia coli* with vectors containing εdC showed that the lesion is only weakly mutagenic in bacteria, with about a 2% mutation frequency, and dA and T again being the most commonly misincorporated bases (Palejwala *et al.*, 1991; Basu *et al.*, 1993; Palejwala *et al.*, 1993; Moriya *et al.*, 1994). Studies with mammalian polymerases α and δ *in vitro* showed that dA and T incorporation predominated, while polymerase β incorporated primarily dC (Shibutani *et al.*, 1996). Furthermore, studies in mammalian cells *in vivo* showed that εdC is highly mutagenic, with T and dA incorporation being most common (Moriya *et al.*, 1994; see also Barbin *et al.*, this volume).

A thorough study of the thermodynamic properties of εdC was performed recently (Gelfand *et al.*, 1998). The presence of the lesion at the centre of a 13-mer duplex of DNA causes an average 70% loss in Gibb's free energy as compared with the same duplex that has an unmodified base pair at the centre. The significant destabilization brought about by the presence of the lesion was found to be mostly independent of the sequence context of the lesion and the base pair opposite the lesion. The enthalpy, however, was shown to be more sequence-specific and dependent on the opposite base, the drop in overall enthalpy for the sequence used in our study being greatest for εdC•dA; εdC•dG and εdC•T were very similar, and εdC•dC showed the lowest change in enthalpy.

The proteins that repair εdC lesions in bacteria and humans have been identified (Hang *et al.*, 1998; Saparbaev & Laval, 1998). In these studies, the *E. coli* mismatch-specific uracil glycosylase and the human mismatch-specific thymine-DNA glycosylase readily removed εdC lesions from double-stranded DNA substrates. Additionally, in the case of the bacterial enzyme, εdC was removed more efficiently than thymine bases from G•T mispairs, which were thought to be its primary substrate. Furthermore, Saparbaev and Laval (1998) compared the efficiency of the excision rates with different residues opposite the lesion site, making it particularly interesting for our own work. For the bacterial mismatch-specific uracil glycosylase, the rates of excision were surprisingly similar, with only a twofold difference between εC•dG, the most active substrate, and εC•dA, the least active.

In this paper, we summarize the structural studies that we have performed on the 11-mer duplexes d(C-G-T-A-C-εC-C-A-T-G-C)•d(G-C-A-T-G-X-G-T-A-C-G), where X is dG, dA, T and dC.

Henceforth, these duplexes are referred to in this paper as εdC•dG, εdC•dA, εdC•T and εdC•dC. These four structures were solved, as we will briefly explain, by molecular dynamics simulations, restrained by high-resolution nuclear magnetic resonance (NMR) distance restraints and full-relaxation matrix back-calculations. The chemical structure of εdC and the numbering scheme of the duplex sequence used in these studies are shown in Figure 1.

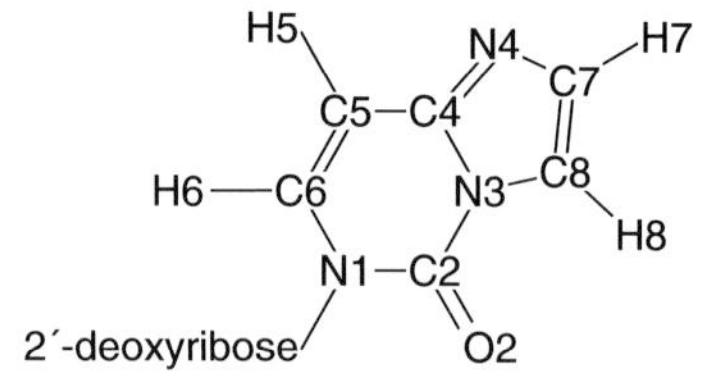

H5 N4 H7
C5–C4 C7
H6—C6 N3–C8 H8
N1–C2
2′-deoxyribose O2

3,N^4-etheno-2′-deoxycytosine

C1	G2	T3	A4	C5	εC6	C7	A8	T9	G10	C11
G22	C21	A20	T19	G18	X17	G16	T15	A14	C13	G12

Figure 1. Chemical structure of 3,N^4-etheno-2′-deoxycytidine and the numbering scheme of the duplexes used in the studies

Methods and materials

The methods used to elucidate three of the four structures presented in this paper were published previously (Cullinan *et al.*, 1996; Korobka *et al.*, 1996; Cullinan *et al.*, 1997); the methods used for εdC•dC were similar to those used for εdC•dG. Therefore, we include only a summary of the procedures herein. εC was synthesized and incorporated into the oligodeoxynucleotide sequence by standard phosphoramidite procedures. The strands were annealed in a 1:1 ratio, and the duplex was then dissolved in 10 mmol/L phosphate buffer, pH 6.9, with 50 mmol/L NaCL and 1 mmol/L EDTA in 99.96% D_2O or 90% H_2O/10% D_2O.

One- and two-dimensional NMR experiments were carried out, primarily concentrating on proton resonances. Distance restraints were calculated from the NMR data for use in simulations of molecular dynamics. Starting structures of A- and B-form DNA were created, and NOE distance-restrained dynamics were performed with the X-PLOR program (Brünger, 1993). Convergence of the structures to a single conformation that satisfied the NMR data was used as the criterion for a final distance-restrained structure. These structures were then passed through an additional round of molecular dynamics, restrained by back-calculation of the full-relaxation matrix.

Results

The results of three of these structural studies have been reported in detail elsewhere (Cullinan *et al.*, 1996; Korobka *et al.*, 1996; Cullinan *et al.*, 1997). We summarize here the points that seem most interesting, with minimal reference to the NMR data. In each of the four cases, it is worth pointing out that the perturbation of the structure is located at the central base pairs. The lesion-containing base pair must make adjustments to accommodate the exocyclic ring, and the flanking base pairs are affected by the changes to the central base pair. The data are consistent, however, in showing that the three base pairs at each end of the duplex are unaffected by whatever disruption of the structure takes place at the centre of the duplex. In the discussion of the structures below, the structural parameters (e.g. Xdisp, rise, shift, slide) were measured by the Curves program (Lavery & Sklenar, 1989). (The values of some parameters presented here differ somewhat from those published previously, since we used more recent files.)

ε**dC•dG** (Cullinan *et al.*, 1997)

The εdC•dG structure adopts a right-handed helix of the B-form family, with no gross perturbation of the sugar phosphate backbone. All of the residues have a glycosidic torsion angle in the *anti* orientation, and all of the sugar puckers are in a south orientation (C2′-endo/C3′-exo), except for the sugar of εdC, which is in the C3′-endo/C4′-exo range. There is a bend in the axis of roughly 14° towards the major groove at the lesion site. Steric clashes at the lesion site are relieved by formation of a 'wobble' structure by the bases, in which the dG opposite the lesion is tucked into the minor groove (normalized Xdisp of –0.17 nm) and the εdC is swung into the major groove (normalized Xdisp of 0.11 nm), such that it is stacked directly underneath the base of C5. For the C5-εC6 base step, this gives an average value of 0.27 nm for the shift and –0.12 nm for the slide, while for the G16-G17 base step, the values were 0.08 nm and

–0.22 nm, respectively. The εdC base is also sharply rolled, the values for the C5-εC6 and εC6-C7 base steps being 14.2° and –23.1°, respectively. This roll causes a much smaller roll of the base of C7, and this central disruption of the structure seems to be the origin of the bend in the axis, with the base pairs following C7•G16 stacking parallel to them. This structure is stabilized by the presence of three hydrogen bonds involving G17, one of which is with the εdC base. The strongest of the hydrogen bonds is between G17(N2H2) and C7(O4′), with weaker bonds present across the base pair between C17(H1) and εC6(O2) and between C17(N2H1) and G18(O4′). Normal Watson-Crick alignment is present in all base pairs other than the lesion site itself. Several stereo views of the central three base pairs are shown in Figure 2, while the full duplex is shown in Figure 6.

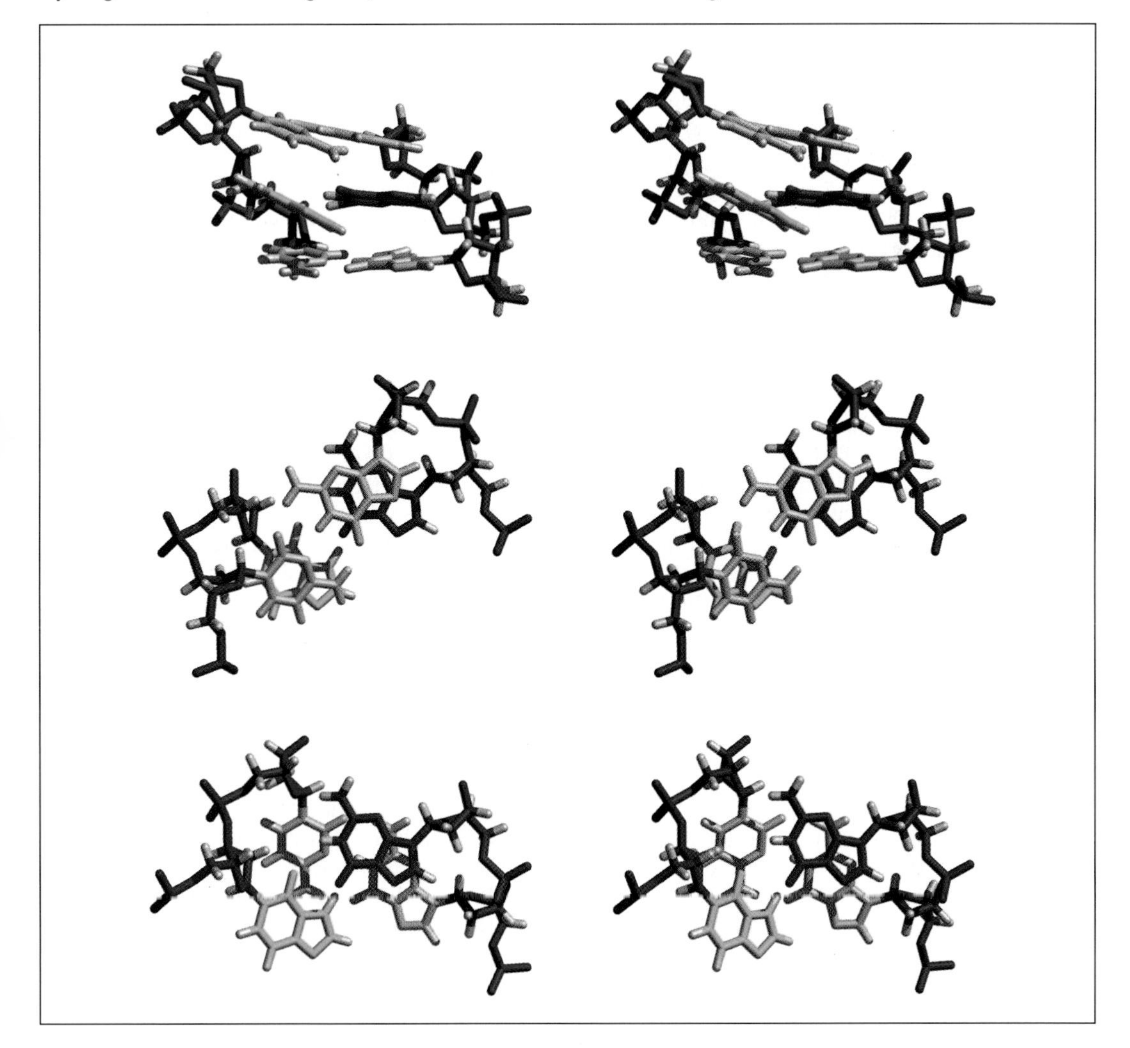

Figure 2. 3,N^4-Etheno-2´-deoxycytidine (εdC)–deoxyguanosine

This steroview shows the central three base pairs of the duplex. The εdC base is in green and the dG opposite it is in red. (Top) View from the major groove. (Centre) Axial view, showing stacking interaction with the C5–G18 base pair above the εC6–G17 base pair. (Bottom) Another view from the same perspective showing the εC6–G17 base pair above the C7–G16 base pair

εdC•dA (Korobka *et al.*, 1996)

The εdC•dA structure is also a right-handed helix of the B-form family, but at first glance appears to be a 12-mer rather than the 11-mer that it is. This is because the steric clashes of the εC6 and the purine base of A17 at the lesion site are relieved by formation of a staggered structure, such that the εC6 base stacks to the 3′ side of the A17 base, and the A17 base stacks to the 3′ side of the εC6 base. Thus, the εC6-A17 base pair exhibits a stagger of –0.34 nm and a stretch of –0.27 nm, allowing them to partially stack on top of one another. The εC6 base is displaced to the minor groove side (normalized Xdisp of –0.11 nm), while the A17 base is slightly displaced to the major groove side (normalized Xdisp of 0.04 nm). This staggered conformation dramatically increases the rise for the εC6-C7 base step and the A17-G18 base step, those being 0.7 nm and 0.68 nm, respectively. This increased rise lengthens the duplex, such that its end-to-end distance is 0.3–0.45 nm greater than that of the other three duplexes we solved. The key NMR data leading to this structure were NOE interactions between εC6(H7,H8) and A17(H2′,H2″). There is an overall bend of the axis of about 16° toward the minor groove. No hydrogen bonding is seen in the lesion-containing base pair, but all other base pairs exhibit normal Watson-Crick alignments.

All glycosidic torsion angles are in the *anti* orientation, and all sugar puckers, including those in the central base pair, are in the C2′endo/C3′-exo range. The central region of the duplex is shown in Figure 3 and the full duplex in Figure 6.

εdC•T (Cullinan *et al.*, 1996)

The εdC•T structure is also a right-handed helix of the B-form family, with no significant perturbation of the sugar phosphate backbone. In this case, the presence of the lesion is accommodated by a *syn* orientation of the εdC about the glycosidic bond, while all other residues remain in *anti*. This allows for the formation of a strong hydrogen bond between εC6(N4) and T17(N3H). The NMR evidence for the *syn* orientation of the εdC was a very strong interaction between εC6(H6) and εC6(H1′). The bases of the central pair remain coplanar, but with a propeller twist of 19°, while the C7•G16 base pair shows a buckle of 12°. The εC6 base and the T17 base are displaced to opposite grooves, with normalized Xdisp values of –0.45 and –0.09 nm, respectively. The bases at the lesion site stack well with the C7•G16 base pair and somewhat with the C5•G18 base pair. There is a slight bend in the helical axis (about 9°) toward the major groove.

All base pairs other than the lesion site retain Watson-Crick hydrogen bonding. All sugar puckers are in the C2′-endo/C3′-exo range. Stereoviews of the central base pairs are shown in Figure 4 and the full-length duplex is shown in Figure 6.

εdC•dC

The εdC•dC structure is a right-handed helix of the B-form family, with no major perturbations of the sugar phosphate backbone but with a kink in the helical axis of almost 27° towards the major groove. This is by far the most bent of the four εdC-containing structures we have solved. In this duplex, the C17 base has swung into the minor groove (normalized Xdisp of –0.34 nm), such that the G18–C17 base step shows a shift of –0.3 nm and a slide of –0.16 nm, while the C17–G16 step shows a shift of 0.25 nm and a slide of 0.12 nm. The εC6 base undergoes a tip of about –10.3°, resulting in a tilt of 12.8° and a roll of 21.4° in the εC6–C7 base step. The εC6 base is displaced to the major groove, with a normalized Xdisp of 0.17 nm. When looking at the relative positions of εC6 and C17, this results in a shear of 0.51 nm, a stretch of 0.15 nm, a slight stagger of –0.04 nm, a buckle of –7.9°, and a propeller twist of –21.4°. Despite this mild misalignment of their base planes, the O2 of εC6 is on the same plane as the C17 base, and the orientations of the bases allow the formation of a hydrogen bond between εC6(O2) and C17(N4H2). All other base pairs exhibit normal Watson-Crick hydrogen bonding. All residues have the glycosidic torsion angle in the *anti* range. The sugars of all residues other than εC6 are in a C2′-endo/C3′-exo range, while the sugar of εC6 is clearly C3′-endo. In this conformation, the H2′ of εC6 points into the area where the base of C7 would normally lie, which may account for the high roll in the εC6–C7 base step. With the C7 and G16 bases in the same plane and the bases following them (A8 through T15) remaining parallel to the C7•G16 base pair,

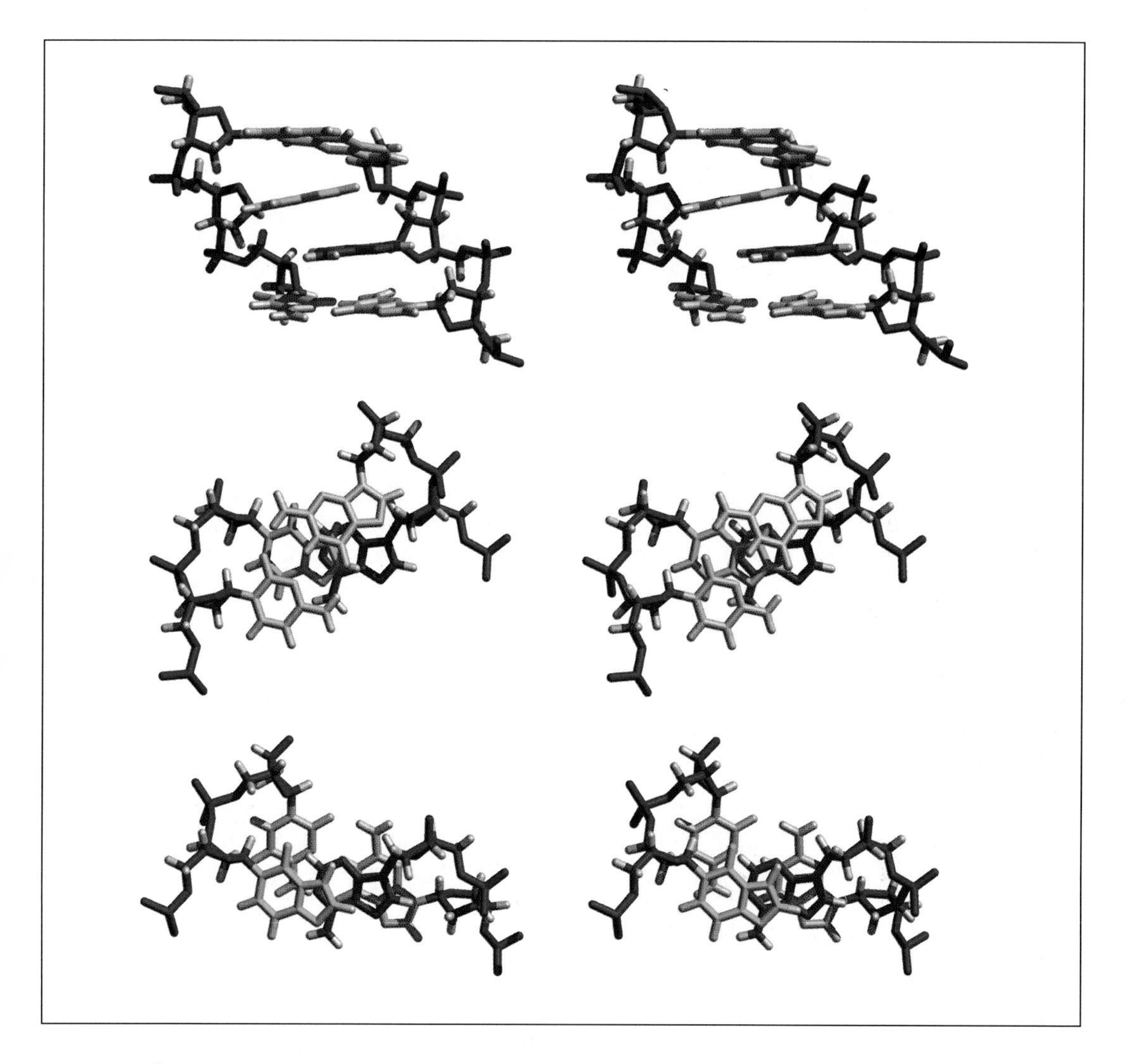

Figure 3. 3,N^4-Etheno-2´-deoxycytidine (εdC)–deoxyadenosine

This steroview shows the central three base pairs of the duplex. The εdC base is in green and the dA opposite it is in red. (Top) View from the major groove. (Centre) Axial view, showing stacking interaction with the C5–G18 base pair above the εC6–A17 base pair. (Bottom) Another view from the same perspective showing the εC6–A17 base pair above the C7–G16 base pair

the roll of C7 seems to be the primary cause of the sharp kink in the helical axis. It is also noteworthy that the duplex is somewhat overwound on one side of the lesion (twist of 58.5° for the εC6–C7 step, 50.5° for the G18–C17 step) and underwound on the other (twist of 13.8° for the C5–εC6 step, 22.0° for the C17–G16 step). Figure 5 shows stereoviews of the central three base pairs, while Figure 6 shows the full-length duplex.

Discussion

When comparing the four εdC-containing duplexes presented here, one of the more remarkable features that emerges is that, while 21 out of 22 residues are the same for all, the structure around the lesion site varies considerably among them. In both cases in which a pyrimidine is opposite the lesion site, both bases remain fairly co-planar, but in one case the εdC base is in

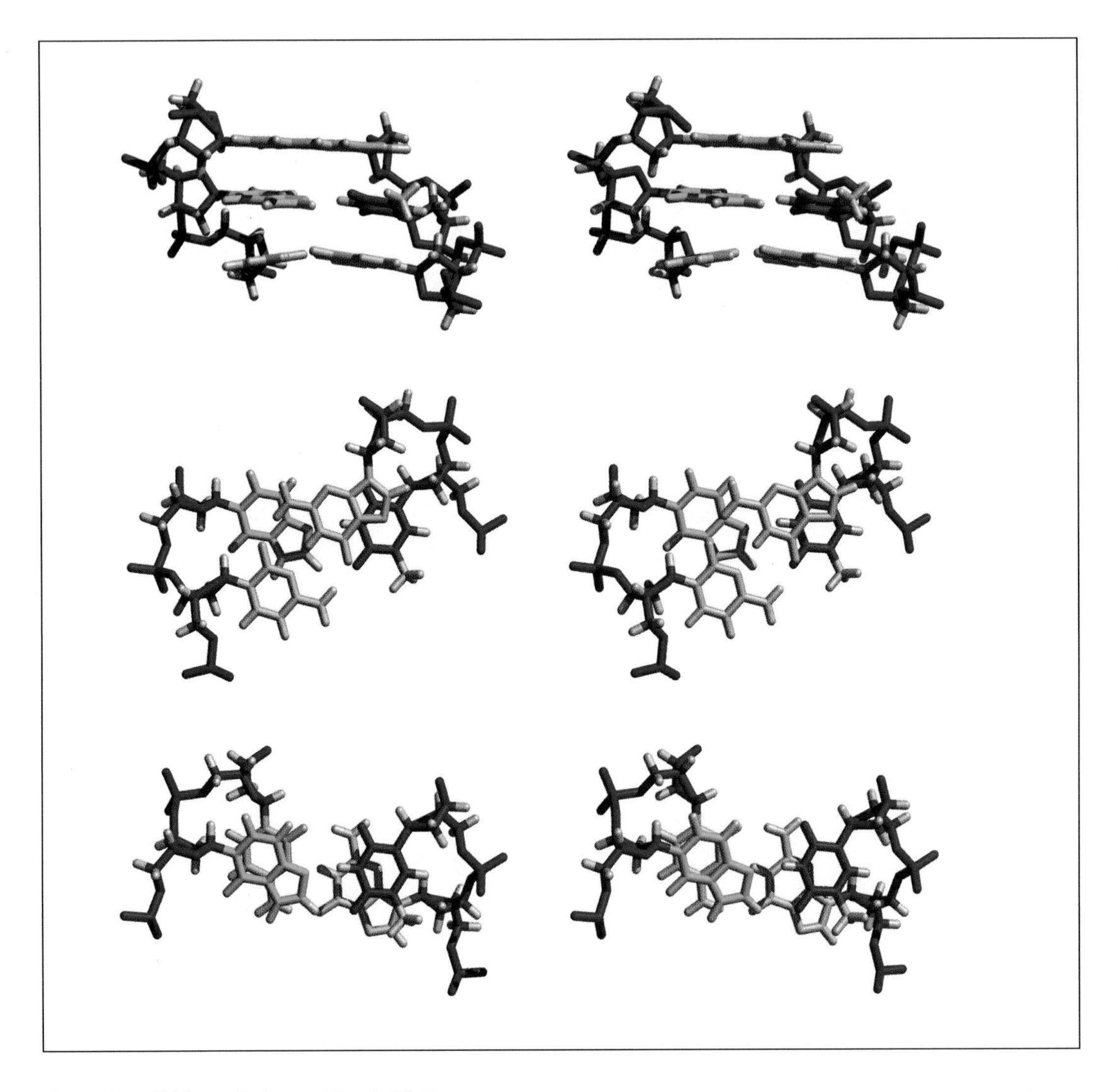

Figure 4. 3,N^4-Etheno-2′-deoxycytidine (εdC)–thymine

This steroview shows the central three base pairs of the duplex. The εdC base is in green and the T opposite it is in red. (Top) View from the major groove. (Centre) Axial view, showing stacking interaction with the C5–G18 base pair above the εC6–T17 base pair. (Bottom) Another view from the same perspective showing the εC6–T17 base pair above the C7–G16 base pair

syn, while in the other it is in an *anti* conformation. In both cases in which a purine is opposite the lesion, there is considerable stagger of the base pairs, but in one case they both point into the helix and stagger partially over one another, while in the other they are displaced into opposite grooves. In addition, these structures are staggered in opposite directions. Yet, there must be some similarity between them all, if repair enzymes are able to recognize all four structures. There are certain similarities among all except the εdC•dA duplex. The three that are similar all show a kink (or bend) of the helical axis towards the major groove, the point of the kink being near the lesion site. Also, in these three structures, the base opposite the lesion is always

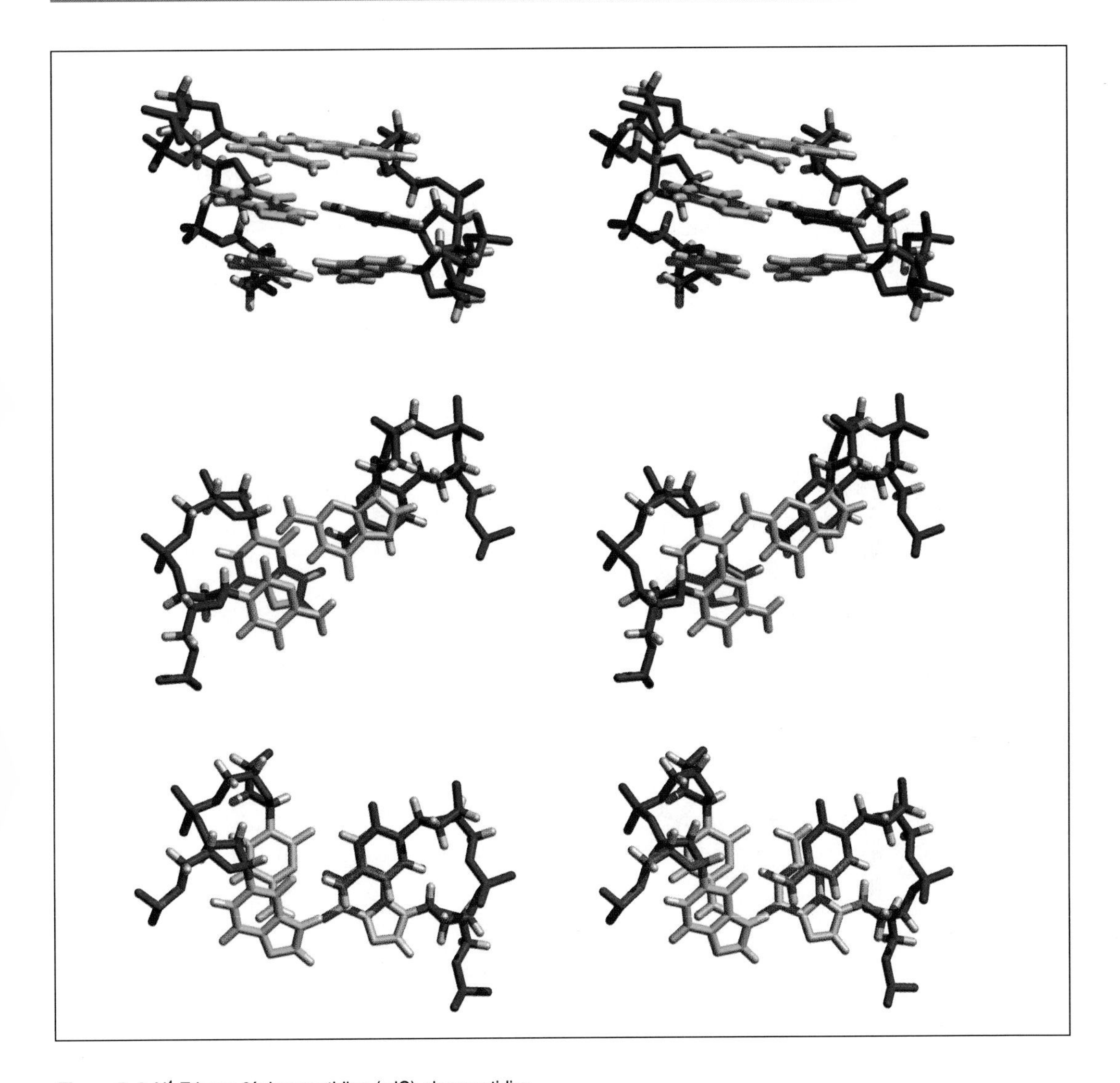

Figure 5. 3,N^4-Etheno-2′-deoxycytidine (εdC)–deoxycytidine

This steroview shows the central three base pairs of the duplex. The εdC base is in green and the dC opposite it is in red. (Top) View from the major groove. (Centre) Axial view, showing stacking interaction with the C5–G18 base pair above the εC6–C17 base pair. (Bottom) Another view from the same perspective showing the εC6–C17 base pair above the C7–G16 base pair

tucked into the minor groove, to a greater or lesser extent, with normalized Xdisp values ranging from –0.09 to –0.335 nm. Whether these similarities are significant remains to be shown. Now that our structural studies of εdC opposite each of the four deoxynucleotides have been completed, data from other studies of εdC should be reexamined for correlations.

The results of the mutagenesis experiments showed clearly that, under the right conditions (polymerase and repair systems), any of the four deoxynucleotides can be incorporated opposite the εdC lesion. Our studies have shown that, in each of the four cases, there is a reasonable way for the duplex to relieve steric hindrance and stabilize the structure. What is more difficult to

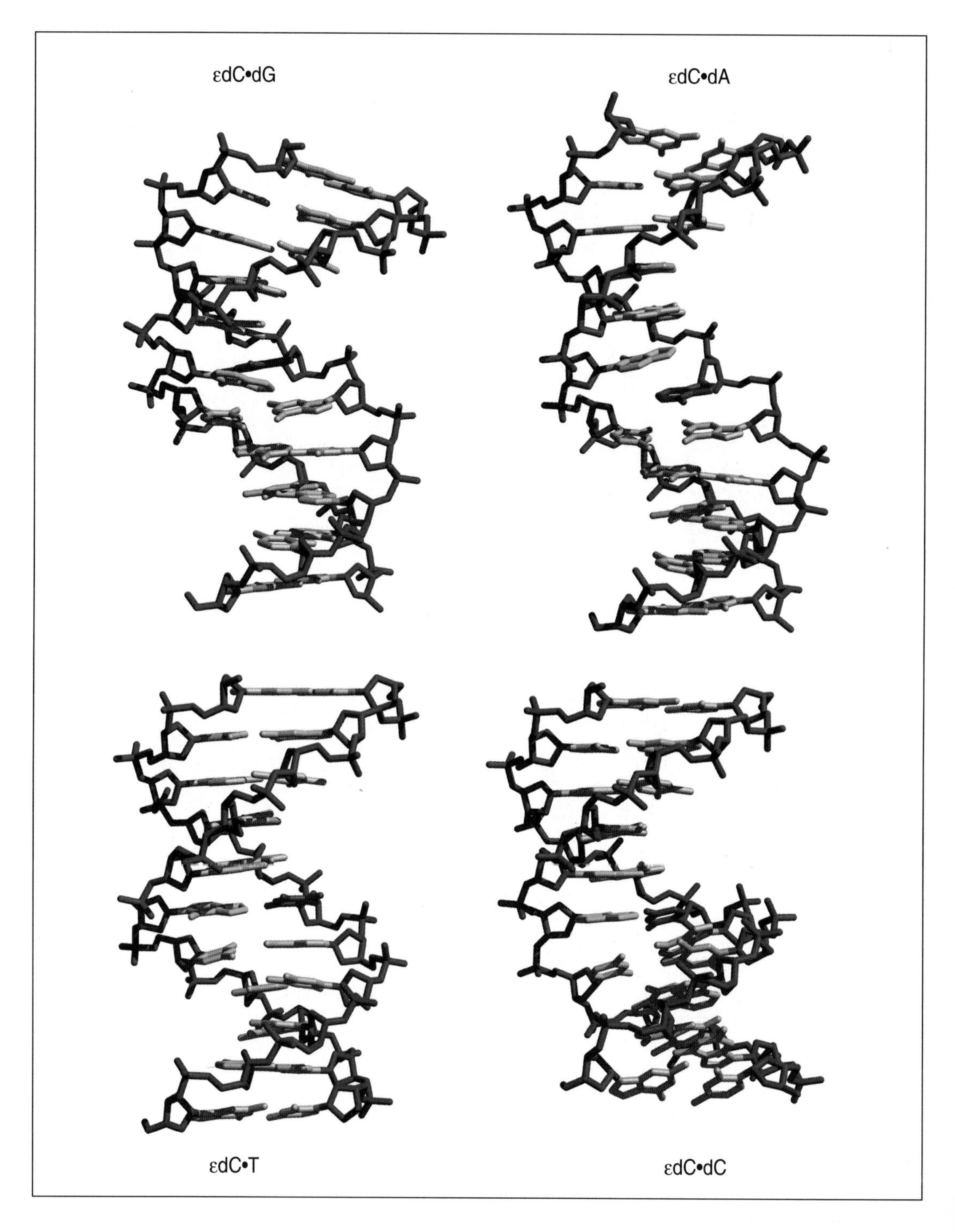

Figure 6. Full-length duplexes. This view shows the major groove around the lesion site and how the distortions in the centre affect the rest of the duplex.

explain, however, is the base preference shown by the polymerases. The Klenow fragment of bacterial polymerase I and mammalian polymerase α and δ all preferentially incorporate dA or T opposite the lesion (Singer & Spengler, 1986; Simha *et al.*, 1991; Zhang *et al.*, 1995; Shibutani *et al.*, 1996). From the structural data presented here, there is no obvious reason why this should be so. It should be noted, however, that any argument that would successfully explain that result would also conflict with the results obtained *in vivo* in bacteria, which show the etheno lesion to be a weak mutagen, implying that dG is being incorporated opposite it, and with results showing that polymerase β incorporates dC (Palejwala *et al.*, 1991; Basu *et al.*, 1993; Palejwala *et al.*, 1993; Moriya *et al.*, 1994; Shibutani *et al.*, 1996). Until more is known about the structures of the various polymerases, preferably bound to DNA, speculation along these lines is likely to be futile.

The most striking and puzzling aspect of the results of the thermodynamics experiments is that in a duplex in which, from our data, only the central few base pairs are disrupted at all, the overall free energy (ΔG) drops by about 70% (Gelfand *et al.*, 1998). This is presumably due in large part to a change in the solvation energy. It is surprising that the energy is so consistently affected in all of the structures, when the local perturbations are quite different. In addition, enthalpy, unlike free energies, is dependent on sequence context and the base opposite the lesion. As the values for ΔG are roughly equal and the enthalpy varies between them, the entropic term must make up the difference. In the case of εdC•dC, where there is relatively little enthalpic change, the large change in free energy must be due to a decrease in entropy, while in the case of εdC•dA, where the loss of enthalpy was the highest, it seems to have been compensated for by a gain in entropy. It is worth noting that εdC•dA shows both the greatest loss of enthalpy and is the most disrupted of the structures, in relation to canonical Watson-Crick base pair alignments. For the most part, though, it is very difficult to correlate the experimentally observed changes in enthalpy and free energy to specific structural details.

Experimental results with repair enzymes, however, show correlations with the structural data. Both bacterial mismatch-specific uracil glycosylase and human mismatch-specific thymine-DNA glycosylase also remove the thymine in a G•T mismatch; the guanine is displaced towards the minor groove, while the thymine is displaced toward the major groove, forming a 'wobble' base pair (Hunter *et al.*, 1987). As mentioned above, in each of our structures except the εdC•dA, the base opposite the εdC is displaced towards the minor groove. This could be a key element in recognition by this enzyme of the εdC lesion. In addition, in the crystal structure of the mismatch-specific uracil glycosylase in a complex with DNA, the DNA can be seen to have been forced into a staggered structure quite similar to that seen in the εdC•dA (Barrett *et al.*, 1998). It is possible that even though the εdC•dA does not have the base opposite the lesion displaced into the minor groove, it is recognized by these enzymes simply because its normal structure resembles that of the bound duplex.

Acknowledgements

This work was made possible thanks to NIH grant CA47995 and NIEHS grant ES04068.

References

Barbin, A. & Bartsch, H. (1986) Mutagenic and promutagenic properties of DNA adducts formed by vinyl chloride metabolites. In: *The Role of Cyclic Nucleic Acid Adducts in Carcinogenesis and Mutagenesis* (Singer, B. & Bartsch, H., eds), IARC Scientific Publications No. 70. Lyon, IARC, pp. 345–358

Barbin, A., Besson, F., Perrard, M.H., Bereziat, J.C., Kaldor, J., Michel, G. & Bartsch, H. (1985) Induction of specific base-pair substitutions in *E. coli* TrpA mutants by chloroethylene oxide, a carcinogenic vinyl chloride metabolite. *Mutat. Res.* 152, 147–156

Barrett, T.E., Savva, R., Panayotou, G., Barlow, T., Brown, T., Jiricny, J. & Pearl, L.H. (1998) Crystal structure of a G:T/U mismatch-specific DNA glycosylase: Mismatch recognition by complementary-strand interactions. *Cell* 92, 117–129

Basu, A.K., Wood, M.L., Niedernhofer, L.J., Ramos, L.A. & Essigmann, J.M. (1993) Mutagenic and genotoxic effects of three vinyl chloride-induced DNA lesions: 1,N^6-Ethenoadenine, 3,N^4-ethenocytosine, and 4-amino-5-(imidazol-2-yl)imidazole. *Biochemistry* 32, 12793–12801

Brünger, A. (1993) *X-PLOR, Version 3.1: A System for X-Ray Crystallography and NMR*. New Haven, Yale University Press, pp. 295–348

Creech, J.L. Jr & Johnson, M.N. (1974) Angiosarcoma of liver in the manufacture of polyvinyl chloride. *J. Occup. Med.* 16, 150–151

Cullinan, D., Korobka, A., Grollman, A.P., Patel, D.J., Eisenberg, M. & de los Santos, C. (1996) NMR solution structure of an oligodeoxynucleotide duplex containing the exocyclic lesion 3,N^4-etheno-2′-deoxycytidine opposite thymidine: Comparison with the duplex containing deoxyadenosine opposite the adduct. *Biochemistry* 35, 13319–13327

Cullinan, D., Johnson, F., Grollman, A.P., Eisenberg, M. & de los Santos, C. (1997) Solution structure of a DNA duplex containing the exocyclic lesion 3,N^4-etheno-2′-deoxycytidine opposite 2′-deoxyguanosine. *Biochemistry* 36, 11933–11943

Gelfand, C.A., Plum, E., Grollman, A.P., Johnson, F. & Beslauer, K.J. (1998) The impact of an exocyclic cytosine adduct on DNA duplex properties: Significant thermodynamic consequences despite modest lesion-induced structural alterations *Biochemistry* 37, 12507–12512

Guengerich, F.P., Crawford, W.M., Jr & Watanabe P.G. (1979) Activation of vinyl chloride to covalently bound metabolites: Roles of 2-chloroethylene oxide and 2-chloroacetaldehyde. *Biochemistry* 18, 5177–5182

Hang, B., Medina, M., Fraenkel-Conrat, H. & Singer, B. (1998) A 55-kDa protein isolated from human cells shows DNA glycosylase activity toward 3,N^4-ethenocytrosine and the G/T mismatch. *Proc. Natl Acad. Sci. USA* 95, 13561–13566

Hunter, W.N., Brown, T., Kneale, G., Anand, N.N., Rabinovich, D. & Kennard, O. (1987) The structure of guanosine–thymidine mismatches in B-DNA at 2.5-A resolution. *J. Biol. Chem.* 262, 9962–9970

Korobka, A., Cullinan, D., Cosman, M., Grollman, A.P., Patel, D.J., Eisenberg, M. & de los Santos, C. (1996) Solution structure of an oligodeoxynucleotide duplex containing the exocyclic lesion 3,N^4-etheno-2′-deoxycytidine opposite 2′-deoxyadenosine, determined by NMR spectroscopy and restrained molecular dynamics. *Biochemistry* 35, 13310–13318

Kusmierik, J.T. & Singer, B. (1992) 1,N^2-Ethenodeoxyguanosine: Properties and formation in chloroacetaldehyde-treated polynucleotides and DNA. *Chem. Res. Toxicol.* 5, 634–638

Lavery, R. & Sklenar, H. (1989) Defining the structure of irregular nucleic acids: Conventions and principles. *J. Biomol. Struct. Dyn.*, 6, 655–667

Maltoni, C., Lefemine, G., Chieco, P. & Carretti, D. (1974) Vinyl chloride carcinogenesis: Current results and perspectives. *Med. Lav.* 65, 421–444

Moriya, M., Zhang, W., Johnson, F. & Grollman, A.P. (1994) Mutagenic potency of exocyclic DNA adducts: Marked differences between *Escherichia coli* and simian kidney cells. *Proc. Natl Acad. Sci. USA* 91, 11899–11903

Nair, J., Barbin, A., Guichard, Y. & Bartsch. H. (1995) 1,N^6-Ethenodeoxyadenosine and 3,N^4-ethenodeoxycytine in liver DNA from humans and untreated rodents detected by immunoaffinity/^{32}P-postlabeling. *Carcinogenesis* 16, 613–617

Palejwala, V.A., Simha, D. & Humayun, M.Z. (1991) Mechanisms of mutagenesis by exocyclic DNA adducts. Transfection of M13 viral DNA bearing a site-specific adduct shows that ethenocytosine is a highly efficient RecA-independent mutagenic noninstructional lesion. *Biochemistry* 30, 8736–8743

Palejwala, V.A., Rzepka, R.W. & Humayun, M.Z. (1993) UV irradiation of *Escherichia coli* modulates mutagenesis at a site-specific ethenocytosine residue on M13 DNA. Evidence for an inducible recA-independent effect. *Biochemistry* 32, 4112–4120

Saparbaev, M. & Laval, J. (1998) 3,N^4-Ethenocytosine, a highly mutagenic adduct, is a primary substrate for *Escherichia coli* double-stranded uracil-DNA glycosylase and human mismatch-specific thymine-DNA glycosylase. *Proc. Natl Acad. Sci. USA* 95, 8508–8513

Scherer, E., Winterwerp, H. & Emmelot, P. (1986) Modification of DNA and metabolism of ethyl carbamate *in vivo*: Formation of 7(2-oxoethyl)guanine and its sensitive determination by reductive tritiation using ^{3}H-sodium borohydride. In: *The Role of Cyclic Nucleic Acid Adducts in Carcinogenesis and Mutagenesis* (Singer, B. & Bartsch, H., eds), IARC Scientific Publications No. 70. Lyon, IARC, pp. 109--125

Shibutani, S., Suzuki, N., Matsumoto, Y. & Grollman, A.P. (1996) Miscoding properties of 3,N^4-etheno-2′-deoxycytidine in reactions catalyzed by mammalian DNA polymerases. *Biochemistry* 35, 14992–14998

Simha, D., Palejwala, V.A. & Humayun, M.Z. (1991) Mechanisms of mutagenesis by exocyclic DNA adducts. Construction and in vitro template characteristics of an oligonucleotide bearing a single site-specific ethenocytosine. *Biochemistry* 30, 8727–8735

Singer, B. & Spengler, S.J. (1986) Replication and transcription of polynucleotides containing ethenocytosine, ethenoadenine and their hydrated intermediates. In: *The Role of Cyclic Nucleic Acid Adducts in Carcinogenesis and Mutagenesis* (Singer, B. & Bartsch, H., eds), IARC Scientific Publications No. 70. Lyon, IARC, pp. 359–371

Zhang, W., Johnson, F., Grollman, A.P. & Shibutani, S. (1995) Miscoding by the exocyclic and related DNA adducts 3,N^4-etheno-2′-deoxycytidine, 3,N^4-ethano-2′-deoxycytidine, and 3-(2-hydroxyethyl)-2′-deoxyuridine. *Chem. Res. Toxicol.* 8, 157–163

Exocyclic DNA Adducts in Mutagenesis and Carcinogenesis
B. Singer & H. Bartsch
IARC Scientific Publications No. 150
International Agency for Research on Cancer, Lyon, 1999

Thermal destabilization of DNA oligonucleotide duplexes by exocyclic adducts on adenine or cytosine depends on both the base and the size of adduct

J. Sági & B. Singer

Numerous carcinogens or their bifunctional metabolites modify DNA bases by forming additional exocyclic rings on the base moiety. These modifications form exocyclic rings between N1 and N^6 of dA, N3 and N^4 of dC or N1 and N^2 of dG, as well as the N^2 and N3 of dG. This study focuses on the reaction products of dA and dC with chloroacetaldehyde, bis-chloroethyl nitrosourea and *para*-benzoquinone, which form etheno, ethano and *para*-benzoquinone derivatives, respectively. The three dC adducts and three dA adducts were each incorporated site-specifically into 25-nucleotide-long deoxyoligonucleotides. All duplexes with a single modified dA or dC adduct opposite the normal complement showed decreased thermal stability, as compared with the unmodified control duplex. The destabilizations ranges from –2 °C to –13 °C, depending on saturation, the size of the adduct and the nature of the base. Energy-minimized molecular models of the duplexes illustrate various degrees of distortions by the adducts, the *para*-benzoquinone adducts showing the greatest distortion.

Introduction

Metabolites of vinyl chloride, benzene and bis-chloroethyl nitrosourea can form etheno (ε), ethano (E) and *para*-benzoquinone (pBQ) exocyclic adducts, respectively, on dA, dC and dG in DNA (Singer & Hang, 1997). These adducts differ in structure and in the number of added rings (Figure 1). The formation of exocyclic rings on carcinogen-modified DNA bases *in vivo* can have biological consequences that may be modulated by the steric structure of the adduct and/or by the effect of the adduct on the secondary structure of DNA. Changes in the secondary structure can be observed by determining thermal and thermodynamic stability and the conformation of duplexes. These determinations have been carried out with the etheno and, in part, the pBQ adducts. The duplex thermal stability of εA was measured with self-complementary hexamers (Basu *et al.*, 1987) and dodecamers (Leonard *et al.*, 1994), both containing two adducts. With a single εA, the thermodynamics of melting was studied with 15-mer duplexes (Hang *et al.*, 1998a); solution conformation was also determined earlier (Kouchakdjian *et al.*, 1991; de los Santos *et al.*, 1991). The detailed thermodynamics of melting for 13-mer duplexes with a single εC has been described recently (Gelfand *et al.*, 1998), and solution conformation was also determined (Cullinan *et al.*, 1997). The effect of pBQ-C on stability was measured with 25-mer duplexes (Sági *et al.*, 1998) and that of pBQ-A with both 25-mer and 22-mer duplexes (Hang *et al.*, 1998b). Differences were found not only in duplex length but also in the sequences flanking the adduct and the salt and the duplex concentrations used for the stability measurements. Therefore, the effects of different adducts are not always comparable. In this work we investigated how the etheno, ethano and benzoquinone adducts of dC and dA influence the thermal stability of 25-mer duplexes under the same conditions. To better understand the

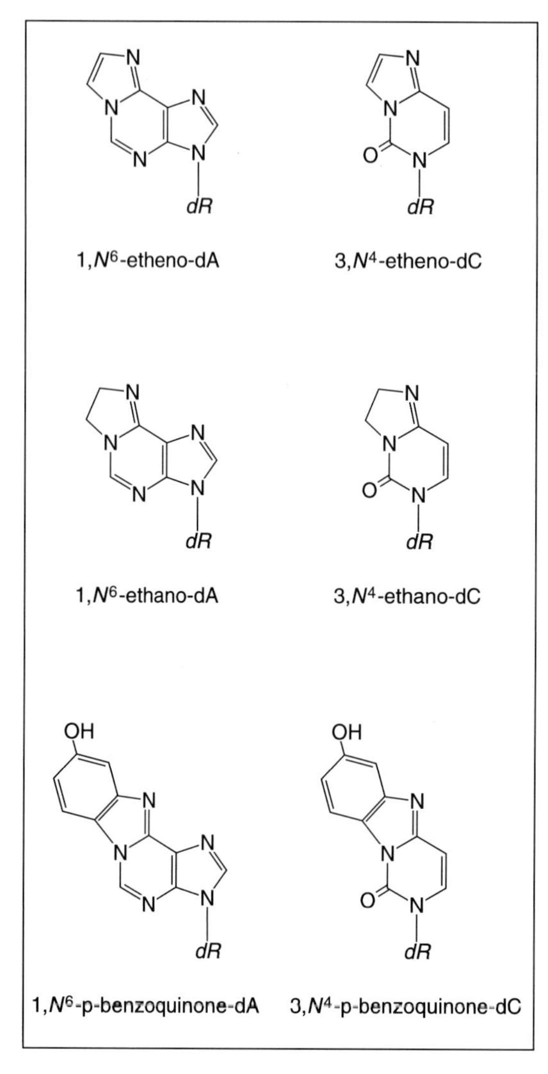

Figure 1. Chemical structures of the adenine and cytosine adducts

results of the observed destabilization, visualization by energy-minimized molecular modelling of the local conformation of the adduct-containing duplexes was also used (see also the Appendix for molecular models).

Experimental methods

Oligodeoxynucleotides

The following 25-nucleotide-long oligodeoxynucleotides were used for thermal denaturation studies:

1 5′ CCGCT**A**G**C**GGGTACCGAGCTCGAAT 3′
2 CCGCT**εA**GCGGGTACCGAGCTCGAAT
3 CCGCT**EA**GCGGGTACCGAGCTCGAAT
4 CCGCT**pBQ-**AGCGGGTACCGAGCTCGAAT
5 CCGCTAG**εC**GGGTACCGAGCTCGAAT
6 CCGCTAG**EC**GGGTACCGAGCTCGAAT
7 CCGCTA**pBQ-**CGGGTACCGAGCTCGAAT
8 3′ GGCGATCGCCCATGGCACGAGCTTA 5′

where εA, EA and pBQ-A signify $1,N^6$-etheno-dA, $1,N^6$-ethano-dA and $1,N^6$-*para*-benzoquinone-dA, respectively, in position 6 of the 5′-to-3′ strand. εC, EC and pBQ-C signify $3,N^4$-etheno-dC, $3,N^4$-ethano-dC and $3,N^4$-*para*-benzoquinone-dC, respectively, in position 8 of the strand. The oligodeoxynucleotides were synthesized as described by Litinski *et al.* (1997) and Chenna *et al.* (this volume). All of the duplexes contained the normal complementary base opposite the adduct, i.e. T opposite A and A adducts and G opposite C and C adducts.

Thermal stability

Determination and analysis of the absorption versus temperature melting (T_m) profiles were carried out essentially as described earlier (Sagi *et al.*, 1998; Hang *et al.*, 1998a).

Molecular modelling

Nine-nucleotide-long segments of the 25-mer duplexes used to determine T_m were selected for molecular mechanics. These contained an unmodified central A or C or the modified A or C adducts with four flanking base pairs on each side. The sequences were: 5′ CGCTA*GCGG/GCGATCGCC for the A adducts and 5′ CTAGC*GGGT/GATCGCCCA for the C adducts, where A* and C*, respectively, indicate the position of the modified base. Original structures were built in canonical B-conformation in the biopolymer module of InsightII (Molecular Simulations, Inc., San Diego, CA) on a Silicon Graphics O2 work station (SiliconGraphics, Mountain View, CA). Terminal phosphate groups were removed, and exocyclic adducts were manually formed on the central A and C bases. The 'all-atom' carbon types were used to assign potentials for the modified nucleosides in the AMBER force field (version 4.1) incorporated into InsightII, and this force field was used in the calculations. Sixteen sodium ions were then placed 0.25 nm from the phosphate oxygens to neutralize anions. Energy minimization was carried in the Discover 3 module of the InsightII package with the distance-dependent dielectric constant set to 4.0 and the cutoff value for non-bonded interactions extended to 1.2 nm to simulate solution conditions. Distance restraints were applied on all H bonds, at an average of 35 kcal/mol. To retain coplanarity of the benzetheno ring, torsion restraints were also applied for pBQ-A. This assembly was energy-minimized first by the Steepest descent method to a gradient < 20 kcal/nm.mol, which was followed by the Polak-Ribiere conjugate gradient method to a gradient < 0.01 kcal/nm.mol. The geometry-optimized structures, representing a local minimum of the system, were used for visualization of the structural distorsion caused by the adduct.

Results

All six adducts were incorporated into the same 25-mer duplexes, previously used in biochemical studies. The cytosine adducts εC, EC and pBQ-C are at position 8 of the 5′-to-3′ strand in the AGCGG sequence, and the adenine adducts εA, EA and pBQ-A are at position 6 in the CTAGC sequence of the 25-mer (see above). The same salt and buffer conditions were used in all experiments (see Table 1). Thermal stability was determined at several duplex concentrations and is compared at a duplex concentration of 1.5 µmol/L in Table 1.

The melting transitions were all monophasic, and the transitions were completely reversible with the 25-mers (curves not shown). All of the adducts studied reduced the T_m value of the unmodified 25-mer control duplex, the degree of destabilization depending on the type of the adduct and the base (Table 1).

Table 1. Effect of etheno (ε), ethano (E) and *para*-benzoquinone (pBQ) adducts of adenine and cytosine on the thermal stability of 25-mer duplexes

Adduct	T_m (°C)	ΔT_m (°C)	H_{280} (%)
None	73.0		37.2
–CT**εA**GC–	70.2	–2.8	36.3
–CT**EA**GC–	70.6	–2.4	39.2
–CT**pBQ-A**GC–	63.8	–9.2	34.1
–AG**εC**GG–	66.6	–6.4	38.5
–AG**EC**GG–	63.8	–9.2	24.8
–AG**pBQ-C**GG–	60.4	–12.6	34.2

Thermal stability was measured in solutions containing 0.1 mol/L NaCl, 10 mmol/L sodium phosphate and 10 mmol/L EDTA, pH adjusted to 7.0, with 1.5 µmol/L of duplex concentration. Duplexes were prepared with equal amounts of each strand. Solutions were degassed by an initial fast heating (5 °C/min) up to 90 °C, then annealed by slow cooling over 30 min. Measurements were carried out with ramp rates of 0.2 °C/min in the range of ± 20 °C of the T_m of the samples and 0.5 °C/min at other temperatures. Absorption data were collected at 260 and 280 nm at 0.5-°C intervals. T_m values are defined as the maximum of the first derivative curve. The values shown in the Table are averages of three to four measuments of separately annealed samples. T_m and thermal hyperchromicity values were reproducible within 0.5 °C and 1%, respectively.

εA had only a small negative effect on the thermal stability of the 25-mer duplex, and the ΔT_m was –2.8 °C. Only a small effect was anticipated since the destabilization depends strongly on duplex length, and 15-mer duplexes had been found to be destabilized by an average of –9 °C (Hang *et al.*, 1998a); however, 25-mers were used since these studies were done in parallel with those on repair (Singer & Hang, this volume).

The etheno ring differs from the ethano ring by its double bond in the five-membered exocyclic ring (Figure 1). A smaller base overlap or stacking can be anticipated for the ethano ring with neighbouring bases as compared with the fully conjugated heteroaromatic etheno ring; however, EA does not cause a different destabilization effect from that of εA (Table 1).

In comparison with the one-ring etheno and ethano adducts of adenine, the two-ring *para*-benzoquinone adduct pBQ-A caused a significantly larger destabilization of the duplex (Table 1). The more than threefold greater effect of pBQ-A as compared with the effects of the single-ring adducts may reflect the differences in size.

A different pattern was found with the cytosine adducts: εC, EC and pBQ-C caused greater destabilization of the 25-mer parent duplex than the corresponding purine adducts, and this effect apparently depends on the saturation and size of the adduct (Table 1). εC decreased the T_m by 6.4 °C, while EC decreased it by 9.2 °C. This difference in ΔT_m between εC and EC is likely to be the result of double-bond saturation. The effect of pBQ-C, –12.6 °C, is the largest among the C adducts and is even larger than that observed with pBQ-A.

Energy-minimized molecular models for the unmodified control duplexes and the A and C adduct-containing duplexes, respectively, are shown in Figure 2. Only the side views are shown in each case for three base pairs of the duplex. This modelling is used only to illustrate the initial accommodation of the adduct into the double helix, starting from one initial structure, and the degree of distortion. Duplexes with εA and εC have been modelled previously on the basis of studies with nuclear magnetic resonance (NMR) (Kouchakdjian *et al.*, 1991; Cullinan *et al.*, 1997); our models of these two adducts are in general agreement with the previous ones.

Discussion

The differences in T_m between the adenine and cytosine series and among the three modifications may result to some extent from differences in neighbouring bases. Sequence generally has an effect on $-\Delta T_m$, as described for εA (Hang *et al.*, 1998a) and εC (Gelfand *et al.*, 1998); however, these differences are minor and much less than the two- to threefold differences observed here between εA and εC or between EA and EC. Therefore, factors other than the effect of flanking sequence may be responsible for the differential effects of the same exocyclic ring structure in adenine and cytosine. It is likely that stacking is the main contributor to the differences. Both εA and εC cause mainly enthalpic destabilization of the duplexes (Gelfand *et al.*, 1998; Hang *et al.*, 1998a), which may be correlated to base stacking. The resulting $\Delta\Delta G^o$ values were, however, very different. εA caused an average $\Delta\Delta G^o_{37}$ of –4 kcal/mol with 15-mer duplexes (Hang *et al.*, 1998a); while εC caused a much stronger decrease in free energy for the formation of 13-mer duplexes, the $\Delta\Delta G^o_{25}$ was an average of about –14 kcal/mol when the opposite base was G (Gelfand *et al.*, 1998). Local conformations, as determined by NMR, were also different: the εA was stacked into the double helix (Kouchakdjian *et al.*, 1991), while εC was displaced and shifted toward the major groove as compared with unmodified C (Cullinan *et al.*, 1997).

There was also some evidence for the different stacking patterns in our 'initial accommodation' molecular models. The purine adducts are generally better stacked with the neighbouring bases than are the pyrimidine adducts. With the purine adducts, the opposite thymine is displaced, although some displacement of the adduct itself towards the major groove can also be observed. With the pyrimidine adducts, the opposite guanine is apparently well stacked, and the adduct is moved towards the major groove to an extent that apparently interrupts the continuity of stacking in the strand. With the pBQ adducts, stacking with bases in the opposite strand can also be seen, as a consequence of the size of the adduct.

Conclusions

Thermal destabilization of 25-mer duplexes by a single site-specific exocyclic adduct of adenine

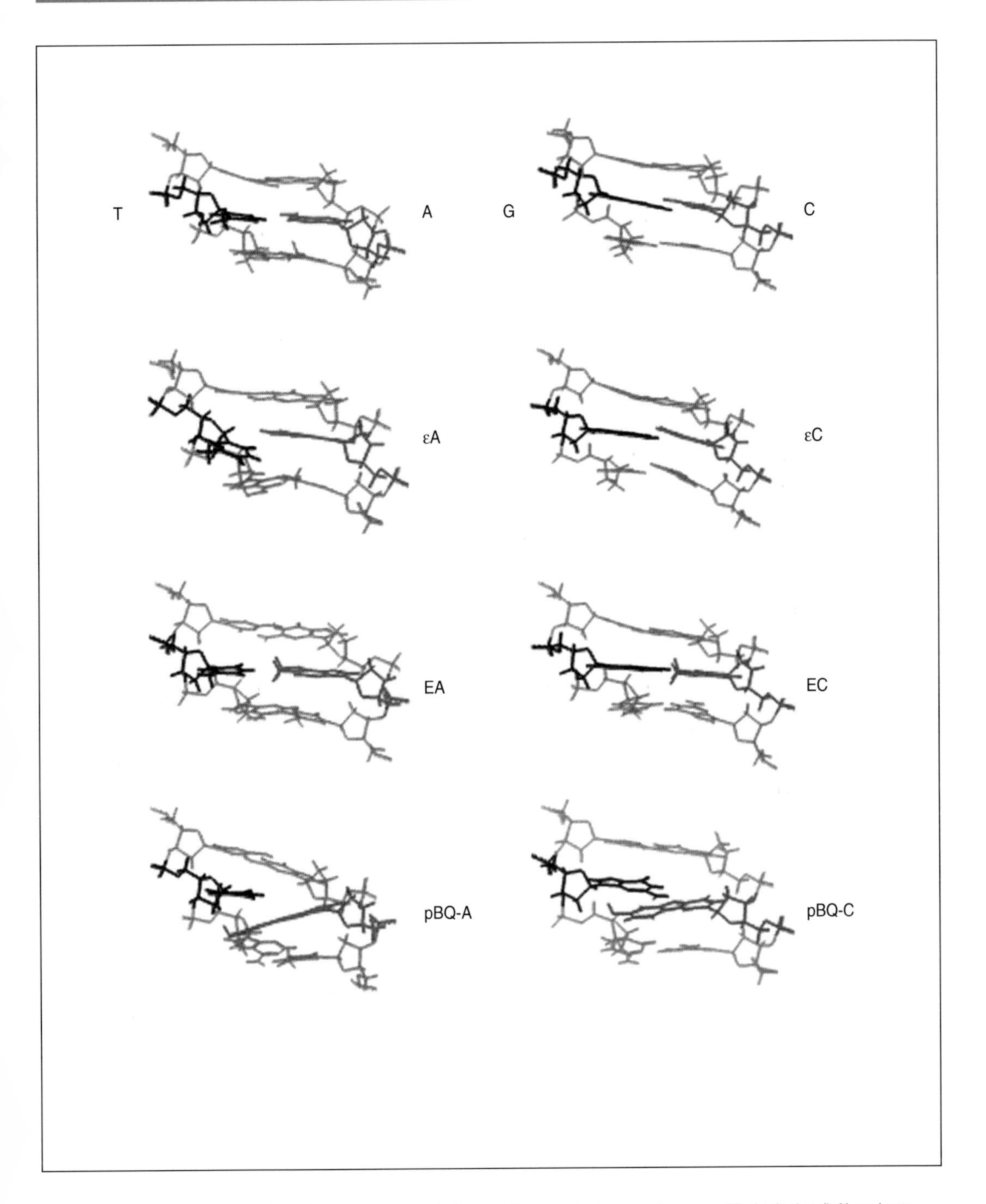

Figure 2. Three-base pair views from the major groove of etheno, ethano and p-benzoquinone modified adenine (left) and cytosine (right) in the energy-minimized oligonucleotide duplexes. The nucleotides of unmodified and modified adenines and cytosines are red and the complementary nucleotides are blue. The central A or A adduct is flanked by T and G. The central C or C adduct is flanked by G and G. The flanking base pairs are shown in grey.

and cytosine appears to depend on both the size of the exocyclic ring and the base that contains the exocyclic adduct. εA and EA reduced stability to only a small but similar extent, by –2.8 and –2.4 °C, respectively, which is probably the result of good stacking. The two-ring exocyclic adduct, pBQ-A, however, caused greater destabilization, by –9.2 °C, which may result in part from the adduct size. All of the cytosine adducts had a larger negative effect on duplex stability than did the adenine adducts. This effect gradually increased with saturation and size. The destabilizations were –6.4, –9.2 and –12.6 °C, respectively, for the etheno, ethano and *para*-benzoquinone adduct of cytosine. These stronger effects may originate from less well stacked local conformations.

References

Basu, A.K., Niedernhofer, L.J. & Essigmann, J.M. (1987) Deoxyhexanucleotide containing a vinyl chloride induced DNA lesion, 1,N^6-ethenoadenine: Synthesis, physical characterization, and incorporation into a duplex bacteriophage M13 genome as part of an amber codon. *Biochemistry* 26, 5626–5635

Cullinan, D., Johnson, F., Grollman, A.P., Eisenberg, M. & de los Santos. C. (1997) Solution structure of a DNA duplex containing the exocyclic lesion 3,N^4-etheno-2′-deoxycytidine opposite 2′-deoxyguanosine. *Biochemistry* 36, 11933–11943

Gelfand, C.A., Plum, G.E., Grollman, A.P., Johnson, F. & Breslauer, K.J. (1998) The impact of an exocyclic cytosine adduct on DNA duplex properties: Significant thermodynamic consequences despite modest lesion-induced structural alterations. *Biochemistry* 37, 12507–12512

Hang, B., Sági, J. & Singer, B. (1998a) Correlation between sequence-dependent glycosylase repair and the thermal stability of oligonucleotide duplexes containing 1,N^6-ethenoadenine. *J. Biol. Chem.* 273, 33406

Hang, B. Chenna, A., Sági, J. & Singer, B. (1998b) Differential cleavage of oligonucleotides containing the benzene-derived adduct, 1,N^6-benzetheno-dA, by the major human AP endonuclease HAP1 and *Escherichia coli* exonuclease III and endonuclease IV. *Carcinogenesis*, 19, 1339–1343

Kouchakdjian, M., Eisenberg, M., Yarema, K., Basu, A., Essigmann, J. & Patel, D.J. (1991) NMR studies of the exocyclic 1,N(6)-ethenodeoxyadenosine adduct (εdA) opposite thymidine in a DNA duplex—Nonplanar alignment of εdA(anti) and dT(anti) at the lesion site. *Biochemistry* 30, 1820–1828

Leonard, G.A., McAuley-Hecht, K.E., Gibson, N.J., Brown, T., Watson, W.P. & Hunter, W.N. (1994) Guanine-1,N^6-ethenoadenine base pairs in the crystal structure of d(CGCGAATT(edA)GCG). *Biochemistry* 33, 4755–4761

Litinski, V., Chenna, A., Sági, J. & Singer, B. (1997) Sequence context is an important determinant in the mutagenic potential of 1,N^6-ethenodeoxyadenosine (εA): Formation of εA basepairs and elongation in defined templates. *Carcinogenesis* 18, 1609–1615

Sági, J., Chenna, A., Hang, B. & Singer, B. (1998) A single cyclic p-benzoquinone adduct can destabilize a DNA oligonucleotide duplex. *Chem. Res. Toxicol.* 11, 329–334

de los Santos, C., Kouchakdjian, Yarema, K.M., Basu, A., Essigmann, J. & Patel, D.J. (1991) NMR studies of the exocyclic 1,N^6-ethenodeoxyadenosine adduct (εdA) opposite deoxyguanosine in a DNA duplex. εdA(syn).dG(anti) pairing at the lesion site. *Biochemistry* 30, 1828–1835

Singer, B. & Hang, B. (1997) What structural features determine repair enzyme specificity and mechanism in chemically modified DNA? *Chem. Res. Toxicol.* 10, 713–732

Exocyclic DNA Adducts in Mutagenesis and Carcinogenesis
B. Singer & H. Bartsch
IARC Scientific Publications No. 150
International Agency for Research on Cancer, Lyon, 1999

Chapter V. Cyclic adducts as biomarkers

A sensitive immunoslot–blot assay for detection of malondialdehyde–deoxyguanosine in human DNA

C. Leuratti, R. Singh, E.J. Deag, E. Griech, R. Hughes, S.A. Bingham, J.P. Plastaras, L.J. Marnett & D.E.G. Shuker

As part of a large programme on food risk assessment, we have become interested in the endogenous production of genotoxic agents from dietary precursors. Malondialdehyde (MDA), a product of lipid peroxidation and prostaglandin biosynthesis, is mutagenic in bacterial and mammalian systems. MDA reacts with DNA, and the major adduct (M_1-dG) has been detected in healthy human liver and leukocyte DNA. Analytical methods used so far for the detection of M_1-dG have not been applied to large numbers of individuals or a large variety of samples. Often, only a few micrograms of DNA from human tissues are available for analysis, and a very sensitive assay is needed to detect background levels of M_1-dG in very small amounts of DNA. In this paper, we describe the development of an immunoslot–blot (ISB) assay for the measurement of M_1-dG in 1 µg of DNA. The limit of detection of the assay is about 5 adducts per 10^8 bases. The advantages of ISB over other assays for DNA adduct detection, such as the possibility of analysing 1 µg DNA per sample and the fact that it is less time-consuming and laborious, mean that it can be more easily used for routine analysis of large numbers of samples in biomonitoring.
A series of human samples was analysed, and levels of 0.3–6.43 M_1-dG per 10^7 normal bases were detected in 42 gastric biopsy samples and 0.7–16.65 M_1-dG per 10^7 normal bases in 28 samples of leukocyte DNA. In an initial study in five human volunteers on standardized diets, the levels of M_1-dG in leukocyte DNA changed in relation to meat, vegetable and tea intake.

Introduction

Diet is a major modulatory factor in cancer risk. For example, epidemiological studies have shown a positive association between consumption of meat and an increased risk for colorectal cancer and have also demonstrated the protective effects of fruit and vegetables against cancers at several sites (World Cancer Research Fund, 1997). The aim of the work reported here was to establish the relationship between specific DNA damage and diet and use the resulting biomarkers in studies in humans.

Malondialdehyde (MDA) is a carbonyl compound generated by lipid peroxidation (Janero, 1990) and prostaglandin biosynthesis (Marnett, 1994a). MDA is mutagenic in bacterial and mammalian systems (Basu & Marnett, 1983; Benamira *et al.*, 1995) and reacts with DNA, forming adducts with deoxyguanosine (dG), deoxyadenosine (dA) and deoxycytidine (dC) (Marnett, 1994b). The main adduct, a highly fluorescent pyrimidopurinone (M_1-dG) (Figure 1), has been detected by gas chromatography and electron capture-negative chemical ionization with mass spectrometry in disease-free human liver and leukocyte DNA at levels of 0.48–11 adducts per 10^7 normal nucleotides (Chaudhary *et al.*, 1994; Rouzer *et al.*, 1997). The adduct has also been measured in

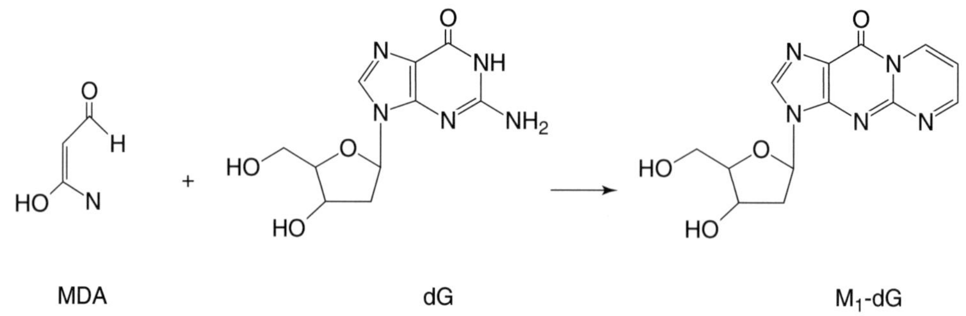

Figure 1. M_1-dG is formed in DNA by reaction of malondialdehyde (MDA) with deoxyguanosine.

human leukocyte, breast tissue and gastric DNA with ^{32}P-postlabelling techniques (Vaca *et al.*, 1995; Wang *et al.*, 1996; Leuratti *et al.*, 1997). There is some evidence that dietary intake of fatty acids affects the levels of M_1-dG (Fang *et al.*, 1996).

In a study of site-specific mutagenicity, M_1-dG was shown to induce mainly G → T transversions and G → A transitions (Fink *et al.*, 1997), indicating a direct involvement of the adduct in causing mutations. Further investigations are needed to elucidate the importance of M_1-dG in human carcinogenesis and other chronic diseases, to evaluate its distribution in various tissues and cell types and to determine the extent to which adduct levels are modulated by dietary composition or other environmental factors. The methods used so far for detecting M_1-dG are sensitive and/or specific but require large amounts of DNA for analysis or are time-consuming and laborious. In the study reported here, an immunoslot–blot (ISB) assay for the measurement of M_1-dG in ≤ 5 µg of DNA was developed. In this assay, we can detect background levels of the adduct in leukocytes and gastric biopsy samples from humans on normal diets.

Materials and methods

Synthesis of 1,N²-M₁-deoxyguanosine

1,N^2-M_1-deoxyguanosine (M_1-dG) was synthesized and characterized as previously described (Chapeau & Marnett, 1991).

DNA modified by malondialdehyde *in vitro*

Calf thymus DNA (CT-DNA, 0.26 mg/ml) was incubated with MDA (2 mmol/L, generated by hydrolysis of tetramethoxypropane) in 0.1 mol/L potassium dihydrogen orthophosphate (pH 4.5) for four days at 37 °C, precipitated with ethanol and redissolved in water. Standard MDA-modified CT-DNA was digested to deoxynucleosides, and the amount of M_1-dG present was measured by high-performance liquid chromatography (HPLC)–fluorescence from a calibration curve obtained with synthetic standard. Total DNA was quantified by HPLC with ultraviolet detection after digestion to monodeoxynucleotides (Leuratti *et al.*, 1998).

Immunoslot–blot

A murine monoclonal antibody (D10A1) specific for the M_1-dG adduct has been prepared and characterized by Sevilla *et al.* (1997). Immunoslot–blots (Figure 2) were carried out essentially as described by Mientjes *et al.* (1996) with modifications (Leuratti *et al.*, 1998). Briefly, CT-DNA containing various amounts of M_1-dG (prepared by diluting modified DNA with control DNA), control CT-DNA and human DNA samples (3.5 µg) were sonicated for 15 min in a water-bath sonicator, heat-denatured for 10 min, cooled on ice and mixed with an equal volume of 2 mol/L ammonium acetate. Single-stranded DNA (1 µg DNA/sample, in triplicate) was immobilized on nitrocellulose filters (0.1 µm, BA79, Schleicher & Schuell, Dassel, Germany) with a Minifold II, 72-well slot–blot microfiltration apparatus (Schleicher & Schuell, Dassel, Germany). The slots were rinsed with 200 µl of 1 mol/L ammoni-

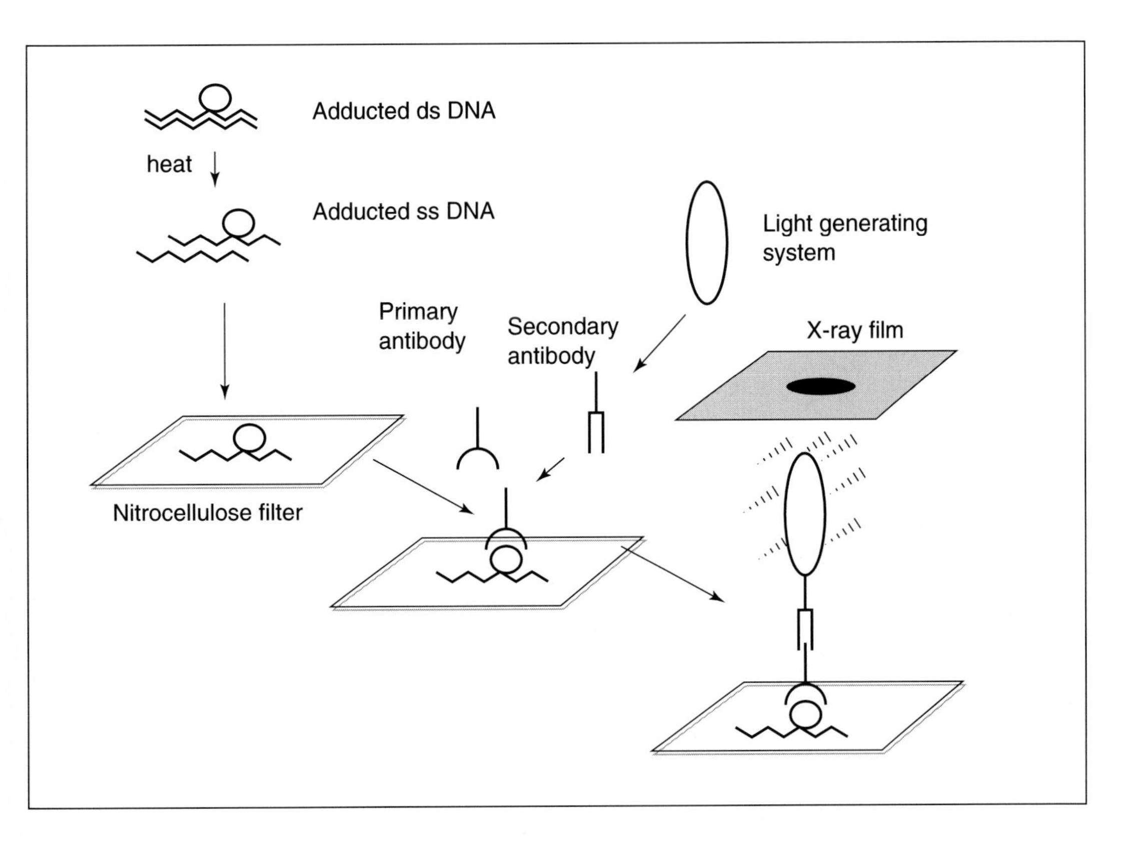

Figure 2. Scheme of the immunoslot–blot procedure

um acetate. The filters were removed from the support, baked at 80 °C for 90 min, washed twice with phosphate-buffered saline (PBS)–0.1%Tween 20 containing 0.5% fat-free milk powder and incubated overnight at 4 °C with the anti-M_1-dG monoclonal antibody D10A1 diluted 1:150000 in 40 ml PBS–0.1% Tween 20 containing 0.5% fat-free milk powder. The nitrocellulose filters were washed with PBS–0.1% Tween 20 and incubated with goat anti-mouse IgG horseradish peroxidase conjugate (Dako, Denmark), diluted 1:4000 in 32 ml PBS–0.1% Tween 20 (0.5% fat-free milk powder), for 2 h at room temperature and washed again with PBS–0.1% Tween 20. Enzymatic activity was visualized by bathing the filter for 5 min in SuperSignal Ultra (Pierce). The filters were exposed to a film (Hyperfilm-ECL; Amersham), and the density of the bands was evaluated directly on the filters with a MultiAnalyst Chemiluminescence Analyser (Biorad) by integrating the whole volume of each band. Standard curves were generated with modified CT-DNA (optical density signal against the amount of M_1-dG). The adduct levels in human DNA were quantified by reference to the standard curve.

High-performance liquid chromatography with ^{32}P-postlabelling

1,N^2-MDA–deoxyguanosine monophosphate (M_1-dGp) was synthesized by the procedure of Seto *et al.* (1985), with modifications, characterized by electrospray–mass spectroscopy and used for the development of a HPLC/^{32}P-postlabelling assay for the detection of the adduct in human samples (Leuratti *et al.*, 1998).

Blood samples were obtained from five human volunteers housed at the MRC Dunn Nutrition Centre (Cambridge, United Kingdom), who were assigned randomly to one of the following diets: 420 g meat; 420 g meat + 400 g vegetables; 420 g meat + 3 g tea or 420 g meat + 400 g vegetables + 3 g tea. The meat was given as roast-beef sandwiches, lasagne, sweet and sour pork or beefsteak. The vegetables were green peas, broccoli and brussels sprouts. Tea was given as six cups of 500 mg black tea. All of the diets were consumed for 15 days. Blood samples were collected on day 0 (when entering the study) and on days 13 and 15 of each diet period. The energy content of each diet was kept constant by addition of a caloric drink (Hycal), and the diets were adjusted to meet the energy requirements of each volunteer. Body weight was recorded daily.

Professor A.T.R. Axon (General Infirmary, Leeds, United Kingdom) kindly provided the human gastric biopsy samples. DNA was extracted from whole blood and biopsy samples with the Qiagen genomic DNA extraction kit (Qiagen Ltd, Crawley, Sussex, United Kingdom) and digested with proteinase K and RNase A. The purity of the DNA was assessed by a 260:280 ratio in a Perkin-Elmer UV spectrophotometer and by reversed-phase HPLC after digestion to deoxynucleotides.

Results and discussion

The availability of a new monoclonal antibody (D10A1) produced and characterized by Sevilla *et al.* (1997) made it possible to develop an ISB assay for the detection of M_1-dG in intact DNA with only 1 μg of DNA per slot. Synthetic primary analytical standards were used to quantify M_1-dG by HPLC–fluorescence in MDA-treated DNA. Thus, MDA-modified CT-DNA containing known amounts of adduct was used to generate calibration curves for each ISB analysis, and the binding of D10A1 antibody was proportional to the amount of M_1-dG in the samples. Figure 3 shows a typical ISB strip with constant amounts of standard (1 μg/slot) containing decreasing amounts of M_1-dG. Lines 9 and 10 contain internal standards, which are commercially available DNAs with either very low (S, ultrapure CT-DNA, Sigma) or very high (BM, human leukocyte DNA, Boehringer Mannheim) levels of M_1-dG. Lines 11–22 contain human samples. The assay has a limit of detection of about 0.156 fmol M_1-dG per μg intact DNA (5 adducts/10^8 normal bases), giving a sensitivity comparable to that of previously described methods but requiring much less DNA. A slightly lower limit of detection of 0.08 fmol/μg (2.5 adducts in 10^8 nucleotides) was obtained when a

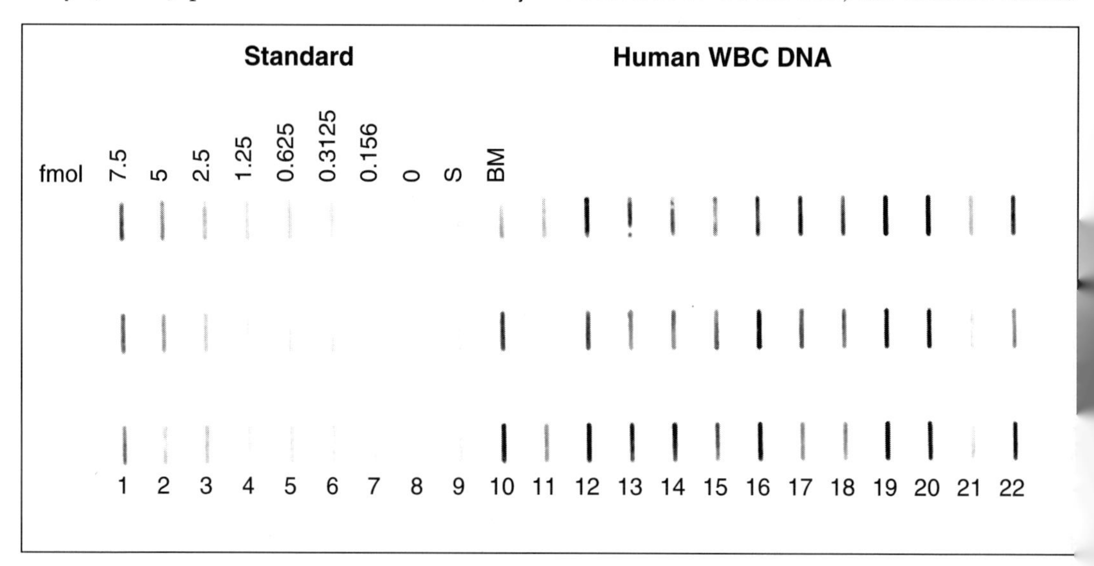

Figure 3. Typical immunoslot–blot analysis of human samples

Lines 1–7: Calf thymus DNA containing decreasing amounts of M_1-dG (10 fmol to 0.156 fmol); line 8: control DNA; lines 9–10: internal standards containing either low (S) or high (BM) amounts of M_1-dG. Lines 11–22: Human leukocyte (WBC) DNA samples isolated from volunteers on controlled diets. All samples are present in triplicate on the filter.

synthetic oligonucleotide containing M_1-dG was used as a standard instead of CT-DNA (Leuratti *et al.*, 1998).

In order to compare the ISB with the previously established HPLC/^{32}P-postlabelling method, four human blood samples in which enough DNA was available were analysed by both methods. A good correlation in the adduct levels measured was found between the two assays (Leuratti *et al.*, 1998).

ISB presents many advantages over other assays, including ^{32}P-postlabelling. As only 1 µg of DNA is needed per well, all samples can be analysed in triplicate on the same ISB. This assay is much less laborious and time-consuming than other assays, allowing routine analysis of a large number of samples in a short time. Quantification of the adduct in human samples was based on a calibration curve obtained by using a standard on the same filter as the real samples, thus standardizing the results obtained on different ISBs.

The possibility of non-specific binding, however, must be taken into consideration when using an assay based on adduct detection by direct antibody binding. The monoclonal antibody used in this work has been completely characterized by Sevilla *et al.* (1997) and shown to recognize M_1-G ribonucleoside with higher specificity than M_1-dG. In the present work, all human DNA samples analysed by ISB were checked by HPLC, and no RNA contamination was detected. Sevilla *et al.* (1997) showed that D10A1 does not bind to unmodified intact RNA or DNA. In our study, no binding was detected with unmodified oligonucleotide (Leuratti *et al.*, 1998). The very low background binding to control DNA was probably due to the presence of endogenous levels of the adduct in the CT-DNA. Previously described competitive inhibition studies showed that cross-reactivity of antibody binding to other related endogenous exocyclic adducts was very low (Sevilla *et al.*, 1997). To assess the specificity of D10A1 antibody binding to M_1-dG in intact DNA in the ISB assay, competitive inhibition studies were conducted. Free M_1-dG added at a concentration of 50 pg/ml PBS resulted in 60–90% signal inhibition. The antibody binding to the membrane was completely inhibited when free M_1-dG was present at 5 ng/ml PBS (data not shown). Addition of free M_1-dG as a competitive inhibitor suppressed the signal completely in all human samples

Levels of 0.31–6.43 (average, 1.93 ± 1.58) adducts per 10^7 normal bases were detected in DNA from 42 human gastric biopsy samples, indicating wide interindividual variation, as reported previously in other tissues (Chaudhary *et al.*, 1994; Fang *et al.*, 1996; Wang *et al.*, 1996). Levels of 0.7–16.65 M_1-dG per 10^7 bases were measured in 28 samples of leukocyte DNA from human volunteers on controlled diets, and the levels increased in three out of five volunteers given a high-meat diet in comparison with a free-diet period (e.g. blood samples collected before starting the study). The levels of adduct were reduced in two non-smoking volunteers after they added vegetables and tea to their diet (Figure 4).

Conclusions

With the ISB assay for the analysis of M_1-dG in DNA, we can detect background levels of the adduct in DNA from human leukocytes and gastric biopsy samples with a sensitivity similar to that of previously published methods, including ^{32}P-postlabelling and gas chromatography and electron capture-negative chemical ionization with mass spectrometry. The advantages of ISB, such as the possibility of analysing 1 µg DNA per well and the fact that it is less time-consuming and laborious than other assays, means that it

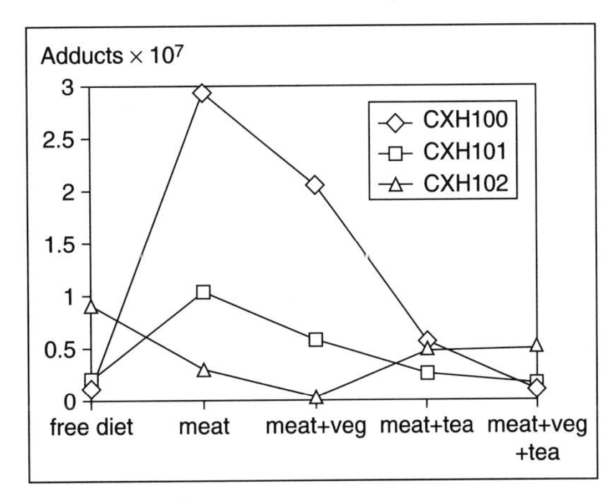

Figure 4. M1-dG levels in DNA isolated from blood of three non-smoking volunteers after different diets

could be used for analysing a large number of samples. For example, studies are currently under way on the relation between M_1-dG and dietary fat and antioxidant intake and on the distribution of M_1-dG in various human tissues. Initial results on human volunteers on controlled diets suggested that the M_1-dG levels in leukocyte DNA vary with the diet.

Acknowledgements

Professor A.T.R. Axon and Dr S. Everett, Leeds General Infirmary, United Kingdom, provided human gastric biopsy samples. This work was supported in part by the United Kingdom Ministry of Agriculture, Fisheries and Food, contracts nos. FS1716 and FS1735.

References

Basu, A.K. & Marnett, L.J. (1983) Unequivocal demonstration that malondialdehyde is a mutagen. *Carcinogenesis* 4, 331–333

Benamira, M., Johnson, K., Chaudhary, A., Bruner, K., Tibbetts, C. & Marnett, L.J. (1995) Induction of mutations by replication of malondialdehyde-modified M13 DNA in *Escherichia coli*: Determination of the extent of DNA modification, genetic requirements for mutagenesis, and types of mutations induced. *Carcinogenesis* 16, 93–99

Chapeau, M-C. & Marnett, L.J. (1991) Enzymatic synthesis of purine deoxynucleoside adducts. *Chem. Res. Toxicol.* 4, 636–638

Chaudhary, A.K., Nokubo, M., Reddy, G.R., Yeola, S.N., Morrow, J.D., Blair, I.A. & Marnett, L.J. (1994) Detection of endogenous malondialdehyde–deoxyguanosine adducts in human liver. *Science* 265, 1580–1582

Fang, J-L, Vaca, C.E., Valsta, L.M. & Mutanen, M. (1996) Determination of DNA adducts of malondialdehyde in humans: Effects of dietary fatty acid composition. *Carcinogenesis* 17, 1035–1040

Fink, S.P., Reddy, G.R. & Marnett, L.J. (1997) Mutagenicity in *Escherichia coli* of the major DNA adduct derived from the endogenous mutagen malondialdehyde. *Proc. Natl Acad. Sci. USA* 94, 8652–8657

Janero, D.R. (1990) Malondialdehyde and thiobarburic acid-reactivity as diagnostic indices of lipid peroxidation and peroxidative tissue injury. *Free Radicals Biol. Med.* 9, 515–540

Leuratti, C., Bingham, S., Hughes, R., Axon, A.T.R., Everett, S., Farmer, P.B. & Shuker, D.E.G. (1997) Detection by HPLC/^{32}P-postlabelling of malondialdehyde–deoxyguanosine monophosphate in human blood and tissue DNA in relation to diet and *H. pylori* infection. *Proc. AACR* 39, 353

Leuratti, C., Singh, R., Lagneau, C., Farmer, P.B., Plastaras, J.P., Marnett, L.J. & Shuker, D.E.G. (1998) Determination of malondialdehyde-induced DNA damage in human tissues using an immunoslot–blot assay. *Carcinogenesis* 19, 1919–1924

Marnett, L.J. (1994a) Generation of mutagens during arachidonic acid metabolism. *Cancer Metastasis Rev.* 13, 303–308

Marnett, L.J. (1994b) DNA adducts of α,β-unsaturated aldehydes and dicarbonyl compounds. In: *DNA Adducts: Identification and Biological Significance*, IARC Scientific Publications No. 125 (Hemminki, K., Dipple, A., Shuker, D.E.G., Kadlubar, F.F., Segerback, D. & Bartsch, H., eds). Lyon, IARC, pp. 151–163

Mientjes, E.J., Hochleitner, K., Luiten-Schuite, A., van Delft, J.H.M., Thomale, J., Berends, F., Rajewski, M.F., Lohman, P.H.M. & Baan, R.A. (1996) Formation and persistence of O^6-ethylguanine in genomic and transgene DNA in liver and brain of λlacZ transgenic mice treated with N-ethyl-N-nitrosourea. *Carcinogenesis* 17, 2449–2454

Rouzer, C.A., Chaudary, A.K., Nokubo, M., Ferguson, D.M., Reddy, G.R., Blair, I.A. & Marnett, L.J. (1997) Analysis of the malondialdehyde–2′-deoxyguanosine adduct pyrimidopurinone in human leukocyte DNA by gas chromatography/electron capture/negative chemical ionization/mass spectrometry. *Chem. Res. Toxicol.* 10, 181–188

Seto, H., Takesue, T. & Ikemura, T. (1985) Reaction of malondialdehyde with nucleic acid. II. Formation of fluorescent pyrimido[1,2-a]purin-10(3H)-one mononucleotide. *Chem. Soc. Jpn.* 58, 3431–3435

Sevilla, C.L., Mahle, N.H., Eliezer, N., Uzieblo, A., O'Hara, S.M., Nokubo, M., Miller, R., Rouzer, C.A. & Marnett, L.J. (1997) Development of monoclonal antibodies to the malondialdehyde–deoxyguanosine adduct, pyrimidopurinone. *Chem. Res. Toxicol.* 10, 172–180

Vaca, C.E., Fang, J.-L., Mutanen, M. & Valsta, L. (1995) ^{32}P-Postlabelling determination of DNA adducts of malondialdehyde in humans: Total white blood cells and breast tissue. *Carcinogenesis* 16, 1847–1851.

Wang, M., Dhingra, K., Hittelman, W.N., Liehr, J.G., de Andrade, M. & Li, D. (1996) Lipid peroxidation-induced putative malondialdehyde–DNA adducts in human breast tissues. *Cancer Epidemiol. Biomed. Prev.* 5, 705–710

World Cancer Research Fund (1997) *Food, Nutrition and the Prevention of Cancer: A Global Perspective.* Washington DC, American Institute for Cancer Research

Exocyclic DNA Adducts in Mutagenesis and Carcinogenesis
B. Singer & H. Bartsch
IARC Scientific Publications No. 150
International Agency for Research on Cancer, Lyon, 1999

^{32}P-Postlabelling with high-performance liquid chromatography for analysis of abundant DNA adducts in human tissues

K. Yang, J.-L. Fang, D. Li, F.-L. Chung & K. Hemminki

Abundant complex DNA adducts can be detected in human tissues by a combined ^{32}P-postlabelling and high-performance liquid chromatography (HPLC) method. The HPLC profiles reveal a panorama of nuclease P1-resistant human adducts, which are not among the known human DNA adducts and are suspected of being endogenous. Lipid peroxidation-induced DNA adducts and I-compounds are two possible candidates for these adducts. Therefore, we performed two experiments: one was to identify chromatographically the lipid peroxidation-induced adducts among other human adducts with two acrolein- and crotonaldehyde-derived propano adduct standards (Acr-dG3 and Cro-dG1&2) and a structurally unknown adduct (Cro-DNA) derived from crotonaldehyde-treated DNA; and the other was to analyse the adducts in breast tissue from patients with breast cancer and from controls and to compare their behaviour with that of I-compounds in cancerous tissues. In the first experiment, Acr-dG3 and Cro-dG1 were detected in three human lung tissues, at levels ranging from 3.4 to 8.9 ($\times 10^{-8}$) and from not detectable to 2.9 ($\times 10^{-8}$), respectively. Acr-dG3 and Cro-DNA were detected in three human colon tissues, at levels of 0.2–0.4 ($\times 10^{-8}$) and 1.2–3.4 ($\times 10^{-8}$), respectively. In the second experiment, adjacent and tumorous breast tissues from 15 patients with breast cancer (of an average age of 33.4 years) and normal breast tissue from 18 controls (of an average age of 57.3) were analysed for the abundant complex adducts. The total adduct levels in the adjacent and tumorous tissues were lower than in the normal tissues (with medians of 8.0, 11.8 and 13.3 ($\times 10^{-7}$), respectively). Significant differences in the adduct levels between adjacent or tumorous tissues and normal tissues were observed in three HPLC peaks, and age was significantly associated with three peaks. These results are consistent with our speculation that the abundant adducts are comprised of lipid peroxidation-induced adducts and human homologues of I-compounds.

Introduction

Abundant lipophilic DNA adducts that can be extracted into 1-butanol have been detected in a variety of human tissues by ^{32}P-postlabelling with high-performance liquid chromatography (HPLC). These adducts resemble bulky DNA adducts in that they can be enriched by nuclease P1 treatment in the ^{32}P-postlabelling assay. Their apparent levels range from 1.8 adducts/10^7 nucleotides in colon tissue to as high as 17 adducts/10^7 nucleotides in skin (Hemminki *et al.*, 1997; Yang *et al.*, 1998). These lipophilic adducts represent only a part of the abundant adducts present in human tissues. Some evidence suggests that these adducts are formed endogenously (Yang *et al.*,1998).

Endogenous DNA adducts induced by lipid peroxidation products have been studied extensively because of their potential role in carcinogenesis (Marnett & Burcham, 1993; Chung *et al.*, 1996). Several major lipid peroxidation products, e.g. malondialdehyde, 4-hydroxynonenal, acrolein and crotonaldehyde, belong to the class of difunctional electrophiles which form exocyclic DNA adducts by reaction with exocyclic and ring nitrogens of the bases (Esterbauer *et al.*, 1991). Some resulting adducts have been detected in DNA from various tissues of unexposed rodents and humans (Chaudhary *et al.*, 1994; Nath & Chung, 1994; Nair *et al.*, 1995). The detection of a putative 2,3-epoxy-4-hydroxynonanal–DNA adduct in our previous study suggested the likelihood of the presence of other exocyclic adducts among the complex adducts (Yang *et al.*, 1998).

I-Compounds are endogenous DNA modifications of unknown structure (for reviews, see Marnett & Burcham, 1993; Randerath *et al.*, 1993a; De Flora *et al.*, 1996). They have been studied extensively in experimental rodents but rarely in humans because of a lack of standards and the practical restrictions of human studies. Some interesting features of I-compounds are that their levels increase with age and decrease during carcinogenesis and tumour promotion.

Since both exocyclic adducts and I-compounds are resistant to nuclease P1 treatment and their levels are expected to fall into the range of our analysis, we focused on their possible association with the abundant human adducts by carrying out two experiments. One was to detect some acrolein- and crotonaldehyde-derived DNA adducts among the complex adducts in human lung and colon tissues. The other was to analyse the abundant adducts in breast tissues of patients with breast cancer and controls and to compare their behaviour with that of I-compounds, in order to shed light on their role in human carcinogenesis. We regarded breast cancer as a challenging subject because in two recent studies, breast cancer was related to increased levels of aromatic (Li *et al.*, 1996) and malondialdehyde-induced DNA adducts (Wang *et al.*, 1996). The same DNA samples used in those two studies were examined in the current investigation by the ^{32}P-postlabelling/HPLC method.

Materials and methods

DNA samples

Human lung and colon tissues were collected from a few individuals, and the DNA was isolated as described elsewhere (Yang *et al.*, 1998). Breast tumours and histologically normal adjacent tissues were collected in the United States from 18 women with breast cancer who were undergoing mastectomy, and normal breast tissue was collected from 18 women undergoing reduction mammoplasty; DNA was isolated as described previously (Li *et al.*, 1996). Personal characteristics of the breast tissue donors are listed in Table 2 below. Owing to practical restriction in sample collection, the breast cancer patients and the control patients were not matched in terms of age or race. The mean age of the breast cancer patients was 57.3 years (SD, 12.9 years) which was significantly higher than that of the control patients, 33.4 years (SD 10.6 years) ($p < 0.00001$). Most of the cancer patients were white, whereas most of the control patients were black.

Standards

Three standards were used to detect the corresponding DNA adducts in human tissues: acrolein-derived propano-dG 3′-monophosphate standards, isomer 3 (Acr-dG3, previously designated AdG3 (see Nath & Chung, 1994)), which might contain two isomers; crotonaldehyde-derived propano-dG 3′-monophosphate standards, diastereomers 1 and 2 (Cro-dG1&2, previously designated CdG1&2 (see Nath & Chung, 1994)) and a standard (designated Cro-DNA) of unknown structure which is a major adduct in crotonaldehyde-modified calf thymus DNA. The chemical structures of Acr-dG3 and Cro-dG1&2 are shown in Figure 1. The standards Acr-dG3 and Cro-dG1&2 and the crotonaldehyde-modified calf thymus DNA were prepared as described elsewhere (Chung *et al.*, 1984).

^{32}P-Postlabelling–high-performance liquid chromatography analysis of DNA adducts in breast tissues

A modification of the dinucleotide/-monophosphate version of the postlabelling assay described by Randerath *et al.* (1989) was used. To analyse DNA adducts in the breast tissues, 5–10 μg DNA were hydrolysed to adducted dinucleotide monophosphates and normal nucleosides at 37 °C for 45 min by prostatic acid phosphodiesterase (0.02 U/μl) and nuclease P1 (0.2 μg/μl) in a volume of 5–10 μl 3.2 mmol/L sodium acetate buffer containing 0.06 mmol/L zinc chloride (pH 5.2) according to the amount of DNA. The proteins were precipitated by adding 100 μl chilled ethanol after DNA digestion, maintained at –20 °C for 20 min and centrifuged at 14 000 rpm for 15 min. The supernatant was evaporated to dryness in a fresh tube and subjected to a postlabelling reaction. The ^{32}P-postlabelling reaction

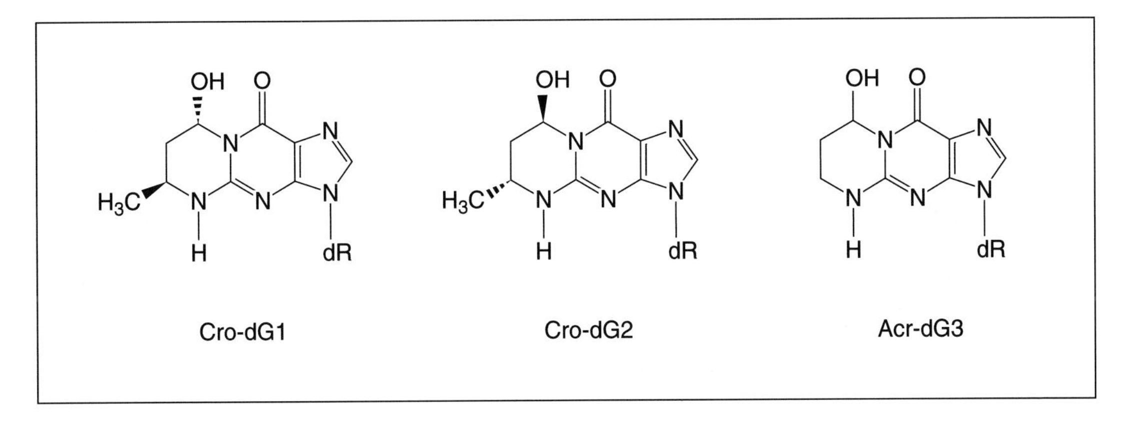

Figure 1. Chemical structures of the adduct standards, acrolein-derived propano-dG 3′-monophosphate, isomer 3 (Acr-dG3), crotonaldehyde-derived propano-dG 3′-monophosphate diastereomers 1 (Cro-dG1) and 2 (Cro-dG2)

was conducted in 2 µl of a solution containing 20 mmol/L 2-(*N*-cyclohexylamino)ethanesulfonic acid buffer (pH 9.6), 20 mmol/L magnesium chloride, 2 mmol/L spermidine, 20 mmol/L dithiothreitol, 3.3 U T4 polynucleotide kinase and 7 µCi [γ-^{32}P]ATP (3000 Ci/mmol). After a 40-min incubation at 37 °C, the labelled modified dinucleotides were converted to modified and normal mononucleotides by treatment with 5 mU (in 1 µl water) snake venom phosphodiesterase at 37 °C for 30 min, then diluted to 8 µl with water and subjected to HPLC analysis.

A Beckman Gold HPLC system with an on-line radioactive detector and a C18 reversed-phase HPLC column (Kromasil 5 C18 100A, 250 × 2.00 mm, Phenomenex, USA) were used for adduct resolution and detection. The HPLC was run at room temperature with a binary system constituted of methanol and buffer A in programme A (see below). The flow rate in the HPLC column was 0.2 ml/min. The individual peaks in the HPLC profile were integrated for peak area, and the individual DNA adduct levels were calculated from the peak areas and the ratio between the HPLC peak area and radioactive counting in a scintillation counter (Beckman, LS 6000TA).

Detection of putative endogenous DNA adducts in human lung and colon tissues

The same postlabelling method described above was used to analyse DNA adducts in human lung and colon tissues and to retrieve the standard adduct (Cro-DNA) from crotonaldehyde-modified calf thymus DNA. The standards Acr-dG3 and Cro-dG1&2 (50 fmol in 1 µl water) were directly labelled and then converted to 5′-monophosphate forms by nuclease P1 (5 µg, in 1 µl 1.5 mmol/L zinc chloride). Removal of the 3′-phosphate took up to 24 h at 37 °C due to the resistance of the standards to nuclease P1 treatment. Both human DNA adducts and standards were separated by HPLC with methanol and buffer A in programme B (see below). The peaks for the standards and for human adducts that had similar retention times to the standards were collected, evaporated to dryness and analysed again, separately and in a mixture, by HPLC to confirm the identity of adducts with the standards and to quantify the adducts. The second HPLC separation was performed with methanol and buffer B in programme C (see below). Since a collected peak from the first HPLC separation was likely to contain several adduct components, the apparent levels of the putative endogenous DNA adducts in human lung and colon tissues were calculated from the peak areas in the second HPLC profiles.

The buffers used for HPLC were: buffer A: 0.5 mol/L ammonium formate, 20 mmol/L phosphoric acid, pH 4.6; buffer B: 0.5 mol/L ammonium formate, 20 mmol/L phosphoric acid, pH 3.2. The programmes used were:

- Programme A: 0–1.5 min, 2–13% methanol; 1.5–12 min, 13–14.5% methanol; 12–21 min, 14.5–26.5% methanol; 21–45 min, 26.5–50% methanol; 45–67 min, 50–65% methanol; 67–75 min, 65–100% methanol

- Programme B: 0–5 min, 2% methanol, isocratic; 5–55 min, 2–54.5% methanol
- Programme C: 0–40 min, 2% methanol, isocratic; 40–65 min, 2–52% methanol.

Statistical analysis

Nemenyi's rank sum test and Mann-Whitney's rank sum test were used to compare the distributions of adduct levels in different groups. Spearman's rank correlation was used to analyse the correlation between adduct level and age.

Results

Putative endogenous DNA adducts in human lung and colon tissues

In our previous studies on lipophilic adducts, butanol extraction was carried out after the labelling reaction (Hemminki *et al.*, 1997). Since many early-eluting adducts, including the propano DNA adducts, cannot be extracted into 1-butanol in substantial proportion (Figure 2), butanol extraction was omitted in this study in order to recover the whole spectrum of the nuclease P1-

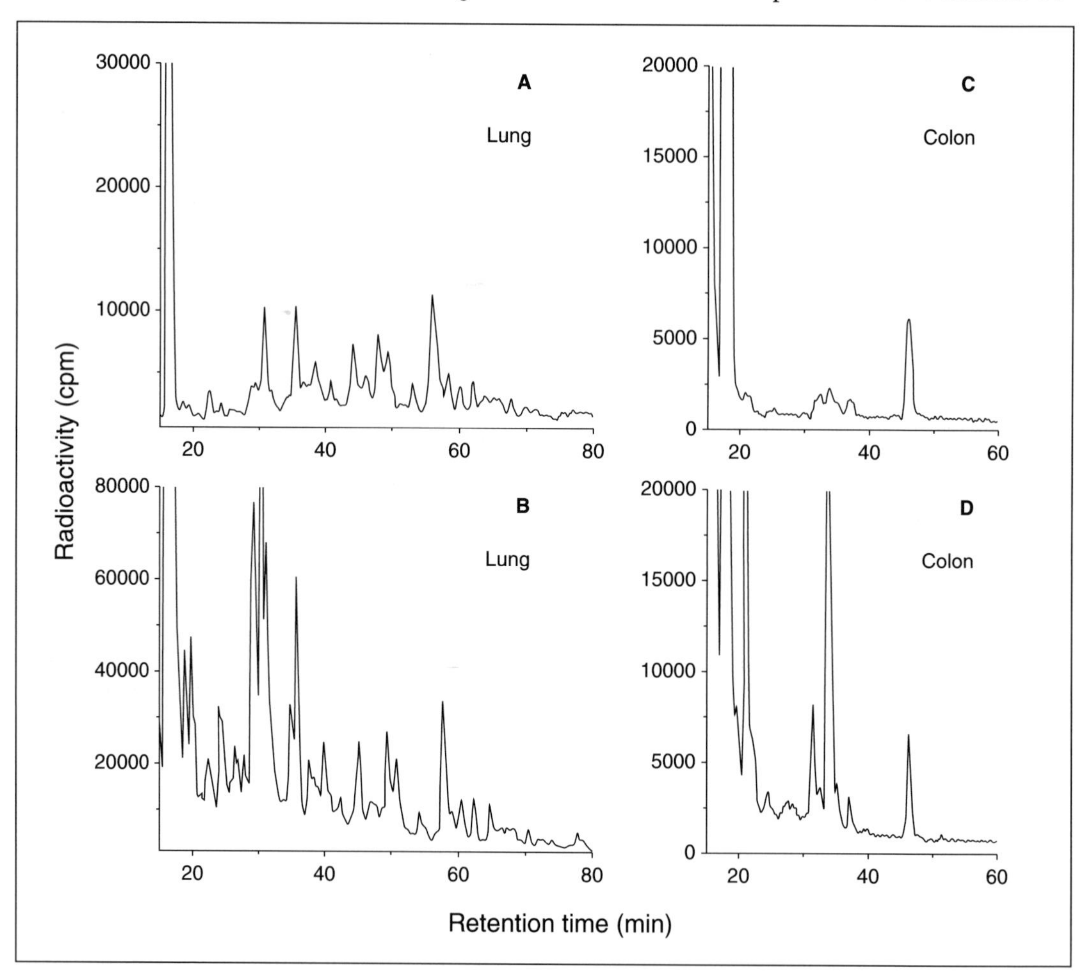

Figure 2. High-performance liquid chromatography profiles of abundant DNA adducts in human tissues determined by monophosphate version of the ^{32}P-postlabelling assay with (A,C) and without (B,D) 1-butanol extraction after labelling reaction

A and C, adducts extracted into butanol phase; B and D, adducts recovered without butanol extraction

resistant adducts for HPLC analysis. The HPLC profiles of the DNA adducts in colon and lung tissue are shown in Figure 3 (AI and BI). Inorganic phosphate, excess ATP and normal nucleotides eluted before 17 min. Colon tissues had a simple DNA adduct pattern, with four major adduct peaks eluted after 20 min when the adducts were separated by programme B (Figure 3AI). The DNA adduct pattern of the lung tissue samples was much more complicated, showing large tandem peaks that were not well separated in programme B (Figure 3BI).

In the HPLC profiles of the colon DNA samples, peaks a and b (Figure 3AI) had the same retention times as the standards Acr-dG3 and Cro-DNA, respectively. These adducts co-migrated with Acr-dG3 or Cro-DNA when eluted in HPLC by programme C after being mixed with the standards. Figure 3AII, III and IV show co-migration of the adduct in peak b with the standard Cro-DNA. In HPLC profiles of the lung DNA samples, peaks c and d (Figure 3BI) had same retention time as the standards Acr-dG3 and Cro-dG1&2. When mixed with Cro-dG1&2 and eluted in programme C, peak c was shown to contain an adduct that co-migrated with Cro-dG1 (Figure 3C). Co-migration of peak d with Acr-dG3 was confirmed by elution in a mixture in programme C (Figure 3BII, III and IV). Only Acr-dG3 was detected in both tissues. Cro-dG1 was found only in lung DNA and Cro-DNA only in colon DNA.

The apparent levels of these acrolein- and crotonaldehyde-related DNA adducts determined in the human colon and lung tissues are listed in Table 1.

DNA adducts in breast tissues

The HPLC profiles showed a complex pattern of DNA adducts in the breast tissues (Figure 4). Inorganic phosphate, excess ATP and normal nucleotides eluted before 13 min. Twelve prominent adduct components eluted between 15 and 50 min, and fraction 13 included small peaks between 53 and 75 min, where the non-polar polycyclic aromatic hydrocarbon adducts were expected to elute. The adducts eluted before 15 min and after 53 min were not studied further. There was no systematic difference between the adduct patterns of normal, tumour and tumour-adjacent breast tissues.

The adduct levels in the 12 HPLC peaks marked in Figure 4 were calculated (Table 2). The highest median adduct level was observed in normal tissues from the control subjects, while the lowest was observed in the tissues adjacent to breast tumours. Correlation analysis of the adduct levels in the adjacent and tumour tissues of the 15 cancer patients for whom both tissues were analysed showed a high correlation, with a coefficient of 0.65 ($p < 0.01$). The adduct levels in the individual HPLC fractions are listed in Table 3. Large interindividual variations were observed in

Table 1. Levels of some endogenous DNA adducts in DNA from human lung and colon tissues (adducts per 10^8 nucleotides)

Tissue	Acr-dG3	Cro-dG1	Cro-DNA
Lung			
	3.4	ND	ND
	8.9	2.9	ND
	3.7	1.0	ND
Colon			
	0.4	ND	1.2
	0.2	ND	3.4
	0.3	ND	2.2

Acr-dG3, acrolein-derived propano-dG, isomer 3; Cro-dG1 and Cro-dG2, crotonaldehyde-derived propano-dG, diastereomers 1 and 2; ND, not detected

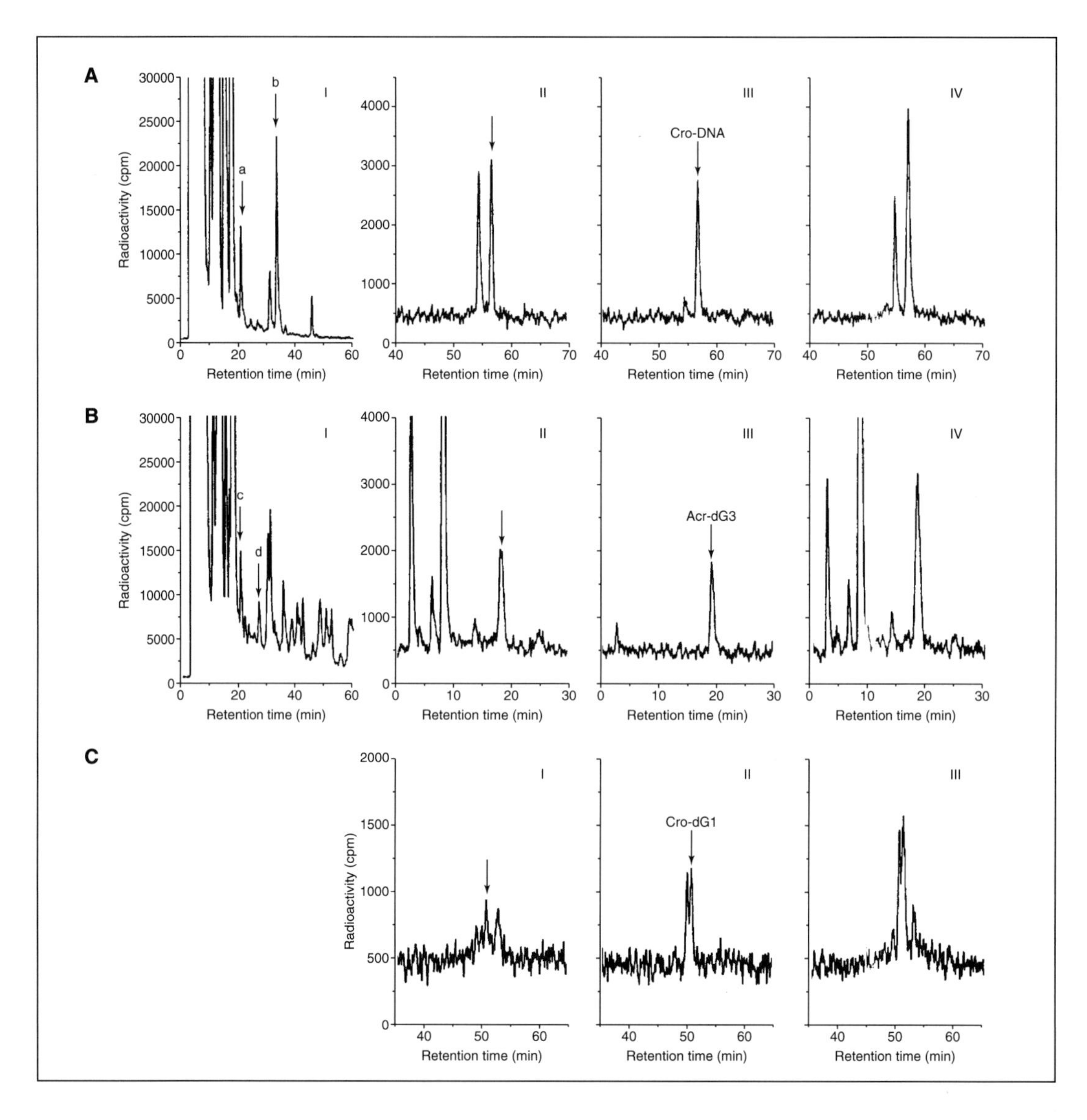

Figure 3. High-performance liquid chromatography (HPLC) chromatograms showing co-migration of the HPLC peaks observed in human colon (A) and lung (B,C) tissues with acrolein- and crotonaldehyde-related adduct standards

In AI and BI, HPLC programme B was used. In AII-IV, BII-IV and CI-III, HPLC programme C was used. AI, DNA adducts in a colon DNA sample; arrows a and b indicate peaks that have the same retention time as acrolein-derived propano-dG, isomer 3 (Acr-dG3) and crotonaldehyde–DNA (Cro-DNA), respectively; AII, collected fraction b in AI; AIII, collected Cro-DNA; AIV, mixture of same amounts of human adducts and Cro-DNA used in AII and AIII. BI, DNA adducts in a lung DNA sample; arrows c and d indicate peaks that have the same retention time as Acr-dG3 and crotonaldehyde-derived propano-dG diastereomers 1 and 2 (Cro-dG1&2), respectively; BII, collected fraction c in BI; BIII, collected Acr-dG3; BIV, mixture of same amounts of human adducts and Cro-DNA used in BII and BIII. CI, collected fraction d in BI; CII, collected Cro-dG1&2; CIII, mixture of same amounts of human adducts and Cro-dG1&2 used in CI and CII. Arrows in AII, BII and CI indicate human adducts that co-migrated with the standards indicated by arrows in AIII, BIII and CII, respectively.

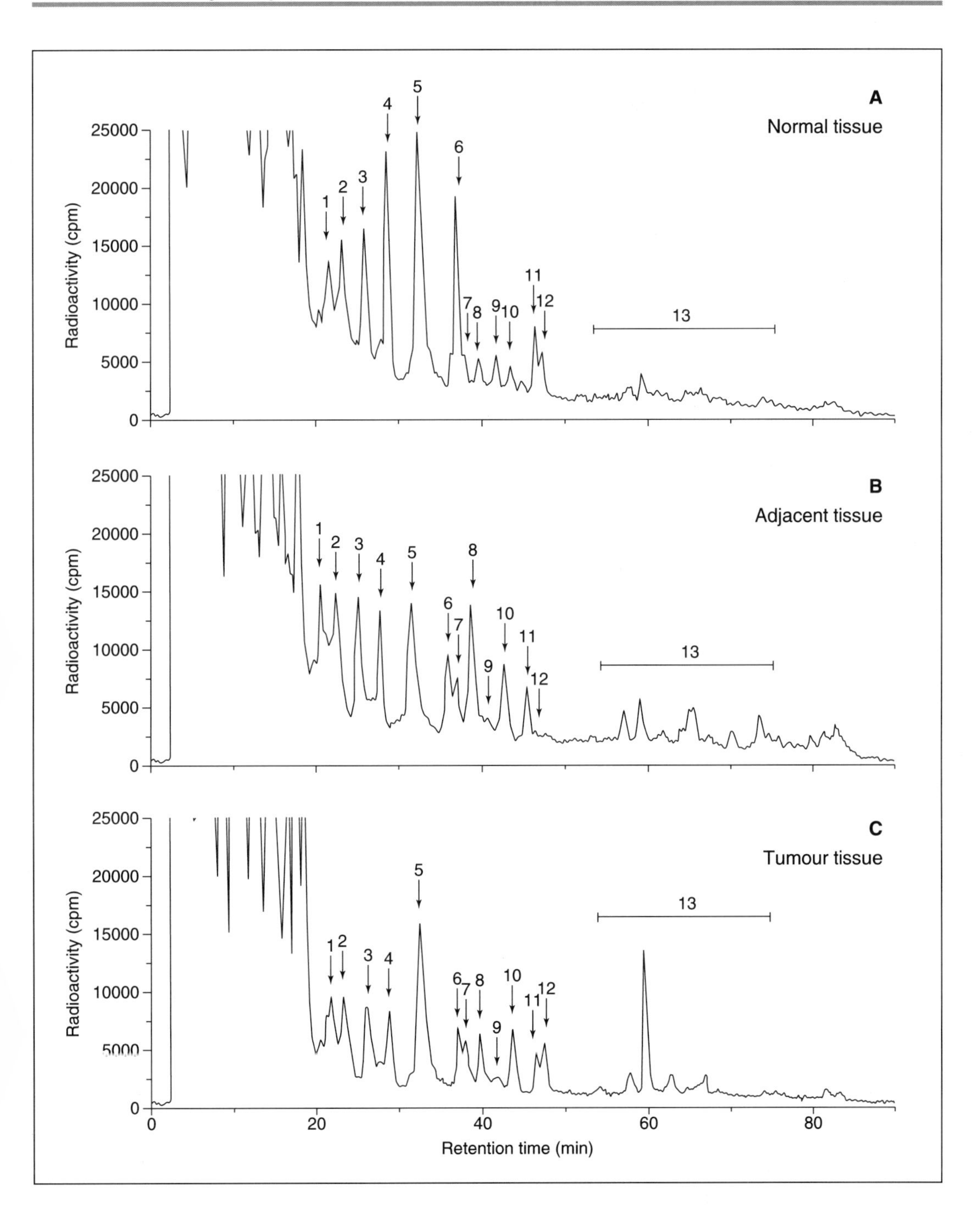

Figure 4. High-performance liquid chromatography (HPLC) chromatograms of adducts in breast tissue DNA from a control woman (A); in DNA of adjacent breast tissue (B) and in tumour tissue (C) from a breast cancer patient

Arrows indicate the HPLC peaks or fractions from which adduct levels were analysed.

Table 2. Personal characteristics of the women from whom breast tissues were collected, and DNA adduct levels in the breast tissues

Cancer patients				Controls		
		Adduct level[a]				
Age (years)	Race	Adjacent	Tumour	Age (years)	Race	Adduct level
55	Hispanic	11.9	36.8	30	White	11.3
44	White	10.9	14.3	42	White	12.7
51	Black	19.8	11.2	49	Black	5.3
66	White	8.0	5.3	19	Black	3.9
53	Hispanic	4.9	2.0	29	Black	21.0
48	White		2.5	17	White	13.9
				46	Black	38.3
46	White	21.0	48.8	20	Black	51.6
41	Hispanic	2.7	4.9	46	White	8.4
53	Black	6.0	12.4	32	Black	20.8
80	White	1.4	2.3	28	Black	22.4
41	White	13.5	8.4	46	Other	4.9
74	White	18.9	12.0	17	Black	8.5
70	White	2.2	1.0	27	Black	7.6
65	Hispanic	23.5	26.0	39	Hispanic	9.8
67	White	1.4	12.4	39	White	15.9
54	White		21.2			
				37	Black	16.2
79	White		1,0			
44	White	1.8	0.8	38	Black	88.7
Median		8.0	11.8			13.3

[a] Adducts per 10^7 nucleotides. Total level of 12 HPLC fractions shown in Figure 3

both total and individual adduct levels. In almost all the fractions, the highest median adduct level was observed in normal tissues from control subjects, and most of the medians were markedly higher than those observed in the adjacent and tumour tissues of the cancer patients. Statistical analysis showed that three peaks (4, 6, 11) were significantly different between normal tissue and adjacent or tumour tissues (Table 3).

No significant difference was found between black and white women in the adduct levels in adjacent and control tissues. The association between age and individual adduct levels was analysed by the nonparametric Spearman test. When the data for adjacent and control tissues were combined for analysis, a weak but significant negative correlation between adduct level and age was found for HPLC fractions 3, 4 and 5 (Table 4). When the data for adjacent and control tissues were analysed separately, the same trend of an association was found, but no significance was detected owing to the decrease in sample size and age range (Table 4).

Discussion

One of the major advantages of using HPLC in studying DNA adducts is that the entire spectrum of adducts labelled in the ^{32}P-postlabelling assay can be analysed after a single injection and an overall estimate of the adducts present in human tissues can be made, while a single postlabelling assay does not cover all adducts and does not recover different adducts with same efficiency. Unlike most studies of DNA adducts, which have

Table 3. DNA adduct levels (adducts per 10^8 nucleotides) in human breast tissues

Tissue	Peak 1	Peak 2	Peak 3	Peak 4[a]	Peak 5	Peak 6[b]
Adjacent						
Median	18.6	6.2	11.7	1.3	10.0	7.9
Range	0.6-51.7	0.3-35.9	0.4-35.2	0.5-38.1	2.2-61.2	1.1-38.6
Tumour						
Median	11.6	8.6	6.4	5.1	7.7	3.4
Range	0.9-297.0	0.6-91.8	0.5-46.9	0.8-23.1	1.7-68.5	0.5-20.0
Normal						
Median	17.0	22.0	11.8	13.8	26.2	13.4
Range	3.0-81.3	3.9-170.0	0.7-100.2	0.9-264.7	6.1-195.0	1.8-59.7

Tissue	Peak 7	Peak 8	Peak 9	Peak 10	Peak 11[c]	Peak 12
Adjacent						
Median	1.8	0.75	0.9	1.5	3.6	1.6
Range	0.4-19.2	0.4-13.7	0.6-10.7	0.5-8.0	0.3-17.3	0.7-4.7
Tumour						
Median	3.1	1.2	0.6	0.6	1.0	1.2
Range	0.4-19.6	0.7-33.7	0.3-4.9	0.5-8.1	0.5-34.1	0.3-99.3
Normal						
Median	3.5	2.8	1.9	1.7	5.7	2.9
Range	0.7-30.3	0.9-22.7	0.4-19.3	0.4-9.3	1.0-39.2	0.7-53.9

[a] A < N, $p < 0.05$
[b] T < N, $p < 0.01$
[c] T < N, $p < 0.05$

Table 4. Correlation coefficients of DNA adduct levels with age

Tissue	Peak 3	Peak 4	Peak 5
Adjacent to tumour	− 0.21	− 0.10	− 0.10
Normal	− 0.42	− 0.17	− 0.24
Adjacent plus normal	− 0.37, $p < 0.05$	− 0.39, $p < 0.05$	− 0.38, $p < 0.05$

focused on a specific adduct or a group of adducts, the present study emphasized the analysis of individual adducts in the panorama of adducts.

Acrolein and crotonaldehyde are generated not only during lipid peroxidation but also by burning fat, by cigarette smoking and in the metabolism of some carcinogens, such as *N*-nitrosopyrrolidine and cyclophosphamide (Nath & Chung, 1994). Thus, both endogenous and exogenous sources may contribute to the formation of Acr-dG and Cro-dG. Yet, because of the low levels of acrolein and crotonaldehyde in the environment and the presence of Acr-dG and Cro-dG in tissues of untreated experimental animals or humans, the view that these propano adducts are generated endogenously is favoured (Chung *et al.*, 1996). The detection of several acrolein- and crotonaldehyde-induced DNA adducts in this study demonstrates that the complex of DNA adducts analysed in human tissues by our ^{32}P-postlabelling–HPLC assay indeed includes more endogenous adducts, in addition to the previously detected 2,3-epoxy-4-hydroxy-nonanal–DNA adduct. Acr-dG and Cro-dG have been detected in relatively large amounts of DNA (about 50 μg DNA or more for each assay) from human liver, mammary gland and leukocytes by a ^{32}P-HPLC method in which a HPLC separation step was used before and a thin-layer chromatography separation step after the labelling reaction to purify the propano adducts before the final HPLC analysis (Nath & Chung, 1994; Nath *et al.*, 1996). Our detection of the propano adducts in human lung and colon tissues confirmed their expected prevalence. The breast tissue DNA was not analysed for the three exocyclic adduct standards owing to insufficient amounts of DNA.

In this study, the analysis of the acrolein- and crotonaldehyde-induced DNA adducts was focused on identification rather than quantification, but the apparent levels of the propano adducts in human lung and colon tissues were comparable to those reported in other human tissues. Moreover, human and rat liver, mammary tissue and leukocytes showed similar ratios of Acr-dG3 to Cro-dG1 (Table 5). Similar trends were observed when the levels of Acr-dG3 and Cro-dG1 in human lung and colon tissues determined in this study were compared with those reported in animal lung and colon tissues (Table 5). As only 10 μg DNA were used in this study, the fact that Cro-dG1&2 was not detected in colon tissues is likely simply to reflect low adduct levels. Our results demonstrate that the propano adducts can be detected by our simple ^{32}P-postlabelling–HPLC method.

When analysed within the background of the complex abundant DNA adducts, the lipid peroxidation-related adducts seemed to be a major constituent in colon tissue, since the acrolein- and crotonaldehyde-induced adducts were detected in two out of the four predominant HPLC fractions which eluted after 20 min (Figure 3AI). In contrast, the propano adducts seemed to represent only a minor proportion of the adducts in lung and breast tissue (Figure 3BI and Figure 4). Several lines of evidence suggest that the unidentified abundant adducts (the main HPLC peaks) might include endogenous DNA adducts. Firstly, the

Table 5. Propano DNA adduct levels in human and animal tissues reported in the literature[a] and detected in this study

Tissue	Acr-dG3	Cro-dG1	Cro-dG2
Human			
Liver	0.6–14.8[b]	2.8–20.6	2.6–6.8
Mammary	0.2–13.2	0.08–1.54	0.12–1.02
Leukocyte	0.06–0.5	0.1–0.5	0.06–0. 22
Lung[c]	3.4–8.9	ND–2.9	ND
Colon[c]	0.2–0.4	ND	ND
Rat			
Liver	0.2–0.8	2.8–14.8	1.0–4.0
Mammary	2.12–2.68	0.08–0.26	0.1–0.2
Leukocyte	3.08–5.8	1.16–1.88	0.32–0.54
Lung	7.4–12.9	0.96–3.32	1.38–6.06
Colon	4.9–18.68	0.2–1.36	0.68–1.34

Acr-dG3, acrolein-derived propano-dG 3′-monophosphate, isomer 3; Cro-dG1 and Cro-dG2, crotonaldehyde-derived propano-dG 3′-monophosphate, diastereomers 1 and 2; ND, not detected

[a] Data from Nath & Chung (1994) and Nath *et al.* (1996) unless otherwise indicated

[b] Adducts per 10^8 nucleotides

[c] Data from present study

patterns of the lipophilic DNA adducts were tissue-specific, distinguished from the patterns of DNA adducts induced by exogenous carcinogens/mutagens that showed consistency in various tissues (Li *et al.*, 1990). Secondly, in lung tissue and lymphocytes, the levels of these DNA adducts were much higher than those of the aromatic DNA adducts measured at the same time by a ^{32}P-postlabelling–thin-layer chromatography method (Hemminki *et al.*, 1997; Yang *et al.*, 1998; unpublished data). Thirdly, some of these adducts were age-dependent (Yang *et al.*, 1998). Fourthly, in lymphocytes, these adducts were not related to occupational exposure to polycyclic aromatic hydrocarbons or to cigarette smoking (Yang *et al.*, 1998). Fifthly, the sample donors were unlikely to have had occupational exposure to other kinds of carcinogens. Finally, these adducts were not likely to arise from carcinogens or mutagens in the diet, because the DNA adducts of some extensively studied carcinogens in food, like alkenylbenzenes and heterocyclic aromatic amines, have not been unequivocally detected in human tissues (Phillips, 1994; Turesky, 1994).

Beside exocyclic adducts, the most frequently studied endogenous DNA adducts include oxidized and alkylated DNA bases and I-compounds. The well-studied DNA oxidation products, such as 7,8-dihydro-8-oxy-2′-deoxyguanosine and 5,6-dihydroxy-5,6-dihydrothymine, and the major DNA alkylation products, *N*7-alkyl-deoxyguanosine and O^6-alkyl-deoxyguanosine, are sensitive to nuclease P1 treatment (Lutgerink *et al.*, 1992; Haque *et al.*, 1994) and are expected to migrate closely with normal nucleotides in HPLC, thus excluding their presence among the unidentified complex DNA adducts that we analysed with the dinucleotide/monophosphate version of the ^{32}P-postlabelling assay. In contrast, I-compounds fall into the range of levels detected in our method and are therefore candidates for the unidentified human adducts.

As the women with cancer and the controls were not matched on age, the reduction in the adduct level in the adjacent and tumorous tissues cannot be explained solely by cancer. The observation of three HPLC fractions in which the adduct levels were negatively associated with age and three fractions in which the adduct levels were decreased significantly in adjacent or tumour tissues indicates that both age and cancer status could have contributed to the decline in the adduct levels; however, we were unable to do a multivariate analysis because of the small number of individuals in this study. If cancer status made the largest contribution, then the decline would be consistent with the assumption that the human adducts are type I I-compounds, since they are present at lower levels in tumour tissues (Randerath *et al.*, 1993b), and this decrease may overcome the age-dependent increase in I-compound levels. In contrast, if age made the largest contribution, it would be more difficult to explain the results, since the levels of most I-compounds increase with age (Randerath *et al.*, 1993a).

Another possible factor in interpreting the results is that the control tissues were from women undergoing reduction mammoplasty, who may have had endocrine disorders. A high fat diet (Li & Randerath, 1990) and a high-cholesterol/high-fat diet (Vulimiri *et al*, 1995) were shown to reduce the levels of I-compounds. Thus, a disturbance in fat metabolism can have opposite effects and may have contributed to the higher adduct levels in the controls. Furthermore, hormonal effects have been suggested to regulate the levels of I-compounds (Randerath *et al.*, 1993a)

The studies carried out on this set of breast DNA samples are unique in that both carcinogenic adducts (i.e. aromatic adducts and malondialdehyde–DNA adducts) and modifications to DNA suspected to be non-carcinogenic have been analysed, and opposite results were observed. In addition, the present study is among the few that have addressed the unidentified abundant human adducts. Considering the difficulties of studying human I-compounds, which are reasonably expected to exist in human tissues, our results are quite informative; however, although these results are consistent with our speculation, a final conclusion can be reached only when the standards become available.

References

Chaudhary, A.K., Nokubo, M., Reddy, G.R., Yeola, S.N., Morrow, J.D., Blair, I.A. & Marnett, L.J. (1994) Detection of endogenous malondialdehyde–deoxyguanosine adducts in human liver. *Science* 265, 1580–1582

Chung, F.-L., Young, R. & Hecht, S.S. (1984) Formation of cyclic 1,N^2-propanodeoxyguanosine adducts in DNA upon reaction with acrolein and crotonaldehyde. *Cancer Res.* 44, 990–995

Chung, F.-L., Chen, H.-J.C. & Nath, R.G. (1996) Lipid peroxidation as a potential endogenous source for the formation of exocyclic DNA adducts. *Carcinogenesis* 17, 2105–2111

De Flora, S., Izzotti, A., Randerath, K., Randerath, E., Bartsch, H., Nair, J., Balansky, R., van Schooten, F., Degan, P., Fronza, G., Walsh, D. & Lewtas, J. (1996) DNA adducts and chronic degenerative diseases. Pathogenetic relevance and implications in preventive medicine. *Mutat. Res.* 366, 197–238

Esterbauer, H., Schaur, R.J. & Zollner, H. (1991) Chemistry and biochemistry of 4-hydroxynonenal, malonaldehyde and related aldehydes. *Free Radical Biol. Med.* 11, 81–128

Haque, K., Cooper, D.P. & Povey, A.C. (1994) Optimization of ^{32}P-postlabelling assays for the quantitation of O6-methyl and N7-methyldeoxyguanosine-3′-monophosphates in human DNA. *Carcinogenesis* 15, 2485–2490

Hemminki, K., Yang, K., Rajaniemi, H., Tyndyk, M. & Likhachev (1997) Postlabelling-HPLC analysis of lipophilic DNA adducts from human lung. *Biomarkers* 2, 341–347

Li, D. & Randerath, K. (1990) Association between diet and age-related DNA modifications (I-compounds) in rat liver and kidney. *Cancer Res.* 50, 3991–3996

Li, D., Xu, D.C. & Randerath, K. (1990) Species and tissue specificities of I-compounds as contrasted with carcinogen adducts in liver, kidney and skin DNA of Sprague-Dawley rats, ICR mice and Syrian hamsters. *Carcinogenesis* 11, 2227–2232

Li, D., Wang, M., Dhingra, K. & Hittelman, W.N. (1996) Aromatic DNA adducts in adjacent tissues of breast cancer patients: Clues to breast cancer etiology. *Cancer Res.* 56, 287–293

Lutgerink, J.T., van den Akker, E., Smeets, I., Pachen, D., van Dijk, P., Aubry, J.M., Joenje, H., Lafleur, M.V. & Retel, J. (1992) Interaction of singlet oxygen with DNA and biological consequences. *Mutat. Res.* 275, 377–386

Marnett, L.J. & Burcham, P.C. (1993) Endogenous DNA adducts: Potential and paradox. *Chem. Res. Toxicol.* 6, 771–785

Nair, J., Barbin, A., Guichard, Y. & Bartsch, H. (1995) 1,N^6-Ethenodeoxyadenosine and 3,N^4-ethenodeoxycytidine in liver DNA from humans and untreated rodents detected by immunoaffinity/32P-postlabelling. *Carcinogenesis* 16, 613–617

Nath, R.G. & Chung, F.-L. (1994) Detection of exocyclic 1,N^2-propanodeoxy-guanosine adducts as common DNA lesions in rodents and humans. *Proc. Natl Acad. Sci. USA* 91, 7491–7495

Nath, R.G., Ocando, J.E. & Chung, F.-L. (1996) Detection of 1,N^2-propanodeoxy-guanosine adducts as potential endogenous DNA lesions in rodent and human tissues. *Cancer Res.* 56, 452–456

Phillips, D.H. (1994) DNA adducts derived from safrole, estragole and related compounds, and from benzene and its metabolites. In: *DNA Adducts: Identification and Biological Significance* (Hemminki, K., Dipple, A., Shuker, D.E.G., Kadlubar, F.F., Segerbäck, D. & Bartsch, H., eds), IARC Scientific Publications No. 125. Lyon, IARC, pp. 131–140

Randerath, K., Randerath, E., Danna, T.F., van Golen, K.L. & Putman, K.L. (1989) A new sensitive ^{32}P-postlabeling assay based on the specific enzymatic conversion of bulky DNA lesions to radiolabeled dinucleotides and nucleotide 5′-monophosphates. *Carcinogenesis* 10, 1231–1239

Randerath, K., Li, D., Moorthy, B. & Randerath, E. (1993a) I-compounds—Endogenous DNA markers of nutritional status, ageing, tumor promotion and carcinogenesis. In: *Postlabelling Methods for Detection of DNA Adducts* (Phillips, D.H., Castegnaro, M. & Bartsch, H., eds), IARC Scientific Publications No. 124. IARC, Lyon, pp. 157–165

Randerath, K., Zhou, G.-D., Hart, R.W., Turturro, A. & Randerath, E. (1993b) Biomarkers of aging: Correlation of DNA I-compound levels with median lifespan of calorically restricted and ad libitum fed rats and mice. *Mutat. Res.* 295, 247–263

Turesky, R.J. (1994) DNA adducts of heterocyclic aromatic amines, arylazides and 4-nitroquinoline 1-oxide. In: *DNA Adducts: Identification and Biological Significance* (Hemminki, K., Dipple, A., Shuker, D.E.G., Kadlubar, F.F., Segerbäck, D. & Bartsch, H., eds), IARC Scientific Publications No. 125. Lyon, IARC, pp. 217–228

Vulimiri, S.V., Shibano, T., Randerath, E. & Randerath, K. (1993) Specific reduction of putative indigenous modifications (I-compounds) in pig liver DNA by high cholesterol/high fat diet. *Proc. Am. Assoc. Cancer Res.* 34, 558

Wang, M., Dhingra, K., Hittelman, W.N., Liehr, J.G., de Andrade, M. & Li, D. (1996) Lipid peroxidation-induced putative malondialdehyde–DNA adducts in human breast tissues. *Cancer Epidemiol. Biomarkers Prev.* 5, 705–710

Yang, K., Fang, J.-L. & Hemminki, K. (1998) Abundant lipophilic DNA adducts in human tissues. *Mutat. Res.* 422, 285–295

Exocyclic DNA Adducts in Mutagenesis and Carcinogenesis
B. Singer & H. Bartsch
IARC Scientific Publications No. 150
International Agency for Research on Cancer, Lyon, 1999

Cancer risk assessment for crotonaldehyde and 2-hexenal: An approach

E. Eder, D. Schuler & Budiawan

Crotonaldehyde and 2-hexenal are bifunctional compounds that form 1,N^2-propanodeoxyguanosine adducts and are mutagenic and genotoxic; crotonaldehyde is carcinogenic. Analysis of the mutations resulting from crotonaldehyde-induced DNA damage revealed the importance of deoxyguanosine adducts. Humans are exposed ubiquitously to these compounds by various routes. The highest daily intake of crotonaldehyde is assumed to be derived from cigarette smoke (31–169 µg/kg body weight), and the highest intake of 2-hexenal is probably from fruit and vegetables (31–165 µg/kg body weight per day). Because these compounds are suspected to play on important role in carcinogenicity, we developed sensitive ^{32}P-postlabelling techniques for DNA adducts of crotonaldehyde and hexenal, in order to improve estimates of cancer risk. The respective standards were also synthesized and characterized spectroscopically. We report here the results of the ^{32}P-postlabelling, e.g. the stability of the adducts in respect of nuclease P1 treatment, their labelling efficiencies, thin-layer chromatography of adduct spots and the recoveries and detection limits.

In untreated male Fischer 344 rats, neither crotonaldehyde nor 2-hexenal adducts were detected, but crotonaldehyde adducts were found in the tissues of rats given single doses of 200 or 300 mg/kg body weight and in the livers of rats after repeated doses of 1 or 10 mg/kg body weight. The adduct levels were higher 20 h after gavage than after 12 h. The adducts persist to a certain extent. 2-Hexenal adducts were detected in tissues of male Fischer 344 rats after gavage with single doses of 50, 200 or 500 mg/kg body weight. The highest adduct levels were measured 48 h after gavage, but no adducts were found 8 h after gavage.

Two approaches for cancer risk estimation are discussed. One is based on the correlation between the covalent binding index, calculated from adduct levels, and the median toxic dose (TD_{50}) (Lutz, 1986) and showed a cancer risk of 1 per 10^7 lives for hexenal, assuming dietary intakes of 31–165 µg/kg body weight per day. The other is based on a cancer incidence of 0.07 at a dose of crotonaldehyde of 4.2 mg/kg body weight per day assessed from the study of Chung *et al.* (1986), which can be interpreted as a risk of 5.8–18 new cases per 10^4 smokers, assuming a consumption of 30 cigarettes per day. The latter approach may, however, lead to an overestimate of the cancer risk associated with exposure to crotonaldehyde; the estimate based on our binding studies resulted in a 20-fold lower estimate of the carcinogenic risk of crotonaldehyde.

Introduction

Crotonaldehyde ($CH_3CH{=}CHCHO$) and 2-hexenal ($C_3H_7CH{=}CHCHO$) are reactive bifunctional α,β-unsaturated carbonyl compounds which form DNA adducts (Lutz *et al.*, 1982; Chung *et al.*, 1984; Hoffman *et al.*, 1989; Eder *et al.*, 1990; Marnett, 1994). These compounds are mutagenic (Eder *et al.*, 1982; Marnett *et al.*, 1985; Eder *et al.*, 1992), genotoxic (Feron *et al.*, 1991; Eder *et al.*, 1993a) and carcinogenic (Bittersohl, 1974; Chung *et al.*, 1986). A selection of the results of tests for genotoxicity is summarized in Table 1. Crotonaldehyde and 2-hexenal belong to the group of β-alkyl-substituted acrolein congeners that form a regioisomer of 1,N^2-propanodeoxyguanosine adducts in which the OH group is adjacent to the N1 of the guanine moiety (Eder *et al.*, 1993b). The structures, configurations and conformations of the adducts are elucidated by spectroscopic methods, in particular, ^{1}H-nuclear magnetic resonance

Table 1. Mutagenicity, genotoxicity and carcinogenicty of crotonaldehyde and 2-hexenal

Compound	Mutagenicity	SOS Chromotest	Strand breaks
Crotonaldehyde	Strain TA104, 940 revertants/μmol (Marnett *et al.*, 1985) Strain TA100, 1952 revertants/μmol Threefold cell density (Eder *et al.*, 1992)	Strain PQ37 SOSIP = 10.6×10^{-3} (with ethanol as solvent) (Eder *et al.*, 1993a)	L1210 cells, measured by alkaline elution, at 800 mmol (cell toxicity) (Eder *et al.*, 1993a)
2-Hexenal	Strain TA104, 460 revertants/μmol (Marnett *et al.*, 1985) Strain TA100, 980 revertants/mmol Threefold cell density (Eder *et al.*, 1992)	Strain PQ37, no activity because of high toxicity (Eder *et al.*, 1993a)	L1210 cells at 250 μmol (no cell toxicity) (Eder *et al.*, 1993a)

Carcinogenicity: Hepatocarcinogenic in Fischer 344 rats (Chung *et al.*, 1986); squamous-cell carcinoma in mouth and bronchi of humans after occupational exposure (Bittersohl, 1974)

(NMR) (Eder & Hoffman, 1992, 1993). Besides 1,N^2-propanoadducts, 7,8-cyclic guanosine adducts and biscyclic adducts are also formed, as shown in Scheme 1. We recently published an analysis of mutations resulting from crotonaldehyde-induced DNA lesions in a plasmid shuttle vector system (Czerny *et al.*, 1998) and showed that 82% of the point mutations were at G:C base pairs, with a hot spot at base pair 133 of the *sup* F gene of plasmid pZ 189, indicating that guanosine adducts play a major role in mutagenesis of crotonaldehyde. The other DNA lesions found also support the hypothesis that DNA adducts with deoxyguanosine play a crucial role in the genotoxicity of crotonaldehyde and probably also of other α,β-unsaturated carbonyl compounds that predominantly form guanine adducts.

Mankind is ubiquitously exposed to this type of compound, as they are found in the environment, in food, at work places, in surface water and in soil and they are produced endogenously (Eder *et al.*, 1990, 1992, 1993a). These genotoxic substances are considered to play a role in carcinogenicity (Nath & Chung, 1994), but the data base is insufficient to assess their role in human cancer. Measurements of DNA adducts as markers for initiation of cancer cells and for other possible events in cancer induction could improve estimates of the carcinogenic hazard of these substances, e.g. via relationships between adduct levels and carcinogenic potency. Such relationships do not presently exist, but DNA adduct levels can be used in a linear extrapolation of carcinogenic risk if relationships between dose and cancer potency are known. We therefore developed highly sensitive ^{32}P-postlabelling techniques to detect the 1,N^2-propanodeoxyguanosine adducts of the important α,β-unsaturated carbonyl compounds crotonaldehyde, as an example of an environmentally and occupationally relevant compound, and 2-hexenal, occurring in plants and food. In our studies, the detection limits for the 1,N^2-propanodeoxyguanosine adducts are 1 adduct per 10^9 nucleotides for crotonaldehyde and 3 adducts per 10^8 for 2-hexenal. The results of the ^{32}P-postlabelling methods are shown for crotonaldehyde and 2-hexenal, and the respective adduct levels in Fischer 344 rats are reported. Approaches to carcinogenic risk assessments for crotonaldehyde and 2-hexenal are discussed on the basis of the adduct levels determined by the ^{32}P-postlabelling technique.

Results and discussion

^{32}P-Postlabelling technique

The development of the ^{32}P-postlabelling technique will be described in detail elsewhere.

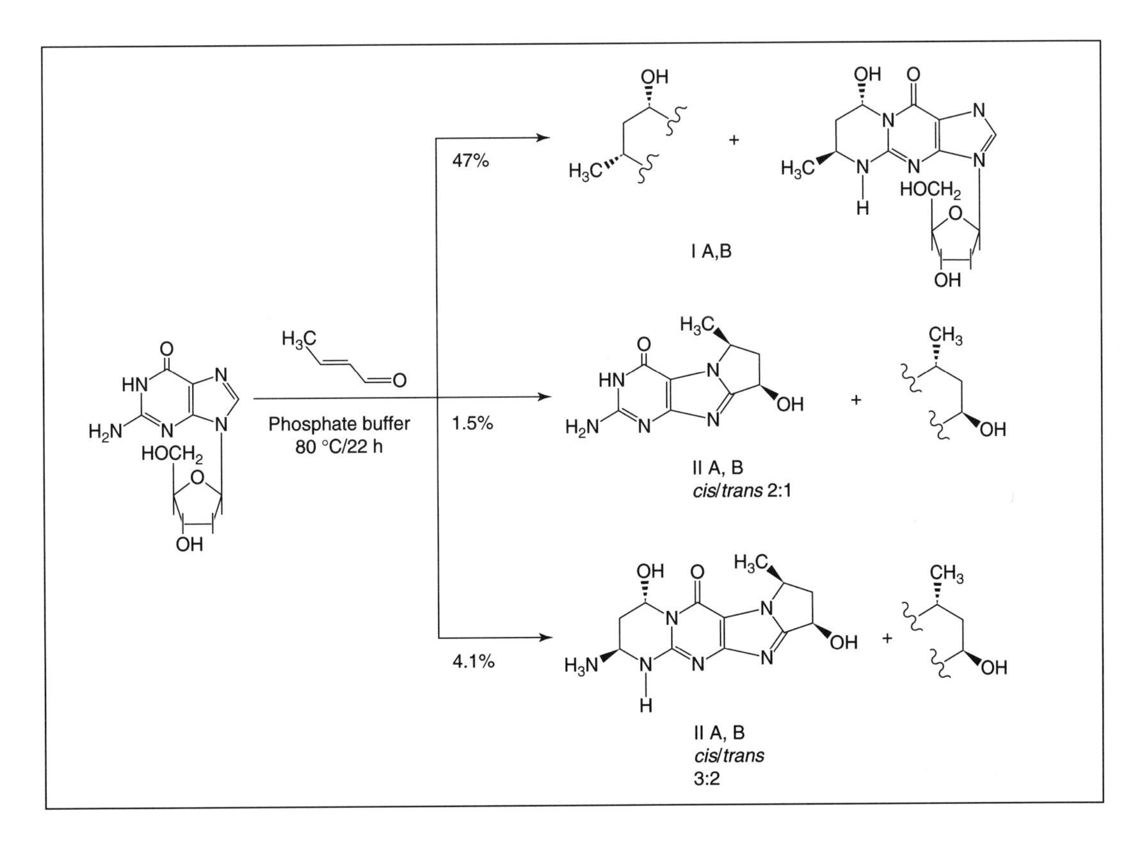

Scheme 1. Reaction of crotonaldehyde with deoxyguanosine

Standards are needed to develop methods, define optimum conditions and determine labelling efficiencies and rates of recovery. Furthermore, these standards can be used for a clear attribution of the thin-layer chromatography (TLC) spots to the respective DNA adducts and thus avoid artefacts. To do this, the high-performance liquid chromatography and ultraviolet spectra of the synthesized diastereomers of the 1,*N*²-propanodeoxyguanosine-3´-monophosphates of crotonaldehyde and 2-hexenal are determined. As already described, only one type of regioisomer and only the *trans* isomers are formed with these two β-alkyl-substituted acrolein congeners. The ^{1}H-NMR, ^{13}C-NMR and mass spectra are also recorded (not shown).

The adducts of crotonaldehyde and 2-hexenal are stable in the range pH 4–11 and are also relatively stable during nuclease P1 treatment at pH 3 (Figure 1). In order to optimize the labelling efficiency, the pH, incubation time, amount of polynucleotide kinase, temperature and concentrations of 1,4-dithiothreitol, spermidin, bicin, $MgCl_2$ and other metal ions were investigated. Figure 2 shows the dependence of a selection of such parameters on the labelling efficiency of crotonaldehyde adducts. The labelling efficiency of crotonaldehyde adducts was about 80% and that of 2-hexenal adducts about 35% (Figure 3).

The adducts were detected and quantified as the respective TLC spots either by Cerenkov counting or with an Instant Imager (Canberra Packard). Figure 4 shows the adduct spots of crotonaldehyde as an example. About 40% of the crotonaldehyde adducts were recovered. The detection limit was about 1 adduct per 10^9 nucleotides, with an absolute detection limit of about 92 amol. In the case of the 2-hexenal adducts, the contact transfer method was used, in addition to nuclease P1 enrichment, in order to eliminate background radioactivity in the region of the adduct spot on the TLC plate. The

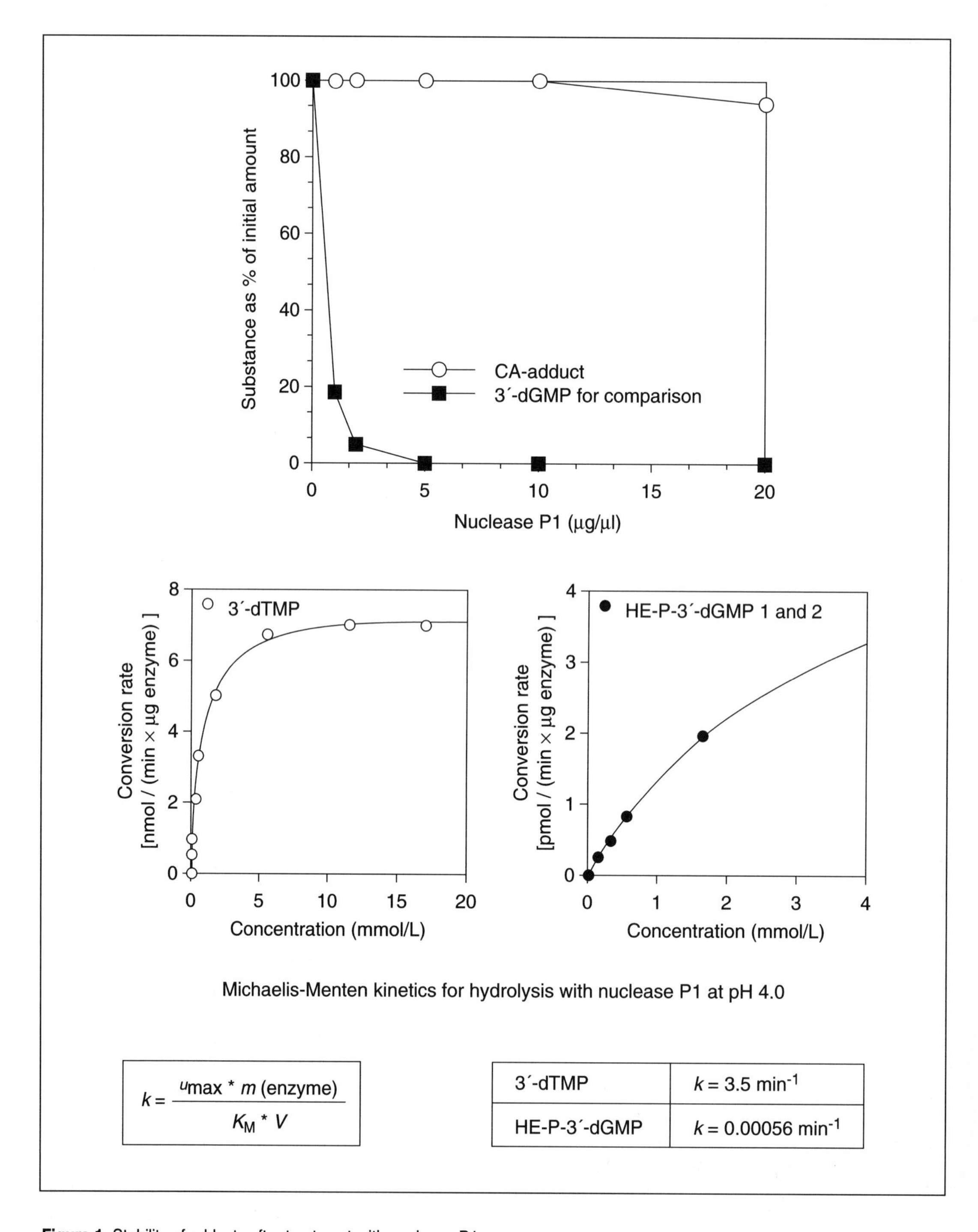

$$k = \frac{u_{max} * m\,(\text{enzyme})}{K_M * V}$$

3´-dTMP	$k = 3.5\ min^{-1}$
HE-P-3´-dGMP	$k = 0.00056\ min^{-1}$

Figure 1. Stability of adducts after treatment with nuclease P1

CA, crotonaldehyde

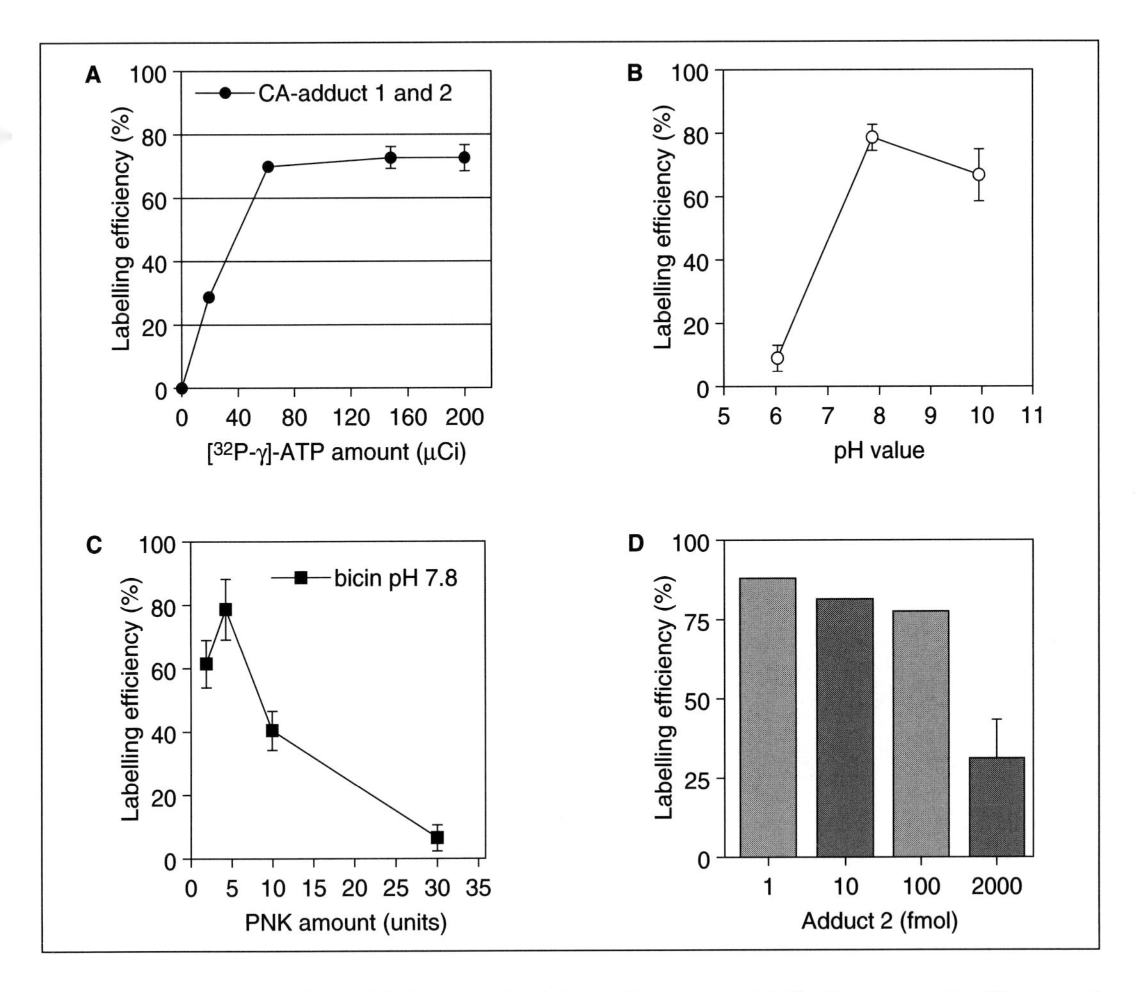

Figure 2. Labelling efficiency of cronaldehyde adducts in relation to (A) amount of ATP, (B) pH of kinase buffer, (C) amount of polynucleotide kinase (PNK) and (D) amount of adduct

recovery was 10% in comparison with 20% without contact transfer. In spite of the low rate of recovery, a detection limit of 3 adducts per 10^8 nucleotides was achieved with a standard deviation of 40%. For comparison, the standard deviation was 27% at an adduct level of 1 adduct per 10^6 nucleotides.

Studies in experimental animals

Various doses of crotonaldehyde and 2-hexenal in corn oil were administered by gavage to male Fischer 344 rats.

No crotonaldehyde adducts were detected in either liver DNA from untreated male rats (Figure 4) or untreated calf thymus DNA (not shown). The crotonaldehyde adduct levels in rat liver were 2.9 ± 0.15 adducts/10^8 nucleotides 20 h after a single gavage dose of 200 mg/kg body weight, 3.4 ± 0.5 adducts/10^8 nucleotides 20 h after a single gavage dose of 300 mg/kg body weight and 1.7 ± 0.6 adducts/10^8 nucleotides 12 h after the dose of 200 mg/kg body weight. The highest adduct levels were found in liver, followed by lung, kidney and colon. Thus, the highest adduct concentrations were found in the organs in which cancer was induced in a long-term study (liver; Chung *et al.*, 1986) and in an epidemiological study at the workplace (lung; Bittersohl, 1974).

DNA adducts of crotonaldehyde were also detected in the livers of rats treated with repeat-

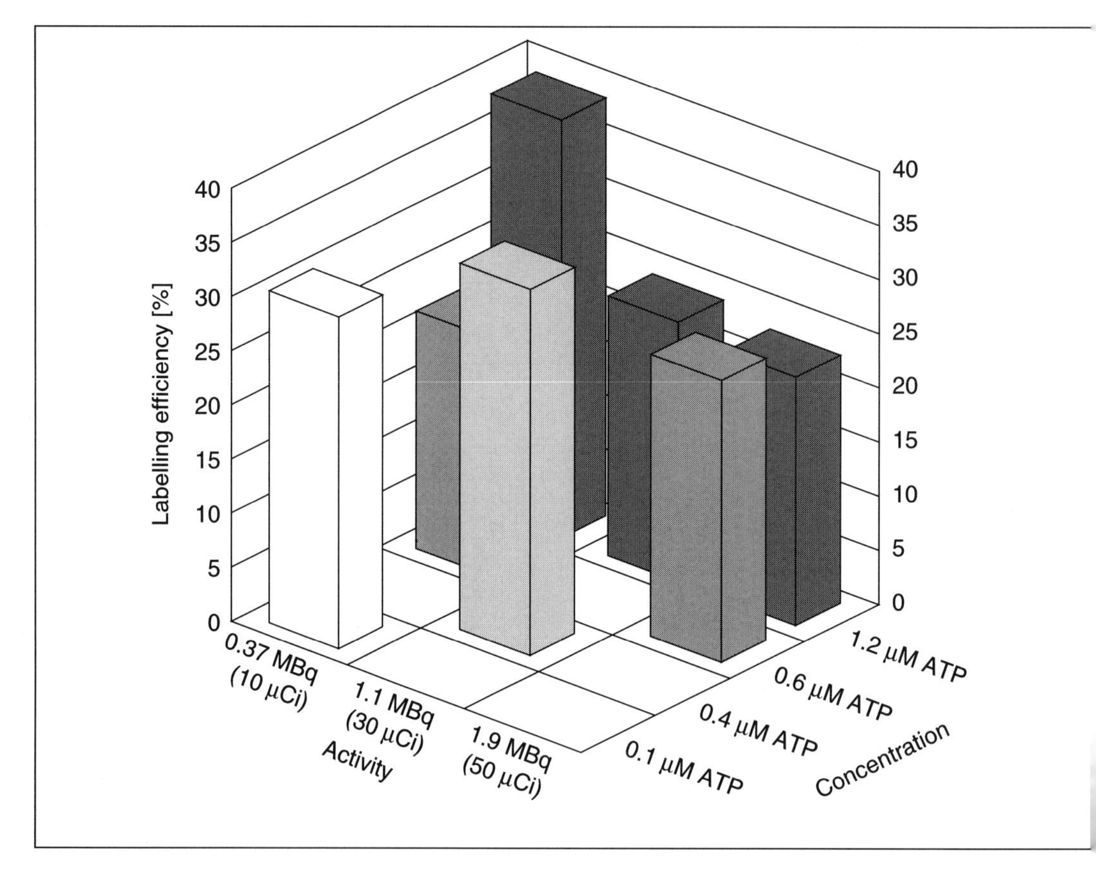

Figure 3. Effect of concentration and radioactivity of ATP on labelling efficiency

ed doses five times a week for six weeks, with levels of 6.2 ± 0.5 adducts/10^8 nucleotides at a dose of 10 mg/kg body weight and 2.0 ± 0.8 adducts/10^8 nucleotides at a dose of 1 mg/kg body weight. The adducts persisted to a certain extent in the rat liver (not shown): one week after the end of four weeks of repeated gavage 69% of the initial level remained, and two weeks after the end of administration 18% was still present. The initial level was measured 24 h after the last gavage.

The results clearly demonstrate that crotonaldehyde forms DNA adducts *in vivo* and that these adducts are stable; however, these adducts are evidently repaired, although slowly. There is no indication in our studies that other adducts are formed in detectable amounts; however, we cannot exclude the possibility, because we did not include standards for other adducts e.g. the 7,8-cyclic adducts. Neverthess, our studies *in vitro* indicate that the polyethylene imine–cellulose chromatographic properties of the other possible adducts, as shown in Scheme 1, should not be very different from those tested here (Eder *et al.*, 1992) and that we would have found them if they were stable and present in detectable amounts.

It is possible that intake of crotonaldehyde by control rats from the environment, drinking-water, food and disinfectants could lead to a detectable steady-state background of adducts. The results after repeated gavage of low doses show that the lowest daily dose that results in a detectable adduct level would be about 50 µg/kg body weight per day. In principle, a crotonaldehyde intake of 50–100 µg/kg body weight and more can occur during animal maintenance. This would explain the discrepancy between our

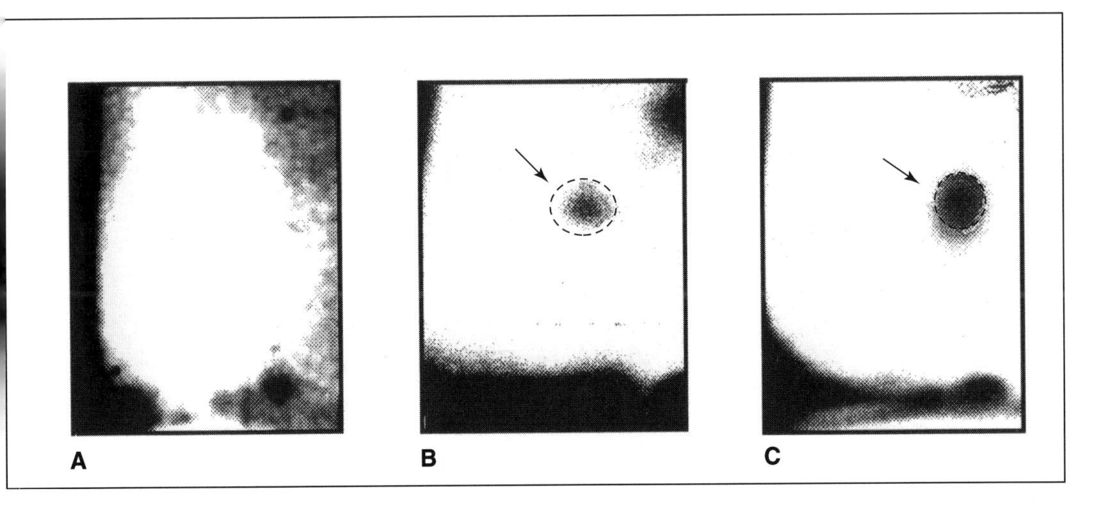

Figure 4. Thin-layer chromatograms of (A) untreated male Fisher 344 rat liver DNA, (B) rat liver DNA after gavage with crotonaldehyde and (C) rat liver DNA after addition of 30 fmol crotonaldehyde adduct standard

results and those of others (Nath & Chung, 1994) who found a background of crotonaldehyde adducts in untreated rats. Adduct levels in the range of 1 adduct per 10^9 to about 10^8 nucleotides can be expected under those circumstances.

After gavage of rats with 500 mg/kg body weight of 2-hexenal, 1,N^2-propanodeoxguanosine adducts were detected at various levels depending on the time after gavage, with none detected at 8 h, 0.61 adducts/10^6 nucleotides after 24 h, 1.8 adducts/10^6 nucleotides 48 h after gavage and 0.7 adducts/10^6 nucleotides 96 h after gavage. The results provide some insight into the pharmacokinetics of 2-hexenal and the formation of 2-hexenal adducts. As already shown with crotonaldehyde, maximum adduct levels are also found relatively late, in view of the fact that crotonaldehyde and 2-hexenal are directly binding compounds which do not need metabolic activation. In this study, we demonstrated formation of 2-hexenal adducts *in vivo* for the first time. Gölzer *et al.* (1996) reported recently that no adducts could be detected after gavage of animals with 2-hexenal, but they looked for adducts as early as 12 h after gavage. We could detect no adducts 8 h after gavage, and none would have appeared by 12 h. The results of Gölzer *et al.* (1996) are therefore in line with ours, showing that no adducts are detected in untreated animals.

Figure 5 shows the organ distribution of adducts two days after gavage with 2-hexenal at 200 or 500 mg/kg body weight. The highest adduct levels were found in the forestomach, liver and oesophagus. 2-Hexenal appears to bind at the site of initial contact, even though adducts are formed relatively late. It appears to be directly resorbed by the tissues of the forestomach, oesophagus and duodenum but to a lesser degree in the glandular stomach; however, systemic distribution also occurs, since the adduct levels in the kidney were also high. In contrast to the situation with crotonaldehyde, very low adduct concentrations were found in lung. Witz (1989) suggested that glutathione conjugation of α,β-unsaturated carbonyl compounds is not necessarily a detoxification process, since glutathione conjugation is in equilibrium and the α,β-unsaturated carbonyl compounds can be released by dissociation of the conjugate in cell compartments of low glutathione content, e.g. next to the DNA. Besides glutathione conjugation, Michael-addition conjugations with other –SH–group containing compounds (e.g. proteins) may also occur. Thus, protein adducts of this type can be considered 'stabilized' transport forms of this class of compounds, which prevent further detoxification of these compounds. Some support for this hypothesis is found in our results. One is the relatively long time until adducts can

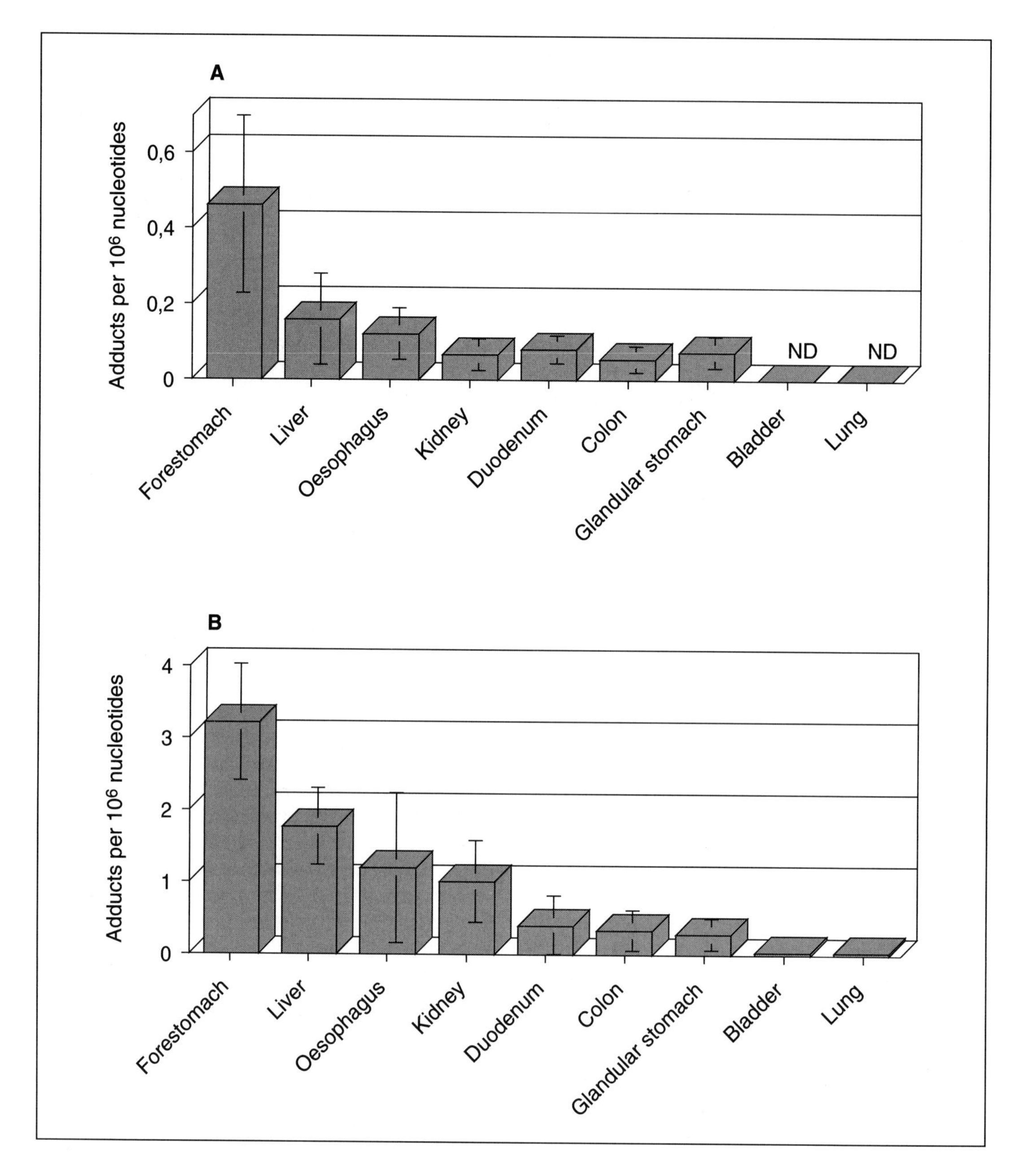

Figure 5. Organ distribution of 2-hexenal adduct levels in male Fischer 344 rat liver DNA two days after gavage with (A) 200 mg/kg body weight and (B) 500 mg/kg body weight; ND, not detected

be detected, and another is the relatively high adduct concentration in the kidney. In this context, it would be interesting to know whether adducts are formed earlier in organs other than in the liver, such as in the forestomach where adduct formation should start directly after gavage. 2-Hexenal is not very reactive (Eder & Hoffman, 1993), however, and it may take some time before sufficient amounts of adducts are formed for detection.

It is also interesting that the ranking of adduct levels differs slightly at the lower dose from that at the higher dose. The adduct levels in the duodenum and glandular stomach are higher than those in the kidney with a dose of 200 mg/kg body weight, whereas with 500 mg/kg body weight the levels in the kidney are clearly higher than those in the duodenum and glandular stomach. Clearly, less 2-hexenal is available after the first pass of the low dose through liver (systemic distribution) than with the high dose. Figure 6 shows the dose dependence of adduct levels in the organs with the highest adduct concentrations.

Cancer risk assessment on the basis of DNA adduct levels

As neither dose–response data from a long-term bioassay nor adduct levels in cancer tissues are available in the case of the 2-hexenal, we attempted to estimate the cancer risk on the basis of the TD_{50} derived from the correlation between the covalent binding index (CBI) and the TD_{50} of Lutz (1986). The CBI of 2-hexenal, calculated from our results, is 0.35 at a dose of 500 mg/kg body weight, 0.08 at 200 mg/kg and 0.06 at 50 mg/kg. Figure 7 shows that the CBI of 2-hexenal is less than 0.06 at doses under 50 mg/kg body weight. The described assessment is valid, however, only if the following are assumed:

- the mutagenic potency of the measured 2-hexenal-propanodeoxyguanosine adducts is the average of the mutagenic potency of all adducts formed;
- liver is the most susceptible target organ, and the CBI is not significanly higher in other organs;
- adduct formation is not essentially higher in cellular sub-populations;
- the CBI is constant in the low dose range;
- the CBI correlates with the TD_{50} according to Lutz (1986);

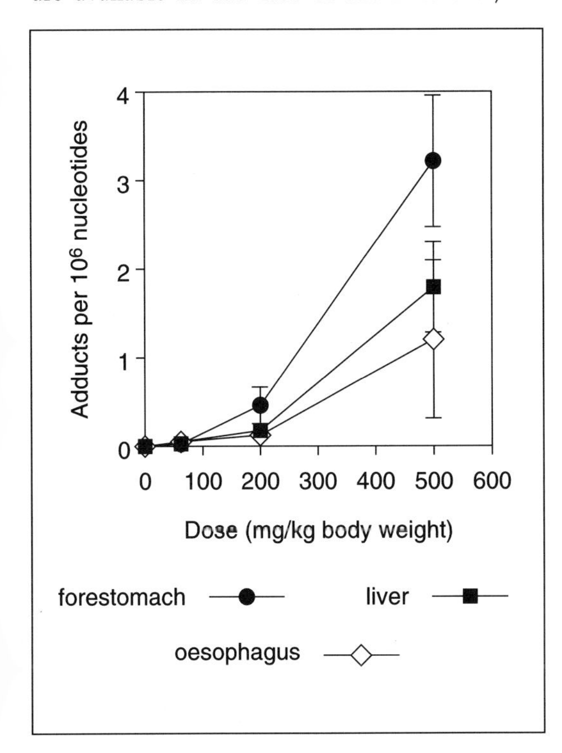

Figure 6. Dose-dependence of 2-hexenal adduct levels in various organs of male Fischer 344 rats

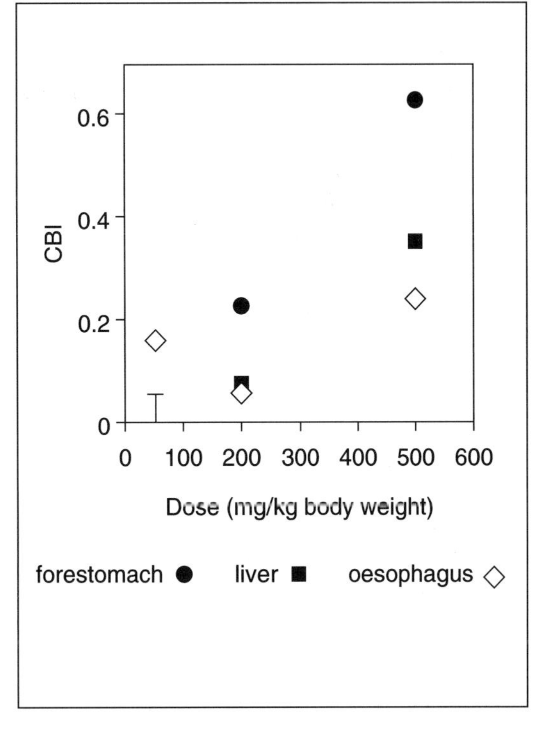

Figure 7. Dependence of covalent binding index (CBI) on dose in various organs of male Fischer 344 rats

- the dose–tumour incidence relationship is linear at low doses and
- the dose–tumour incidence relationship is similar in rats and humans.

A TD_{50} of 1.57 mol/kg per day or 154 g/kg per day can be derived from the correlation of Lutz (1986) with a CBI of 0.06 and a dose of 50 mg 2-hexenal per kg body weight. The TD_{50} value is, however, only a hypothetical value used for these calculations, since it is far higher than the LD_{50}. The mean intake of 2-hexenal in the German population is estimated from epidemiological data to be about 0.03 mg/kg body weight per day (Table 2), and a maximum intake of 0.16 mg/kg can be calculated for the 5th percentile (5% of the people have the highest consumption). A linear extrapolation on the basis of the TD_{50} would lead to cancer risks of 1 per 10^7 lives in the case of an intake of 0.03 mg/kg and 5 per 10^7 lives at an intake of 0.16 mg/kg body weight per day (5th percentile). Since the carcinogenic potency of the adducts measured is unknown, introduction of a safety factor of 10 is recommended. The risk for the group with the highest intake would then be $5/10^6$ lives, which would be acceptable. In any case, the benefits derived from eating fruits and vegetables are much higher than the estimated cancer risk.

Another approach to cancer risk assessment can be used for crotonaldehyde. Chung *et al* (1986) reported a 7% incidence of hepatocellular carcinogenicity at a daily dose of approximately 4 mg/kg body weight. Table 3 gives a selection of important sources of crotonaldehyde which are relevant for intake. The highest intake, 33–105 µg/kg body weight, is assumed to come from cigarette smoke. This leads to a cancer risk of 5.8–18 new cases per 10^4 smokers on the basis of 0.07 cases per 4 mg/kg body weight (Chung *et al.*, 1986). Assuming a dietary intake of about 1 µg/kg body weight, a cancer risk of 1.8 incidences per 10^5 humans would result. Under certain circumstances, however, i.e. enzymatic and abiotic degradation processes in food, a higher intake may occur. Food contamination at a level of about 1–2 mg/kg has been observed occasionally (BUA-Stoffbericht 98, 1993). Under these conditions, a maximum intake of about 15 µg/kg body weight per day may be anticipated and would lead to a cancer risk of 2.6 per 10^4 humans. Endogenous formation of crotonaldehyde is, in general, considered to be low (Esterbauer *et al.*, 1982) and is therefore not taken into consideration in this risk estimation.

The cancer risk estimated by the above approach appears to be relatively high; however, a linear dose–cancer relationship has been extrapolated from the study in experimental animals. Furthermore, the daily dose is estimated roughly on the basis of the average drinking-water consumption of the test animals in the study of Chung *et al.* (1986), and the dose–cancer incidence relationship is based on only one value because the rats at the high dose did not develop hepatocellular carcinoma, but only one neoplastic nodule. This approach may well have led to an overestimate of the cancer risk due to limited data.

A comparison of our dose–adduct level relationship with this cancer risk assessment also indicates that the risk is overestimated, since the lowest dose at which detectable adduct levels occur is 50 µg/kg body weight with our detection limit. This is in the range of the highest intake from cigarette smoke but two to three magnitudes higher than the intake from food. Thus, a cancer risk of 2.6 per 10^4 humans is very high if the underlying dose is two magnitudes below the dose that would lead to detectable DNA adduct levels.

Our studies of binding with crotonaldehyde *in vivo* provide a possibility for proving the validity of this approach. In view of the correlation between CBI and TD_{50}, a CBI of 15 can be calculated at a TD_{50} of 0.4 mmol/kg per day or 29 mg/kg per day. On the basis of our results for binding, we calculated a CBI of 2. The two CBIs are in the same range and differ only by a factor of 7. A TD_{50} of 600 mg/kg per day is calculated according to the correlation of Lutz (1986). The carcinogenic risk of crotonaldehyde would thus be 20-fold lower if it were estimated on the basis of adduct levels, as shown above for 2-hexenal (1 per 10^6 lives as an average dietary risk and 2.9–9 per 10^5 arising from cigarette smoking). The higher CBI of 15 estimated from the carcinogenesis bioassay than that found from our results can be explained by a greater carcinogenic potency of crotonaldehyde than estimated from adduct formation. The higher potency can be accounted for either by

Table 2. Fruit and vegetable consumption relevant for intake of 2-hexenal in Germany

Food	Consumption (g/day)		Intake of hexenal (mg/day)	
	Mean	Maximum	Mean	Maximum
Bananas	17	98	0.68	3.9
Pulses	16	73	0.54	2.5
Apples	41	163	0.41	1.6
Cabbage	14	74	0.20	1.0
Grapes	4.8	48	0.077	0.77
Cauliflower	3.8	38	0.053	0.53
Tomatoes	19	86	0.032	0.15
Strawberries	3.6	33	0.023	0.21
Total			**2.0**	**10.7**

From Behörde für Arbeit (1995)

Table 3. Significant sources of crotonaldehyde intake in humans

Source	Concentration	Estimated daily intake (μg/65-kg body weight)	References
Air 1 m from highway	1–2 μg/m³	0.5 (24-h inhalation)	Zweidinger *et al.* (1988)
Air, work place (production)	300–600 μg/m³	22–44	Hoechst (personal communication)
Tobacco smoke	72–228 μg/cigarette	33–105 (30 cigarettes/day)	Vickroy (1976); Kuwata *et al.* (1979)
Fruit and vegetables	1.4–100 μg/kg	0.01–0.77 (500 g/day)	Winter & Willhalm (1964); Linko *et al* (1978)
Fish	71.4–1000 μg/kg	0.28–3 (200 g/day)	Yurkowski & Bordeleau (1995); Yoshida *et al* (1984)
Meat	10–270 μg/kg	0.03–0.83 (200 g/day)	Noleau & Toulemonde (1986), Cantoni *et al.* (1969)
Beer	0.8–20 μg/L	0.01–0.31 (1 L/day)	Greenhoff & Wheeler (1981); Hashimoto & Eshima (1977)
Wine	300–700 μg/L	2.3–5.4 (0.5 L/day)	Sponholz (1982)
Whisky	30–210 μg/L	0.05–0.33 (0.1 L/day)	Miller & Danielson (1988)

high cytotoxicity or by additional adduct formation. Indeed, there is evidence for both (Holmberg & Malmfors, 1974; Eder & Hoffman, 1992). The additional adducts, e.g. cyclic 7,8-deoxyguanosine adducts and biscyclic 1,N^2,7,8-adducts, are, however, unstable, leading to depurination, and are therefore not detectable by ^{32}P-postlabelling. The discrepancy can also be explained by the uncertainty in the estimate that a dose of 4 mg/kg per day leads to a 7% incidence in experimental animals, as discussed above.

The introduction of an additional safety factor of 10 into the estimate of the carcinogenic risk of hexenal (see above) is justified, since a sevenfold higher CBI, or a 20-fold higher cancer risk, is found with crotonaldehyde on the basis of the TD_{50} than on the basis of the *in-vivo* binding studies. In principle, the above considerations for crotonaldehyde are also valid for the structurally related 2-hexenal, which forms the same adducts although with lower chemical reactivity (Eder & Hoffman, 1992, 1993). The difference in reactivity is also expressed in the CBIs of crotonaldehyde and 2-hexenal.

References

Behörde für Arbeit (1995) Standards zur Expositionsabschätzung. Hamburg, Arbeitsgemeinschaft der leitenden Medizinalbeamtinnen und -beamten der Länder, Gesundheit und Soziales

Bittersohl, G. (1974) Epidemiologische Untersuchungen über Krebserkrankungen bei Arbeiten mit Aldol und aliphatischen Aldehyden. *Arch. Geschwulstforsch.* 43, 172–176

BUA-Stoffbericht 98 (1993) Ökologischer Aspekt: Vorkommen und Verteilung auf die Kompartimente. In: Crotonaldehyd (2-Butenal), BUA-Stoffbericht 98 (GDCH ed). S. Hirzel, Wissenschaftliche Verlagsgesellschaft

Cantoni, C., Bianchi, M.A., Renon, P. & Calcinardi, C. (1969) Studi sulle alterazioni delle coppe. *Atti Soc. Ital. Sci. Vet.* 23, 752–756

Chung, F.L., Young, R. & Hecht, S.S. (1984) Formation of cyclic 1,N^2-propanodeoxy-guanosine adducts in DNA upon reaction with acrolein or crotonaldehyde. *Cancer Res.* 44, 990–995

Chung, F.L., Tanaka, T. & Hecht, S.S. (1986) Induction of liver tumors in F 344 rats by crotonaldehyde. *Cancer Res.* 46, 1285–1289

Czerny, C., Eder, E. & Rünger, T.M. (1998) Genotoxicity and mutagenicity of the α,β-unsaturated carbonyl compound crotonaldehyde (butenal) on a plasmid shuttle vector. *Mutat. Res.* 407, 125–134

Eder, E. & Hoffman, C. (1992) Identification and characterization of deoxyguanosine crotonaldehyde adducts. Formation of 7,8 cyclic adducts and 1,N^2,7,8- biscyclic adducts. *Chem. Res. Toxicol.* 5, 802–808

Eder, E. & Hoffman, C. (1993) Identification and characterization of deoxyguanosine adducts of mutagenic β-alkyl substituted acrolein congeners. *Chem. Res. Toxicol.* 6, 486–494

Eder, E., Henschler, D. & Neudecker, T. (1982) Mutagenic properties of allylic and α,β-unsaturated compounds: Consideration of alkylating mechanisms. *Xenobiotica* 12, 831–848

Eder, E., Hoffman, C., Bastian, H., Deininger, C. & Scheckenbach, S. (1990) Molecular mechanisms of DNA damage initiated by α,β-unsaturated carbonyl compounds as criteria for genotoxicity and mutagenicity. *Environ. Health Perspectives* 88, 99–106

Eder, E., Deininger, C., Neudecker, T. & Deininger, D. (1992) Mutagenicity of β-alkyl substituted acrolein congeners in the *Salmonella typhimurium* strain TA100 and genotoxicity-testing in the SOS chromotest. *Environ. Mol. Mutag.* 19, 338–345

Eder, E., Scheckenbach, S., Deininger, C. & Hoffman, C. (1993a) The possible role of α,β-unsaturated carbonyl compounds in mutagenesis and carcinogenesis. *Toxicol. Lett.* 67, 87–103

Eder, E., Hoffman, C., Sporer, S. & Scheckenbach, S. (1993b) Biomonitoring studies and susceptibility markers for acrolein congeners and allylic and benzylic compounds. *Environ. Health Perspectives* 99, 245–247

Esterbauer, H., Cheesman, K.H., Dianzani, M.U., Poli, G. & Slater, T.F. (1982) Separation and characterization of the aldehydic products of lipid peroxidation stimulated by ADP-Fe^{2+} in rat liver microsomes. *Biochem. J.* 208, 129–140

Feron, V.J., Til, H.P., de Vrijer, F., Woutersen, R.A., Carsee, F.R. & van Bladeren, P.J. (1991) Aldehydes: Occurrence, carcinogenic potential, mechanism of action and risk assessment. *Mutat. Res.* 259, 363–385

Gölzer, P., Janzowski, C., Pool-Zobel, B.L. & Eisenbrand, G. (1996). (E)-2-Hexenal-induced DNA damage and formation of cyclic 1,N^2-(1,3-propano)-2′-deoxyguanosine adducts in mammalian cells. *Chem. Res. Toxicol.* 9, 1207–1213

Greenhoff, K. & Wheeler R.E. (1981) Analysis of beer carbonyls at the part per billion level by combined liquid chromatography and high pressure liquid chromatography. *J. Inst. Brew.* 86, 35–41

Hashimoto, N. & Eshima, T. (1977) Composition and pathway of formation of stale aldehydes in bottled beer. *.J. Am. Soc. Brew. Chem.* 35, 145–150

Hoffman, C., Bastian, H., Wiedenmann, M., Deininger, C. & Eder, E. (1989) Detection of acrolein congener DNA adducts isolated from cellular systems. *Arch. Toxicol.* 13, 219–223

Holmberg, B. & Malmfors, T. (1974) The cytotoxicity of some organic solvents. *Environ. Res.* 7, 183–192

Kuwata, K., Uebori, M. & Yamasaki, Y. (1979) Determination of aliphatic and aromatic aldehydes in polluted airs as their 2,4-dinitrophenyl hydrazones by high performance liquid chromatograph. *J. Chromatogr. Sci.* 17, 264–268

Linko, R.R., Kallio, H., Pyysalo, T. & Raimo, K. (1978) Volatile monocarbonyl compounds of carrot roots at various stages of maturity. *Z. Lebensm. Unters. Forsch.* 166, 208–211

Lutz, W.K. (1986) Quantitative evaluation of DNA binding data for risk estimation and classification of direct and indirect carcinogens. *J. Cancer Res. Clin. Oncol.* 112, 85–91

Lutz, D., Eder, E., Neudecker, T. & Henschler, D. (1982) Structure–mutagenicity relationship in α,β-unsaturated carbonyl compounds and their corresponding allylic alcohols. *Mutat. Res.* 93, 305–315

Marnett, L.J. (1994) DNA adducts of α,β-unsaturated aldehydes and dicarbonyl compounds. In: *DNA Adducts, Identification and Biological Significance* (Hemminki, K., Dipple, A., Shuker, D.E.G., Kadlubar, F.F., Segerbäck, D. & Bartsch, H., eds), IARC Scienitifc Publications No. 125. Lyon, IARC, pp. 151–163

Marnett, L.J., Hurd, H.K., Hollstein, M.C., Levin, D.E., Esterbauer, H. & Ames, B.N. (1985) Naturally occurring carbonyl compounds are mutagens in Salmonella-tester-strain TA 104. *Mutat. Res.* 148, 25–34

Miller, B.E. & Danielson, N.D. (1988) Derivatization of vinyl aldehydes with anthrone prior to high-performance liquid chromatography with fluorimetric detection. *Anal. Chem.* 60, 622–626

Nath, R.G. & Chung, F.L. (1994) Detection of exocyclic 1,N^2-propanodeoxy-guanosine adducts as common DNA lesions in rodents and humans. *Proc. Natl Acad. Sci. USA* 91, 7491–7495

Noleau, I. & Toulemonde, B. (1986) Quantitative study of roasted chicken flavour. *Lebensm.-Wiss. Technol.* 19, 122–125

Sponholz, W.-R. (1982) Analyse und Vorkommen von Aldehyden in Weinen. *Z. Lebensm. Unters. Forsch.* 174, 458–462

Vickroy, D.G. (1976) The characterization of cigarette smoke from Cytrel®, smoking products and its comparison to smoke from fluecured tobacco. I Vapor phase analysis. *Beitr. Tabakforsch.* 8, 415–421

Winter, M. & Willhalm, B. (1964) [Studies on aromas. X. Fresh strawberry aroma. Analysis of volatile carbonyl, ester and alcohol components.] *Helv. Chim. Acta* 47, 1215–1227 (in French)

Witz, G.C. (1989) Biological interactions of α,β-unsaturated aldehydes. *Free Radical Biol. Med.* 7, 333–349

Yoshida, A., Sasaki, K. & Ohshiba, K. (1984) Suppressing experiments for fishy odor. II. Effects of spices, seasonings and food additives for fishy odor and volatile carbonyl compounds. *Seikatsu Esei* 28, 211–218

Yurkowski, M. & Bordeleau, M.A. (1965) Carbonyl compounds in salted cod. II. Separation and identification of volatile monocarbonyl compounds from heavily salted cod. *J. Fish. Res. Board Can.* 22, 27–32

Zweidinger, R.B., Sigsby, J.J.E., Tejada, S.B., Stumpf, F.D., Dropkin, D.L., Ray, W.D. & Duncan, J.W. (1988). Detailed hydrocarbon and aldehyde mobile source emissions from roadway studies. *Environ. Sci. Technol.* 22, 956–962

Exocyclic DNA Adducts in Mutagenesis and Carcinogenesis
B. Singer & H. Bartsch
IARC Scientific Publications No. 150
International Agency for Research on Cancer, Lyon, 1999

Chapter VI. DNA repair and effects on replication

Mammalian enzymatic repair of etheno and *para*-benzoquinone exocyclic adducts derived from the carcinogens vinyl chloride and benzene

B. Singer & B. Hang

Two human carcinogens that have been extensively studied are vinyl chloride and benzene. The active metabolites used in this study are chloroacetaldehyde (CAA) and *para*-benzoquinone (pBQ). Each forms exocyclic adducts between the N1 and N^6 of A, the N3 and N^4 of C and the N1 and N^2 of G. Only CAA has been found to form the N^2,3 adduct of G. CAA and pBQ adducts differ structurally in size and in the number of added rings, pBQ adding two rings to the base, while etheno bases have a single five-membered ring. The mechanism of repair of these two types of adducts by human enzymes has been studied in our laboratory with defined oligodeoxynucleotides and a site-specific adduct. The etheno derivatives are repaired by DNA glycosylase activity; two mammalian glycosylases are responsible: alkylpurine-DNA-*N*-glycosylase (APNG) and mismatch-specific thymine-DNA glycosylase. The former repairs 1,N^6-ethenoA (εA) as rapidly as the original substrate, 3-methyladenine, while the latter repairs 3,N^4-ethenoC (εC) more efficiently than the G/T mismatch. Our finding that there are separate enzymes for εA and εC has been confirmed by the use of tissue extracts from an APNG knockout mouse. As pBQ is much less efficient than CAA in modifying bases, the biochemical studies required total synthesis of the nucleosides. Furthermore, the pBQ adduct-containing oligomers are cleaved, to various extents by a different class of enzyme: human and bacterial *N*-5′-alkylpurine (AP) endonucleases. The enzyme incises such oligomers 5′ to the adduct and generates 3′-hydroxyl and 5′-phosphoryl termini but leaves the modified base on the 5′-terminus of the 3′ cleavage fragment ('dangling base'). Using active-site mutants of the human AP endonuclease, we found that the active site for the primary substrate, abasic (AP) site, is the same as that for the bulky pBQ adducts. There appears to be no clear rationale for the widely differing recognition and repair mechanisms for these exocyclic adducts, as shown for the repair of the same types of modification on different bases (e.g. εA and εC) and for completely unrelated lesions (e.g. AP site and pBQ adducts). Another important variable that affects the rate and extent of repair is the effect of neighbouring bases. In the case of εA, this sequence-dependent repair correlates with the extent of double-strandedness of the substrate, as demonstrated by thermal stability studies.

Introduction

Many commonly used chemicals have been found to cause mutation or cancer in animals and humans. Some hazardous chemicals act directly, but most need metabolic activation to exert their biological effects (Guengerich, 1992). Two unrelated chemicals that have been identified as human carcinogens are vinyl chloride and benzene (IARC, 1987). The active metabolites that have been investigated primarily are chloroethylene oxide and chloroacetaldehyde (CAA) for vinyl chloride (Barbin *et al.*, 1975; O'Neill *et al.*, 1986) and hydroquinone and *para*-benzoquinone (pBQ) formed from benzene (Snyder *et al.*, 1987; Huff *et al.*, 1989). This paper focuses on two ultimate metabolites, CAA and pBQ, that have been studied extensively in terms of the chemical modification of DNA and individual bases.

In vitro, CAA forms exocyclic adducts on dA, dC and dG, which have been studied *in vitro* for their chemical and biochemical properties (Kusmierek & Singer, 1982; Singer *et al.*, 1988; Kusmierek & Singer, 1992; reviewed by Leonard, 1992). Four cyclic etheno adducts have been characterized (1,N^6-εA (εA), 3,N^4-εC (εC), 1,N^2-εG and N^2,3-εG), which have also been identified and quantified *in vivo* (Ribovich *et al.*, 1982; Swenberg *et al.*, 1992; Nair *et al.*, 1995; Guichard *et al.*, 1996; Müller *et al.*, 1997). Their effects in replication have been studied *in vitro* and *in vivo*. *In vitro*, only N^2,3-εG is highly mutagenic (Singer *et al.*, 1987, 1991). *In vivo* there are conflicting reports on the frequency of mutation for εC and εA (Palejwala *et al.*, 1991; Basu *et al.*, 1993; Palejwala *et al.*, 1993; Moriya *et al.*, 1994; Pandya & Moriya, 1996). The variation apparently depends on the systems used, which include both bacterial and mammalian hosts. For example, there are reports of up to 70 and 80% mutants for εA and εC, respectively, with mammalian vectors (Moriya *et al.*, 1994; Pandya & Moriya, 1996). In the case of N^2,3-εG, the one published study conducted *in vivo* showed about 13% G→A mutations (Cheng *et al.*, 1991).

The *para*-benzoquinone adducts (1,N^6-pBQ-A, 3,N^4-pBQ-C and 1,N^2-pBQ-G) are also formed *in vitro* (Jowa *et al.*, 1990; Pongracz *et al.*, 1990; Chenna & Singer, 1995, 1997) but are bulkier than the etheno adducts owing to the two-ring addition to the base (see Figure 1, and see the Appendix for molecular models). Replication

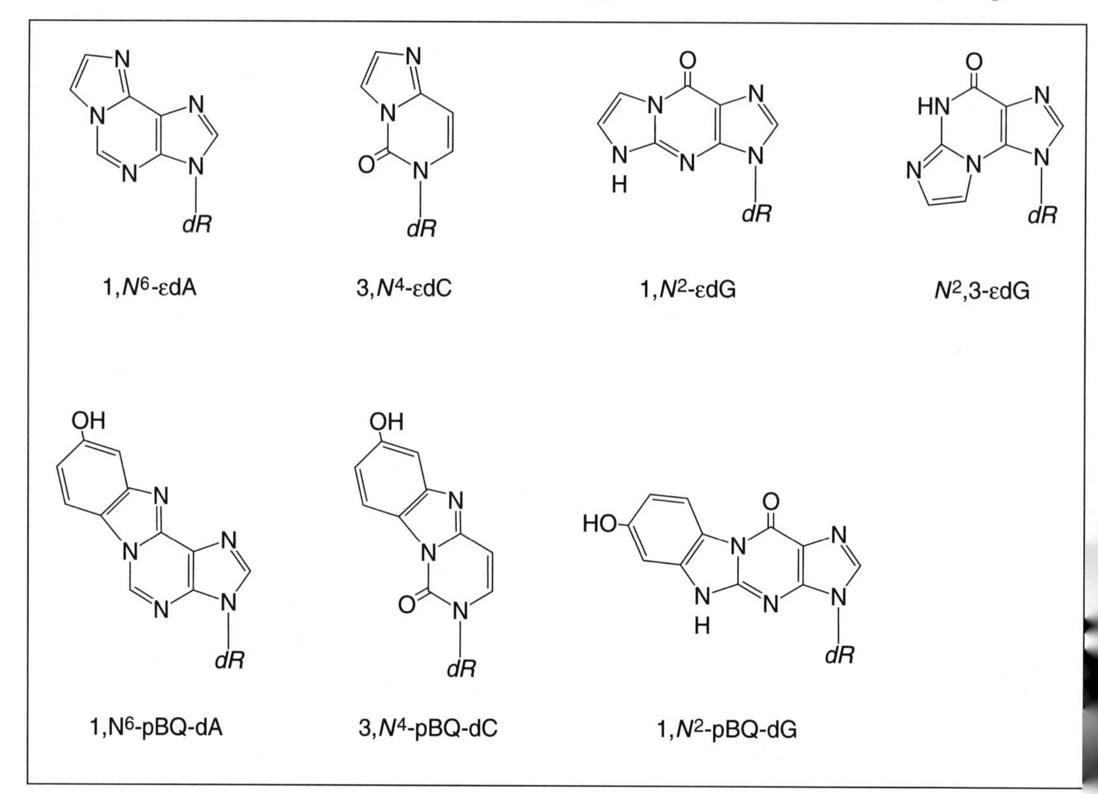

Figure 1. Structures of etheno (ε) derivatives (top) and *para*-benzoquinone (pBQ) derivatives (below)

studies *in vitro* indicate that the major effect of all these modifications is primarily blockage before or at the position of the adduct (see Chenna *et al.*, this volume). It is not clear whether some by-pass occurs, particularly as a result of sequence context. This was shown directly with εA-containing oligomers (Litinski *et al.*, 1997).

Repair of the etheno and pBQ adducts has been a major focus in our laboratory, since in order to preserve genetic integrity and cell survival such adducts must be removed from DNA so that they do not exert their potential mutagenic and/or lethal effects. In addition, the repair of exocyclic adducts with various structural changes has been useful as a model system for understanding the structure–function relationship in enzyme recognition in conjunction with other structural studies, such as nuclear magnetic resonance and X-ray crystallography.

Materials and methods

Substrates

Oligomers containing etheno adducts and pBQ adducts (Figure 2) were synthesized on a DNA synthesizer with phosphoramidites made in this laboratory (Dosanjh *et al.*, 1994; Chenna & Singer, 1995, 1997), with the exception of 1,N^6-εA which was purchased from Glen Research. The purity and composition of the oligomers were determined by enzymatic digestion followed by high-performance liquid chromatography (HPLC) with reversed-phase columns.

The 5′ ^{32}P-end labelling with T4 polynucleotide kinase and annealing to the complementary strand containing a selected base opposite the adduct was performed as described by Rydberg *et al.* (1991). 3′-End labelling with [α-^{32}P]ddATP and terminal deoxynucleotidyltransferase was carried out as described by Hang *et al.* (1996a).

Globally modified DNA containing etheno adducts was prepared by reaction with CAA (Singer *et al.*, 1992). The DNA was purified by repeated ethanol precipitation and stored in 10 mmol/L Tris-HCl/1 mmol.L EDTA, pH 7.8. DNA containing m^{3}A and m^{7}G was prepared by reaction with ^{3}H-*N*-methyl-*N*-nitrosourea or ^{14}C-dimethyl sulfate. Purification was done by the procedure described by Singer *et al.*(1992). Assays for simultaneous detection of etheno and alkyl bases released by glycosylases were performed after mixing the two substrates in equimolar amounts. Fluorescence was used to determine the release of εA, and radioactivity counting was used to determine the release of methylated bases.

Enzymatic assays

Binding assay: A band shift assay was used to detect and quantify the affinity of protein binding to a ^{32}P-end-labelled oligonucleotide containing a site-specifically placed adduct (Rydberg *et al.*, 1991). This is illustrated with a εA-containing 25-mer and human cell extracts in Figure 3.

Nicking assay: Cell-free extracts or purified proteins were incubated with the ^{32}P 5′-end-labelled oligomer substrates at the indicated times and temperatures. After heat denaturation, the samples were separated by denaturing 12% polyacrylamide gel electrophoresis (PAGE). A ^{32}P-labelled oligomer of the expected size of the cleavage product was run concurrently as a size marker. The bands were quanitified by X-ray film autoradiography followed by densitometry with a Molecular Dynamics PhosphorImager (Rydberg *et al.*, 1991; Hang *et al.*, 1998a).

DNA glycosylase assay: Protein, from cell extracts or purified, was reacted with adduct-containing oligomers or globally modified DNA with CAA or ^{14}C- or ^{3}H-labelled alkylating agents (Singer *et al.*, 1992). 3-Methyl-A and 7-methyl-G release was determined by isotope dilution after HPLC separation on an A-8 cation exchange column. Fluorescence for εA detection was determined after HPLC separation on an in-line Perkin-Elmer Fluorescence Detector.

Purification of repair enzymes

Nuclear extracts of human placenta and human lymphoblasts were used for isolation and purification of proteins that excise εA and m^{3}A, respectively. The protocols, involving conventional column chromatography in purification, are described by Singer *et al.* (1992). To purify the pBQ-dC activity from HeLa cells, additional column steps were used, which included two runs of Mono-S HR 5/5 and a Superdex 75 column with a fast protein liquid chromatography system from Pharmacia Biotech (Hang *et al.*, 1996b). The peak active frac-

To study repair of etheno derivatives:

1. 5´-CCGCTε**A**GCGGGTACCGAGCTCGAAT (εA: 1,N^6-etheno-A)
2. 5´-CCGCTAGε**C**GGGTACCGAGCTCGAAT (εC: 3,N^4-etheno-C)
3. 5´-CCGCTAGCε**G**GGTACCGAGCTCGAAT (εG: 1,N^2-etheno-G)
4. 5´-AGCGGGGε**A**GGGAGCT-3´
5. 5´-AGCGGCCε**A**CCGAGCT-3´
6. 5´-AGCGGAAε**A**AAGAGCT-3´
7. 5´-AGCGGTTε**A**TTGAGCT-3´

To study repair of pBQ derivatives:

8. 5´-CCGCT-*p*BQ-**A-**GCGGGTACCGAGCTCGAAT (*p*BQ-A: 1,N^6-benzetheno-A)
9. 5´-CCGCTAG-*p*BQ-**C-**GGGTACCGAGCTCGAAT (*p*BQ-C: 3,N^4-benzetheno-C)
10. 5´-CCGCTAGC-*p*BQ-**G-**GGTACCGAGCTCGAAT (*p*BQ-G: 1,N^2-benzetheno-G)
11. 5´-*p*BQ-**C-**GGGTACCGAGCTCGAAT

Other oligomers used:

12. 5´-CCGCT**Hx**GCGGGTACCGAGCTCGAAT (Hx: hypoxanthine)
13. 5´-CCGCTAGC**oxoG**GGTACCGAGCTCGAAT (oxoG: 7,8-dihydro-8-oxo-G)
14. 5´-CCGCTAG**T**GGGTACCGAGCTCGAAT (T/G mismatch)
GGCGATC**G**CCCATGGCTCGAGCTTA-5´
15. 5´-ATTACGAATGCCCACACCGC**T**GGCGCCCACCACCACTAGCTGGCC
TAATGCTTACGGGTGTGGCG**G**CCGCGGGTGGTGGTGATCGACCGG-5´
16. 5´-CCGCTAG**U**GGGTACCGAGCTCGAAT (U/G mismatch)
GGCGATC**G**CCCATGGCTCGAGCTTA-5´

Figure 2. Synthetic oligonucleotides used in the study

For oligonucleotides 1–10 and 12–13, the double strand was used. The base opposite the adduct is always the normal complementary base (e.g. εA/T, pBQ–C/G). Oligomer 11 was used in single-stranded form as a marker. Oligonucleotides 14–16 are shown in double-stranded form with the mismatch in bold.

tion from the last step was shown to be homogeneous on a silver-stained sodium dodecyl sulfate (SDS)-PAGE gel. To purify the εC-DNA glycosylase activity, also from the HeLa cells, a εC-DNA affinity matrix (Sepharose CL-4B) was prepared, and the column was used for the two final steps in the purification (Table 1) (Hang *et al.*, 1998b).

Other methods

The purity of isolated proteins and the relative molecular mass were assessed by SDS-PAGE and silver staining. The purified proteins isolated by SDS-PAGE were microsequenced by the Edman method. The protein concentration was determined by the method of Bradford (1976). Eluted

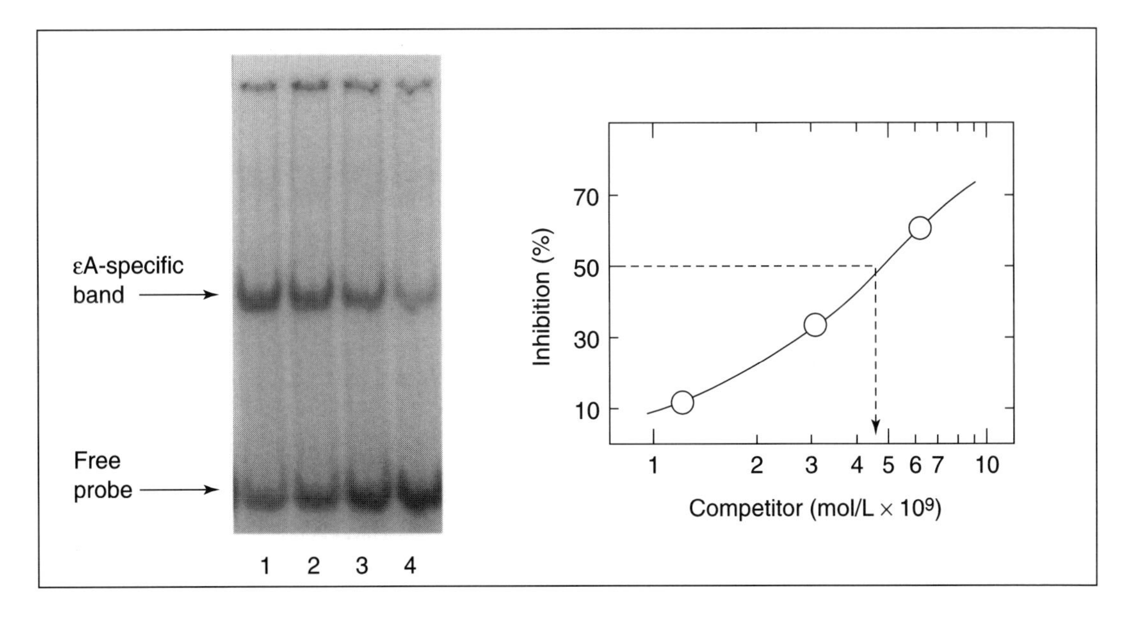

Figure 3. DNA binding protein from HeLa cells specific for 1,N^6-ethenoadenine (εA)

Left, competition assay with increasing concentrations of the εA-containing 25-mer. The εA binding band is indicated. Right, determination of binding constant from data on left (Rydberg *et al.*, 1991)

protein (εC-DNA glycosylase) was denatured and renatured essentially as described by Hager and Burgess (1980). Thermal denaturation of oligomers was performed with a Beckman DU 7400 diode-array spectrophotometer at 260 or 280 nm (Sági *et al.*, 1998).

Repair of etheno adducts

1,N^6-Ethenoadenine is a substrate for mammalian alkylpurine-DNA-N-glycosylases

In 1991, we found that a human cell-free extract contained a specific protein that bound to an εA-containing 25-mer oligomer (Figure 3A)

Table 1. Purification of 3,N^4-ethenocytosine (εC)–DNA glycosylase from HeLa cells

Purification step	Protein (mg)	Specific activity[a] (units/mg)	Yield (%)
Whole cell extract	5785	8.8	
Ammonium sulfate precipitate	3639	11.2	80.4
Phosphocellulose (P11)	444	38.6	33.7
Blue Sepharose	63	85.3	10.6
Mono-S	9	327.2	5.8
εC–DNA affinity (1)	0.2	9190	3.6
εC–DNA affinity (2)	None[b]	–	–

[a] One unit of εC–DNA glycosylase is defined as 1 fmol of the εC-oligomer cleaved after 10 min at 37 °C.

[b] No measurable protein was recovered, but there was sufficient εC activity to determine that G/T mismatch activity remained.

(Rydberg *et al.*, 1991). Competition with excess unlabelled εA-containing 25-mer showed a high binding constant (Figure 3B). This protein appeared to have associated glycosylase activity towards the εA, as shown by release of the free base (Figure 4) followed by 5′ nicking at the adduct after incubation with a HeLa cell-free extract or partially purified 'εA-binding protein' (Singer *et al.*, 1992).

Subsequently, both m^3A and εA in DNA were treated with both the partially purified human 'εA binding protein' and the cell-free extract of cloned human alkylpurine-DNA-*N*-glycosylase (APNG, a gift from Dr T.R. O'Connor and Dr J. Laval (O'Connor & Laval, 1991)). It was unexpected that both preparations bound and released εA and m^3A in the same manner. Furthermore, purification of enzyme from human placenta showed that the two activities co-eluted and the ratio remained approximately constant (Singer *et al.*, 1992).

Saparbaev *et al.* (1995) used purified cloned APNGs from humans, rats, yeast and *Escherichia coli* to test the species specificity of the excision of both εA and m^3A from either globally modified DNA or the same εA containing 25-mer oligomer used in the binding experiments (Rydberg *et al*, 1991). They found increasing amounts of repair of εA with evolutionary progress: m^3A-DNA glycosylase II from *E. coli* had the least activity and APNG from humans was three orders of magnitude more effective towards εA (Saparbaev *et al.*, 1995). The authors concluded that at least two chemically different adducts, εA and m^3A, were released from DNA by the same enzyme and mechanism.

The other three etheno adducts, εC, 1,N^2-εG and N^2,3-εG, are also released by HeLa cell-free extracts with differing efficiencies (Dosanjh *et al.*, 1994b). A closer examination showed that εC is not repaired by human APNG but by a separate DNA glycosylase (see below). Both of the etheno dG adducts are released by the HeLa cell-free extract but at a much lower rate than is εA or εC (Figure 5). Since a cell extract was used, it is unclear whether the two dG adducts are repaired

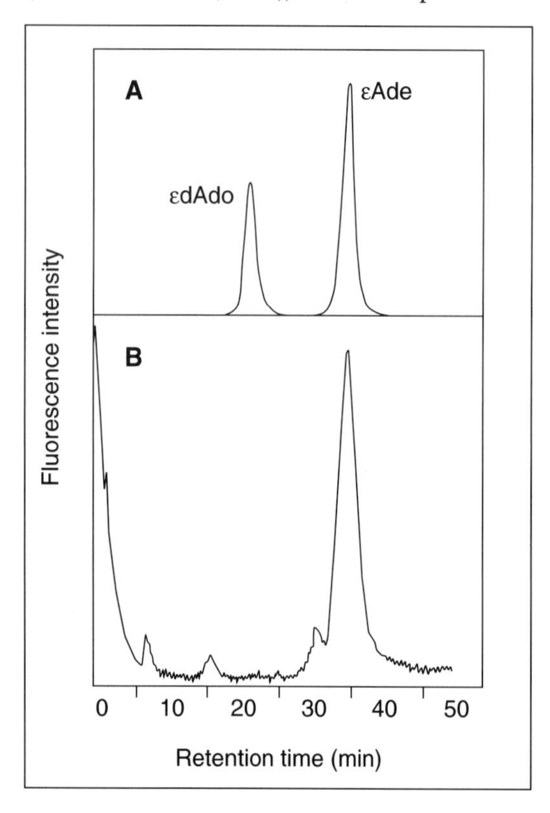

Figure 4. Detection by fluorescence of release of ethenoadenine (εAde) but not ethenodeoxyadenosine (εAdo) by cell-free HeLa extracts

(A) Retention time of markers; (B) retention time of released εAde by HeLa cell extract from chloroacetaldehyde-treated DNA (Rydberg *et al.*, 1992)

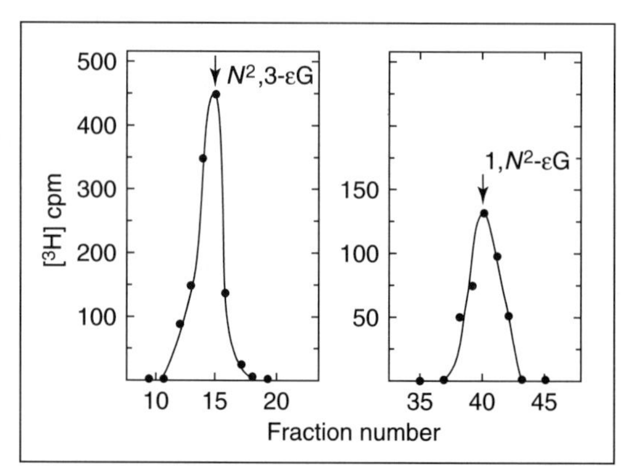

Figure 5. Release of free ^{3}H-N^2,3-ethenoguanine (εG) (left) and 1,N^2-εG (right) by HeLa cell extract with high-performance liquid chromatography to separate the free bases for quantification

From Dosanjh *et al.* (1994b)

by APNG, although m³A-DNA glycosylase II protein from *E. coli* is reported to release N^2,3-εG (Matijasevic *et al.*, 1992).

3,N⁴-Ethenocytosine is repaired by a separate enzyme, the human G/T(U) mismatch-DNA glycosylase

After extensive purification of HeLa cell-free extracts to isolate APNG, it was found that the specific enzyme fraction did not cleave εC-containing oligonucleotides. Separation of εA and εC activities could be demonstrated after three column steps in purification (Figure 6). By further use of a gel filtration column chromatography (Superdex 75), the size of the native protein was found to be about 66 kDa (Singer & Hang, 1997), in contrast to the human APNG which has an apparent molecular mass of 32 kDa (O'Connor, 1993).

Efforts were then made to further purify and identify the 'εC-DNA glycosylase', as shown in Table 1. The use of εC-DNA affinity columns enabled us to separate the target protein to a high degree of purity (Figure 7). The protein had an apparent molecular mass of 55 kDa on SDS-PAGE (Hang *et al.*, 1998b), which is close to that found for the human mismatch-specific thymine-DNA glycosylase (TDG), purified and cloned by Nedderman and colleagues (Nedderman & Jiricny, 1993; Nedderman *et al.*, 1996). On testing our protein for cleavage of a G/T mismatch containing oligomer (sequence 15, Figure 2), we showed that the two activities, εC and G/T mismatch, co-eluted in the last three column steps. Moreover, the isolated and renatured 55-kDa protein showed activity towards both substrates, although εC was more efficiently cleaved. Another proof that the two activities reside in the same protein was obtained by use of a TDG from the unrelated thermophilic archaeon *Methanobacterium thermoautotrophicum* THF (Horst & Fritz, 1996).

At the same time as we were conducting our experiments, Saparbaev and Laval (1998) purified εC-DNA glycosylase activity from *E. coli*. By microsequencing, the enzyme was found to be the previously described double-stranded uracil-DNA glycosylase. In addition, a recombinant human TDG, which is the homologue of the *E. coli* enzyme, was found to act on εC in oligomers.

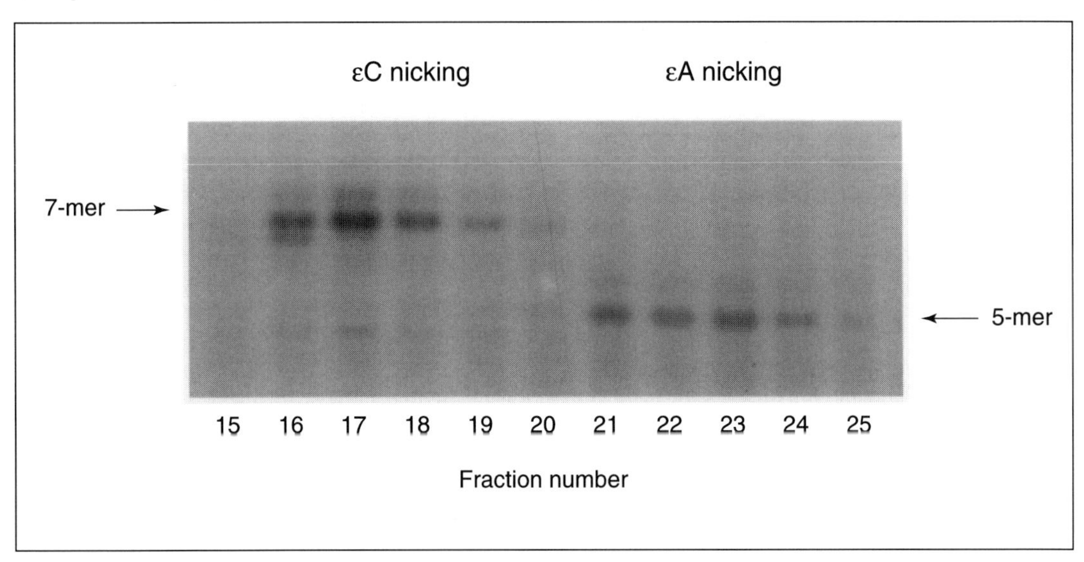

Figure 6. Separation on denaturing polyacrylamide gel electrophoresis of 3,N^4-ethenocytosine (εC) and 1,N^6-ethenoadenine (εA) activity

Fractions are from Mono-S FPLC column chromatography. Note the complete separation of the two activities: εC nicking produces a 7-mer product, and εA nicking produces a 5-mer (Hang *et al.*, 1996a).

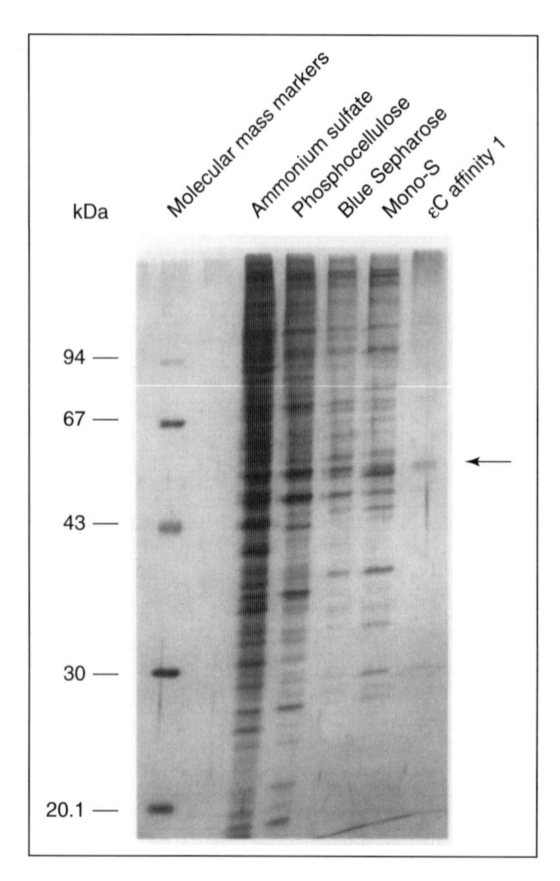

Figure 7. Silver-stained sodium dodecyl sulfate–polyacrylamide gel electrophoresis of steps in the purification of the 3,*N*4-ethenocytosine (εC)-DNA glycosylase

See Table 1 for analytical data. The arrow indicates the position of the 55-kDa εC activity, as proved by isolation and renaturation (Hang *et al.*, 1998b).

Use of knockout mice lacking alkylpurine-DNA-N-glycosylase for studying substrate specificity

Recent developments in the techniques of gene deletion in mice have added a powerful tool for studying the biological and biochemical features of targeted enzymes. We used the 'knockout' mouse developed by Elder *et al* (1997, 1998), which has a deleted *apng* gene, to study the substrate specificity of this enzyme (Hang *et al.*, 1997a). Various mouse tissues, including those from testes, liver and lung, were used to prepare cell-free extracts. In a DNA nicking assay to test the oligonucleotides εA, εC, hypoxanthine and 7,8-dihydro-8-oxoguanine (Figure 2), it was found that εA and εC are repaired by separate gene products, that the previously identified 'εA-DNA glycosylase' (Rydberg *et al.*, 1992) and 'hypoxanthine-DNA glycosylase' (Karren & Lindahl, 1978, 1980; Myrnes *et al.*, 1982) are both functions of mouse APNG, and that mouse APNG does not measurably contribute to the repair of 7,8-dihydro-8-oxoguanine (Figures 8 and 9).

An experiment conducted simultaneously with an independently developed knockout mouse lacking *apng* gave similar results for εA, hypoxanthine and 7,8-dihydro-8-oxoguanine (Engelward *et al.*, 1997).

Sequence dependence of 1,N^6-ethenoadenine repair

Many biochemical functions are influenced by the immediate sequence context of adducts in DNA or oligomers. Repair is not an exception. In the experiments described here, involving εA repair by APNG, which requires a double-stranded substrate, our design was to flank εA by tandem bases in a 15-mer in order to maximize the effects of purines or pyrimidines in stabilizing double-strandedness. Sequences 4–7 in Figure 2 show the complete sequences used; they are referred to below by the central pentamer.

Repair with human APNG was studied as a function of both enzyme concentration and time. Figure 10 shows the time-dependence of εA repair flanked by four sequences: –CCεACC–, –GGεAGG–, –TTεATT– and –AAεAAA–. The effects of flanking purines and pyrimidines were clearly different and could be related to the greater stability of G/C pairs compared with A/T pairs. This hypothesis was tested in a series of experiments. It was clearly shown by thermal denaturation of the oligonucleotides used that –GGεAGG– (or –CCεACC–) had a significantly higher T_m than did –TTεATT– (–AAεAAA–) (see Table 1 of Hang *et al.*, 1998a for thermodynamic parameters). A plot of two melting curves was used to determine that the difference conferred by sequence was highly significant (Figure 11).

In addition, studies of the effect of temperature shift on repair (Figure 12) confirmed that the specific sequence dependence of repair was parallel to the degree of denaturation of these duplexes. A direct correlation between double-strandedness and repair efficiency had not been previously demonstrated but only hypothesized.

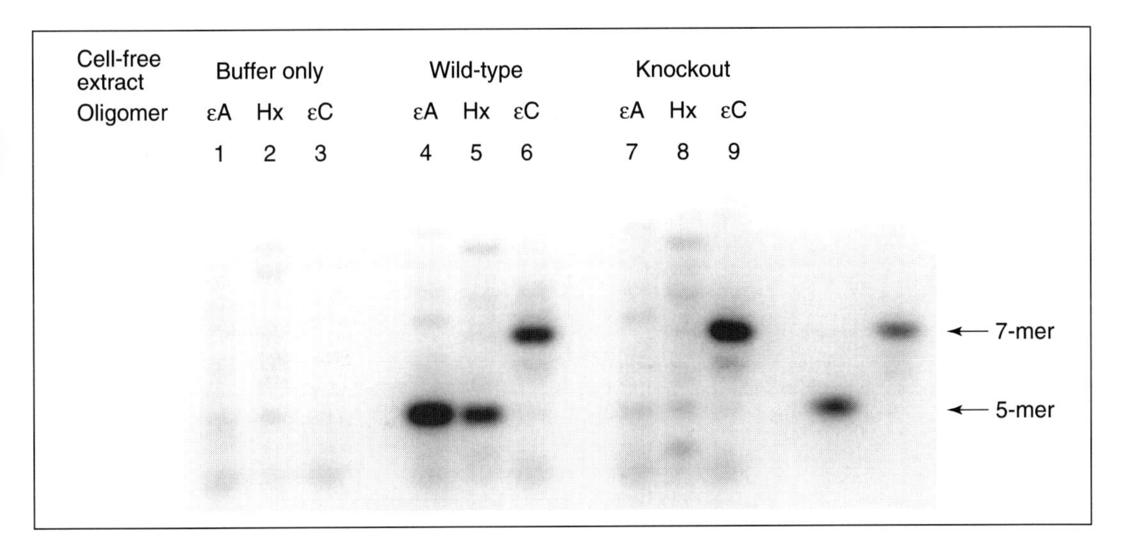

Figure 8. Autoradiogram of denaturing polyacrylamide gel electrophoresis comparing extent of nicking by wild-type and knock-out mouse testis cell-free extracts of 1,N^6-ethenoadenine (εA), hypoxanthine (Hx) and 3,N^4-ethenocytosine (εC)

The third panel shows that the alkylpurine-DNA-*N*-glycosylase knockout mouse extract acts only on εC (Hang *et al.*, 1997a).

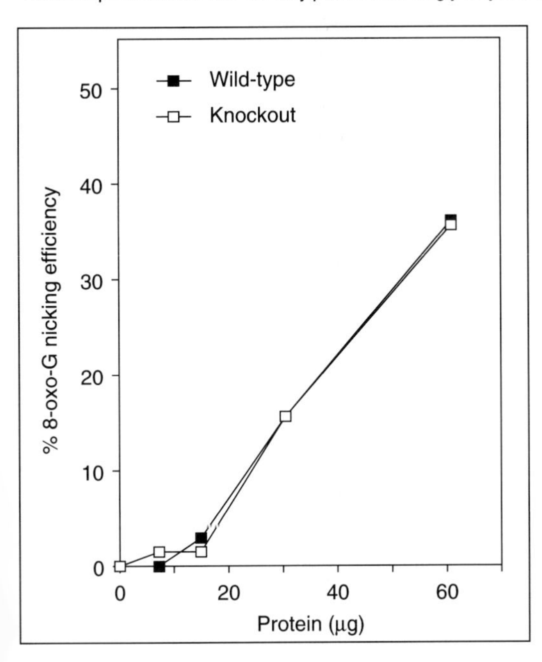

Figure 9. Comparison of 7,8-dihydro-8-oxoguanine (8-oxo-G) nicking activity in wild-type and knockout mouse testis extract, showing no detectable difference in nicking between mice with (+) and without (–) alkylpurine-DNA-*N*-glycosylase after 1-h incubation (Hang *et al.*, 1997a)

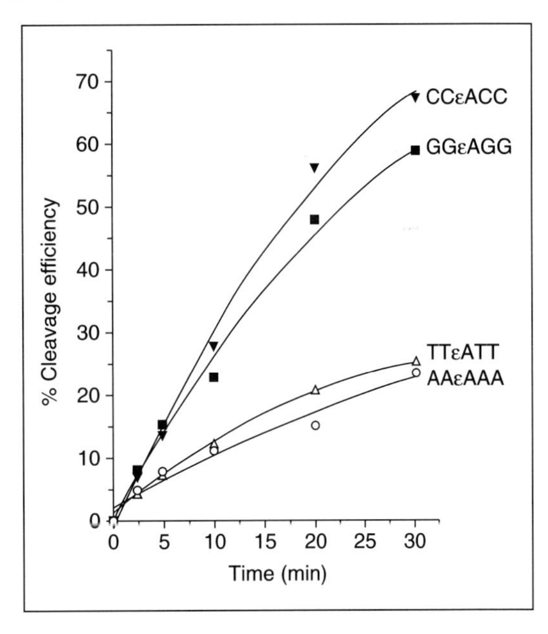

Figure 10. Difference in rate of cleavage of 1,N^6-ethenoadenine as a function of purine or pyrimidine flanking bases with human alkylpurine-DNA-*N*-glycosylase on 15-mer substrates

ε, etheno; for complete sequences, see Figure 2 (Hang *et al.*, 1998a).

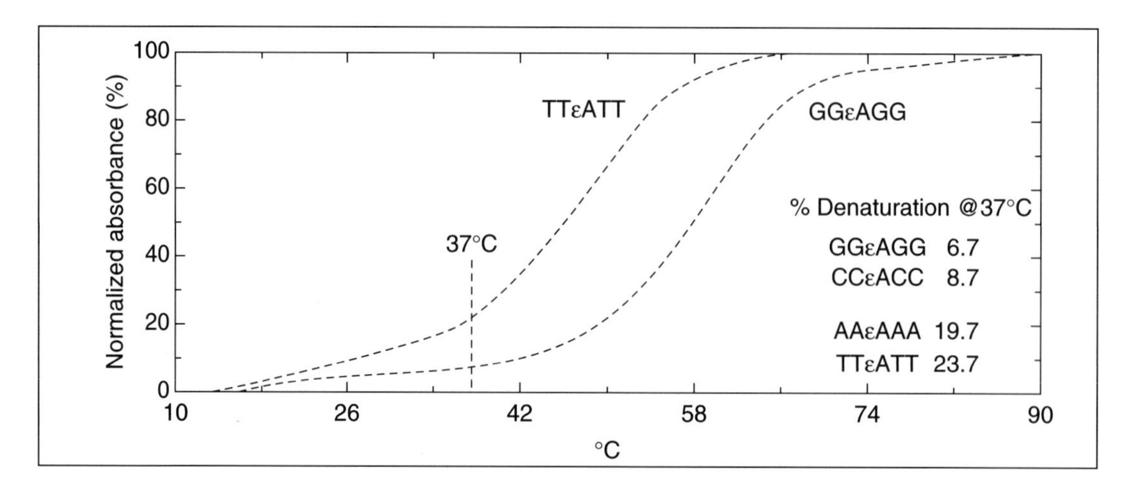

Figure 11. Superimposed melting curves of two representative 1,N^6-ethenoadenine (εA)-containing 15-mer duplexes with the same flanking sequences as in Figure 10

Calculated percent denaturation at 37 °C for all four εA oligomers is given. Note the large difference between purine and pyrimidine neighbours affecting the G/C and A/T content. For detailed thermodynamic parameters of these dupluxes, see Hang *et al.* (1998a).

Repair of *para*-benzoquinone adducts

All three pBQ adducts, pBQ-dC, pBQ-dA and pBQ-dC, were synthesized as described by Chenna and Singer (1995, 1997) and site-specifically incorporated into 25-mer oligomers. On initial testing, all were cleaved by a HeLa cell extract at 5′ to the adduct (Chenna *et al.*, 1995; Singer & Hang, 1997). This new activity was purified from HeLa cells to apparent homogeneity and partially microsequenced. It was then found that the enzyme was identical to the major human apurinic/apyrimidinic (AP) endonuclease (variously termed Ape, HAP1, APEX and Ref 1: Demple *et al.*, 1991; Robson & Hickson, 1991; Seki *et al.*, 1992; Xanthoudakis *et al.*, 1992). Thus, we had a 'new' substrate for an 'old' enzyme. This was confirmed by using a recombinant human AP endonuclease, Ape, a gift from Dr Bruce Demple. In addition, two bacterial 5′ AP endonucleases, exonuclease III and endonuclease IV (a gift from Dr David Wilson III), were much more efficient in cleaving all three pBQ adducts (Figure 13). Cross-comparisons with adducts with different neighbour groups (Figure 2, sequences 8–10) are not indicative of relative efficiency of repair.

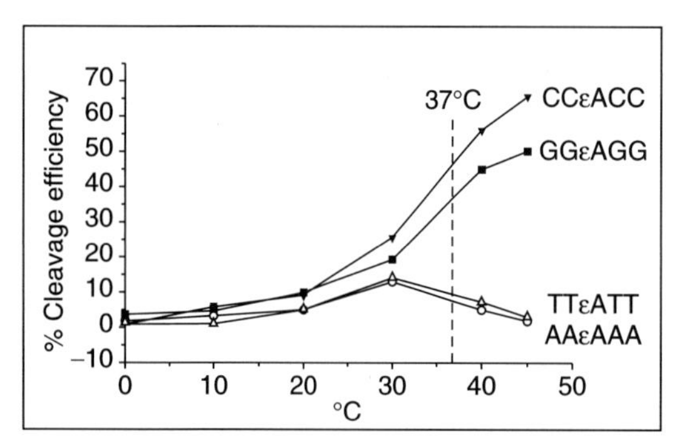

Figure 12. Effect of reaction temperature on cleavage efficiency of 1,N^6-ethenoadenine (εA)-containing 15-mers with different tandem neighbour bases

Each point represents incubation of an independent reaction mixture with human alkylpurine-DNA-*N*-glycosylase at the indicated temperatures. The reaction at normal physiological tepmerature (37 °C) is shown by the dashed vertical line (Hang *et al.*, 1998a).

The mechanism by which an AP endonuclease cleaves an existing AP site is well described (Doetsch & Cunningham, 1990; Friedberg *et al.*, 1995). In the case of pBQ adducts, the enzyme recognizes the bulky adduct, cleaves at the 5′ site, and leaves the adduct as a 'dangling base' on the 3′ fragment (Figure 14). The difference in size and structure between an AP site and a pBQ adduct is enormous, yet both are efficiently repaired, and use of active site mutants suggested that the same

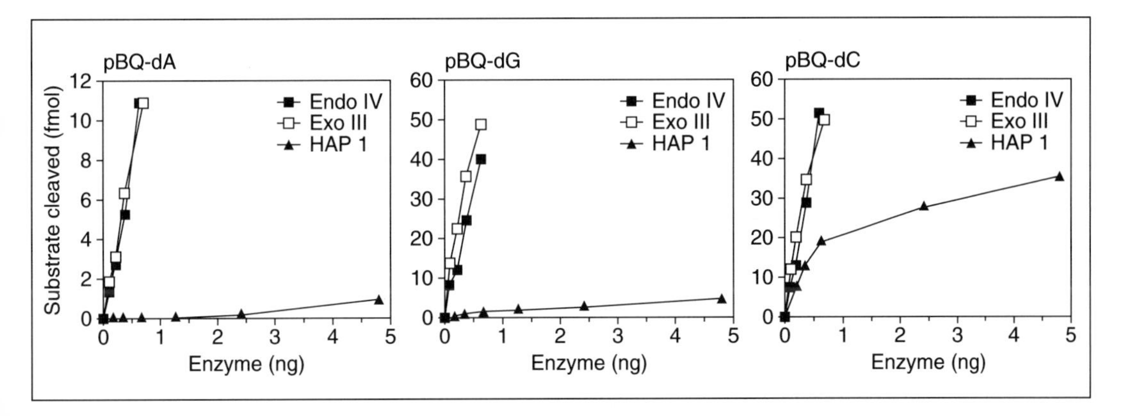

Figure 13. Cleavage of *para*-benzoquinone (pBQ)–dA (left), pBQ–dG (middle) and pBQ–dC by three 5′ apurinic/apyrimidinic endonucleases: HAP1, exonuclease III (Exo III) and endonuclease IV (Endo IV)

Except for the enzyme, all reaction conditions were identical (Hang *et al.*, 1998c).

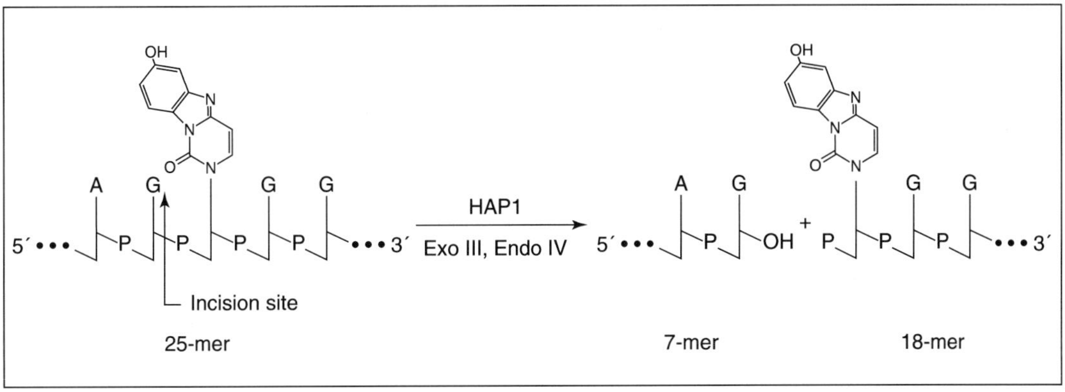

Figure 14. Proposed scheme for cleavage by 5′ apurinic/apyrimidinic endonucleases of 3,N^4-benzetheno–2′-deoxycytidine in a 25-mer oligomer with the adduct at position 8 from the 5′ side

The 5′ cut yields a 7-mer fragment with a 3′-OH group plus an 18-mer with a 5′ terminal phosphoryl group with the 3,N^4-benzetheno–2′-deoxycytidine residue attached as shown (Hang *et al.*, 1996b).

active site is used (Hang *et al.*, 1997b). Co-crystallization of the human AP endonuclease with these substrates would lead to a better understanding of the structural requirements for recognition and the catalytic mechanism(s).

References

Barbin, A., Bresil, H., Croisy, A., Jacquignon, P., Malaveille, C., Montesano, R. & Bartsch, H. (1975) Liver-microsome-mediated formation of alkylating agents from vinyl bromide and vinyl chloride. *Biochem. Biophys. Res. Commun.* 67, 596–603

Basu, A.K., Wood, M.L., Niedernhofer, L.J., Ramos, L.A. & Essigmann, J.M. (1993) Mutagenic and genotoxic effects of three vinyl chloride-induced DNA lesions: 1,N^6-ethenoadenine, 3,N^4-ethenocytosine, and 4-amino-5-(imidazol-2-yl)imidazole. *Biochemistry* 32, 12793–12801

Bradford, M.M. (1976) A rapid and sensitive method for the quantitation of microgram quantities of protein utilizing the principle of protein–dye binding. *Anal. Biochem.* 72, 248–254

Cheng, K.C., Preston, B.D., Cahill, D.S., Dosanjh, M.K., Singer, B. & Loeb, L.A. (1991) The vinyl chloride DNA derivative, N^2,3-ethenoguanine, produces G→A transitions in *E. coli*. *Proc. Natl Acad. Sci. USA* 88, 9974–9978

Chenna, A. & Singer, B. (1995) Large scale synthesis of p-benzoquinone-2´-deoxycytidine and p-benzoquinone-2´-deoxyadenosine adducts and their site-specific incorporation into DNA oligonucleotides. *Chem. Res. Toxicol.* 8, 865–874

Chenna, A. & Singer, B. (1997) Synthesis of a benzene metabolite adduct, (3´´-hydroxyl)-1,N^2-benzetheno-2´-deoxyguanosine, and site-specific incorporation of the phosphoramidite into DNA oligonucleotides. *Chem. Res. Toxicol.* 10, 165–171

Chenna, A., Hang, B., Rydberg, B., Kim, E., Pongracz, K., Bodell, W.J. & Singer, B. (1995) The benzene metabolite p-benzoquinone forms adducts with DNA bases that are excised by a repair activity from human cells that differs from an ethenoadenine glycosylase. *Proc. Natl Acad. Sci. USA* 92, 5890–5894

Demple, B., Herman, T. & Chen, D.S. (1991) Cloning and expression of APE, the cDNA encoding the major human apurinic endonuclease: Definition of a family of DNA repair enzymes. *Proc. Natl Acad. Sci. USA* 88, 11450–11454

Doetsch, P.W. & Cunningham, R.P. (1990) The enzymology of apurinic/apyrimidinic endonucleases. *Mutat. Res.* 2-3, 173–201

Dosanjh, M.K., Roy, R., Mitra, S. & Singer, B. (1994a) 1,N^6-Ethenoadenine is preferred over 3-methyladenine as substrate by a cloned human N-methylpurine-DNA glycosylase (3-methyladenine-DNA glycosylase). *Biochemistry* 33, 1624–1628

Dosanjh, M.K., Chenna, A., Kim, E., Fraenkel-Conrat, H., Samson, L. & Singer, B. (1994b) All four known cyclic adducts formed in DNA by the vinyl chloride metabolite chloroacetaldehyde are released by a human DNA glycosylase. *Proc. Natl Acad. Sci. USA* 91, 1024–1028

Elder, R.H., Weeks, R.J., Willington, M.A., Cooper, D.P., Rafferty, J.A., Deans, B., Hendry, J.H. & Margison, G.P. (1997) Alkylpurine-DNA-*N*-glycosylase (APNG) knockout mice. *Proc. Am. Assoc. Cancer. Res.* **38**, 360

Elder, R.H., Jansen, J.G., Weeks, R.J., Willington, M.A., Deans, B., Watson, P.J., Mynett, K.J., Bailey, J.A., Cooper, D.P., Rafferty, J.A., Heeran, M.C., Wijnhoven, S.W.P., van Zeeland, A.A. & Margison, G.P. (1998) Alkylpurine-DNA-*N*-glycosylase knockout mice show increased susceptibility to induction of mutations by methyl methanesulfonate. *Mol. Cell. Biol.* 18, 5828–5837

Engelward, B.P., Weeda, G., Wyatt, M.D., Broekhof, J.L., del Wit, J., Donker, I., Allan J.M., Gold, B., Hoejimakers, J.H. & Samson, L.D. (1997) Base excision repair deficient mice lacking the AAG alkyladenine DNA glycosylase. *Proc. Natl Acad. Sci. USA* 94, 13087–13092

Friedberg, E., Walker, G. C. & Seide, W. (1995) *DNA Repair and Mutagenesis*. Washington DC, ASM Press

Guengerich, F.P. (1992) Metabolic activation of carcinogens. *Pharmacol. Ther.* 54, 17–61

Guichard, Y., El Ghissassi, F., Nair, J., Bartsch, H. & Barbin, A. (1996) Formation and accumulation of DNA ethenobases in adult Sprague-Dawley rats exposed to vinyl chloride. *Carcinogenesis* 17, 1553–1559

Hager, D.A. & Burgess, R.R. (1980) Elution of proteins from sodium dodecyl sulfate-polyacrylamide gels, removal of sodium dodecyl sulfate, and renaturation of enzymatic activity: Results with sigma subunit of *Escherichia coli* RNA polymerase, wheat germ DNA topoisomerase, and other enzymes. *Anal. Biochem.* 109, 76–86

Hang, B., Chenna, A., Rao, S. & Singer, B. (1996a) 1,N^6-Ethenoadenine and 3,N^4-ethenocytosine are excised by separate human DNA glycosylases. *Carcinogenesis* 17, 155–157

Hang, B., Chenna, A., Fraenkel-Conrat, H. & Singer, B. (1996b) An unusual mechanism for the major human apurinic/apyrimidinic (AP) endonuclease involving 5´ cleavage of DNA containing a benzene-derived exocyclic adduct in the absence of an AP site. *Proc. Natl Acad. Sci. USA* 93, 13737–13741

Hang, B., Singer, B., Margison, G.P. & Elder, R.H. (1997a) Targeted deletion of alkylpurine-DNA N-glycosylase in mice eliminates repair of 1,N^6-ethenoadenine and hypoxanthine but not of 3,N^4-ethenocytosine or 8-oxoguanine. *Proc. Natl Acad. Sci. USA* 94, 12869–12874

Hang, B., Rothwell, D.G., Sági, J., Hickson, I.D. & Singer, B. (1997b) Evidence for a common active site for cleavage of an AP site and the benzene-derived exocyclic adduct, 3,N^4-benzetheno-dC, in the major human AP endonuclease. *Biochemistry* 36, 15411–15418

Hang, B., Sági, J. & Singer, B. (1998a) Correlation between sequence-dependent glycosylase repair and the thermal stability of oligonucleotide duplexes containing 1,N^6-ethenoadenine. *J. Biol. Chem.* 273, 33406–33413

Hang, B., Medina, M., Fraenkel-Conrat, H. & Singer, B. (1998b) A 55-kDa protein isolated from human cells shows DNA glycosylase activity toward 3,N^4-ethenocytosine and the G/T mismatch. *Proc. Natl Acad. Sci. USA* 95, 13561–13566

Hang, B., Chenna, A., Sági, J. & Singer, B. (1998c) Differential cleavage of oligonucleotide containing the benzene-derived adduct, 1,N^6-benzethano dA, by the major human AP endonuclease HAP1 and *Escherichia coli* exonuclease III and endonuclease IV. *Carcinogenesis* 19, 1339–1343

Huff, J.E., Haseman, J.K., DeMarini, D.M., Eustis, S., Maronpot, R.R., Peters, A.C., Persing, R.L., Chrisp, C.E. & Jacobs, A.C. (1989) Multiple-site carcinogenicity of benzene in Fischer 344 rats and B6C3F1 mice. *Environ. Health Perspectives* 82, 125–163

Horst, J.P. & Fritz, H.J. (1996) Counteracting the mutagenic effect of hydrolytic deamination of DNA 5-methylcytosine residues at high temperature: DNA mismatch N-glycosylase Mig.Mth of the thermophilic archaeon *Methanobacterium thermoautotrophicum* THF. *EMBO J.* 15, 5459–5469

IARC (1987) *IARC Monographs on the Evaluation of the Carcinogenic Risks to Humans*, Suppl. 7. Lyon

Jowa, L., Witz, G., Snyder, R., Winkle, S. & Kalf, G.F. (1990) Synthesis and characterization of deoxyguanosine–benzoquinone adducts. *J. Appl. Toxicol.* 10, 47–54

Karran, P. & Lindahl, T. (1978) Enzymatic excision of free hypoxanthine from polydeoxynucleotides and DNA containing deoxyinosine monophosphate residues. *J. Biol. Chem.* 253, 5877–5879

Karran, P. & Lindahl, T. (1980) Hypoxanthine in deoxyribonucleic acid: Generation by heat-induced hydrolysis of adenine residues and release in free form by a deoxyribonucleic acid glycosylase from calf thymus. *Biochemistry* 19, 6005–6011

Kusmierek, J.T. & Singer, B. (1982) Chloroacetaldehyde-treated ribo- and deoxyribopolynucleotides. 1. Reaction products. *Biochemistry*, 21, 5717–5722

Kusmierek, J.T. & Singer, B. (1992) 1,N^2-ethenodeoxyguanosine: Properties and formation in chloroacetaldehyde-treated polynucleotides and DNA. *Chem. Res. Toxicol.*, 5, 634–638

Leonard, N.J. (1992) Etheno-bridged nucleotides in structural diagnosis and carcinogenesis. *Biochem. Mol. Biol. Chemtracts* 3, 273–297

Litinski, V., Chenna, A., Sági, J. & Singer, B. (1997) Sequence context is an important determinant in the mutagenic potential of 1,N^6-etheno-deoxyadenosine (εA): Formation of εA basepairs and elongation in defined templates. *Carcinogenesis* 18, 1609–1615

Matijasevic, Z., Sekiguchi, M. & Ludlum, D.B. (1992) Release of N^2,3-ethenoguanine from chloroacetaldehyde-treated DNA by *Escherichia coli* 3-methyladenine DNA glycosylase II. *Proc. Natl Acad. Sci. USA* 89, 9331–9334

Moriya, M., Zhang, W., Johnson, F. & Grollman, A.P. (1994) Mutagenic potency of exocyclic DNA adducts: Marked differences between *Escherichia coli* and simian kidney cells. *Proc. Natl Acad. Sci. USA* 91, 11899–11903

Müller, M., Belas, F.J., Blair, I.A. & Guengerich, F.P. (1997) Analysis of 1,N^2-ethenoguanine and 5,6,7,9-tetrahydro-7-hydroxy-9-oxoimidazo[1,2-a]purine in DNA treated with 2-chlorooxirane by high performance liquid chromatography/electrospray mass spectrometry and comparison of amounts to other DNA adducts. *Chem. Res. Toxicol.* 10, 242–247

Myrnes, B., Guddal, P.H. & Krokan, H. (1982) Metabolism of dITP in HeLa cell extracts, incorporation into DNA by isolated nuclei and release of hypoxanthine from DNA by a hypoxanthine–DNA glycosylase activity. *Nucleic Acids Res.* 10, 3693–3701

Nair, J., Barbin, A., Guichard, Y. & Bartsch, H. (1995) 1,N^6-Ethenodeoxyadenosine and 3,N^4-ethenodeoxycytine in liver DNA from humans and untreated rodents detected by immunoaffinity/^{32}P-postlabeling. *Carcinogenesis* 16, 613–617

Neddermann, P. & Jiricny, J. (1993) The purification of a mismatch-specific thymine-DNA glycosylase from HeLa cells. *J. Biol. Chem.* 268, 21218–21224

Neddermann, P., Gallinari, P., Lettieri, T., Schmid, D., Truong, O., Hsuan, J.J., Wiebauer, K. & Jiricny, J. (1996) Cloning and expression of human G/T mismatch-specific thymine-DNA glycosylase. *J. Biol. Chem.* 271, 12767–12774

O'Connor, T.R. (1993) Purification and characterization of human 3-methyladenine-DNA glycosylase. *Nucleic Acids Res.* 21, 5561–5569

O'Connor, T.R. & Laval, J. (1991) Human cDNA expressing a functional DNA glycosylase excising 3-methyladenine and 7-methylguanine. *Biochem. Biophys. Res. Commun.* 176, 1170–1177

O'Neill, I.K., Barbin, A., Friesen, M. & Bartsch, H. (1986) Reaction kinetics and cytosine adducts of chloroethylene oxide and chloroacetaldehyde: Direct observation of intermediates by FTNMR and GC–MS. In: *The Role of Cyclic and Nucleic Acid Adducts in Carcinogenesis and Mutagenesis* (Singer, B. & Bartsch, H., eds), IARC Scientific Publications No. 70, Lyon, IARC, pp. 57–73

Palejwala, V.A., Simha, D. & Humayun, M.Z. (1991) Mechanisms of mutagenesis by exocyclic DNA adducts. Transfection of M13 viral DNA bearing a site-specific adduct shows that ethenocytosine is a highly efficient RecA-independent mutagenic noninstructional lesion. *Biochemistry* 30, 8736–8743

Palejwala, V.A., Rzepka, R.W., Simha, D. & Humayun, M.Z. (1993) Quantitative multiplex sequence analysis of mutational hot spots. Frequency and specificity of mutations induced by a site-specific ethenocytosine in M13 viral DNA. *Biochemistry* 32, 4105–4111

Pandya, G.A. & Moriya, M. (1996) 1,N^6-Ethenodeoxyadenosine, a DNA adduct highly mutagenic in mammalian cells. *Biochemistry* 35, 11487–11492

Pongracz, K., Kaur, S., Burlingame, A.L. & Bodell, W.J. (1990) Detection of (3′-hydroxy)-3,N^4-benzetheno-2′-deoxycytidine-3′-phosphate by ^{32}P-postlabeling of DNA reacted with p-benzoquinone. *Carcinogenesis* 11, 1469–1472

Ribovich, M.L., Miller, J.A., Miller, E.C. & Timmins, L.G.(1982) Labeled 1,N^6-ethenoadenosine and 3,N^4-ethenocytidine in hepatic RNA of mice given[ethyl-1,2(–3)H or ethyl-1(–14)C]ethyl carbamate (urethan). *Carcinogenesis* 3, 539–546

Robson, C.N. & Hickson, I.D. (1991) Isolation of cDNA clones encoding a human apurinic/apyrimidinic endonuclease that corrects DNA repair and mutagenesis defects in *E. coli* xth (exonuclease III) mutants. *Nucl. Acids Res.* 19, 5519–5523

Rydberg, B., Dosanjh, M.K. & Singer, B. (1991) Human cells contain protein specifically binding to a single 1,N^6-ethenoadenine in a DNA fragment. *Proc. Natl Acad. Sci. USA* 86, 6839–6842

Rydberg, B., Qiu, Z.-H., Dosanjh, M.K. & Singer, B. (1992) Partial purification of a human DNA glycosylase acting on the cyclic adduct, 1,N^6-ethenodeoxyadenosine. *Cancer Res.* 52, 1377–1379

Sági, J., Chenna, A., Hang, B. & Singer, B. (1988) A single cyclic p-benzoquinone adduct can destabilize a DNA oligonucleotide duplex. *Chem. Res. Toxicol.* 11, 329–334

Saparbaev, M., Kleibl, K. & Laval, J. (1995) *Escherichia coli, Saccharomyces cerevisiae,* rat and human 3-methyladenine DNA glycosylases repair 1,N^6-ethenoadenine when present in DNA. *Nucl. Acids Res.* 23, 3750–3755

Saparbaev, M. & Laval, J. (1998) 3,N^4-Ethenocytosine, a highly mutagenic adduct, is a primary substrate for *Escherichia coli* double-stranded uracil-DNA glycosylase and human mismatch-specific thymine-DNA glycosylase. *Proc. Natl Acad. Sci. USA* 95, 8508–8513

Seki, S., Hatsushika, M., Watanabe, S., Akiyama, K., Nagao, K. & Tsutsui, K. (1992) cDNA cloning, sequencing, expression and possible domain structure of human APEX nuclease homologous to *Escherichia coli* exonuclease III. *Biochim. Biophys. Acta* 1131, 287–299

Singer, B. & Grunberger, D. (1983) *Molecular Biology of Mutagens and Carcinogens*. New York, Plenum Press

Singer, B. & Hang, B. (1997) Perspective: What structural features determine repair enzyme specificity and mechanism in chemically modified DNA? *Chem. Res. Toxicol.* 10, 713–732

Singer, B., Spengler, S.J., Chavez, F. & Kusmierek, J.T. (1987) The vinyl chloride-derived nucleoside, N^2,3-ethenoguanosine, is a highly efficient mutagen in transcription. *Carcinogenesis* 8,745–747

Singer, B., Spengler, S.J. & Kuśmierek, J.T. (1988) Reactions of vinyl chloride and its metabolites with bases in nucleic acids and the potential biological consequences. In: *Chemical Carcinogens: Activation Mechanisms, Structural and Electronic Factors, and Reactivity* (Politzer, P.A. & Martin, F., eds). Amsterdam, Elsevier, pp.188–207

Singer, B., Kuśmierek, J.T., Folkman, W., Chavez, F. & Dosanjh, M.K. (1991) Evidence for the mutagenic potential of the vinyl chloride induced adduct, N^2,3-ethenodeoxyguanosine, using a site-directed kinetic assay. *Carcinogenesis* 12, 745–747

Singer, B., Antoccia, A., Basu, A.K., Dosanjh, M.K., Fraenkel-Conrat, H., Gallagher, P.E., Kuśmierek, J.T., Qiu, Z.H. & Rydberg, B. (1992) Both purified human 1,N^6-ethenoadenine-binding protein and purified human 3-methyladenine-DNA glycosylase act on 1,N^6-ethenoadenine and 3-methyladenine. *Proc. Natl Acad. Sci. USA* 89, 9386–9390

Snyder, R., Jowa, L., Witz, G., Kalf, G. & Rushmore, T. (1987) Formation of reactive metabolites from benzene. *Arch. Toxicol.* 60, 61–64

Swenberg, J.A., Fedtke, N., Ciroussel, F., Barbin, A. & Bartsch, H. (1992) Etheno adducts formed in DNA of vinyl chloride-exposed rats are highly persistent in liver. *Carcinogenesis* 13, 727–729

Xanthoudakis, S., Miao, G., Wang, F., Pan, Y.C. & Curran, T. (1992) Redox activation of Fos-Jun DNA binding activity is mediated by a DNA repair enzyme. *EMBO J.* 11, 3323–3335

Exocyclic DNA Adducts in Mutagenesis and Carcinogenesis
B. Singer & H. Bartsch
IARC Scientific Publications No. 150
International Agency for Research on Cancer, Lyon, 1999

Enzymology of the repair of etheno adducts in mammalian cells and in *Escherichia coli*

M. Saparbaev & J. Laval

Exocyclic adducts are generated in cellular DNA by reaction with epoxides that are formed metabolically from various industrial pollutants and by reaction with activated aldehydes that arise during membrane lipid peroxidation. The etheno (ε) derivatives of purine and pyrimidine bases, e.g. 3,*N*⁴-ethenocytosine, 1,*N*⁶-ethenoadenine, *N*²,3-ethenoguanine and 1,*N*²-ethenoguanine, are probably involved in carcinogenesis because they are highly mutagenic and genotoxic. Therefore, the repair processes that eliminate exocyclic adducts from DNA should play a crucial role in maintaining the stability of the genetic information.

The DNA glycosylases implicated in the repair of etheno adducts have been identified. Human and *Escherichia coli* 3-methyladenine-DNA-glycosylases excise 1,*N*⁶-ethenoadenine residues. We have identified two homologous proteins present in human cells and *E. coli* that remove 3,*N*⁴-ethenocytosine residues by DNA glycosylase activity. The human enzyme is an activity of the mismatch-specific thymine-DNA glycosylase, while the bacterial enzyme is an activity of the double-stranded uracil-DNA glycosylase, i.e., the homologue of the human enzyme. The fact that 1,*N*⁶-ethenoadenine and 3,*N*⁴-ethenocytosine are recognized and efficiently excised by DNA glycosylases *in vitro* suggests that these enzymes may be responsible for the repair of these mutagenic lesions *in vivo* and may contribute importantly to genetic stability.

Introduction

Some compounds that are widespread in the environment, such as ethyl carbamate and vinyl chloride, a chemical carcinogen used in plastic manufacture (Green & Hathway, 1978), are genotoxic (Miller & Miller, 1983; Guengerich & Kim, 1991). They are oxidized in mammalian cells by cytochrome P450 to reactive electrophilic species, which react with RNA and DNA bases to form etheno (ε)-bridged adducts (Dahl *et al.*, 1978; Leithäuser *et al.*, 1990; Park *et al.*, 1990). Etheno adducts such as 1,N^6-ethenoadenine (εA), 3,N^4-ethenocytosine (εC), N^2,3-ethenoguanine (N^2,3-εG) and 1,N^2-ethenoguanine (1,N^2-εG) are produced by reaction of DNA with two oxidized metabolites of vinyl chloride, 2-chloroacetaldehyde and 2-chloroethylene oxide (Bolt, 1988; Guengerich *et al.*, 1994). Moreover, the generation of exocyclic DNA adducts by products of membrane lipid peroxidation has been demonstrated (El Ghissassi *et al.*, 1995; Chung *et al.*, 1996). The level of εC residues present in human liver has been found to be 2.8 ± 0.9 per 10^7 bases (Marnett & Burcham, 1993).

The deleterious effects of etheno bases are due to the mutagenic potential of these cyclic DNA lesions and are summarized in Table 1. In *Escherichia coli* and simian kidney cells, εC produces mainly εC•G→A•T transversions and εC•G→T•A transitions (Basu *et al.*, 1993; Moriya *et al.*, 1994). In a single-stranded shuttle vector containing a single εC residue, the targeted mutation frequency yield was 2% in *E. coli*, 32% in SOS-induced *E. coli* cells and 81% in simian kidney cells (Moriya *et al.*, 1994). εA residues are highly mutagenic in mammalian cells, where they cause mainly εA•T→G•C transitions (Pandya & Moriya, 1996), but are weakly mutagenic in *E. coli*. In bacterial systems, N^2,3-εG has miscoding properties, producing N^2,3-εG•C→A•T transitions (Cheng *et al.*, 1991), and 1,N^2-εG causes G•C→A•T transitions (Langouët *et al.*, 1997).

Table 1. Mutagenic properties of etheno adducts *in vivo*

Etheno adduct	Mutagenic properties				Reference
	Mammalian cells		*E. coli*		
	Targeted event	Frequency (%)	Targeted event	Frequency (%)	
εA	εA → G,T,C	70	εA → D,T,C	0.18–0.53	Pandya & Moriya (1996)
εC	εC → A,T	81	εC → A,T	2–32	Moriya *et al.* (1994)
N^2,3-εG	ND	ND	N^2,3-εG → A	13	Cheng *et al.* (1991)
1,N^2-εG	ND	ND	1,N^2-εG → A,T,C	16	Langouët *et al.* (1997)

The repair of etheno adducts has been investigated *in vitro*, and the repair of εA and N^2,3-εG adducts in DNA by DNA glycosylases present in crude extracts of mammalian cells has been described (Oesch *et al.*, 1986; Singer *et al.*, 1992). Purified *E. coli* 3-methyladenine (3-MeA)-DNA glycosylase (AlkA protein) releases N^2,3-εG when present in DNA (Matijasevic *et al.*, 1992). The excision of εA by pure 3-MeA-DNA glycosylases of various origins has shown that the human enzyme is by far the most efficient (Saparbaev *et al.*, 1995). The enzyme that repairs εC was identified and partially purified from human cells and shown to be different from the alkyl-*N*-purine glycosylase (ANPG) protein (Hang *et al.*, 1996, 1997), but it was not characterized at the molecular level. We have shown that two homologous proteins present in *E. coli* and human cells remove εC residues by DNA glycosylase activity (Saparbaev & Laval, 1998). The human enzyme was identified as the mismatch-specific thymine-DNA glycosylase (hTDG) (Neddermann *et al.*, 1996), whereas the bacterial enzyme is the double-stranded uracil-DNA glycosylase (dsUDG), the homologue of hTDG (Gallinari & Jiriçny, 1996). The dsUDG protein also repairs T in G/T mismatches (Saparbaev & Laval, 1998), a substrate specificity that it shares with hTDG.

In the study reported here, the enzymic properties of DNA glycosylases that repair etheno adducts were further characterized *in vitro*. The fact that etheno adducts are recognized and excised with good efficiency by various DNA-glycosylases *in vitro* suggests that these enzymes may be responsible for repair of these mutagenic lesions *in vivo* and contribute importantly to genetic stability.

Materials and methods

Materials

3,N^4-Ethenocytosine and 3,N^4-ethenocytidine were purchased from Sigma (USA). The oligonucleotides containing a single εA residue at position Y were purchased from Eurogentec (Angers, France):

εA25 5′-CCGCT**Y**GCGGGTACCGAGCTCGAAT-3′

NC42 5′- AGAGATTAGAGAGAGATTAGA**Y**GA-GATTAGAGAGAGATTAGA-3′

CV42 5′-AGAGATTTTTAGAGATTTTT**Y**GA-GATTTTTAGAGATTTTT-3′

These sequences are referred to below as εA25, NC42 and CV42, respectively. The sequences NC42 (non-curved DNA) and CV42 (curved DNA) were previously used to study the properties of curved DNA (Bonnefoy & Rouvière-Yaniv, 1991). All of the complementary oligonucleotides were synthesized by Dr E. Lescot in our laboratory. The 34-mer oligonucleotide: 5′-AAATACATCGTCAC-CTGGG**X**CATGTTGCAGATCC-3′, in which X = εC, U or T at position 20, was purchased from Genset, Paris, and these sequences are referred to as (εC-34), (U-34) and (T-34a). This sequence was previously used to identify εA and the hypoxanthine-DNA glycosylases (Saparbaev *et al.*, 1995; Saparbaev & Laval, 1998). It should be noted that the G/T mismatch in that sequence is within the 5′-GpG-3′/3′-CpT-5′ context (Sibghat-Ullah *et al.*, 1996).

The oligonucleotides were labelled as described previously (Saparbaev & Laval, 1998). The duplex oligonucleotides, were obtained by annealing (εC-34), (U-34) or (T-34) with the complementary oligonucleotides containing dA, dG, dC or T opposite X, as described previously (Saparbaev & Laval, 1998), and are referred to as εC-34/G, εC-34/A, εC-34/T, εC-34/C, U-34/G, U-34/A, U-34/T, U-34/C and T-34/G, T-34/A, T-34/T, T-34/C. In addition, for determination of G/T glycosylase activity, we used 30-mer oligonucleotides containing thymine in various sequence contexts: (i) a 30-mer in which T at position 11 (T-30) is within the 5′-CpG-3′/3′-GpT-5′ context: 5′-(TGACTG-CATATGCATGTAGACGATGTGCAT); (ii) a 34-mer oligonucleotide containing thymine at position 19 (T-34b) within the 5′-TpG-3′/3′-ApT-5′ context: 5′-d(CGGTATCCACCAGGTCATTAATAAC-GATGAAGCC) and (iii) a 40-mer oligonucleotide containing thymine at position 20 (T-40) within the 5′-GpG-3′/3′-CpT-5′ context: 5′-d(AATTGC-TATCTAGCTCCGCTCGCTGGTACCCATCTCAT-GA) annealed to a complementary oligonucleotide containing dG opposite **T**. These duplex oligonucleotides are referred to as T-30/G, T-34b/G and T-40/G, respectively.

Xth protein, terminal transferase and molecular biological products were purchased from Boehringer (Mannheim). T4 polynucleotide kinase was purchased from New England BioLabs. The *E. coli* formamido pyrimidine DNA glycosylase (FPG) (Boiteux *et al.*, 1990), Tag and AlkA proteins (O'Connor & Laval, 1990; Tudek *et al.*, 1998) were purified as described elsewhere. The ANPG40 (O'Connor & Laval, 1991), ANPG70 (O'Connor, 1994), the rat alkylpurine DNA glycosylase (APDG60) (O'Connor & Laval, 1990) and methyladenine DNA glycosylase (MAG) (Saparbaev *et al.*, 1995) proteins were purified to apparent homogeneity from extracts of *E. coli* BH290 (*tag, alkA*) harbouring plasmids containing ANPG40, ANPG70, APDG60 and MAG cDNA, respectively. Human hTDG and *E. coli* εC-DNA glycosylases were purified as described previously (Saparbaev & Laval, 1998). The activity of the various proteins was tested with their classical substrates and was checked just before use.

DNA glycosylase assays

The release of etheno bases was measured by one of two approaches: (i) by the cleavage of the oligonucleotide containing a single adduct at a defined position and analysis of the products of the reaction in sequencing gels, as described elsewhere (Saparbaev & Laval, 1998) or (ii) by high-performance liquid chromatography (HPLC) in a model System Gold (Beckman). The products of the reaction were detected by monitoring ultraviolet absorption at 270 and 280 nm. The species were separated on a C_{18} μ Bondapak column (Waters) eluted isocratically at 1 ml/min with 50 mmol/L $NH_4H_2PO_4$ (pH 4.5) containing 10% methanol (v/v).

Results

Kinetics of excision of ethenoadenine and other modified bases by human, yeast and E. coli 3-methyladenine-DNA glycosylases

Since εA and hypoxanthine residues are excised by human, yeast and *E. coli* 3-MeA-DNA glycosylases, we compared the enzymic activities of these proteins towards different base lesions and established the kinetics for excision of the modified bases. The kinetics constants are listed in Table 2. For the human (ANPG40), yeast (MAG) and *E. coli* (AlkA) proteins, DNA containing εA, hypoxanthine, 3-MeA or 7-meGua residues was used as the substrate. The best substrate for all three enzymes was 3-MeA. Comparison of the kinetics constants for the excision of εA and hypoxanthine showed dramatic differences in the efficiency of the three enzymes, the human one being by far the most efficient and *E. coli* AlkA and yeast MAG proteins being extremely inefficient. This results may imply that εA and hypoxanthine are not physiologically important DNA lesions for rapidly propagating monocellular organisms, perhaps due to the low cellular concentration of reactive epoxides in the case of εA and a slow rate of adenine deamination in duplex DNA for hypoxathine.

Comparison of the k_{cat}/K_m values of the human ANPG40 protein for the release of εA and hypoxathine residues suggests that the specificity of the reaction is comparable to that for release of 7-MeGua residues but it is much less efficient than that for release of 3-MeA residues.

Table 2. Kinetics constants for human, yeast and bacterial proteins for excision of 3-methyladenine (3-meAde), 7-methylguanine (7-meGua), 1, N^6-ethenoadenine (εA) and hypoxanthine from DNA

Substrate	DNA glycosylase								
	Human			Yeast			E. coli		
	K_m	k_{cat}	K_m/k_{cat}	K_m	k_{cat}	K_m/k_{cat}	K_m	k_{cat}	K_m/k_{cat}
3-meAde	8[a]	11	1.4				5[a]	0.5	0.1
7-meGua	25	0.35	0.014						
Hypoxanthine	6	0.21	0.035	~ 10^3			420	0.84×10^{-3}	2×10^{-6}
εA	24	~0.05	~0.002	200			800	~ 10^{-3}	~10^{-6}

The units for the constants are: K_m, nmol/L; k_{cat}, min^{-1}; k_{cat} / K_m, $(nmol/L \times min)^{-1}$

[a] The apparent K_m of the 3-meAde-DNA glycosylases were determined by incubating the respective protein with increasing concentrations of appropriate substrate. The data were treated as described by Lineweaver and Burk.

E. coli double-stranded uracil-DNA glycosylase protein acts on DNA containing 3,N⁴-ethenocytosine residues like a DNA glycosylase

The dsUDG protein recognizes εC residues when present in a duplex oligonucleotide. The products excised from εC-34/G by the repair enzymes were characterized by HPLC as described in Materials and methods. Figure 1A shows the elution profile of authentic samples of εC and ethenocytidine which eluted at 11.9 and 14.5 min, respectively. HPLC under the same conditions of the products released from the εC-34/G duplex oligonucleotide after incubation with heat-inactivated dsUDG protein showed two peaks eluting at 3.1 and 5.5 min, and no material eluted at 11.9 min (Figure 1B). When the oligonucleotide was incubated with native dsUDG protein, a single peak eluting at 11.9 min was observed (Figure 1C). Furthermore, this product co-chromatographed with an authentic sample of εC (Figure 1D). No material was detected at the position of the nucleoside (Figure 1C). These results show that the εC residues are excised from the oligonucleotide by dsUDG as a free base.

The products of the reaction were further characterized by polyacrylamide gel electrophoresis. AP site-specific endonucleases do not incise the εC-34/G duplex oligonucleotide. After treatment of εC-34/G with the dsUDG and hTDG proteins, no cleavage was detected, but subsequent treatment with enzymes that incise DNA at AP sites generated a band migrating at the position of the 19-mer (data not shown). These results indicate that the bacterial and human enzymes generate AP sites in DNA containing εC residues and therefore act like DNA glycosylases.

Kinetics of excision of 3,N⁴-ethenocytosine (εC), uracil and thymine from duplex oligonucleotides containing either εC/G, U/G or G/T mispairs by double-stranded uracil-DNA glycosylase and human thymine-DNA glycosylase

Since εC residues are excised in *E. coli* by the dsUDG protein and in human cells by the hTDG protein, in order to evaluate the relative substrate specificity of the two enzymes for their various substrates, the kinetics of the excision of εC, uracil and thymine residues opposite G in a double strand were measured (Saparbaev & Laval, 1998). By far the best substrate for the *E. coli* dsUDG enzyme was εC as compared with uracil, which had a 50-fold lower specificity constant (Table 2). Table 3 shows the kinetics constants measured for the hTDG protein on the same set of oligonucleotides. Comparison of the specificity constants for the enzyme with εC/G or U/G mismatches as substrate shows that uracil and εC are equally good substrates for the human enzyme and better substrates than the T/G mismatch. The latter is believed to be the physiological substrate of this enzyme.

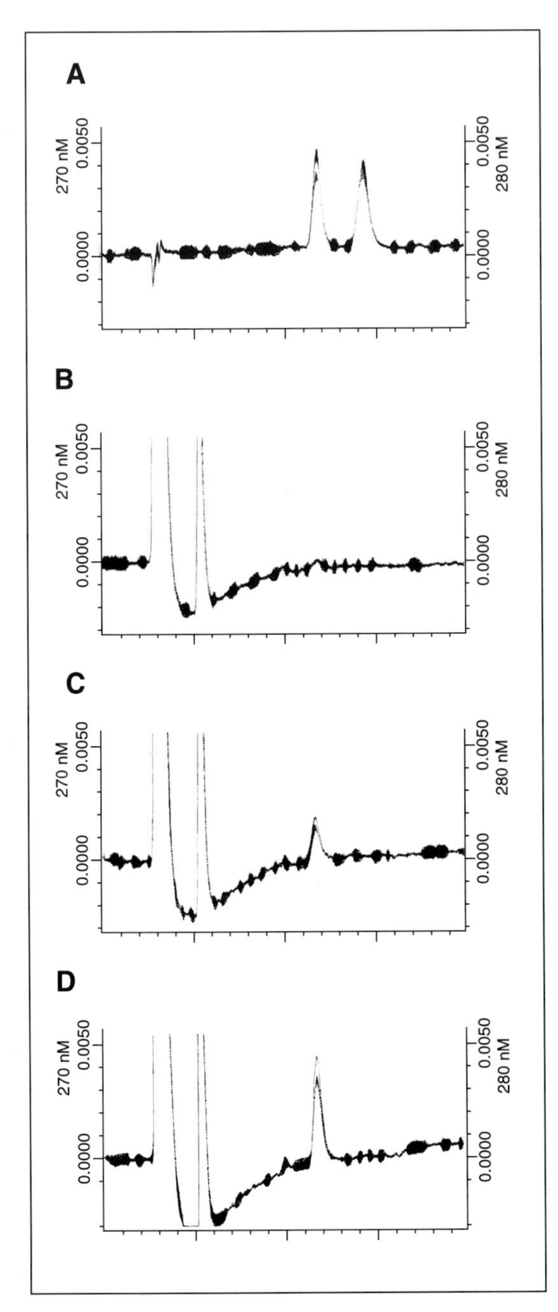

Figure 1. High-performance liquid chromatography (HPLC) of the products released by double-stranded uracil-DNA glycosylase (dsUDG) protein acting on duplex oligonucleotide containing 3,N^4-ethenocytosine (εC) residue

(A) Elution profile of ethenocytosine and ethenocytidine dissolved in water, with retention times of 11.9 and 14.5 min, respectively; (B) duplex oligonucleotide εC-34/G (1000 pmol) was incubated with 4.8 µg of previously heat-denatured (90 °C, 5 min) dsUDG enzyme (total volume, 50 µl) for 1 h at 37 °C. A 20-µl aliquot of the reaction mixture was analysed directly by HPLC as described in Materials and methods. Two peaks eluted at 3.1 and 5.5 min, and no material was detected at 11.9 or 14.5 min. (C) Same experiment as in B, but with native dsUDG enzyme. In addition to the two peaks observed in B, eluting at 3.1 and 5.5 min, there is a single peak eluting at 11.9 min at the position expected for ethenocytosine. (D) Same experiment as in C, but before HPLC analysis an aliquot of incubated reaction mixture was supplemented with 130 pmol of pure ethenocytosine residue. A single symmetrical peak eluted at 11.9 min.

Effect of curvature of DNA on excision of 1,N⁶-ethenoadenine by AlkA, MAG and ANPG40 proteins

The curvature of the DNA duplex has proved to be an important feature in its interactions with many proteins. Bent DNA has been associated with promoters, origins of replication and matrix-associated regions (Caddle *et al.*, 1990; Hagerman, 1990; Von Kries *et al.*, 1990). Moreover, sections of bent DNA have been shown to be preferential substrates for topoisomerases and DNA nicking enzymes (Linial & Shlomai, 1988; Howard *et al.*, 1991). These observations raise the possibility that curvature in the DNA substrate could modulate the efficiency of DNA glycosylases in excising damaged bases. Therefore, the efficiency of excision of εA residues by 3-MeA-DNA glycosylases of various origins was investigated with duplex oligonucleotides NC42/T (noncurved) and CV42/T (curved) as substrate. The results presented in Table 4 show that the rates of excision of εA residues from curved and noncurved DNA by AlkA protein were identical; with ANPG40, the rate was slightly inhibited in curved DNA; whereas with the MAG protein, the rate from curved DNA was twofold higher than from noncurved DNA.

Effect of neighbouring bases on G/T mismatch DNA-glycosylase activity of E. coli double-stranded uracil-DNA glycosylase and human thymine-DNA glycosylase proteins

As the bacterial dsUDG protein excises thymine from an oligonucleotide containing a G/T mismatch (Barrett *et al.*, 1998; Saparbaev & Laval,

Table 3. Kinetic constants of the *E. coli* double-stranded uracil-DNA glycosylase (dsUDG) and human thymine-DNA glycosylase (hTDG) proteins for the excision of 3,N^4-ethenocytosine (εC), uracil and thymine (opposite guanine)

Substrate[a]	Protein					
	hTDG			dsUDG		
	K_m	k_{cat}	K_m/k_{cat}	K_m	k_{cat}	K_m/k_{cat}
εC	24	9 x 10^{-3}	0.4 x 10^{-3}	2.5[b]	1.0	0.4
Uracil	12	21 x 10^{-3}	1.7 x 10^{-3}	23	0.2	0.8 x 10^{-2}
Thymine	13	0.9 x 10^{-3}	0.07 x 10^{-3}	26	0.4 x 10^{-3}	0.2 x 10^{-4}

[a] Substrate concentration, 2.5–100 nmol/L (Saparbaev & Laval, 1998)
[b] Substrate concentration, 0.1–6 nmol/L (Saparbaev & Laval, 1998)

Table 4. Influence of noncurved and curved DNA conformation on initial velocity of excision of 1,N^6-ethenoadenine (εA) by human (ANPG40), yeast (MAG) and *E. coli* (AlkA) proteins

Origin of 3-MeAde DNA glycosylase (1 unit[a])	εA released (fmol/min)		Relative efficiency
	NC42/T oligonucleotide	CV42/T oligonucleotide	
ANPG40 protein	214	169	0.79
MAG protein	8.8	17.0	1.93
AlkA protein	0.12	0.12	1.00

[a] Activity of 3-MeAde-DNA glycosylases measured with ^{3}H-dimethyl sulfate-DNA substrate

1998), the human and *E. coli* enzymes have common substrate specificity. Sibghat-Ullah *et al.* (1996) demonstrated that the thymine-DNA glycosylase activity of hTDG protein strongly depends upon the neighbouring bases. In order to study the substrate specificity of *E. coli* dsUDG enzyme, thymine-DNA glycosylase activity was measured with duplex oligonucleotides containing mismatched thymine in various sequences. Oligonucleotides T-30, T-34a, T-34b and T-40 were annealed to four complementary oligonucleotides containing dA, dG, dC or T opposite thymine. The results (Table 5) show that the hTDG protein preferentially excises thymine opposite G rather than opposite C and T. The efficiency of excision depends on the base 5′ to the mismatched guanine, in the following order: CpG > GpG > TpG. These results are in good agreement with those obtained previously (Sibghat-Ullah *et al.*, 1996; Waters & Swann, 1998). The removal of thymine by dsUDG protein also depended on the base 5′ to the mismatched guanine, in the following order: GpG > CpG > TpG. This order is different from that for the specificity of the hTDG protein.

Base-pair specificity of the ethenocytosine-DNA glycosylases

dsUDG and the hTDG recognized εC only when present in a double-stranded oligonucleotide, and no detectable excision of εC was observed when the lesion was present in a single-stranded oligonucleotide. The base-pair specificity of both enzymes was measured with duplex oligonucleotides containing mismatches generated by each of the four bases opposite εC or uracil. In each case, the initial velocity of the excision of

Table 5. Influence of the base opposite the dTMP residue and sequence context on the initial velocity of excision of thymine by human thymine-DNA glycosylase (TDG) and *E. coli* double-stranded uracil-DNA glycosylase (dsUDG) proteins

Thymine-containing oligonucleotide	Opposite base	Relative efficiency (%)	
		E. coli dsUDG protein	Human TDG protein
T-30	G	27[a] (27[b])	100[a] (100[b])
5′-CpN-3′/3′-GpT-5′	A	0	0
context	C	100	35
	T	10	12
T-34a 5′-GpG-3′/3′-CpT-5′ context	G	100[b]	56
T-34b 5′-TpG-3′/3′-ApT-5′ context	G	0	4
T-40	G	57[a] (53[b])	100[a] (47[b])
5′-GpN-3′/3′-CpT-5′	A	0	0
context	C	100	2
	T	26	0

[a] Relative efficiency as a function of the opposite base

[b] Relative efficiency as a function of the sequence context

εC or uracil was measured (Saparbaev & Laval, 1998).

The excision of εC by dsUDG or hTDG did not show any strict preference, although the εC/G mismatch was the best substrate. In contrast, the repair of uracil residues showed marked preferences with regard to the opposite base. In the case of the *E. coli* protein, uracil was excised from U/G, U/T and U/C mismatches with similar efficiency, whereas its excision from U/A was negligible. The hTDG protein strongly preferred U/G as a substrate; repair of uracil in U/C was negligible, and it was not excised from U/T or U/A. It is striking that εC residues are excised from all four mismatches with comparable efficiency by the two proteins.

Effect of nonspecific DNA on 3,N⁴-ethenocytosine and uracil-DNA glycosylase activity of E. coli double-stranded uracil-DNA glycosylase protein

During the course of investigations on the specific and nonspecific inhibition of the glycosylase activity of dsUDG protein by nucleic acids, we observed that uracil-DNA glycosylase activity was increased when the reaction mixture was supplemented with the 34-mer duplex oligonucleotide containing the regular G/C base pair. As shown in Figure 2, about twice as much induction was observed in the presence of an 8–60-fold molar excess of G/C duplex as compared with the concentration of G/U substrate. In parallel experiments, there was no effect when εC-DNA glycosylase activity was tested in the presence of an excess of nonspecific G/C duplex oligonucleotide (data not shown).

Discussion

Since etheno bases lead to misincorporation after replication or transcription, they have been proposed as critical candidates in the etiology of human cancer. The molecular structures of the various mismatches that involve εC and εA are being solved (Kouchakdjian *et al.*, 1991; de los

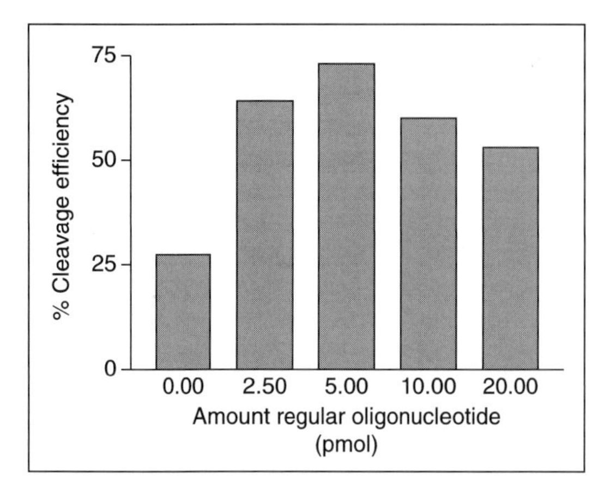

Figure 2. Effect of nonspecific DNA on 3,N^4-ethenocytosine and uracil-DNA glycosylase activity of *E. coli* double-stranded uracil-DNA glycosylase (dsUDG) protein

Increasing amounts of regular duplex oligonucleotide were added to reaction mixture containing 0.3 pmol of ^{32}P-labelled U-34/G duplex oligonucleotide and 25 ng of pure dsUDG protein and incubated for 30 min at 37 °C. The products of the reaction were analysed and quantified as described in Materials and methods.

Santos *et al.*, 1991; Korobka *et al.*, 1996; Cullinan *et al.*, 1996, 1997). It has been shown that etheno adducts in DNA adapt B-helix conformation, which is stabilized by formation of hydrogen bonds. This explains the highly mutagenic properties of etheno bases *in vivo*. Etheno adducts could be formed in DNA either by exogenous sources (vinyl chloride, ethyl carbamate) or by products of lipid peroxidation such as *trans*-4-hydroxy-2-nonenal generated during cellular metabolism (El Ghissassi *et al.*, 1995; Chung *et al.*, 1996).

Since etheno bases are known to be promutagenic and genotoxic (Cheng *et al.*, 1991; Basu *et al.*, 1993; Moriya *et al.*, 1994; Pandya & Moriya, 1996; Langouët *et al.*, 1997), they must be removed from genomic DNA. In *E. coli* and in *Saccharomyces cerevisiae*, the enzymic activity that excises εA has been identified as belonging to 3-MeAde-DNA glycosylase, the AlkA and Mag proteins, respectively (Saparbaev *et al.*, 1995). Enzymic activity in the excision of εC and εA has been identified in crude extracts of mammalian cells (Oesch *et al.*, 1986). The human protein that binds to and excises εA has been partially purified (Singer *et al.*, 1992) and identified as the ANPG protein (Singer *et al.*, 1992; Saparbaev *et al.*, 1995). It has been shown with partially purified proteins from HeLa cells and *anpg* knockout mice that the enzymes that excise εA and εC are different (Hang *et al.*, 1996, 1997). The kinetics constants for excision of εA residues from DNA duplex oligonucleotide listed in Table 2 show that 3-MeAde residues are the best substrates for the 3-MeAde-DNA glycosylases of various origins that we tested. These results are at variance with those of Dosanjh *et al.* (1994). In the present report, comparison of the rate of excision of εA from curved and noncurved DNA duplex oligonucleotides by 3-MeAde-DNA glycosylases of different origins showed that the bacterial AlkA protein excises εA from both curved and noncurved oligonucleotides with the same efficiency, whereas excision of εA from curved DNA by the human ANPG40 protein was slightly inhibited. In contrast, the yeast MAG protein gave a twofold increase in the rate of excision of εA from curved DNA. The different effects of DNA curvature on excision of εA by bacterial, yeast and human enzymes probably reflects differences in the fine tertiary structure of the enzymes.

In a search for proteins involved in the repair of εC residues, our strategy was, first, to identify DNA glycosylase activity in *E. coli* cell extracts that could recognize εC residues. This requirement was fulfilled, the protein was purified to homogeneity (Saparbaev & Laval, 1998), and it was identified as the dsUDG protein described by Gallinari & Jiriçny (1996). It was established that the hTDG protein (Neddermann *et al.*, 1996), the human homologue of dsUDG, also excises εC residues (Saparbaev & Laval, 1998). In order to compare the substrate specificity of these two enzymes, the kinetics were measured for both enzymes with three duplex oligonucleotides containing different mismatches. The dsUDG protein acted on DNA substrates in the order εC/G >> U/G >> G/T, but by far the preferred substrate for dsUDG was εC (Table 3). The enzyme has an extremely efficient kinetic constant for εC/G (k_{cat}/K_m = 0.4 per min/nmol per L), which is 52-fold higher than for the U/G mismatch (k_{cat}/K_m = 0.77 x 10^{-2} per min/nmol per L).

Moreover, the human protein hTDG, previously characterized as an enzyme that repairs T in G/T mismatches, also very efficiently repairs εC residues in DNA. The order of preference of hTDG for different DNA substrates was U/G > εC/G > T/G. The difference in the kinetics for the excision of εC residues is smaller for the human enzyme than for that of *E. coli*, and the situation is different from that for excision of εA, the mammalian enzymes (human and rat) being much more efficient than the prokaryotic enzymes (Table 2). Comparison of the k_{cat}/K_m values for the hTDG protein for each substrate show that uracil and εC are the preferred substrates. Therefore, the human and *E. coli* enzymes show the same substrate specificity.

The lower k_{cat}/K_m value measured for hTDG on an εC substrate as compared with the constant measured for dsUDG suggests that the activity of the human enzyme is stimulated *in vivo* by other factors, such as those involved in the base excision repair pathway. In order to ascertain the role of hTDG *in vivo,* further investigations will be needed with genetic approaches. The highly promutagenic properties of εC in mammalian cells (Moriya *et al.*, 1994) and its repair by hTDG suggest the possibility that a human genetic disease associated with deficient hTDG activity will be identified.

Excision of T in a G/T mismatch by dsUDG has been reported (Barrett *et al.*, 1998; Saparbaev & Laval, 1998). The substrate specificity of this bacterial enzyme was further investigated with duplex oligonucleotides containing mismatched thymine in various sequence contexts. *E. coli* dsUDG protein preferentially excised thymine when it was opposite C, and the order of preference was T/C > T/G > T/T. This pattern is different from that of the human homologue which preferentially removes T opposite G, and the order of the reaction rate is G/T > T/C > T/T. Furthermore, the rate of thymine excision from G/T mismatches by *E. coli* dsUDG protein depends on the neighbouring bases in the order GpG/T > CpG/T > TpG/T. To counteract the mutagenic effects of 5-methylcytosine deamination, thus generating a G/T mismatch, *E. coli* evolved a Vsr DNA mismatch endonuclease which introduces a site-specific nick at a G/T mismatch (Hennecke *et al.*, 1991). In addition, the extremely low constant measured (k_{cat}/K_m = 1.7 x 10^{-5}) suggests that the G/T mismatch activity of the dsUDG protein has no real biological significance. Nevertheless, this new substrate of the bacterial enzyme could be of interest for structural investigations of the molecular mechanisms involved in the catalytic action of dsUDG.

Investigation of the catalytic properties of the dsUDG protein towards a G/U substrate showed that the uracil-DNA glycosylase activity of the dsUDG protein is stimulated by addition to the reaction mixture of a duplex oligonucleotide with regular base pairs containing G/C instead of G/U. However, we observed no such stimulation of the εC-DNA glycosylase activity of the dsUDG protein after addition of nonspecific DNA. These results suggest that the mechanism of action of dsUDG protein could be different on uracil and εC residues.

Recently, the crystal structures of the *E. coli* dsUDG protein and of a DNA–dsUDG protein complex were solved (Barrett *et al.*, 1998). The dsUDG structure contains five-stranded β sheets at the centre, flanked by α helices, thus forming a β–α–β topology. The dsUDG protein contains a narrow pocket which penetrates back into the core of the enzyme, and the authors proposed that this active-site pocket could serve to accommodate a flipped-out pyrimidine base, such as uracil or thymine. In order to fit the data presented above with the crystal structure of dsUDG protein, it must first be assumed that the active-site pocket is quite large in order to accommodate bulky lesions such as εC residues. In contrast to uracil and thymine residues, there may be specific interactions between εC residues and some amino-acid side-chains inside the enzyme pocket that would account for a discriminatory effect and specific binding. Second, the repair of thymine and uracil residues by dsUDG and hTDG proteins strongly depends on the nature of the opposite base (Neddermann & Jiriçny, 1993; Saparbaev & Laval, 1998). In contrast, repair of the εC residue is not markedly influenced by the opposite base (Saparbaev & Laval, 1998). Furthermore, the efficiency of excision of εA residues by AlkA, MAG and ANPG proteins is also not dependent on the nature of the opposite base (Saparbaev *et al.*, 1995). These facts raise the possibility that the processes involved in recognition of modified bases are different: the εC residue may be recognized *per se*, whereas the uracil and thymine residues are recognized from the structure of the mismatch. In fact, U/A and the T/A base pairs are not repaired (Neddermann & Jiriçny, 1993; Saparbaev & Laval, 1998). It is possible that the dsUDG and the hTDG proteins directly recognize the εC residue by flipping it out into a specific pocket within their active site, by a mechanism similar to that described for the uracil DNA glycosylase protein (Savva *et al.*, 1995).

The crystal structure of the dsUDG–DNA complex shows that the protein inserts into the DNA helix an intercalating 'wedge' formed by Gly-143, Leu-144, Arg-146 and the associated water molecule. This 'wedge' penetrates the base stack of the DNA from the minor groove (Barrett *et al.*, 1998). Insertion of the wedge into the duplex induces local unwinding of the DNA backbone and the formation of strong hydrogen bonds between the widowed guanine and Gly-143.

The following model is proposed to accommodate the results presented above and the known structure of the dsUDG protein in order to account for the recognition of uracil and thymine residues by dsUDG protein. The specific binding of the dsUDG protein to G/U or G/T mismatch causes local denaturation of the DNA duplex at the site of the mismatches by intercalating the amino-acid chain into the base stack of the DNA. This open DNA complex could be stabilized by specific hydrogen-bond interactions between the base in the complementary strand and the intercalating 'wedge'. Local unwinding of the duplex DNA at the site of the lesion could bring the C-1′ carbon of the pyrimidine nucleotide in close proximity to the water molecule and a yet unidentified side-chain carboxyl of a Glu or Asp. That carboxyl group could then act as a general base and abstract a proton from the water, thus generating a nucleophile to attack the C-1′ carbon. The main difference between this model and the flip-out model proposed by Barrett *et al.* (1998) is that the enzyme recognizes the modified base inside an open DNA duplex rather than flipping it out of the helix. By intercalating the PSGLSR motif into the DNA helix as a 'hook', the dsUDG protein could achieve two goals: specific recognition of the mismatch and hydrolysis of the *N*-glycosidic bond.

After this manuscript was completed, Hang *et al.* (1998) suggested that the human εC-DNA glycosylase was the hTDG protein. Those authors purified activity from HeLa cells that excised εC residues, co-eluted with the specific activity of a G/T mismatch and had a molecular mass of 55 kDa (estimated by polyacrylamide gel electrophoresis). However, neither the amino-acid sequence of the protein nor identification of the human gene coding for this activity were presented.

Acknowledgements

We thank Dr Derancourt (UPR 9008 CNRS, Montpellier) for determining the amino-acid sequence of the εCDG/dsUDG protein. This work was supported by grants from the European Commission (ENV4-CT97-0505), the Comité Radioprotection-EDF and the Fondation Franco-Norvegienne. M.S. was the recipient of a grant from the Fondation pour la Recherche Médicale.

References

Barrett, T.E., Savva, R., Panayotou, G., Barlow, T., Brown, T., Jiriçny, J. & Pearl, L.H. (1998) Crystal structure of a G:T/U mismatch-specific DNA glycosylase: Mismatch recognition by complementary-strand interactions. *Cell* 92, 117–129

Basu, A.K., Wood, M.L., Niedernhofer, L.J., Ramos, L.A. & Essigmann, J.M. (1993) Mutagenic and genotoxic effects of three vinyl chloride-induced DNA lesions: 1,N^6-Ethenoadenine, 3,N^4-ethenocytosine, and 4-amino-5-(imidazol-2-yl)imidazole. *Biochemistry* 32, 12793–12801

Boiteux, S., O'Connor, T.R., Lederer, F., Gouyette, A. & Laval, J. (1990) Homogeneous *Escherichia coli* FPG protein. A DNA glycosylase which excises imidazole ring-opened purines and nicks DNA at apurinic/apyrimidinic sites. *J. Biol. Chem.* 265, 3916–3922

Bolt, H.M. (1988) Roles of etheno-DNA adducts in tumorigenicity of olefins. *Crit. Rev. Toxicol.* 18, 299–309

Bonnefoy, E. & Rouvière-Yaniv, J. (1991) HU and IHF, two homologous histone-like proteins of

Escherichia coli, form different protein–DNA complexes with short DNA fragments. *EMBO J.* 10, 687–696

Caddle, M.S., Lussier, R.H. & Heintz, N.H. (1990) Intramolecular DNA triplexes, bent DNA and DNA unwinding elements in the initiation region of an amplified dihydrofolate reductase replicon. *J. Mol. Biol.* 211, 19–33

Cheng, K.C., Preston, B.D., Cahill, D.S., Dosanjh, M.K., Singer, B., & Loeb, L.A. (1991) The vinyl chloride DNA derivative N^2,3-ethenoguanine produces G→A transitions in *Escherichia coli. Proc. Natl Acad. Sci. USA* 88, 9974–9978

Chung, F.-L., Chen, H.-J.C. & Nath, R.G. (1996) Lipid peroxidation as a potential endogenous source for the formation of exocyclic DNA adducts. *Carcinogenesis* 17, 2105–2111

Cullinan, D., Korobka, A., Grollman, P.A., Patel, D.J., Eisenberg, M., & de los Santos, C. (1996) NMR solution structure of an oligodeoxynucleotide duplex containing the exocyclic lesion 3,N^4-etheno-2'-deoxycytidine opposite thymidine: Comparison with the duplex containing deoxyadenosine opposite the adduct. *Biochemistry* 35, 13319–13327.

Cullinan, D., Johnson, F., Grollman, P.A., Eisenberg, M., & de los Santos, C. (1997) Solution structure of a DNA duplex containing the exocyclic lesion 3,N^4-etheno-2'-deoxycytidine opposite 2'-deoxyguanosine. *Biochemistry* 36, 11933–11943

Dahl, G.A., Miller, J.A. & Miller E.C. (1978) Vinyl carbamate as a promutagen and a more carcinogenic analog of ethyl carbamate. *Cancer Res.* 38, 3793–3804

Dosanjh, M.K., Roy, R., Mitra, S., Singer, B. (1994) 1,N^6-Ethenoadenine is preferred over 3-methyladenine as substrate by a cloned human N-methylpurine-DNA glycosylase (3-methyladenine-DNA glycosylase). *Biochemistry* 33, 1624–1628

El Ghissassi, F., Barbin, A., Nair, J., & Bartsch, H. (1995) Formation of 1,N^6-ethenoadenine and 3,N^4-ethenocytosine by lipid peroxidation products and nucleic acid bases. *Chem. Res. Toxicol.* 8, 278–283

Gallinari, P. & Jiriçny, J (1996) A new class of uracil–DNA glycosylases related to human thymine–DNA glycosylase. *Nature* 383, 735–738

Green, T. & Hathway, D.E. (1978) Interactions of vinyl chloride with rat-liver DNA *in vivo. Chem. Biol. Interactions* 22, 211–224

Guengerich, F.P. & Kim, D.-H. (1991) Enzymatic oxidation of ethyl carbamate to vinyl carbamate and its role as an intermediate in the formation of 1,N^6-ethenoadenosine. *Chem. Res. Toxicol.* 4, 413–421

Guengerich, F.P., Min, K.S., Persmark, M., Kim, M.S., Humphreys, W.G. & Cmarik, J.L. (1994) Dihaloalkanes and polyhaloalkenes. In: *DNA Adducts: Identification and Significance* (Hemminki, K., Dipple, A., Shuker, D.E.G., Kadlubar, F.F., Segerbäck, D. & Bartsch, H., eds), IARC Scientific Publications No. 125. Lyon, IARC, pp 57–72

Hagerman, P.J. (1990) Sequence-directed curvature of DNA. *Annu. Rev. Biochem.* 59, 755–781

Hang, B., Chenna, A., Rao, S. & Singer, B. (1996) 1,N^6-Ethenoadenine and 3,N^4-ethenocytosine are excised by separate human DNA glycosylases. *Carcinogenesis* 17, 155–157

Hang, B., Singer, B., Margison, G.P. & Elder, R.H. (1997) Targeted deletion of alkylpurine-DNA-N-glycosylase in mice eliminates repair of 1,N^6-ethenoadenine and hypoxanthine but not of 3,N^4-ethenocytosine or 8-oxoguanine. *Proc. Natl Acad. Sci. USA* 94, 12869–12874

Hang, B., Medina, M., Fraenkel-Conrat, H. & Singer, B. (1998) A 55-kDa protein isolated from human cells shows DNA glycosylase activity toward 3,N^4-ethenocytosine and the G/T mismatch. *Proc. Natl Acad. Sci.USA* 95, 13561–13566

Hennecke, F., Kolmar, H., Brundl, K. & Fritz, H.-J. (1991) The *vsr* gene product of *E. coli* K-12 is a strand- and sequence-specific DNA mismatch endonuclease. *Nature* 353, 776–778

Howard, M.T., Lee, M.P., Hsieh, T.S. & Griffith, J.D. (1991) Drosophila topoisomerase II–DNA interactions are affected by DNA structure. *J. Mol. Biol.* 217, 53–62

Korobka, A., Cullinan, D., Cosman, M., Grollman, P.A., Patel, D.J., Eisenberg, M. & de los Santos, C. (1996) Solution structure of an oligodeoxynucleotide duplex containing the exocyclic lesion 3,N^4-etheno-2′-deoxycytidine opposite 2′-deoxyadenosine, determined by NMR spectroscopy and restrained molecular dynamics. *Biochemistry* 35, 13310–13318

Kouchakdjian, M., Eisenberg, M., Yarema, K., Basu, A., Essigmann, J. & Patel, D.J. (1991) NMR studies of the exocyclic 1,N^6-ethenodeoxyadenosine adduct (epsilon dA) opposite thymidine in a DNA duplex. Nonplanar alignment of epsilon dA(anti) and dT(anti) at the lesion site. *Biochemistry* 30, 1820–1828

Langouët, S., Müller, M. & Guengerich, F.P. (1997) Misincorporation of dNTPs opposite 1,N^2-ethenoguanine and 5,6,7,9-tetrahydro-7-hydroxy-9-oxoimidazo[1,2-a]purine in oligonucleotides by *Escherichia coli* polymerases I exo- and II exo-, T7 polymerase exo-, human immunodeficiency virus-1 reverse transcriptase, and rat polymerase beta. *Biochemistry* 36, 6069–6079

Leithäuser, M.T., Liem, A., Steward, B.C., Miller, E.C. & Miller J.A. (1990) 1,N^6-Ethenoadenosine formation, mutagenicity and murine tumor induction as indicators of the generation of an electrophilic epoxide metabolite of the closely related carcinogens ethyl carbamate (urethane) and vinyl carbamate. *Carcinogenesis* 11, 463–473

Linial, M. & Shlomai, J. (1988) A unique endonuclease from *Crithidia fasciculata* which recognizes a bend in the DNA helix. Specificity of the cleavage reaction. *J. Biol. Chem.* 263, 290–297

Marnett, L.J. & Burcham, P.C. (1993) Endogenous DNA adducts: Potential and paradox. *Chem. Res. Toxicol.* 6, 785–771

Matijasevic, Z., Sekiguchi, M. & Ludlum, D.B. (1992) Release of N^2,3-ethenoguanine from chloroacetaldehyde-treated DNA by *Escherichia coli* 3-methyladenine DNA glycosylase II. *Proc. Natl Acad. Sci. USA* 89, 9331–9334

Miller, J.A. & Miller, E.C. (1983) The metabolic activation and nucleic acid adducts of naturally-occurring carcinogens: Recent results with ethyl carbamate and the spice flavors safrole and estragole. The 1983 Walter Hubert Lecture. *Br. J. Cancer* 48, 1–15

Moriya, M., Zhang, W., Johnson, F. & Grollman, A.P. (1994) Mutagenic potency of exocyclic DNA adducts: Marked differences between *Escherichia coli* and simian kidney cells. *Proc. Natl Acad. Sci. USA* 91, 11899–11903

Neddermann, P. & Jiriçny, J. (1993) The purification of a mismatch-specific thymine-DNA glycosylase from HeLa cells. *J. Biol. Chem.* 268, 21218–21224

Neddermann, P., Gallinari, P., Lettieri, T., Schmid, D., Truong, O., Hsuan, J.J., Wiebauer, K. & Jiriçny, J. (1996) Cloning and expression of human G/T mismatch-specific thymine-DNA glycosylase. *J. Biol. Chem.* 271, 12767–12774

O'Connor, T.R. (1994) Purification and characterization of human 3-methyladenine-DNA glycosylase. *Nucleic Acids Res.* 21, 5561–5569

O'Connor, T.R. & Laval, F. (1990) Isolation and structure of a cDNA expressing a mammalian 3-methyladenine-DNA glycosylase. *EMBO J.* 9, 3337–3342

O'Connor, T. & Laval, J. (1991) Human cDNA expressing a functional DNA glycosylase excising 3-methyladenine and 7-methylguanine. *Biochem. Biophys. Res. Commun.* 176, 1170–1177

Oesch, F., Adler, S., Rettelbach, R. & Doerjer, G. (1986) Repair of etheno DNA adducts by *N*-glycosylases. In: *The Role of Cyclic and Nucleic Acid Adducts in Carcinogenesis and Mutagenesis* (Singer, B. & Bartsch, H., eds), IARC Scientific Publications No. 70. Lyon, IARC, pp. 373–379

Pandya, G.A. & Moriya, M. (1996) 1,N^6-Ethenodeoxyadenosine, a DNA adduct highly mutagenic in mammalian cells. *Biochemistry* 35, 11487–11492

Park, K.K., Surh, Y.J., Stewart, B.C. & Miller, J.A. (1990) Synthesis and properties of vinyl carbamate epoxide, a possible ultimate electrophilic and carcinogenic metabolite of vinyl carbamate and ethyl carbamate. *Biochem. Biophys. Res. Commun.* 169, 1094–1098

de los Santos, C., Kouchakdjian, M., Eisenberg, M., Yarema, K., Basu, A., Essigman, J. & Patel, D.J. (1991) NMR studies of the exocyclic 1,N^6-ethenodeoxyadenosine adduct (epsilon dA) opposite deoxyguanosine in a DNA duplex. Epsilon dA(syn).dG(anti) pairing at the lesion site. *Biochemistry* 30, 1828–1835

Saparbaev, M., & Laval, J. (1998) 3,N^4-Ethenocytosine, a highly mutagenic adduct, is a primary substrate for *Escherichia coli* double-stranded uracil-DNA glycosylase and human mismatch-specific thymine-DNA glycosylase. *Proc. Natl Acad. Sci. USA* 95, 8508–8513

Saparbaev, M., Kleibl, K. & Laval, J. (1995) *Escherichia coli, Saccharomyces cerevisiae,* rat and human 3-methyladenine DNA glycosylases repair 1,N^6-ethenoadenine when present in DNA. *Nucleic Acids Res.* 23, 3750–3755

Savva, R., McAuley-Hecht, K., Brown, T. & Pearl, L. (1995) The structural basis of specific base-excision repair by uracil-DNA glycosylase. *Nature* 373, 487–493

Sibghat-Ullah, Gallinari, P., Xu, Y-Z., Goodman, M.F., Bloom, L.B., Jiriçny, J. & Day, R.S., III (1996) Base analog and neighboring base effects on substrate specificity of recombinant human G:T mismatch-specific thymine DNA-glycosylase. *Biochemistry* 35, 12926–12932

Singer, B., Antoccia, A., Basu, A.K., Dosanjh, M.K., Fraenkel-Conrat, H., Gallagher, P.E., Kusmierek, J.T., Qiu, Z.H. & Rydberg, B. (1992) Both purified human 1,N^6-ethenoadenine-binding protein and purified human 3-methyladenine-DNA glycosylase act on 1,N^6-ethenoadenine and 3-methyladenine. *Proc. Natl Acad. Sci. USA* 89, 9386–9390

Tudek, B., Van Zeeland, A.A., Kusmierek, J.T. & Laval, J. (1998) Activity of *Escherichia coli* DNA-glycosylases on DNA damaged by methylating and ethylating agents and influence of 3-substituted adenine derivatives. *Mutat. Res.* 407, 169–176

Von Kries, J.P., Phi-Van, L., Diekmann, S. & Stratling, W.H. (1990) A non-curved chicken lysozyme 5′ matrix attachment site is 3′ followed by a strongly curved DNA sequence. *Nucleic Acids Res.* 18, 3881–3885

Waters, T.R. & Swann, P.F. (1998) Kinetics of the action of thymine DNA glycosylase. *J. Biol. Chem.*, 273, 20007–20014

Exocyclic DNA Adducts in Mutagenesis and Carcinogenesis
B. Singer & H. Bartsch
IARC Scientific Publications No. 150
International Agency for Research on Cancer, Lyon, 1999

Cellular response to exocyclic DNA adducts

M. Moriya, G.A. Pandya, F. Johnson & A.P. Grollman

The mutagenic potential of three exocyclic DNA adducts was studied in *Escherichia coli* and simian kidney cells by incorporating them into single-stranded DNA. Differences in the mutagenic potency of the adducts were observed between hosts: 1,*N*^{6}-ethenodeoxyadenosine and 3,*N*^{4}-ethenodeoxycytidine were more mutagenic in simian cells, whereas 1,*N*^{2}-(1,3-propan-1,3-diyl)-2′-deoxyguanosine was more mutagenic in *E. coli*. To investigate the cellular response to DNA adducts, a double-stranded DNA vector system was developed. Use of this system showed that 1,*N*^{6}-ethenodeoxyadenosine blocks DNA synthesis strongly, and DNA synthesis past this adduct was highly accurate in *E. coli*. The blockage of DNA synthesis was overcome in an error-free manner by the recombination repair mechanism (daughter-strand gap repair).

Introduction

Exocyclic DNA adducts are produced by endogenous and exogenous agents. Vinyl chloride is a prototype of this reaction: this human carcinogen is metabolized to reactive species which react with DNA to form 1,*N*^{6}-ethenodeoxyadenosine (εdA), 3,*N*^{4}-ethenodeoxycytidine (εdC) and *N*^{2},3-ethenodeoxyguanosine (εdG) (Barbin & Bartsch, 1986). Exocyclic DNA adducts also have been detected in the DNA of animals and human tissues that have not been exposed to exogenous sources (Chaudhary *et al.*, 1994; Nath & Chung, 1994; Nair *et al.*, 1995). Lipid peroxidation products are suspected to be the source of some of these lesions (El Ghissasi *et al.*, 1995; Chung *et al.*, 1996), with α,β-unsaturated aldehydes, such as malondialdehyde and acrolein, acting as intermediates in this process (Marnett, 1994).

Exocyclic DNA adducts are removed from DNA by base excision repair (Dosanjh *et al.*, 1994; Saparbaev *et al.*, 1995; Hang *et al*, 1998; Saparbaev & Laval, 1998) and by nucleotide excision repair (Johnson *et al.*, 1997). Judging by their persistence in genomic DNA, this process is incomplete. Many of these lesions miscode (Simha *et al.*, 1991; Singer *et al.*, 1991; Zhang *et al.*, 1995; Lantouët *et al.*, 1998), and some have been shown to be mutagenic (Cheng *et al.*, 1991; Palejwala *et al.*, 1991; Basu *et al.*, 1993; Moriya *et al.*, 1994; Pandya & Moriya, 1996; Fink *et al.*, 1997; Hashim *et al.*, 1997; Lantouët *et al.*, 1998). In this paper, we describe the mutagenic potential of εdA, εdC and 1,*N*^{2}-(1,3-propan-1,3-diyl)-2′-deoxyguanosine (PdG) (Figure 1) in *Escherichia coli* and mammalian cells, using εdA as the prototype exocyclic adduct. PdG serves as a model for malondialdehyde- and acrolein-derived deoxyguanosine adducts in DNA.

Mutagenicity of exocyclic DNA adducts

Using a single-stranded (ss) shuttle vector system (Moriya, 1993), we evaluated the mutagenic potential of three exocyclic DNA adducts, εdA, εdC and PdG, in *E. coli* and simian kidney cells (Moriya *et al.*, 1994; Pandya & Moriya, 1996). Use

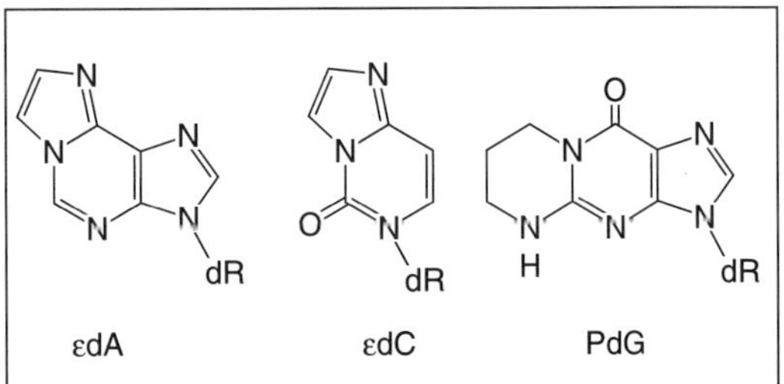

Figure 1. Structures of exocyclic DNA adducts

εdA, 1,*N*^{6}-ethenodeoxyadenosine; εdC, 3,*N*^{4}-ethenodeoxycytidine; PdG, 1,*N*^{2}-(1,3-propan-1,3-diyl)-2′-deoxyguanosine

of ssDNA minimizes DNA repair and permits us to determine the fidelity of *trans*-lesion synthesis in cells. The experimental strategy involves site-specific incorporation of a defined adduct into ssDNA, introduction of modified ssDNA into host cells and analysis of progeny plasmid by differential oligonucleotide hybridization (Pandya & Moriya, 1996).

In *E. coli*, εdA was essentially non-mutagenic (mutation frequency < 0.7%), while εdC showed a mutation frequency of 2% (Table 1). In contrast, PdG was highly mutagenic, directing the exclusive incorporation of dAMP to form PdG→dT transversions. When the host cells were exposed to ultraviolet irradiation before transformation, the mutation frequency for εdC increased while that for εdA was essentially unaffected. The predominant mutations observed were εdC→dT and εdC→dA. The specificity of PdG→dT diminished in irradiated cells, as PdG→dA transitions and PdG→dG non-mutagenic events were also observed. The number of transformants recovered increased in the irradiated hosts, indicating an increased level of trans-lesion synthesis.

The 3′→5′ exonuclease activity associated with the ε subunit of DNA polymerase III catalyses removal of bases incorporated during DNA replication. When NR9232, a mutant strain of *E. coli* carrying a nonfunctional etheno subunit was used as the host cell, the frequency of targeted mutations for εdC increased to 33% (Moriya *et al.*, 1994). In unirradiated NR9232 cells, this value is equal to that observed after irradiation of wild-type

Table 1. Mutagenic potency and specificity of exocyclic DNA adducts

DNA adduct	Host	Targeted events (%) DNA adduct → C, A, T or G				Frequency of targeted mutations (%)
		C	A	T	G	
εdA	AB1157[a]	0	100	0	0	< 0.7
	AB1157/UV[b]	0	99.3	0.7	0	0.7
	COS7	1	31	6	63	70
εdC	AB1157	98	1	1	0	2
	AB1157/UV	67	14	18	0.4	32
	NR9232[c]	67	30	3	0	33
	NR9232/UV	45	50	5	0	55
	COS7	19	50	29	2	81
PdG	AB1157	0	0	100	0	100
	AB1157/UV	0.4	17	51	32	68
	COS7	1	0	7	92	8

εdA, 1,N^6-ethenodeoxyadenosine; εdC, 3,N^4-ethenodeoxycytidine; PdG, 1,N^2-(1,3-propan-1,3-diyl)-2′-deoxyguanosine

[a] Wild type in DNA repair

[b] Ultraviolet radiation pretreatment (20 J/m^2) of host cells

[c] A *mutD5* strain

AB1157. The increased mutation frequency reflected mainly εdC→dA transversions. This marked effect of the etheno subunit on mutagenesis indicates that *trans*-lesional synthesis is catalysed by DNA polymerase III. These results also suggest that dT, as well as non-mutagenic dG, is incorporated opposite εdC, dT being removed preferentially by the 3′→5′ exonuclease activity of the etheno subunit. In contrast, the mutation frequency did not increase with or without ultraviolet irradiation when plasmids containing εdA were introduced into NR9232, indicating that dTMP is inserted exclusively opposite this adduct by DNA polymerase III.

When the same constructs were transfected into simian kidney (COS7) cells, the targeted mutation frequencies of the two etheno adducts were high: 70% for εdA and 81% for εdC (Moriya *et al.*, 1994; Pandya & Moriya, 1996). εdA→dG transitions predominated, followed by εdA→dA (non-mutagenic), εdA→dT and εdA→dC. εdC→dA transversions were most commonly observed, followed by εdC→dT, εdC→dC and εdC→dG. Thus, the preferences for bases inserted opposite exocyclic adducts in COS cells are dC > dT > dA > dG for εdA and dT > dA >> dG > dC for εdC. In spite of its powerful miscoding properties in *E. coli*, PdG was not highly mutagenic in COS cells, the frequency of targeted mutations being 8%. The relative frequency of bases incorporated opposite PdG was dC >> dA > dG. It is interesting to note that PdG has also been shown to induce frameshift mutations in *E. coli* in a repetitive sequence (Benamira *et al.*, 1992).

Structural studies reveal that the non-mutagenic εdA(*anti*):dT(*anti*) pair forms a nonplanar alignment without hydrogen bonds (Kouchakdjian *et al.*, 1991). In contrast, εdA(*syn*):dG(*anti*), which generates the rare εdA→dC transversion, is stabilized by two hydrogen bonds (de los Santos *et al.*, 1991). The structure of εdA:dC that represents the major event in COS cells has not been established. Although εdC is incapable of forming more than one hydrogen bond in B-DNA, in either *syn* or *anti* conformation (Cullinan *et al.*, 1996; Korobka *et al.*, 1996; Cullinan *et al.*, 1997), dG, dT and dA are readily incorporated opposite the lesion during DNA synthesis. Remarkably, the fidelity of DNA synthesis in *E. coli* remains high.

At physiological pH, approximately half of the population of PdG molecules adopts the *syn* conformation, forming a PdG(*syn*):dA(*anti*) Hoogstein pair with two hydrogen bonds (Kouchakdjian *et al.*, 1989). The stability of this structure may account for the exclusive PdG→dT transversions observed in unirradiated *E. coli*, even though PdG represents a strong block to DNA synthesis. However, PdG:dC is the dominant pairing event in COS cells. All of these results suggest that the incorporation of dNTPs opposite an exocyclic DNA adduct is not simply determined by the formation of stable structures between the adduct and incoming nucleotide.

In summary, we have shown that exocyclic DNA adducts generated from endogenous and exogenous sources are mutagenic in mammalian cells. These lesions have the potential of generating mutations in oncogenes and/or tumour suppressor genes and may play a role in carcinogenesis. Endogenous exocyclic adducts contribute to the burden of cellular mutations and may contribute to cellular ageing as well.

Cellular response to exocyclic DNA adducts

The experimental approach described in the previous section was useful for studying the fidelity of *trans*-lesion DNA synthesis. However, *trans*-lesion synthesis is only one of several cellular responses to DNA damage, as DNA adducts are also handled by other mechanisms such as excision repair and recombination repair. To explore these possibilities, we developed a novel experimental approach with strand-specific tags to identify the origin of progeny generated from various pathways *in vivo* (Figures 2 and 3). This strategy involves use of a set of two vectors that are identical except in four regions. Circular ssDNA is prepared from one of these vectors. The other double-stranded vector DNA was digested with *Eco*RV, mixed with circular ssDNA and then denatured. Annealing of ssDNA with its complementary strand forms a gapped heteroduplex DNA. An oligodeoxynucleotide containing a single εdA residue was hybridized, filling the gap, and ligated to the DNA. The resulting closed circular double-stranded DNA was purified. There are four mismatched regions in this heteroduplex:

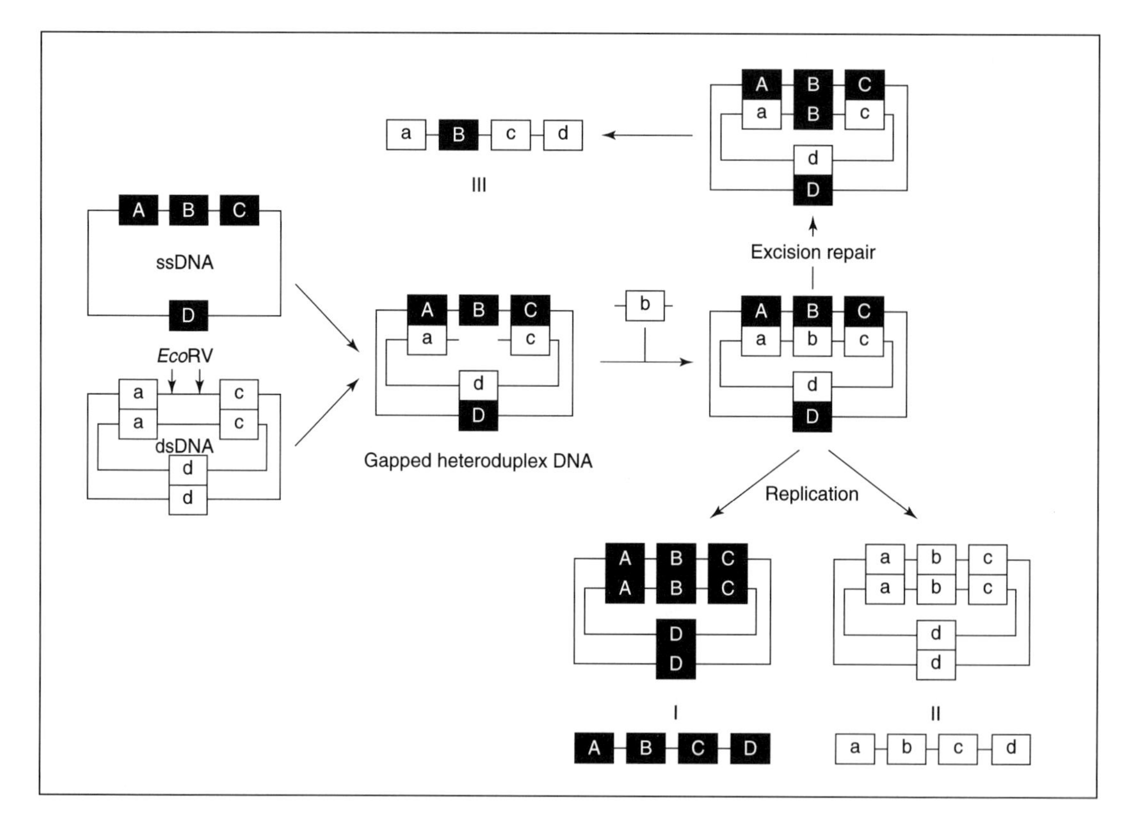

Figure 2. Scheme showing construction of mismatch-containing adducted DNA and progeny expected in cells

Sequences of A, B, C, D, a, b, c, and d are shown in Figure 3. Site b contains 1,N^6-ethenodeoxyadenosine. Progeny I, II and III correspond to those listed in Table 2.

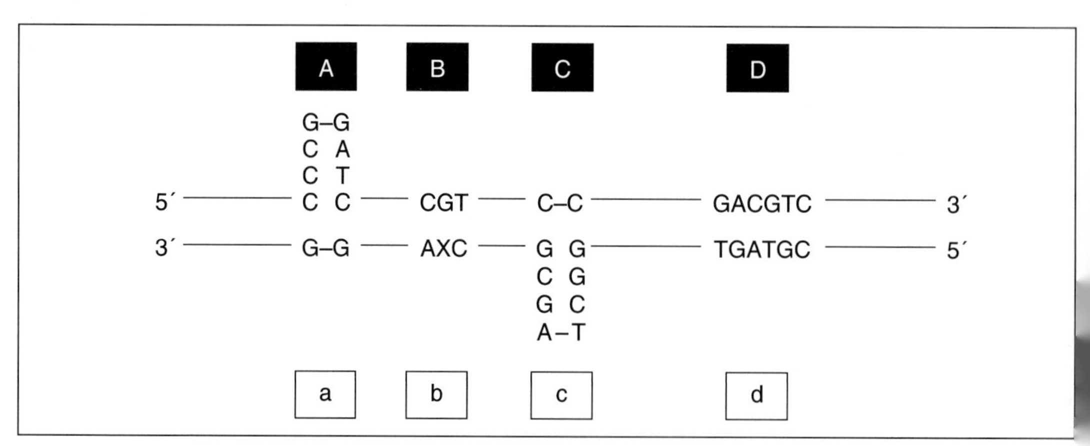

Figure 3. Local sequences of mismatched regions in a DNA construct

X represents 1,N^6-ethenodeoxyadenosine.

Table 2. Linkage analysis for four marker sequences

Progeny	Linkage of markers	No. of plasmid (%)			
		MM934		MM937	
		Control	εdA	Control	εdA
I	A-B-C-D	36	67	35	90
II	a-b-c-d	62	8	63	7
III	a-B-c-d		2		1
IV	a-B-C-d		6		
V	a-B-C-D	0.3	10		
VI	A-B-C-d	1	6		

εdA, 1,N^6-ethenodeoxyadenosine

one strand contains six extra nucleotides at the A/a site, while the other strand has six extra nucleotides at the C/c site (Figure 3). In addition, there are six contiguous mismatches at the D/d site and three mismatches at the B/b site, where εdA (shown as X) was incorporated. The 5'CGT3' in site B is part of the unique *Sna*BI site in the unmodified strand.

Purified heteroduplex DNA was introduced into mismatch repair-deficient *E. coli* cells. Progeny plasmid was used to transform *E. coli*. In the second step, progeny derived from each strand were segregated. Progeny plasmids were analysed for marker sequences by oligonucleotide hybridization.

Several types of progeny are produced from the heteroduplex DNA: replication of the unmodified strand yields progeny containing the A-B-C-D linkage, and the successful *trans*-lesion DNA synthesis across εdA produces progeny with the a-b-c-d linkage. In site b, εdA directs the incorporation of one of the four bases or induces a deletion. When εdA is removed by excision repair, the flanking mismatched bases dA and dC are also removed; a filling reaction converts the sequence of site b to that of site B. Therefore, progeny derived from excision repair contain the a-B-c-d linkage (Figure 2).

We used this approach to determine the base incorporated opposite εdA. As described above, site B contains the unique *Sna*BI site. Therefore, progeny derived from the unmodified strand and repaired molecules are sensitive to this enzyme, whereas progeny derived from *trans*-lesion synthesis are resistant to cleavage. The *Sna*BI-enriched plasmids were used to categorize events targeted to the site of εdA by oligonucleotide hybridization. This analysis showed that all *trans*-lesional events were accompanied by εdA→dA non-mutagenic events, confirming results obtained with ssDNA.

To examine other cellular responses to εdA, progeny plasmids recovered from the first *E. coli* host were used for a second transformation without *Sna*BI digestion. The progeny were analysed with oligonucleotide probes specific to A, B, C, D, a, b, c and d (Table 2). When a control construct was used for transformation of MM934 (*uvrA, alkA, tag1, mutS*), the majority of the progeny plasmids were derived from each strand. The ratio of progeny I to II was not 1:1, probably due to the f1 origin in the vector DNA, which may cause preferential replication of the minus strand. When the εdA-containing construct was used, the percentage of progeny II was markedly lower than that of progeny I, indicat-

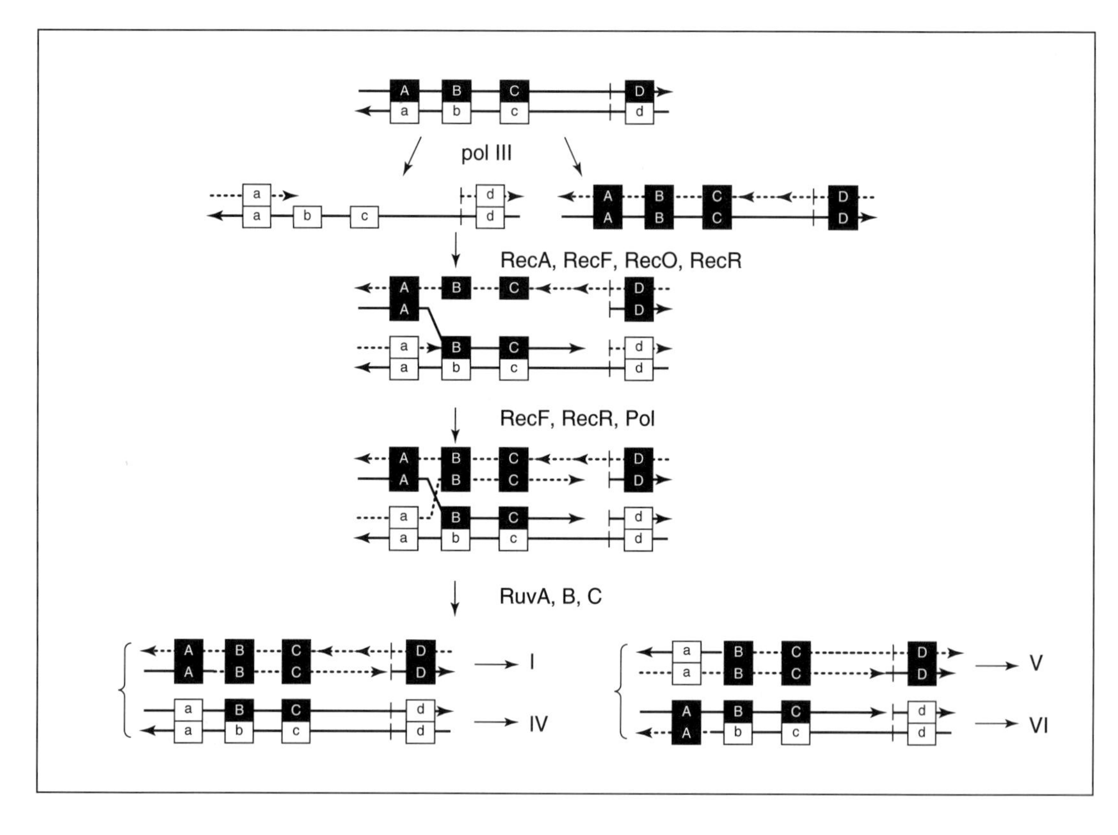

Figure 4. Daughter-strand gap repair mechanism and genes predicted to be involved in each step

A, B, C, D, a, b, c, and d correspond to those shown in Table 2 and Figures 2 and 3, respectively. I, IV, V and VI correspond to those shown in Table 2. Newly synthesized DNA is shown by dotted lines.

ing that εdA blocks DNA synthesis. Progeny III derived from the repaired plasmid accounted for 2% of the total. In addition to these three types of plasmid, three other types of recombinants were recovered. Figure 4 shows how these recombinants are generated. When the modified construct was introduced into a *ΔrecA* derivative (MM937) of MM934, these three recombinants were not observed, these results providing support for the pathway shown in Figure 4. A mutation in the *recF, recO* or *recR* gene also markedly decreased the incidences of these recombinants. Therefore, a likely scenario for *E. coli* is as follows: when εdA-containing DNA replicates before the initiation of repair, DNA polymerase III holoenzyme is strongly blocked at or near εdA. A low level of *trans*-lesional synthesis occurs, which is highly accurate (> 99% error-free), with the insertion of non-mutagenic dTMP opposite the adduct. A fraction of the blocked synthesis is rescued by damage-tolerance mechanism(s), probably daughter strand gap repair, which is mechanistically error-free (Friedberg *et al.*, 1995).

Several exocyclic DNA adducts are formed endogenously in genomic DNA; they are suspected to play a role in ageing and cancer; however, the cellular responses to exocyclic adducts in human cells, including their mutagenicity, remain to be determined.

Acknowledgements

This study was supported by grants CA47995 and CA17395 (APG) and CA17613 (Cancer Center Support Grant) from the National Institutes of Health, USA.

References

Barbin, A. & Bartsch, H. (1986) Mutagenic and promutagenic properties of DNA adducts formed by vinyl chloride metabolites. In: *The Role of Cyclic Nucleic Acid Adducts in Carcinogenesis and Mutagenesis* (Singer, B. & Bartsch, H., eds), IARC Scientific Publications No. 70. Lyon, IARC, pp. 345–358

Basu, A.K., Wood, M.L., Niederhofer, L.J., Ramos, L.A. & Essigmann, J.M. (1993) Mutagenic and genotoxic effects of three vinyl chloride-induced DNA lesions: 1,N^6-ethenoadenine, 3,N^4-ethenocytosine, and 4-amino-5-(imidazol-2-yl)imidazole. *Biochemistry* 32, 12793–12801

Benamira, M., Singh, U. & Marnett, L.J. (1992) Site-specific frameshift mutagenesis by a propanodeoxyguanosine adduct positioned in the $(CpG)_4$ hotspot of *Salmonella typhimurium hisD3052I* carried on an M13 vector. *J. Biol. Chem.* 267, 22392–22400

Chaudhary, A.K., Nokubo, M., Reddy, G.R., Yeola, S.N., Morrow, J.D., Blair, I.A. & Marnett, L.J. (1994) Detection of endogenous malondialdehyde- deoxyguanosine adducts in human liver. *Science.* 265,1580–1582

Cheng, K.C., Preston, B.D., Cahill, D.S., Dosanjh, M.K., Singer, B. & Loeb, L.A. (1991) The vinyl chloride DNA derivative N^2,3-ethenoguanine produces G→A transitions in *Escherichia coli*. *Proc. Natl Acad. Sci. USA* 88, 9974–9978

Chung, F.L., Chen, H.J. & Nath, R.G. (1996) Lipid peroxidation as a potential endogenous source for the formation of exocyclic DNA adducts. *Carcinogenesis* 17, 2105–2111

Cullinan, D., Korobka, A., Grollman, A.P., Patel, D.J., Eisenberg, M. & de los Santos C (1996) NMR solution structure of an oligodeoxynucleotide duplex containing the exocyclic lesion 3,N^4-etheno-2′-deoxycytidine opposite thymidine: Comparison with the duplex containing deoxyadenosine opposite the adduct. *Biochemistry* 35, 13319–13327

Cullinan, D., Johnson, F., Grollman, A.P., Eisenberg, M. & de los Santos C (1997) Solution structure of a DNA duplex containing the exocyclic lesion 3,N^4-etheno-2′-deoxycytidine opposite 2′-deoxyguanosine. *Biochemistry* 36, 11933–11943

Dosanjh, M.K., Chenna, A., Kim, E., Fraenkel-Conrat, H., Samson, L. & Singer, B. (1994) All four known cyclic adducts formed in DNA by the vinyl chloride metabolite chloroacetaldehyde are released by a human DNA glycosylase. *Proc. Natl Acad. Sci. USA* 91, 1024–1028

El-Ghissassi, F., Barbin, A., Nair, J. & Bartsch, H. (1995) Formation of 1,N^6-ethenoadenine and 3,N^4-ethenocytosine by lipid peroxidation products and nucleic acid bases. *Chem. Res. Toxicol.* 8, 278–283

Fink, S.P., Reddy, G.R. & Marnett ,L.J. (1997) Mutagenicity in *Escherichia coli* of the major DNA adduct derived from the endogenous mutagen malondialdehyde. *Proc. Natl Acad. Sci. USA* 94, 8652–8657

Friedberg, E.C., Walker, G.C. & Siede, W. (1995) *DNA Repair and Mutagenesis*. Washington DC, ASM Press

Hang, B., Medina, M., Fraenkel-Conrat, H. & Singer, B. (1998) A 55-kDa protein isolated from human cells shows DNA glycosylase activity toward 3,N^4-ethenocytosine and the G/T mismatch. *Proc. Natl Acad. Sci. USA* 95, 13561–13566.

Hashim, M.F., Schnetz-Boutaud, N. & Marnett, L.J. (1997) Replication of template-primers containing propanodeoxyguanosine by DNA polymerase beta. Induction of base pair substitution and frameshift mutations by template slippage and deoxynucleoside triphosphate stabilization. *J. Biol. Chem.* 272, 20205–20212

Johnson, K.A., Fink, S.P. & Marnett, L.J. (1997) Repair of propanodeoxyguanosine by nucleotide excision repair *in vivo* and *in vitro*. *J. Biol. Chem.* 272, 11434–11438.

Korobka, A., Cullinan, D., Cosman, M., Grollman, A.P., Patel, D.J., Eisenberg, M. & de los Santos C (1996) Solution structure of an oligodeoxynucleotide duplex containing the exocyclic lesion 3,N^4-etheno-2′-deoxycytidine opposite 2′-deoxyadenosine, determined by NMR spectroscopy and restrained molecular dynamics. *Biochemistry* 35, 13310–13318

Kouchakdjian, M., Marinelli, E., Gao, X.L., Johnson, F., Grollman, A. & Patel, D. (1989) NMR studies of exocyclic 1,N^2-propanodeoxyguanosine adducts (X)

opposite purines in DNA duplexes: Protonated X(syn).A(anti) pairing (acidic pH) and X(syn).G(anti) pairing (neutral pH) at the lesion site. *Biochemistry* 28, 5647–57.

Kouchakdjian, M., Eisenberg, M., Yarema, K., Basu, A.K., Essigmann, J. & Patel, D.J. (1991) NMR studies of the exocyclic 1,N^6-ethenodeoxyadenosine adduct (epsilon dA) opposite thymidine in a DNA duplex. Nonplanar alignment of εdA(anti) and dT(anti) at the lesion site. *Biochemistry* 30, 1820–1828

Lantouët, S., Mican, A.N., Müller, M., Fink, S.P., Marnett, L.J., Muhle, S.A. & Guengerich, F.P. (1998) Misincorporation of nucleotides opposite five-membered exocyclic ring guanine derivatives by *Escherichia coli* polymerases in vitro and in vivo: 1,N^2-Ethenoguanine, 5,6,7,9-tetrahydro-9-oxoimidazo[1,2-a]purine, and 5,6,7,9-tetrahydro-7-hydroxy-9-oxoimidazo[1,2-a]purine. *Biochemistry* 37, 5184–5193

Marnett, L.J. (1994) DNA adducts of α, β-unsaturated aldehydes and dicarbonyl compounds. In: *DNA Adducts: Identification and Biological Significance* (Hemminki, K., Dipple, A., Shuker, D.E.G., Kadlubar, F.F., Segerbäck, D. & Bartsch, H., eds), IARC Scientific Publications No. 125. Lyon, IARC, pp. 151–163

Moriya, M. (1993) Single-stranded shuttle phagemid for mutagenesis studies in mammalian cells: 8-Oxoguanine in DNA induces targeted G:CT:A transversions in simian kidney cells. *Proc. Natl Acad. Sci. USA* 90, 1122–1126

Moriya, M., Zhang, W., Johnson, F. & Grollman, A.P. (1994) Mutagenic potency of exocyclic DNA adducts: Marked differences between *Escherichia coli* and simian kidney cells. *Proc. Natl Acad. Sci. USA* 91, 11899–11903

Nair, J., Barbin, A., Guichard, Y. & Bartsch, H. (1995) 1,N^6-Ethenodeoxyadenosine and 3,N^4-ethenodeoxycytidine in liver DNA from humans and untreated rodents detected by immunoaffinity/^{32}P-postlabelling. *Carcinogenesis* 16, 613–617.

Nath, R.G. & Chung, F.L. (1994) Detection of exocyclic 1,N^2-propanodeoxyguanosine adducts as common DNA lesions in rodents and humans. *Proc. Natl Acad. Sci. USA 91*, 7491–7495

Palejwala, V.A., Simha, D. & Hymayun, M.Z. (1991) Mechanisms of mutagenesis by exocyclic DNA adducts. Transfection of M13 viral DNA bearing a site-specific adduct shows that ethenocytosine is a highly efficient RecA-independent mutagenic noninstructional lesion. *Biochemistry* 30, 8736–8743

Pandya, G. & Moriya, M. (1996) 1,N^6-Ethenodeoxyadenosine, a DNA adduct highly mutagenic in mammalian cells. *Biochemistry* 35, 11487–11492

de los Santos, C., Kouchakdjian, M., Yarema, K., Basu, A.K., Essigmann, J. & Patel, D.J. (1991) NMR studies of the exocyclic 1,N^6-ethenodeoxyadenosine adduct (εdA) opposite deoxyguanosine in a DNA duplex. εdA(syn).dG(anti) pairing at the lesion site. *Biochemistry* 30, 1828–1835

Saparbaev, M. & Laval, J. (1998) 3,N^4-Ethenocytosine, a highly mutagenic adduct, is a primary substrate for *Escherichia coli* double-stranded uracil-DNA glycosylase and human mismatch-specific thymine-DNA glycosylase. *Proc. Natl Acad. Sci. USA* 95, 8508–8513.

Saparbaev, M., Kleibl, K. & Laval, J. (1995) *Escherichia coli, Saccharomyces cerevisiae,* and rat and human 3-methyladenine DNA glycosylases repair 1,N^6-ethenoadenine when present in DNA. *Nucleic Acids Res.* 23, 3750–3755

Simha, D., Palejwala, V.A. & Humayun, M.Z. (1991) Mechanisms of mutagenesis by exocyclic DNA adducts. Construction and in vitro template characteristics of an oligonucleotide bearing a single site-specific ethenocytosine. *Biochemistry* 30, 8727–8735

Singer, B., Kusmierek, J.T., Folkman, W., Chavez, F. & Dosanjh, M.K. (1991) Evidence for the mutagenic potential of the vinyl chloride induced adduct, N^2,3-etheno-deoxyguanosine, using a site-directed kinetic assay. *Carcinogenesis* 12, 745–747

Zhang, W., Johnson, F., Grollman, A.P. & Shibutani, S. (1995) Miscoding by the exocyclic and related DNA adducts 3,N^4-etheno-2′-deoxycytidine, 3,N^4-ethano-2′-deoxycytidine, and 3-(2-hydroxyethyl)-2′-deoxyuridine. *Chem. Res. Toxicol.* 8, 157–163

Exocyclic DNA Adducts in Mutagenesis and Carcinogenesis
B. Singer & H. Bartsch
IARC Scientific Publications No. 150
International Agency for Research on Cancer, Lyon, 1999

Role of base excision repair in protecting cells from the toxicity of chloroethylnitrosoureas

D.B. Ludlum, Q. Li & Z. Matijasevic

The chloroethylnitrosoureas react extensively with cellular DNA to produce a variety of DNA adducts, including a deoxycytidine–deoxyguanosine (dC–dG) cross-link that is clearly cytotoxic. It is now well established that O^6-alkylguanine-DNA-alkyltransferase can prevent formation of this dC–dG cross-link and thereby diminish the toxicity of the chloroethylnitrosoureas. Besides alkyltransferase, DNA glycosylases from various species can also contribute to cellular resistance to the chloroethylnitrosoureas, but the mechanism for this increased resistance has not been established. It is known, however, that several chloroethylnitrosourea-modified DNA bases, including the exocyclic adduct, N^2,3-ethanoguanine, are released by *Escherichia coli* 3-methyladenine DNA glycosylase II. In the study described here, we examined the possibility that this enzyme might act on the exocyclic intermediate in dC–dG formation, 1,O^6-ethanodeoxyguanosine, and prevent dC–dG cross-linking in this way. However, the presence of *E. coli* 3-methyladenine DNA glycosylase II does not decrease the amount of dC–dG cross-link formed when chloroethylnitrosourea reacts with DNA, and we conclude that this enzyme does not recognize 1,O^6-ethanodeoxyguanosine. Therefore, its contribution to resistance probably resides in its action on other nitrosourea-induced DNA modifications.

Introduction

The chloroethylnitrosoureas are antitumour agents that produce their cytotoxic action by reacting with cellular DNA (Ludlum, 1997). They form many different DNA adducts, including a cross-link between deoxycytidine (dC) and deoxyguanosine (dG) that is clearly cytotoxic (Tong *et al.*, 1982). Formation of this cross-link can be prevented by O^6-alkylguanine DNA alkyltransferase, so that tumour cells with high levels of this repair enzyme are resistant to treatment with the chloroethylnitrosoureas (Erickson *et al.*, 1980; Pegg *et al.*, 1995).

Studies of differences between sensitive and resistant cells in the distribution of chloroethylnitrosourea-induced DNA modifications have indicated, however, that other DNA repair enzymes besides O^6-alkylguanine-DNA-alkyltransferase probably contribute to the resistance phenomenon (Bodell *et al.*, 1988). By analogy with the adaptive response in *Escherichia coli*, we have suggested that alkylpurine glycosylases might be involved. Matijasevic *et al.* have shown that a resistant tumour cell line has enhanced glycosylase activity (Matijasevic *et al.*, 1991) and that introduction of a cloned yeast glycosylase into glycosylase-deficient *E. coli* cells increases their resistance to chloroethylnitrosourea (Matijasevic *et al.*, 1993). Engelward *et al.* (1996) created alkylpurine glycosylase-deficient mouse cells and showed that these cells have increased sensitivity to *N*,*N'*-bis(2-chloroethyl)-*N*-nitrosourea.

Previously, we investigated the action of cloned and purified *E. coli* 3-methyladenine DNA glycosylase II (Gly II) directly on chloroethylnitrosourea-modified DNA. These studies, which are reviewed here, show that Gly II recognizes several chloroethylnitrosourea adducts, including the exocyclic N^2,3-ethanodeoxyguanosine. The experiments reported in this paper show that this enzyme does not reduce the amount of dC–dG cross-link formation, indicating that bacterial Gly II does not act on 1,O^6-ethanodeoxyguanosine or on the cross-link itself.

Materials and methods

Materials

N-(2-Chloro-1,2-^{3}H-ethyl)-*N*-nitrosourea (^{3}H-CNU) at a specific activity of 1.3 Ci/mmol, was custom-synthesized by Moravek Biochemicals (La Brea, CA, USA). *N*-(2-Chloro-1,2-^{14}C-ethyl)-*N'*-cyclohexyl-*N*-nitrosourea (^{14}C-CCNU), with a specific activity of 24.4 mCi/mmol, was obtained from the Division of Cancer Treatment, National Cancer Institute, Bethesda, MD, USA. Unlabelled CNU and CCNU were also provided by the Division of Cancer Treatment. Professor Mutsuo Sekiguchi (Fukuoka, Japan) kindly provided the *E. coli* MS23 strain carrying the plasmid pYN1000 (Nakabeppu *et al.*, 1984).

Optical markers used in the high-performance liquid chromatography (HPLC) separations were synthesized in our laboratory, as described previosuly (Habraken *et al.*, 1990, 1991a), except for the cross-link 1-[N^3-(2'-deoxycytidyl)]–2-[N^1-(2'-deoxyguanosinyl)]ethane, which was kindly provided by Dr William J. Bodell of the University of California at San Francisco (Bodell & Pongracz, 1993). DNase I, snake venom phosphodiesterase and bacterial alkaline phosphatase were obtained from Worthington; calf thymus DNA and spleen phosphodiesterase came from Sigma. Other chemicals were reagent-grade materials.

Methods

Gly II was isolated from *E. coli* MS23 which harbours the pYN1000 plasmid containing the *AlkA* gene (Nakabeppu *et al.*, 1984). Lysed cells were treated with DEAE cellulose to remove DNA, and the enzyme was purified in the phosphocellulose and DNA cellulose chromatography steps described by Nakabeppu *et al.* (1984). The final product migrated as a single band with a relative molecular mass of 30 000, as visualized by silver staining on a 12.5% polyacrylamide gel with 0.1% sodium dodecyl sulfate. The enzyme was assayed with ^{3}H-*N*-methyl-*N*-nitrosourea (MNU)–DNA as described previously (Habraken *et al.*, 1990). One unit of enzyme activity was defined as the amount of enzyme that released 1 pmol of base from ^{3}H-MNU-DNA in 10 min at 37 °C.

DNAs used as substrates for Gly II were prepared by reacting calf thymus DNA, 5 mg/ml, with either ^{3}H-CNU or ^{14}C-CCNU in 50 mmol/L sodium cacodylate buffer, pH 7.5. The reaction was allowed to proceed for 16 h at 37 °C; modified DNA was then washed free of non-covalently bound radiolabel by repeated ethanol precipitation and re-dissolution in 50 mmol/L sodium cacodylate buffer.

To examine the effect of Gly II on the reaction of ^{3}H-CNU with DNA, the stability of Gly II was first tested under the conditions used for the ^{3}H-CNU reaction. Gly II (0.4 unit) was incubated in the presence or absence of 0.6 mmol/L CNU in 50 mmol/L sodium cacodylate buffer, pH 7, at 37 °C, and aliquots were withdrawn at hourly intervals for assay against ^{3}H-MNU-DNA. The presence of CNU had no discernible effect on the half-life of Gly II activity, which was 10.5 h under these conditions.

The effect of Gly II on the reaction of ^{3}H-CNU with DNA was determined by first thoroughly mixing 75 μCi (0.06 μmol) of ^{3}H-CNU with calf thymus DNA (4 mg/ml) in 50 μl of 50 mmol/L sodium cacodylate buffer, pH 7 and then mixing 15 μl of this mixture with Gly II (0.2 unit) in 20 μl buffer or an equivalent amount of boiled enzyme or buffer alone. The reaction with ^{3}H-CNU was allowed to proceed for 24 h at 37 °C. After 8 h of incubation, an additional 0.1 unit of enzyme in 10 μl of buffer, or the corresponding amount of inactivated enzyme or buffer, was added to the appropriate tube. After 24 h, each ^{3}H-CNU–DNA was precipitated with alcohol and then redissolved and reprecipitated six times until constant specific activity was achieved.

The spectrum of products in each DNA was determined by a method that has been described in detail (Ludlum, 1987). Briefly, the DNA was digested with a combination of DNase I, spleen and venom phosphodiesterases and alkaline phosphatase. HPLC analysis was performed on a 5-μm C_{18} column (4.6 x 250 mm) eluted at 1 ml/min with a slow acetonitrile–phosphate buffer system chosen to separate the dC–dG peak from other derivatives. Optical markers were added during each HPLC separation to verify the identity of radiolabelled peaks. Fractions were collected every 0.5 min and counted in a Beckman LS6500 scintillation counter.

The profiles of radioactivity *versus* fraction number were plotted by a computer program which automatically calculates the background and total radioactivity in each peak. DNA digest containing approximately 15 000 counts per min of adducts was injected for each profile, and the results were normalized to 20 000 counts per min of adducts.

Results

Studies of the action of Gly II on ^{14}C-CCNU–DNA and ^{3}H-CNU–DNA have shown that this enzyme acts on all of the chloroethylnitrosourea-modified deoxynucleosides shown in Figure 1 (Habraken *et al.*, 1991a,b). Release of these modified bases from DNA followed by successful endonuclease, polymerase and ligase action could result in increased resistance to the chloroethylnitrosoureas.

Although any one of the chloroethylnitrosourea-induced modifications shown in Figure 1 could be lethal, the presence of dideoxyguanosinylethane and N^2,3-ethanodeoxyguanosine might be particularly disruptive. If dideoxyguanosinylethane were present, the secondary structure of DNA would almost certainly be distorted. Also, by analogy to the mutagenic effects of N^2,3-ethenodeoxyguanosine (Cheng *et al.*, 1991; Mroczkowska & Kusmierek, 1991), N^2,3-ethanodeoxyguanosine would probably be mutagenic and potentially lethal.

The possibility remains, however, that Gly II may act on other chloroethylnitrosourea-induced DNA modifications besides those shown in Figure 1. The finding that N^2,3-ethanodeoxyguanosine is a substrate for Gly II raised the question of whether other exocyclic adducts produced by the chloroethylnitrosoureas might also be subject to glycosylase action. Of particular interest is the possibility that 1,O^6-ethanodeoxyguanosine might be a substrate. The importance of this possibility is shown in Figure 2. It is well established that the dC–dG cross-link shown at the bottom of this figure is cytotoxic and that its formation can be prevented by the action of O^6-alkylguanine-DNA-alkyltransferase; alkyltransferase acts on O^6-chloroethyldeoxyguanosine, removing the chloroethyl group and restoring deoxyguanosine to its original configuration. After O^6-

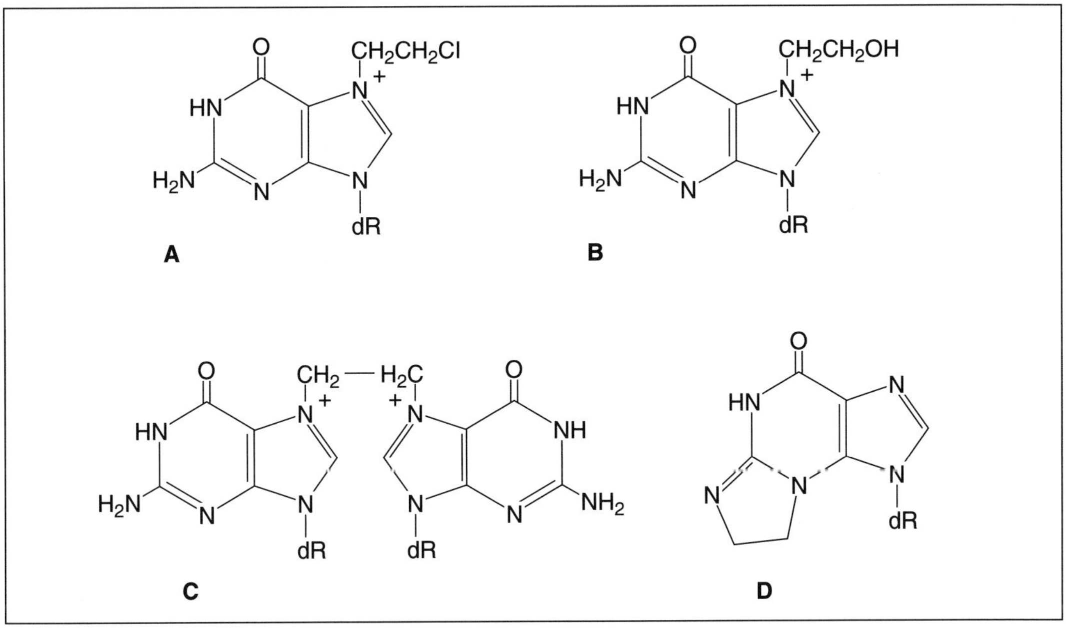

Figure 1. Substrates for *Escherichia coli* 3-methyladenine DNA glycosylase II

Bases are released from DNA by cleavage of their bonds to deoxyribose (dR). A, 7-chloroethyldeoxyguanosine; B, 7-hydroxyethyldeoxyguanosine; C, dideoxyguanosinylethane; and D, N^2,3-ethanodeoxyguanosine

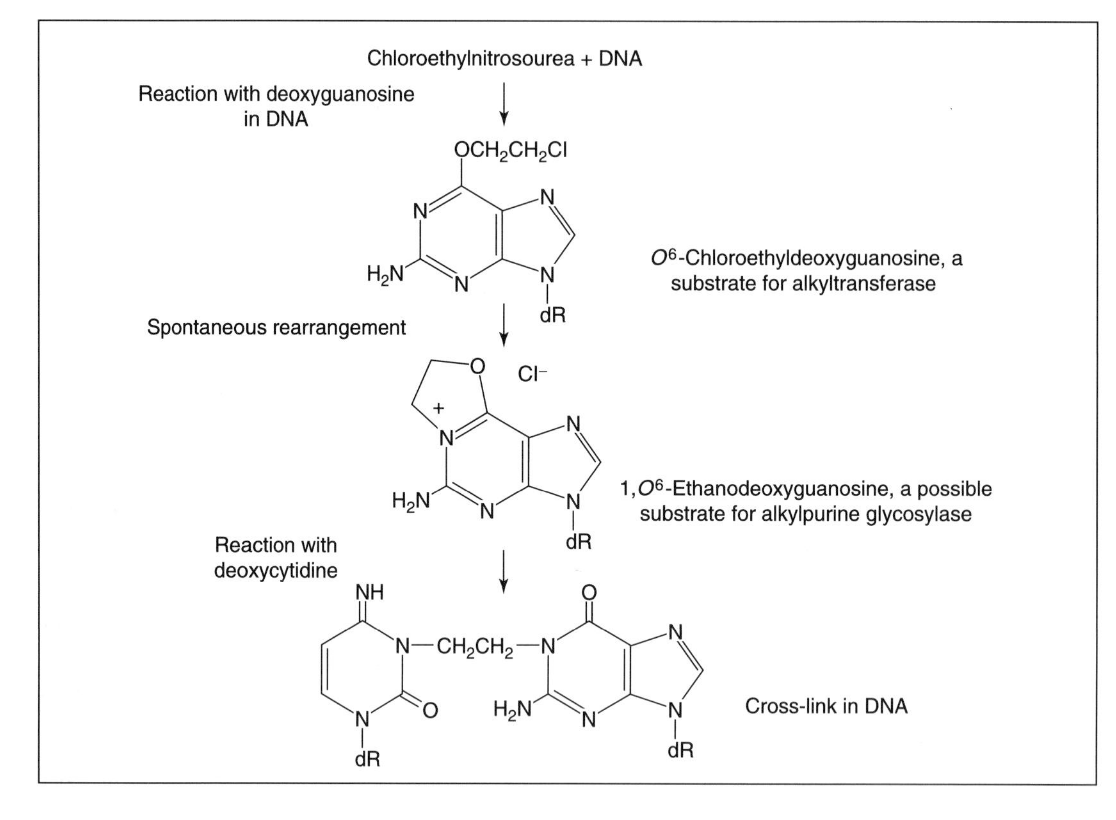

Figure 2. Cross-link formation by the chloroethylnitrosoureas

The formation of cross-links is prevented by repair of O^6-chloroethyldeoxyguanosine and would be prevented if 1,O^6-ethanodeoxyguanosine were a substrate for alkylpurine glycosylase.

chloroethyldeoxyguanosine cyclizes to 1,O^6-ethanodeoxyguanosine, however, dC–dG cross-link formation could still be prevented if this cyclic deoxynucleoside were a substrate for glycosylase action.

We tested this possibility by studying dC–dG cross-link formation in the presence and absence of Gly II. Control studies have shown that Gly II degrades slowly at 37 °C and pH 7. In fact, its half-life under these conditions is 10.5 h, and its activity is not affected by the presence of CNU at a concentration as high as 0.6 mmol/L. Since cross-link formation occurs over several hours (Kohn, 1977), additional enzyme was added after 8 h of incubation to maintain activity, as described in the Methods.

The HPLC profiles of the adducts found in the DNA reacted with ^{3}H-CNU in the absence of enzyme, in the presence of inactivated enzyme or in the presence of active enzyme are shown in Figure 3. Optical density markers for all of the substrates shown in Figure 1 were added to these profiles, and their retention times are indicated in Figure 3. Although the adduct N^2,3-ethanoguanine appears as a shoulder in the first two profiles, it is clearly absent in the DNA that was reacted with ^{3}H-CNU in the presence of active enzyme. The other peak that is clearly reduced in size by glycosylase action is 7-chloroethylguanine. These results are summarized in Table 1, which shows that there was no effect on the amount of dC–dG cross-link present. We conclude that neither the intermediate 1,O^6-ethanodeoxyguanosine nor the cross-link itself is released from DNA by Gly II during its reaction with ^{3}H-CNU.

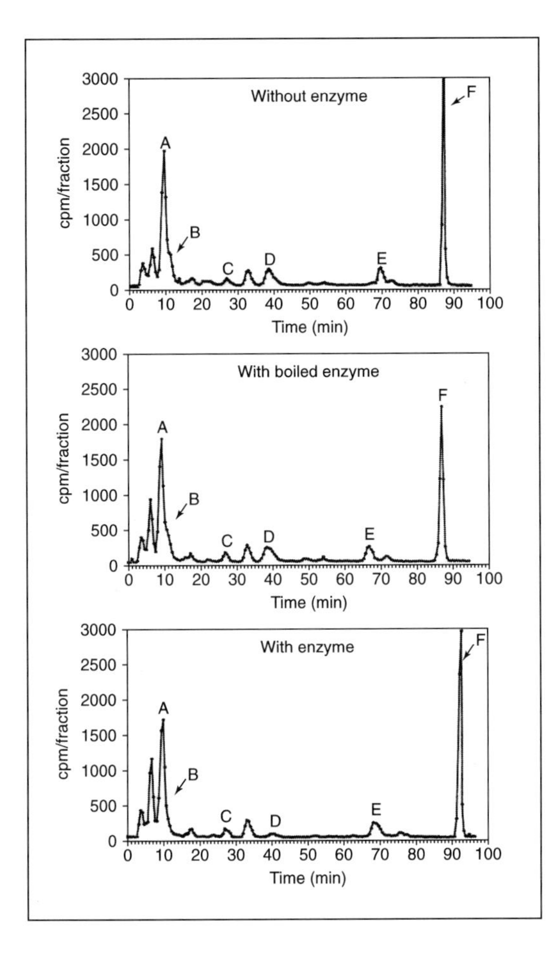

Figure 3. Effect of *Escherichia coli* 3-methyladenine DNA glycosylase II on DNA adduct formation by the chloroethylnitrosoureas

Adduct profiles, normalized to 20 000 counts per minute (cpm) of adducts, are shown for DNA incubated with *N*-(2-chloro-1,2-^{3}H-ethyl)-*N*-nitrosourea in the absence of enzyme (top panel), in the presence of inactivated enzyme (middle panel) or in the presence of active *E. coli* 3-methyladenine DNA glycosylase II (bottom panel). A, 7-hydroxyethylguanine; B, N^2,3-ethanoguanine; C, diguanylethane; D, 7-chloroethylguanine; E, dC–dG cross-link; and F, phosphotriesters

Discussion

Increased expression of Gly II is part of the adaptive response of *E. coli* to simple methylating agents (Karran *et al.*, 1982). Release of 3-methyladenine and 3-methylguanine by this enzyme almost certainly contributes to resistance to methylating agents; however, there is no direct evidence as to which chloroethylnitrosourea-induced lesions are removed by this enzyme *in vivo* to increase cellular resistance to these agents.

As reviewed by Memisoglu and Samson (1996), overexpression of glycosylases in mammalian cells does not always protect them from the toxic effects of alkylating agents. This may be because, as those authors point out, the presence of glycosylase alone does not ensure complete repair of alkylated DNA: the participation of additional repair enzymes is necessary to complete the process. Studies with *Aag* null mice showed clearly that the loss of glycosylase activity increases sensitivity to chloroethylnitrosoureas (Engelward *et al.*, 1996), so that the importance of glycosylase action in protecting cells from the toxicity of these agents is clearly established.

The question of how glycosylase protects cells from the toxicity of chloroethylnitrosoureas at the molecular level remains. Since there is strong evidence that the dC–dG cross-link is cytotoxic, the ability of glycosylase to prevent its formation would explain a protective effect; however, the presence of this enzyme did not decrease the amount of dC–dG that was formed in the experiments reported here. Accordingly, we have concluded that glycosylase does not recognize the exocyclic deoxynucleoside, 1,O^6-ethanodeoxyguanosine, and that some other action of glycosylase provides protection. Any of the lesions shown in Figure 1 could be cytotoxic, but the possibility remains that additional cytotoxic lesions that are substrates for glycosylase action exist.

Acknowledgements

The authors would like to express their appreciation to Professor Mutsuo Sekiguchi for supplying the plasmid containing cloned *E. coli* 3-methyladenine-DNA glycosylase and to Dr William J. Bodell for supplying the optical marker for the dC–dG cross-link. This publication was supported by grant CA-44499 from the National Cancer Institute. Its contents are solely the responsibility of the authors and do not necessarily represent the official views of that Institute.

Table 1. Distribution of alkylation products in DNA[a]

Adduct	No enzyme present	Boiled enzyme present	Active enzyme present
7-Hydroxyethylguanine	34.6	32.6	29.5
Diguanylethane	1.9	2.3	2.3
7-Chloroethylguanine	6.4	6.6	1.2
dC–dG	5.4	4.8	5.8

[a] Percent of total adducts recovered in high-performance liquid chromatography

References

Bodell, W.J. & Pongracz, K. (1993) Chemical synthesis and detection of the cross-link 1-[N^3-(2′-deoxycytidyl)]-2-[N^1-(2′-deoxyguanosinyl)]ethane in DNA reacted with 1-(2-chloroethyl)-1-nitrosourea. *Chem. Res. Toxicol.* 6, 434–438

Bodell, W.J., Tokuda, K. & Ludlum, D.B. (1988) Differences in DNA alkylation products formed in sensitive and resistant human glioma cells treated with N-(2-chloroethyl)-N-nitrosourea. *Cancer Res.* 48, 4489–4492

Cheng, K.C., Preston, B.D., Cahill, D.S., Dosanjh, M.K., Singer, B. & Loeb, L.A. (1991) The vinyl chloride DNA derivative N^2,3-ethenoguanine produces G→A transitions in *Escherichia coli*. *Proc. Natl Acad. Sci. USA* 88, 9974–9978

Engelward, B.P., Dreslin, A., Christensen, J., Huszar, D., Kurahara, C. & Samson, L. (1996) Repair deficient 3-methyladenine DNA glycosylase homozygous mutant mouse cells have increased sensitivity to alkylation induced chromosome damage and cell killing. *EMBO J.* 15, 945–952

Erickson, L.C., Laurent, G., Sharkey, N.A. & Kohn, K.W. (1980) DNA cross-linking and monoadduct repair in nitrosourea-treated human tumour cells. *Nature* 288, 727–729

Habraken, Y., Carter, C.A., Kirk, M.C., Riordan, J.M. & Ludlum, D.B. (1990) Formation of N^2,3-ethanoguanine in DNA after *in vitro* treatment with the therapeutic agent, N-(2-chloroethyl)-N′-cyclohexyl-N-nitrosourea. *Carcinogenesis* 11, 223–228

Habraken, Y., Carter, C.A., Kirk, M.C. & Ludlum, D.B. (1991a) Release of 7-alkylguanines from N-(2-chloroethyl)-N′-cyclohexyl-N-nitrosourea-modified DNA by 3-methyladenine DNA glycosylase. *Cancer Res.* 51, 499–503

Habraken, Y., Carter, C.A., Sekiguchi, M. & Ludlum, D.B (1991b) Release of N^2,3-ethanoguanine from haloethylnitrosourea-treated DNA by *Escherichia coli* 3-methyladenine DNA glycosylase II. *Carcinogenesis*,12, 1971–1973

Karran, P., Hjelmgen, T. & Lindahl, T. (1982) Induction of a DNA glycosylase for N-methylated purines is part of the adaptive response to alkylating agents. *Nature* 296, 770–773

Kohn, K.W. (1977) Interstrand cross-linking of DNA by 1,3-bis(2-chloroethyl)-1-nitrosourea and other 1-(2-haloethyl)-1-nitrosoureas. *Cancer Res.* 37, 1450–1454

Ludlum, D.B. (1987) High performance liquid chromatographic separation of DNA adducts induced by cancer chemotherapeutic agents. *Pharmacol. Ther.* 34, 145–153

Ludlum, D.B. (1997) The chloroethylnitrosoureas: Sensitivity and resistance to cancer chemotherapy at the molecular level. *Cancer Invest.* 15, 588–598

Matijasevic, Z., Bodell, W.J. & Ludlum, D.B (1991) 3-Methyladenine DNA glycosylase activity in a glial cell line sensitive to the haloethylnitrosoureas in comparison with a resistant cell line. *Cancer Res.* 51, 1568–1570

Matijasevic, Z., Boosalis, M., Mackay, W., Samson, L. & Ludlum, D.B. (1993) Protection against chloroethylnitrosourea cytotoxicity by eukaryotic 3-methyladenine DNA glycosylase. *Proc. Natl Acad. Sci. USA* 90, 11855–11859

Memisoglu, A. & Samson, L. (1996) DNA repair functions in heterologous cells. *Crit. Rev. Biochem. Mol. Biol.* 31, 405–447

Mroczkowska, M.M. & Kusmierek, J.T. (1991) Miscoding potential of N^2,3-ethenoguanine studied in an *Escherichia coli* DNA-dependent RNA polymerase *in vitro* system and possible role of this adduct in vinyl chloride-induced mutagenesis. *Mutagenesis* 6, 385–390

Nakabeppu, Y., Kondo, H. & Sekiguchi, M. (1984) Cloning and characterization of the *alkA* gene of *Escherichia coli* that encodes 3-methyladenine DNA glycosylase II. *J. Biol. Chem.* 259, 13723–13729

Pegg, A.E., Dolan, M.E. & Moschel, R.C. (1995) Structure, function, and inhibition of O^6-alkylguanine-DNA alkyltransferase. *Prog. Nucleic Acid Res. Mol. Biol.* 51, 167–223

Tong, W.P., Kirk, M.C. & Ludlum, D.B. (1982) Formation of the cross-link 1-[N^3-deoxy-cytidyl],2-[N^1-deoxyguanosinyl]ethane in DNA treated with N,N'-bis(2-chloroethyl)-N-nitrosourea). *Cancer Res.* 42, 3102–3105

Exocyclic DNA Adducts in Mutagenesis and Carcinogenesis
B. Singer & H. Bartsch
IARC Scientific Publications No. 150
International Agency for Research on Cancer, Lyon, 1999

Localization of chloroacetaldehyde-induced DNA damage in human *p53* gene by DNA polymerase fingerprint analysis

B. Tudek, P. Kowalczyk & J.M. Cieśla

Chloroacetaldehyde (CAA) reacts with DNA bases, forming hydroxyethano derivatives of different stability, which are subsequently converted into etheno (ε) adducts: εA, εC, εG. DNA polymerase fingerprint analysis was used to study the distribution of CAA-induced modifications in the *p53* sequence. A plasmid bearing cDNA containing the human *p53* gene was reacted *in vitro* with CAA, then dehydrated for conversion of hydroxyethano into etheno adducts, and primer extension by T7 DNA polymerase in the presence of four dNTPs was performed. The DNA repair enzymes methylpurine-DNA glycosylase and *Escherichia coli* exonuclease III were used to convert εA residues in the template into DNA strand breaks, which enabled precise localization of the εA residues within the *p53* gene. Hydroxyethano derivatives of adenine and cytosine in a template blocked T7 DNA polymerase and caused premature chain termination opposite adenine or one base before cytosine. After dehydration, both εA and εC were much more easily bypassed by T7 DNA polymerase. Formation of εG was identified as 'stop bands' one base before guanine residues. Modification of cytosine and guanine was additionally recognized by weakening or disappearance of non-specific stops on an undamaged template, probably due to steric hindrance by the tertiary DNA structure for polymerase. Etheno adduction of cytosine and guanine relaxed the compact DNA structure and enabled DNA polymerase to by-pass. In exons 5–8 of *p53*, 143 out of 500 sites appeared to be damaged by CAA, with four particularly densely modified regions between codons 135–147, 218–222, 234–255 and 284–292. The pattern of modification followed the pattern of *p53* mutations found in vinyl chloride-associated liver angiosarcomas in humans and rats, but only in regions that showed 100% homology with the human sequence. The factors that influence DNA damage and induction of mutations in the *p53* gene by CAA and vinyl chloride are discussed.

Introduction

Mutations in the *p53* tumour suppressor gene are commonly found in human cancers. Since loss of protein function may follow a mutational event at any site in the 500-nucleotide evolutionarily conserved region between exons 5 and 8, the spectrum of *p53* mutations provides clues to the etiology and molecular pathogenesis of neoplasia (Greenblat *et al.*, 1994). For example, cases of hepatocellular carcinomas in South Africa and Asia are linked to both hepatitis virus infection and consumption of aflatoxin B_1 in food contaminated with *Aspergillus flavis* (Wogan, 1992). One characteristic of these tumours is frequent AGG→AGT transversions targeted to codon 249 of the *p53* gene (Bressac *et al.*, 1991; Hollstein *et al.*, 1993). The causative role of aflatoxin B_1 in the developement of these tumours was demonstrated by the preferential development of the same mutations in a human liver cell line exposed to this carcinogen (Aguillar *et al.*, 1993) and by binding *in vitro* of the active intermediate, aflatoxin 8,9-epoxide, to *p53* cDNA targeted to the third guanine residue in codon 249 (Puisieux *et al.*, 1991). Furthermore, a 96% correlation was found between the sites of G→T mutations in cases of lung cancer and the sites of benzo[*a*]pyrene adduct formation within the *p53* gene sequence, both when *p53* cDNA was reacted with benzo[*a*]pyrene 7,8-diol-9,10-epoxide *in vitro* (Puisieux *et al.*, 1991) and when HeLa cells were exposed to this epoxide (Denissenko *et al.*, 1996). These data suggest that interaction of carcinogens with DNA is sequence-specific and that the sites of DNA damage are subsequently transformed into sites of mutations; however, mutational events might derive from both DNA damage and lack of repair. Preferential repair of the transcribed DNA strand (Evans *et al.*, 1993) often results in *p53* mutations from the non-transcribed strand (Greenblat *et al.*, 1994). Even in the transcribed strand, a differential rate of repair of specific sequences may occur. Slower repair of ultraviolet radiation-induced DNA damage in diploid human fibroblasts in culture was observed in all but one mutational hot spot found in the *p53* gene in ultraviolet radiation-associated skin cancer (Tornaletti & Pfeifer, 1994).

Vinyl chloride is an environmental carcinogen that is a causative factor in liver angiosarcomas in humans and causes liver angiosarcomas, hepatocellular carcinomas and other cancers in rodents (IARC, 1979). In cases of human liver angiosarcomas associated with exposure to vinyl chloride, very rare A:T→T:A transversions were found in codons 179, 249 and 255 of the *p53* gene (Hollstein *et al.*, 1994). In contrast, in angiosarcomas that were not associated with vinyl chloride, mutations targeted to other sites within the *p53* gene were found: two G:C→A:T transitions in codon 141 and a C:G→T:A transition in codon 136 (Soini *et al.*, 1995). In rats exposed to vinyl chloride, *p53* mutations were also targeted mainly to A:T base pairs (Barbin *et al.*, 1997).

In mammalian cells, vinyl chloride is activated into chloroethylene oxide by cytochrome P450-dependent microsomal monooxygenases (Bartsch *et al.*, 1994). Chloroethylene oxide binds directly to nitrogen atoms on the DNA bases, forming adducts in the following quantitative order: *N*7-(2-oxoethyl)guanine >> 1,N^6-εA > hydroxyethanoguanine > N^2,3-εG > 3,N^4-εC > 1,N^2-εG (Guengerich *et al.*, 1993; Langouët *et al.*, 1997; Müller *et al.*, 1997). Hydroxyethanoguanine seems to be stable and undergoes rearrangement to N^2-(2-oxoethyl)guanine rather than dehydration to the etheno derivative (Langouët *et al.*, 1997). Alternatively, chloroethylene oxide rearranges to form chloroacetaldehyde (CAA). CAA binds to adenine and cytosine in DNA, forming mainly hydroxyethano derivatives, which subsequently dehydrate to 1,N^6-εA and 3,N^4-εC. In polyribonucleotides, the conversion of hydroxyethanocytosine to εC occurs with a calculated half-life of 15 h, and conversion of hydroxyethanoadenine to εA occurs with a half-life of 1.4 h at 37 °C at pH 7.25 (Kuśmierek & Singer, 1982). These hydrated forms might also be formed in DNA of animals exposed to vinyl compounds (Swenberg *et al.*, 1992), but their biological significance is poorly elucidated. Reaction of CAA with guanine in DNA favours formation of N^2,3-εG; 1,N^2-εG is also formed but with at least 100-fold lower efficiency (Kuśmierek & Singer, 1992). If N1 is not blocked by hydrogen bonding in free nucleosides, then formation of 1,N^2-εG prevails over that of N^2,3-εG after modification with CAA (Guengerich & Persmark, 1994). The quantitative relationships among etheno adducts induced by CAA in double-stranded DNA were reported to be the following: 3,N^4-εC ≥ 1,N^6-εA > N^2,3-εG >>> 1,N^2-εG (Kuśmierek & Singer, 1992).

Only one study has so far been performed *in vitro* on single-stranded M13 DNA by a sequencing technique. This showed that DNA damage by CAA is non-random and probably sequence-specific (Premaratne *et al.*, 1993). Here, we show that CAA damage to double-stranded DNA of the *p53* gene is also sequence-dependent.

Materials and methods

Modification of plasmid DNA

Plasmid pSP65 bearing full-length cDNA from the human *p53* gene (a kind gift of Dr P. Hainaut, IARC) was grown in *E. coli* strain DH5α (*recA1, endA1, relA1, thi-1, gyrA96, hsdR17, ΔlacU169, supE44/[F(φ80d(lacZ)ΔM15]*) and isolated by alkaline lysis with final purification by cetyldimethylethylammonium bromide precipitation, as described elsewhere (Sambrook *et al.*, 1989). The plasmid was reacted with 360 mmol/L CAA in 0.3 mol/L cacodylate buffer pH 7.6 at 37 °C for 3 h. After reaction, DNA was ethanol-precipitated, washed and resuspended in 10 mmol/L cacodylate buffer pH 6.5 for dehydration of hydroxyethano into etheno adducts, which was performed for 72 h at 37 °C. After dehydration, the DNA was ethanol-precipitated, resuspended in 10 mmol/L Tris–HCl buffer pH 8.0 with 1 mmol/L EDTA, as were samples containing hydrated cyclic acids, and frozen. The integrity of the modified plasmids was verified by agarose gel electrophoresis. CAA modification did not increase the number of DNA strand breaks in plasmids containing hydrated cyclic adducts or in plasmids containing dehydrated or etheno adducts, as judged by the ratio of covalently closed circular to open circular form of the plasmid (data not shown).

Selective excision of modified bases from DNA

εA residues in the template were converted into DNA strand breaks by digestion of the plasmid with 3-methyladenine-DNA glycosylases from humans (ANPG; a kind gift of Dr J. Laval, Institut Gustave Roussy, France) or from *E. coli* (AlkA; purified from an overproducing strain as described by Tudek *et al.*, 1998) and subsequently with *E. coli* exonuclease III (Promega). The standard reaction mixture for the AlkA protein contained 50 µg plasmid DNA; 70 mmol/L Hepes-KOH, pH 7.6; 1 mmol/L EDTA; 1 mmol/L β-mercaptoethanol and 5% glycerol. That for ANPG contained 70 mmol/L Hepes-KOH, pH 7.6; 1 mmol/L EDTA; 100 µmol/L KCl. Pure enzymes, as judged by sodium dodecyl sulfate–polyacrylamide gel electrophoresis, were added at concentrations of 0.09 µg AlkA and 0.14 µg ANPG. The mixtures were incubated for 30 min at 37 °C; then, Mg^{2+} (final concentration, 1 mmol/L), which are necessary for the activity of exonuclease III from *E.* coli (Xth), were added and, if necessary, KCl (final concentration, 100 mmol/L) and β-mercaptoethanol (5 mmol/L). The plasmid was then incubated with 1 U Xth for a further 30 min at 37 °C. The enzymes were removed by chloroform extraction, and the DNA was ethanol-precipitated and used as a template with selectively removed base derivatives for DNA synthesis *in vitro.*

DNA polymerase fingerprint analysis

The plasmid DNA was denatured with NaOH for 30 min, as described elsewhere (Sambrook *et al.*, 1989), and annealed with one of the primers complementary to exons 5–8 of the following sequences:

R3, 5′-CTGGAGTCTTCCAGTGTGAT-3′
R10, 5′-AAATATTCTCCATCCAGT-3′
R11, 5′-AAATTTCCTTCCACTCGG-3′
R13, 5′-TTCCGTCCCAGTAGATTACC-3′
F3, 5′-GTTGGCTCTGACTGTACCAC-3′
F10, 5′-TCATCTTCTGTCCCTTCCC-3′
F12, 5′-ACTCCCCTGCCCTCAACAAG-3′
F11, 5′-GTTGATTCCACACCCCCG-3′.

The primers were labelled by addition of excess (1–2 µl) of [α-^{35}S]dATP5′ (1000 Ci/mmol) in labelling mix (Amersham) for sequencing reactions and T7 DNA polymerase (4 U per sample). Labelling was performed for 5 min at room temperature (17–20 °C). Subsequently, four dNTPs, each at final concentration of 50 µmol/L, were added, and the samples were incubated for 10 min at 37 °C for chain elongation. The reaction was terminated by addition of formamide, and the synthesis products were analysed by electrophoresis on a sequencing gel. Unmodified plasmid DNA, used as a reference ladder, was sequenced by the Sanger method with an Amersham Sequenase Version 2.0 Kit (Sambrook *et al.*, 1989).

The sites of DNA modification were identified as sites of premature chain termination appearing only on modified templates or as weakening or disappearance on the modified template of the 'stop bands' that had been observed on unmodified DNA, due to a specific tertiary DNA structure (for explanations, see Results and Discussion). Because of low resolution of the 'stop bands', their intensity was estimated arbitrarily on a scale of 0–40, the most intense bands given the value of 40. This provided an approximate assessment of the degree of inhibition of DNA polymerization by the lesion found at that site. When DNA repair enzymes were used, which cleaved the DNA at the site of damage and arrested DNA synthesis, the intensity of the 'stop bands' reflected the relative strength of the modification, for instance for εA and hydroxyethanocytosine. If the DNA repair enzymes did not recognize the lesion, the intensity of the 'stop bands' reflected a compromise value between the number of lesions present at that particular site and their ability to inhibit DNA synthesis.

Results and discussion

Identification of chloroacetaldehyde-induced base lesions in the p53 gene by DNA polymerase fingerprint analysis

Damage to DNA bases in a template may result in arrest of DNA synthesis one base before, opposite the impaired base or one base after; alternatively, DNA polymerase may by-pass the damaged site. We studied the profile of double-stranded DNA damage induced *in vitro* by CAA in the human *p53* gene by investigating arrest of DNA synthesis by T7 DNA polymerase, either directly or after conversion of the damaged bases into DNA strand breaks by DNA glycosylases specific to εA and apurinic/apyrimidinic endonuclease. Premature chain terminations due to damage to the template were observed in CAA-modified DNA, but when full conversion of hydrated adducts into etheno adducts took place during dehydration, the intensity and sometimes the position of the 'stop bands' was changed. In DNA containing hydrated cyclic adducts, in which the adenine hydrates were partially converted into εA, 'stop bands' were observed opposite the adenine residues. Additional bands appeared one base before adenine after cleavage of the template with DNA repair enzymes. Thus, DNA polymerase introduced the base opposite an adenine hydrate, but its extension was limited.

The hydrated form of εA did not appear to be recognized by AlkA, ANPG or Xth proteins, since digestion of DNA containing hydrated cyclic adducts with these enzymes resulted in the appearance of new chain termination sites one base before adenine, due to fragmentation of the template, but the bands opposite adenine were still present (Figure 1). In DNA containing dehydrated or etheno adducts, discrete 'stop bands' were observed one base before the adenine residues and were enhanced by cleavage of the template with AlkA, ANPG and Xth proteins (Figure 1). εA appears to be a much weaker block for T7 DNA polymerase than its hydrated form and arrests DNA synthesis one base before the

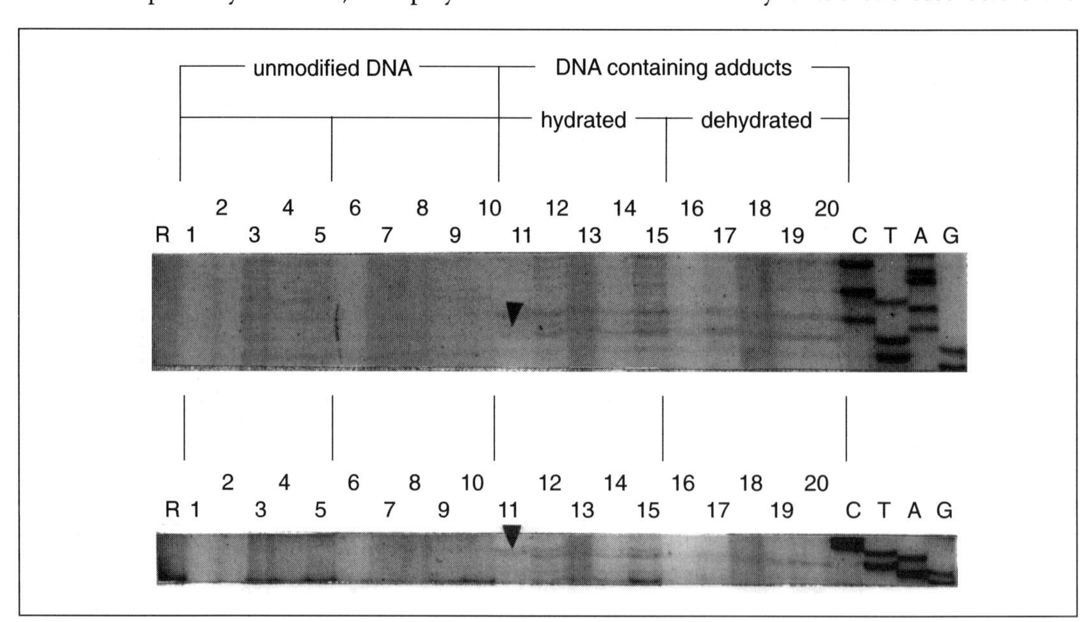

Figure 1. DNA polymerase fingerprint analysis of effect of adenine modification on DNA synthesis *in vitro* by T7 DNA polymerase

Lanes C, T, A, G: sequence of template. Lane R: primer extension on unmodified plasmid. Lanes 1–10: primer extension on unmodified plasmid, which was processed as the modified plasmid but without chloroacetaldehyde (CAA) (lanes 1–5, processed as DNA containing hydrated adducts; lanes 6–10 processed as DNA containing dehydrated adducts). Lanes 11–20: primer extension on CAA-modified plasmid (lanes 11–15, hydrated; lanes 16–20, dehydrated). In lanes 1, 6, 11, 12, no DNA repair enzymes were used; in lanes 2, 7, 12, 17, plasmid was digested with human 3-methyladenine-DNA-glycosylase (ANPG) alone; in lanes 3, 8, 13, 18, with ANPG and exonuclease from *Escherichia coli* (Xth); in lanes 4, 9, 14, 19, with 3-methyladenine-DNA-glycosylase from *E. coli* (AlkA) and Xth; in lanes 5, 10, 15, 20, with Xth alone. Arrows indicate 'stop bands' appearing due to modification of adenine.

lesion under the conditions used here. With higher concentrations of dNTP and longer DNA polymerization, a different pattern of chain termination by εA was observed: one base before, opposite or one to three bases after, or the chain was elongated, depending on the DNA polymerase used and the nucleotides 5′ adjacent to the εA residue (Litinski *et al.*, 1997). We cannot, however, exclude the possibility that the weak 'stop bands' observed one base before adenine derived from ring-opened εA, which could be formed during alkaline denaturation of the plasmid in sequencing reactions. 4-Amino-5-(imidazol-2-yl)imidazole (β), which is formed from εA in alkali (Yip & Tsou, 1973), efficiently inhibits DNA synthesis in *E. coli* when incorporated into a replicating template (Basu *et al.*, 1993). The time for full conversion of εA to β is well over the 30 min needed for plasmid denaturation, since during 2 h of incubation of a hexamer containing a single εA residue at pH 13.0 only 25% of εA was converted into the β product (Basu *et al.*, 1993).

Modification of cytosine residues by CAA was identified by one of two means. A 'stop band' that appeared one base before cytosine on DNA containing hydrated cyclic adducts was enhanced by cleavage with the DNA repair enzymes and had disappeared on templates containing dehydrated or etheno adducts (Figure 2). This was surprising, since it has been suggested (Borys *et al.*, 1994) that ANPG and AlkA can excise cytosine hydrate but not εC. This suggestion was made after the observation that induction of an adaptive response to alkylating agents in *E. coli* markedly reduced the number of C→T mutations in M13*glyU* phage when phage containing hydroxyethanocytosine was transfected into *E. coli*, and the adaptive response had a much less pronounced effect on mutation induction when εC residues prevailed over the hydrated form in phage DNA (Borys *et al.*, 1994). The potential contribution of AlkA and ANPG to the repair of hydroxyethanocytosine needs further research. No repair of εC by AlkA and ANPG was reported in oligonucleotides containing a single εC residue (Hang *et al.*, 1996; Saparbaev & Laval, 1998). Inhibition of DNA synthesis before the cytosine residues on the template containing hydrated cyclic adducts suggests that cytosine hydrate inhibits DNA synthesis one base before the lesion, while εC is easily by-passed by T7 DNA polymerase. Like the hydrated forms of εA and εC, hydroxyethanoguanine is a stronger block for DNA synthesis by various prokaryotic DNA polymerases and rat polymerase β than 1,N^2-εG (Langouët *et al.*, 1997).

Some sites rich in cytosine and guanine constituted a block for polymerase, probably due to their peculiar tertiary structure, and strong 'stop bands' were seen at these sites in unmodified DNA. CAA modification weakened this inhibition in DNA containing hydrated cyclic adducts and on the template containing dehydrated or etheno adducts, and even weaker or no 'stop bands' were

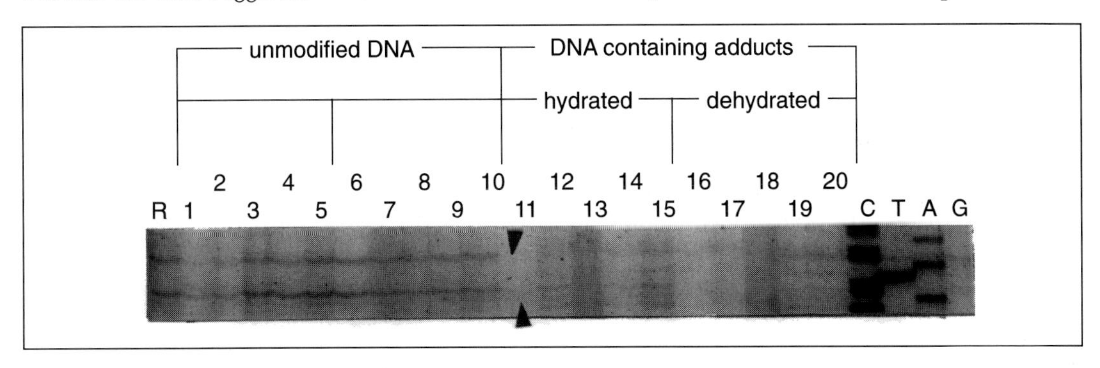

Figure 2. Effect of cytosine modification by chloroacetaldehyde (CAA) on DNA synthesis *in vitro* by T7 DNA polymerase

For explanations, see legend to Figure 1. Lower arrow indicates appearance of 'stop bands' one base before cytosine on template with hydrated adducts, their enhancement by cleavage with DNA repair glycosylases and their disappearance in DNA with dehydrated adducts. The upper arrow indicates disappearance of 'structural stops' after cytosine modification with CAA.

found at these positions (Figure 2). Addition of an etheno group to the 3,N^4 position of cytosine relaxed the compact structure of C:G-rich regions, which were thus by-passed more readily by polymerase, allowing identification of CAA-damaged cytosine residues. Gelfand *et al.* (1998) found that the presence of a single εC residue in DNA causes thermodynamic destabilization of the double helix, which is far from local. Apparently, the energetic consequences of the lesion propagate well beyond the site of structural modification.

Modification of guanine caused the appearance of chain termination sites one base before the guanine residues in both DNA containing hydrated and dehydrated adducts. T7 DNA polymerase was arrested one base before 1,N^2-εG, the minor product of the reaction of CAA with DNA, and before hydroxyethanoguanine (Langouët *et al.*, 1997). Formation of apurinic/apyrimidinic sites that block DNA synthesis might also be expected, owing to the relatively labile *N*-glycosidic bond of N^2,3-εG (Kuśmierek *et al.*, 1989). It was because of the lability of the glycosyl bond, however, that the effect of the presence of N^2,3-εG in the template on the rate of DNA synthesis was not studied. Modification of guanine residues by CAA also caused weakening of non-specific 'structural stop bands' in G:C-rich regions in front of guanine sites in the template (Figure 3). As expected, almost no stops were induced by thymine residues on the modified template (Figure 3), since thymine lacks an exo-

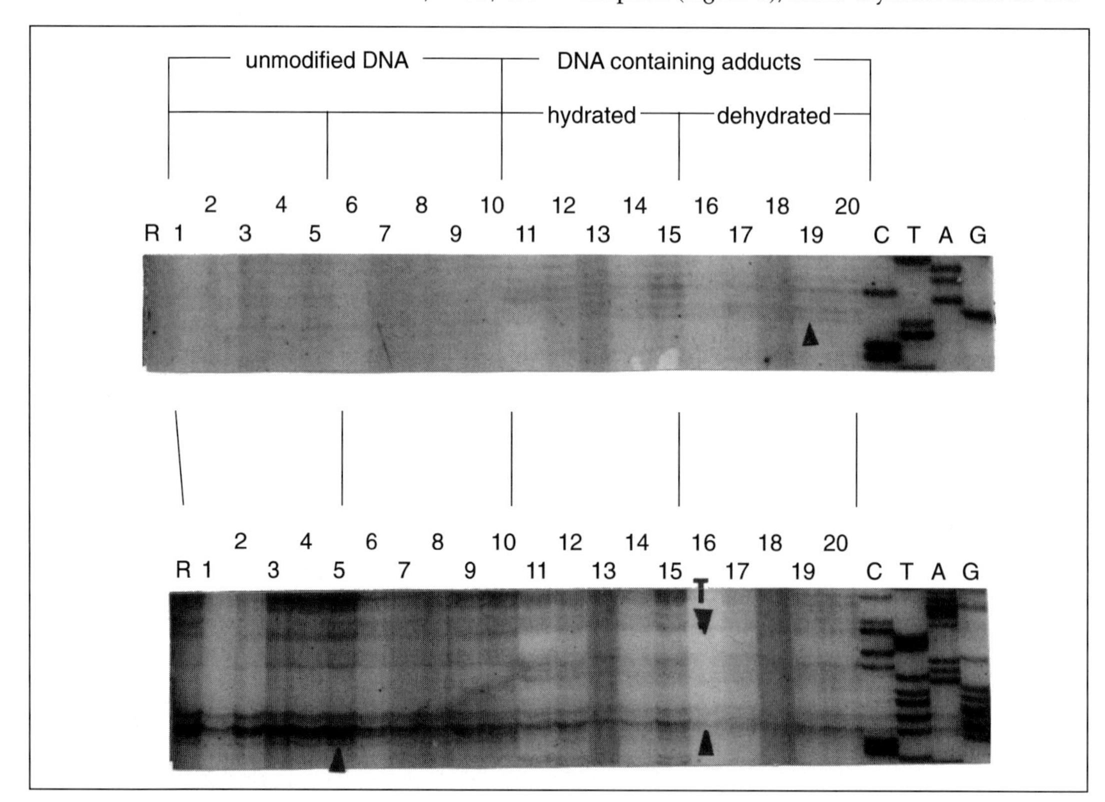

Figure 3. Effect of guanine modification by chloroacetaldehyde (CAA) on DNA synthesis *in vitro* by T7 DNA polymerase

For explanations, see legend to Figure 1. Upper photo: arrow indicates the band one base before guanine on CAA-modified DNA. Lower photo: left arrow on unmodified DNA indicates region of strong structural hindrance for polymerase T7. Right arrow in the same base position indicates weakening of non-specific 'structural stop' after CAA modification of guanine residue. Upper arrow (T) indicates lack of stop band opposite thymine residue.

cyclic nitrogen atom and does not react with CAA (Bartsch *et al.*, 1994). However, about 3% premature chain terminations were found one base before thymine sites on the template of CAA-modified DNA. Generally, 39% premature chain terminations associated with modification of bases were found one base before the adenine residues, 30% before the guanine residues, 28% before the cytosine residues and 3% before the thymine residues.

Localization of modifications due to chloroacetaldehyde in the human p53 gene

DNA polymerase fingerprint analysis, ligation-mediated polymerase chain reaction (PCR) and terminal transferase-dependent PCR have been used to study the sequence-dependence of DNA damage and repair. Each of these methods has its limitations and all probably require supplementation with another one. The ligation-mediated PCR requires DNA repair enzymes that recognize the lesion and cleave the phosphoester bond at or near the site of the damage. Because of the selected substrate specificity of repair proteins, however, this method might result in an underestimate of the extent of DNA damage (Pfeifer *et al.*, 1993). Terminal transferase-dependent PCR (Komura & Riggs, 1998) and DNA polymerase fingerprint analysis sense any DNA irregularity that stops replication, and, since the effect of a damaged base can extend beyond the site of damage (Langouët *et al.*, 1997; Litinski *et al.*, 1997), both methods can give false-positive results or, when the lesion does not affect replication, false-negative results. They can, however, be very useful in assessing the effect of a tertiary DNA structure on DNA damage and repair and the induction of mutations.

DNA polymerase fingerprint analysis has been used extensively to study the effect of DNA damage on replication *in vitro*. These studies revealed marked differences in the mode of interaction with DNA between different classes of chemical carcinogens. While alkylating agents like dimethyl sulfate cause a relatively random alkylation pattern (O'Connor *et al.*, 1988), as subsequently also demonstrated in tissue cultures (Bouziane *et al.*, 1998), compounds that form bulky adducts, like aflatoxin B_1 and benzo[*a*]pyrene show a tendency to bind to specific sequences (Puisieux *et al.*, 1991), and a similar sequence-specific reaction of CAA with single-stranded DNA was suggested (Premaratne *et al.*, 1993).

Using this method, we have found that CAA-induced modifications are distributed non-randomly within the *p53* gene. Several clusters of modifications were found in codons 135–147, 218–222, 234–255 and 284–292. Other regions were modified less densely, e.g. codons 174–181 and 190–204, and no modification could be detected in some, e.g. codons 214–217 and 256–259 (Figure 4). Base modifications were detected at sequences that probably resulted in steric hindrance for DNA polymerase, since 'stop bands' at these positions were present on unmodified template; 93 such 'structural stops' were found in exons 5–8 of the human *p53* gene (Figure 5). Out of 143 sites found to be modified by CAA, 71 were positioned exactly at or adjacent to such sites, suggesting that these sequences interact preferentially with CAA due to the tertiary DNA structure. Structural perturbations in supercoiled DNA, cruciform DNA structures, Z-DNA zones and misaligned purines were shown to be preferential sites for the interaction of haloaldehydes with DNA (Lilley, 1986; Bartsch *et al.*, 1994). We found several inverted repeats within the *p53* gene sequence, which were potentially capable of forming cruciform structures, although not all of these sequences were found to be modified by CAA (Figure 6). However, all of the densely modified regions, except codons 218–220, contained sequences that were either able to form cruciforms or were rich in –GC– bases, thus confirming the contribution of a tertiary DNA structure to the spectrum of DNA modifications by CAA.

The spectrum of CAA-induced modifications was quite different from the general spectrum of *p53* mutations (Figure 4), although extensive modification was observed in three mutational 'hot spots', between codons 245 and 250. In human liver angiosarcomas associated with exposure to vinyl chloride, A:T→T:A transversions were found in codons 179, 249 and 255, and these positions were readily modified by exposure of the *p53* gene to CAA *in vitro* (Figure 7). Clusters of CAA modifications were also found, however, in mutated regions in angiosarcomas that were not associated with exposure to vinyl chloride, namely codons 136, 141 and 146.

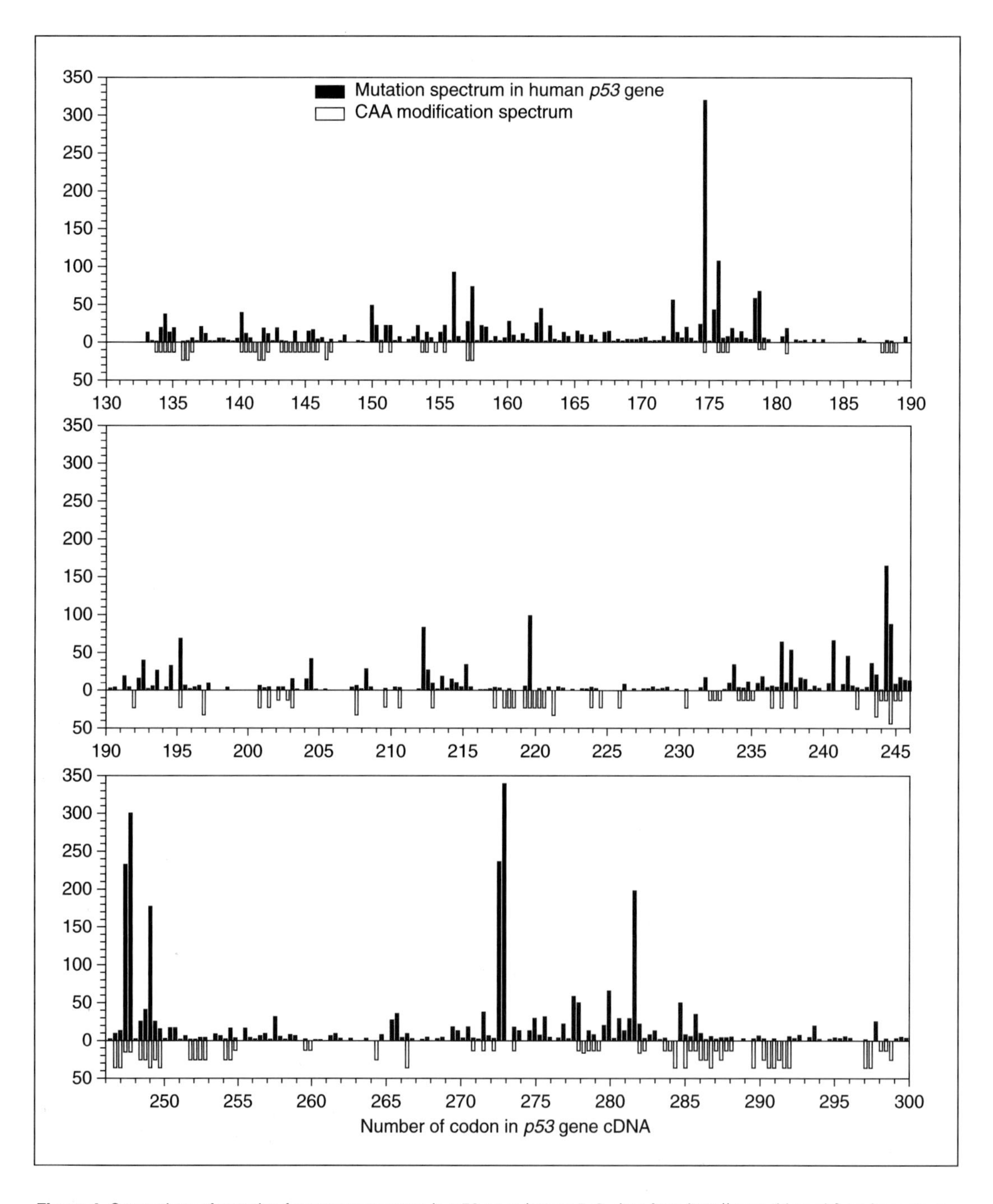

Figure 4. Comparison of mutation frequency spectrum in *p53* gene (exons 5–8, data from http://www.ebi.ac.uk/) and spectrum of chloroacetaldehyde (CAA) modifications

Black columns: position and number of mutations found in human cancers. White columns: position and relative strength of modification, as judged by the apparent intensity of 'stop bands' (for explanations, see Materials and methods)

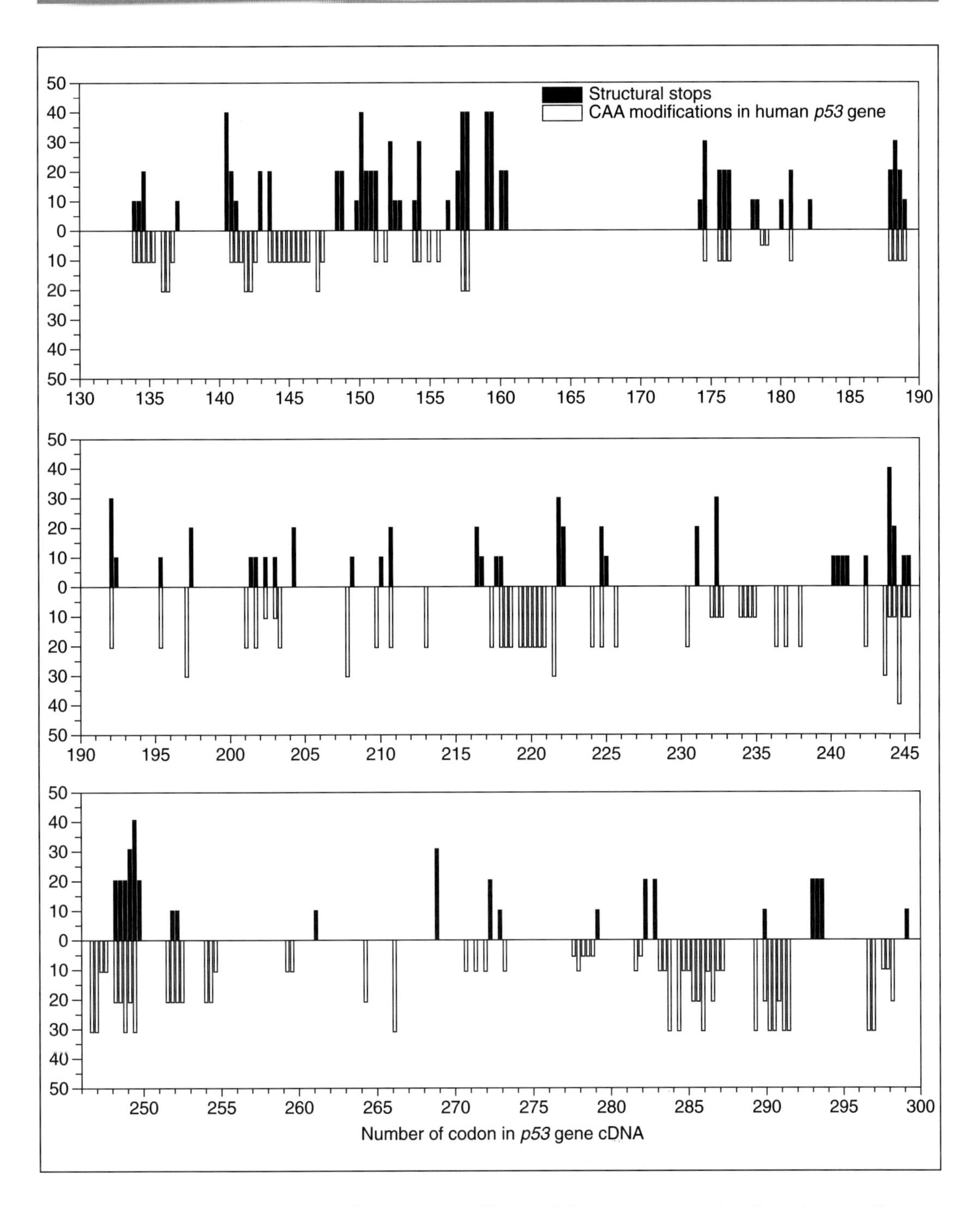

Figure 5. Spectrum of chloroacetaldehyde (CAA)-induced modifications in human *p53* gene and positions of non-specific 'structural stops' within exons 5–8

Black columns: position and relative strength of structural stops. White columns: position and relative strength of CAA modifications

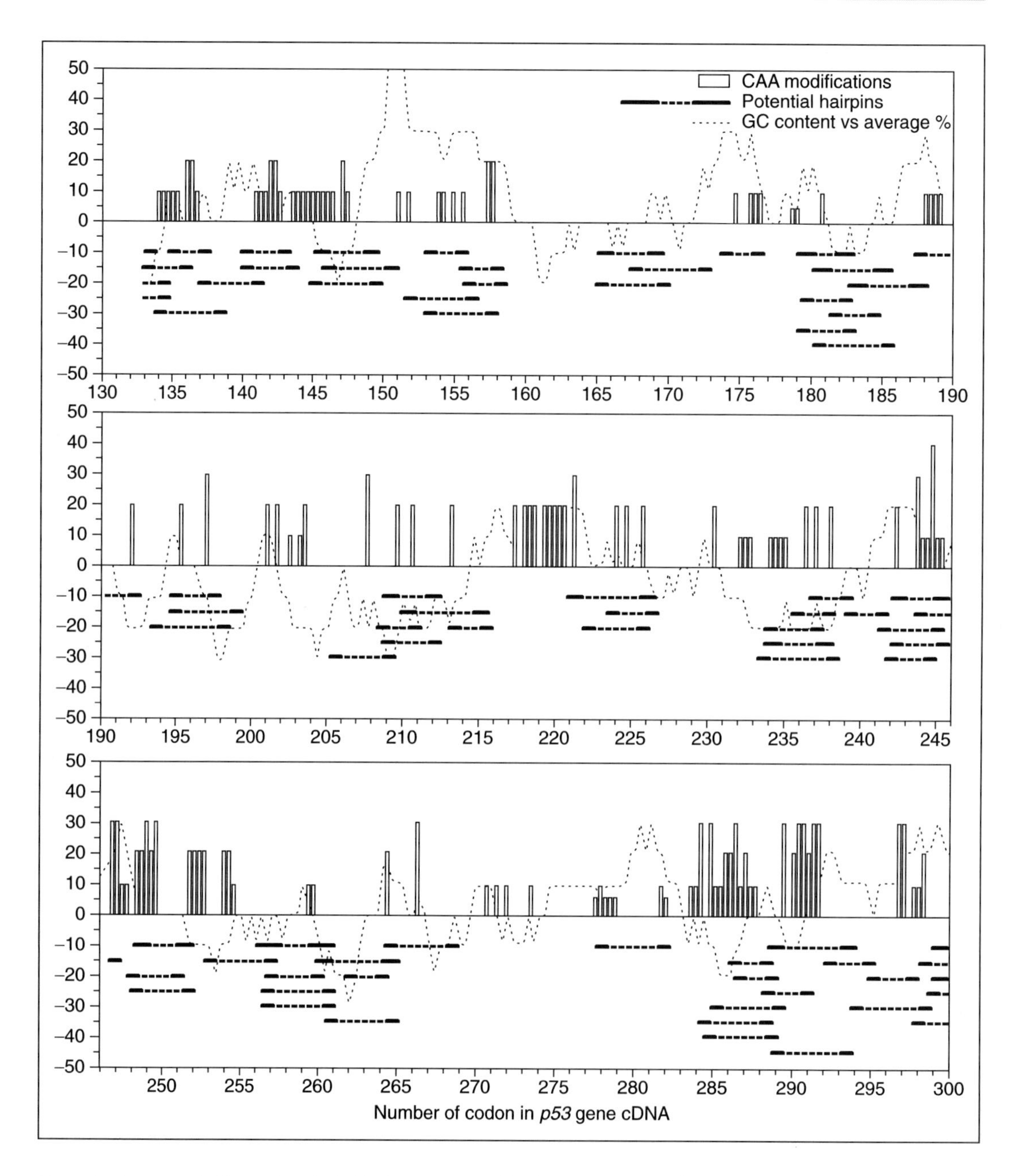

Figure 6. Spectrum of chloroacetaldehyde (CAA)-induced modifications in exons 5–8 of human *p53* gene (white columns) with positions of inverted repeats in the same region that could form cruciform structures and percentage of guanine and cytosine residues in given DNA fragment

G:C content was calculated for nine-nucleotide fragments, and its percentage was compared with that of guanine and cytosine in the whole fragment of *p53* gene containing exons 5–8. Sequences containing only G:C pairs are indicated as +50, and those containing only A:T base pairs as –50.

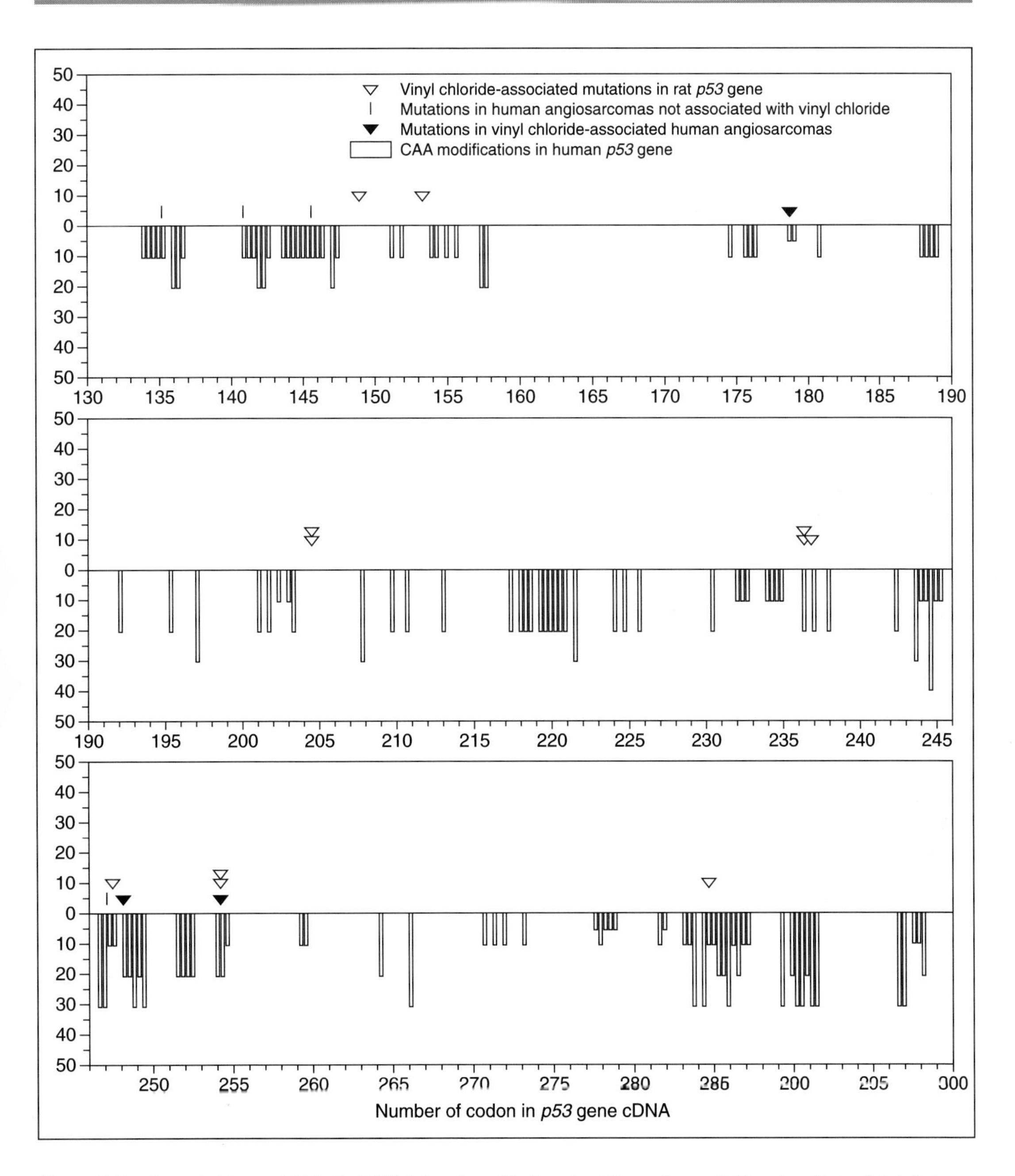

Figure 7. Spectrum of chloroacetaldehyde (CAA)-induced modifications in *p53* gene (exons 5–8) and positions of mutations found in liver angiosarcomas of humans and rats

White columns: position and relative strength of modification. Black triangles: position of mutations found in vinyl chloride-associated human liver angiosarcomas. White triangles: position of mutations found in vinyl chloride-associated rat liver tumours. Vertical lines: position of mutations found in human liver angiosarcomas not associated with exposure to vinyl chloride. Each symbol represents one reported case

Since the human data are very limited, we compared the spectrum of CAA modifications with vinyl chloride-induced mutations in the rat *p53* gene. Eight of 11 mutations that were present in the sequences with strong homology to the human gene correlated with the sites of CAA modifications *in vitro*. Particularly strong homology was found in codons 237, 248, 255 and 285, which correspond to rat codons 235, 246, 253 and 283. In codon 237, no modification was found of the transcribed strand in the second position mutated in the rat *p53* gene, ATG→AGG. This mutation is probably derived from modification of a complementary sequence. The sites of four other mutations in rat codons 149, 154, 162 and 205 were not modified by CAA in the human *p53* gene, but the human and rat sequence in this region shows poor homology (Bienz-Tadmor *et al.*, 1985; Soussi *et al.*, 1988).

These observations strongly support the sequence-specific interaction of vinyl compounds with DNA, which subsequently determines the spectrum of mutations. An intriguing question arising from the analysis of the spectrum of CAA modifications of the *p53* gene and the spectrum of mutations induced by vinyl chloride is the discrepancy between the preferential targeting of mutations to A:T base pairs and the almost equal numbers of adenine, guanine and cytosine sites modified by CAA. All human (Hollstein *et al.*, 1994) and eight of 11 rat *p53* mutations were associated with A:T and only three with G:C base pairs (Barbin *et al.*, 1997). Furthermore, over 50% of *ras* proto-oncogene mutations in rat tumours induced by vinyl chloride were targeted to A:T base pairs (Froment *et al.*, 1994). Several explanations are possible: (i) The spectrum of chloroethylene oxide-induced DNA damage is similar to that of CAA, but the ratio between the base derivatives is different. The amount of εA in chloroethylene oxide-modified DNA is about one order of magnitude higher than that of mutagenic guanine derivatives (Barbin *et al.*, 1985; Cheng *et al.*, 1991; Langouët *et al.*, 1997; Müller *et al.*, 1997) and about two orders of magnitude higher than that of εC (Müller *et al.*, 1997). In CAA-modified DNA, the amount of εA is lower or equal to that of εC and is less than twice as high as that of εG (Kuśmierek & Singer, 1992). In cells, where there is a limited degree of rearrangement due to chloroethylene oxide in comparison with CAA, εA might be the quantitatively predominant lesion. (ii) A total of 56 adenine sites in the *p53* gene interacted readily with CAA, and only a few were mutated in vinyl chloride-associated human and rat liver angiosarcomas (Figure 7). Although the data on carcinogenicity are very limited, preferential repair of some εA residues in various *p53* regions cannot be excluded. Up to 200-fold differences in the efficiency of *N*7-methylguanine repair by the ANPG protein was observed in mammalian cells (T.R. O'Connor, personal communication). (iii) Finally, some *p53* mutations might be favoured because of selected expansion of mutated clones. Although we now know that modification of DNA by CAA is sequence-specific and overlaps with the spectrum of mutations induced by vinyl chloride in the *p53* gene, the mechanism of transformation of DNA damage into mutations is still not well understood.

Acknowledgements

We would like to thank Dr J. Laval (Institut Gustave Roussy, Villejuif, France) for the kind gift of ANPG protein and helpful discussions and Dr J.T. Kuśmierek (Institute of Biochemistry and Biophysics, Polish Academy of Sciences, Warsaw, Poland) for helpful discussions and interest in this work. This study was supported by European Commission grant No. ENV4-CT97-0505 through a collaborative research agreement between IARC and the Institute of Biochemistry and Biophysics, Polish Academy of Sciences, Warsaw (GI/43/4), and by the Polish–French Center of Plant Biotechnology, grant No. C-2/V/8.

References

Aguillar, F., Hussain, S.P. & Cerutti, P. (1993) Aflatoxin B_1 induces the transversion of G to T in codon 249 of the *p53* tumor suppressor gene in human hepatocytes. *Proc. Natl Acad. Sci. USA* 90, 8586–8590

Barbin, A., Laib, R.J. & Bartsch, H. (1985) Lack of miscoding properties of 7-(2-oxoethyl)guanine, the major vinyl chloride–DNA adduct. *Cancer Res.* 45, 2440–2444

Barbin, A., Froment, O., Boivin, S., Marion, M.-J., Belpoggi, F., Maltoni, C. & Montesano, R. (1997) *p53* gene mutation pattern in rat liver tumors induced by vinyl chloride. *Cancer Res.* 57, 1695–1698

Bartsch, H., Barbin, A., Marion, M.-J., Nair, J. & Guichard, Y. (1994) Formation, detection, and role in carcinogenesis of ethenobases in DNA. *Drug Metab. Rev.* 26, 349–371

Basu, A.K., Wood, M.L., Niedernhofer, L.J., Ramos, L.A. & Essigmann, J.M. (1993) Mutagenic and genotoxic effects of three vinyl chloride-induced DNA lesions: 1,N^6-ethenoadenine, 3,N^4-ethenocytosine and 4-amino-5-(imidazol-2-yl)imidazole. *Biochemistry* 32, 12793–12801

Bienz-Tadmor, B., Zakut-Houri, R., Libresco, S., Givol, D. & Oren, M. (1985) The 5′ region of *p53* gene: Evolutionary conservation and evidence for negative regulatory element. *EMBO J.* 4, 3209–3213

Borys, E., Mroczkowska-Slupska, M.M. & Kuśmierek, J.T. (1994) The induction of adaptive response to alkylating agents in *Escherichia coli* reduces the frequency of specific C→T mutations in chloroacetaldehyde-treated M13 *glyU* phage. *Mutagenesis* 9, 407–410

Bouziane, M., Miao, F., Ye, N., Holmquist, G., Chyzak, G. & O'Connor, T.R. (1998) Repair of DNA alkylation damage. *Acta Biochim. Pol.* 45, 191–202

Bressac, B., Kew, M., Wands, J. & Ozturk, M. (1991) Selective G to T mutations of *p53* gene in hepatocellular carcinoma from southern Africa. *Nature* 350, 429–431

Cheng, K.C., Preston, B.D., Cahill, D.S., Dosanjh, M.K., Singer, B. & Loeb, L.A. (1991) The vinyl chloride DNA derivative N^2,3-ethenoguanine produces G→A transitions in *Escherichia coli*. *Proc. Natl Acad. Sci. USA* 88, 9974–9978

Denissenko, M.F., Pao, A., Tang, M. & Pfeifer, G.P. (1996) Preferential formation of benzo[*a*]pyrene adducts at lung cancer mutational hotspots in *p53*. *Science* 274, 430–432

Evans, M.K., Taffe, B.G., Harris, C.C. & Bohr, V.A. (1993) DNA strand bias in the repair of the *p53* gene in normal human and xeroderma pigmentosum group C fibroblasts. *Cancer Res.* 53, 5377–5381.

Froment, O., Boivin, S., Barbin, A., Bancel, B., Trepo, C. & Marion, M.-J. (1994) Mutagenesis of *ras* proto-oncogenes in rat liver tumors induced by vinyl chloride. *Cancer Res.* 54, 5340–5345

Gelfand, C.A., Plum, G.E., Grollman, A.P., Johnson, F. & Breslauer K.J. (1998) The impact of an exocyclic cytosine adduct on DNA duplex properties: Significant thermodynamic consequences despite modest lesion-induced structural alterations. *Biochemistry* 37,12507–12512

Greenblatt, M.S., Bennett, W.P., Hollstein, M. & Harris, C.C. (1994) Mutations in the *p53* gene: Clues to cancer etiology and molecular pathogenesis. *Cancer Res.* 54, 4855–4878

Guengerich, F.P. & Persmark, M. (1994) Mechanism of formation of ethenoguanine adducts from 2-haloacetaldehydes: ^{13}C-labelling patterns with 2-bromoacetaldehyde. *Chem. Res. Toxicol.* 7, 205–208

Guengerich, F.P., Persmark, M. & Humphreys, W.G. (1993) Formation of 1,N^2- and N^2,3-ethenoguanine from 2-halooxiranes: isotopic labelling studies and isolation of hemiaminal derivative of N^2-(2-oxoethyl)guanine. *Chem. Res. Toxicol.* 6, 635–648

Hang, B., Chenna, A., Rao, S. & Singer, B. (1996) 1,N^6-Ethenoadenine and 3,N^4-ethenocytosine are excised by separate human DNA glycosylases. *Carcinogenesis* 19, 155–157

Hang, B., Medina, M., Fraenkel-Conrat, H. & Singer, B. (1998) A 55-kDa protein isolated from human cells shows DNA glycosylase activity toward 3,N^4-ethenocytosine and the G/T mismatch. *Proc. Natl Acad. Sci. USA* 95, 13561–13566

Hollstein, M.C., Wild, C.P., Bleicher, F., Chutimataewin, S., Harris, C.C., Srivatanakul, P. & Montesano, R. (1993) *p53* mutations and aflatoxin B_1 exposure in hepatocellular carcinoma patients from Thailand. *Int. J. Cancer* 53, 51–55

Hollstein, M., Marion, M.-J., Lehman, T., Welsh, J., Harris, C.C., Martel-Planche, G., Kusters, I. & Montesano, R. (1994) *p53* mutations at A:T base pairs in angiosarcomas of vinyl chloride-exposed factory workers. *Carcinogenesis* 15, 1–3

IARC (1979) *IARC Monographs on the Evaluation of Carcinogenic Risk of Chemicals to Humans*, Vol. 19. Lyon, IARC, pp. 377–438

Komura, J.-I. & Riggs, A.D. (1998) Terminal transferase-dependent PCR: A versatile and sensitive method for *in vivo* footprinting and detection of DNA adducts. *Nucleic Acids Res.* 26, 1807–1811

Kuśmierek, J.T. & Singer, B. (1982) Chloroacetaldehyde-treated ribo- and deoxyribopolynucleotides. 1. Reaction products. *Biochemistry* 21, 5717–5722

Kuśmierek, J.T. & Singer, B. (1992) 1,N^2-Ethenodeoxyguanosine: Properties and formation in chloroacetaldehyde-treated polynucleotides and DNA. *Chem. Res. Toxicol.* 5, 634–638

Kuśmierek, J.T., Folkman, W. & Singer, B. (1989) Synthesis of N^2,3-ethenodeoxyguanosine, N^2,3-ethenodeoxyguanosine 5′-phosphate, and N^2,3-ethenodeoxyguanosine 5′-triphosphate. Stability of the glycosyl bond in the monomer and in poly(dG,(dG–dC). *Chem. Res. Toxicol.* 2, 230–233

Langouët, S., Müller, M. & Guengerich, F.P. (1997) Misincorporation of dNTPs opposite 1,N^2-ethenoguanine and 5,6,7,9-tetrahydro-7-hydroxy-9-oxoimidazo[1,2-α]purine in oligonucleotides by *Escherichia coli* polymerases I exo- and II exo-, human immunodeficiency virus-1 reverse transcriptase and rat polymerase β. *Biochemistry* 36, 6069–6079

Lilley, D.M.J. (1986) Cyclic adduct formation at structural perturbations in supercoiled DNA. In: *The Role of Cyclic Nucleic Acid Adducts in Carcinogenesis and Mutagenesis* (Singer, B. & Bartsch, H., eds), IARC Scientific Publications No. 70. Lyon, IARC, pp. 83–99

Litinski, V., Chenna, A., Sagi, J. & Singer, B. (1997) Sequence context is an important determinant in the mutagenic potential of 1,N^6-ethenodeoxyadenosine (εA): Formation of εA basepairs and elongation in defined templates. *Carcinogenesis* 18, 1609–1615

Müller, M., Belas, F.J., Blair, I.A. & Guengerich, F.P. (1997) Analysis of 1,N^2-ethenoguanine and 5,6,7,9-tetrahydro-7-hydroxy-9-oxoimidazo[1,2-α]purine in DNA treated with 2-chlorooxirane by high performance liquid chromatography/electrospray mass spectrometry and comparison of amounts of other DNA adducts. *Chem. Res. Toxicol.* 10, 242–247

O'Connor, T.R., Boiteux, S. & Laval, J. (1988) Ring-opened 7-methylguanine residues in DNA are a block to *in vitro* DNA synthesis. *Nucleic Acids Res.* 16, 5879–5894

Pfeifer, G.P., Singer-Sam, J. & Riggs, A.D. (1993) Analysis of methylation and chromatin structure. *Meth. Enzymol.* 225, 567–583

Premaratne, S., Mandel, M. & Mower, H.F. (1993) Identification of DNA adducts at specific locations by sequencing techniques. *Int. J. Biochem.* 25, 1669–1672

Puisieux, A., Lim, S., Groopman, J. & Ozturk, M. (1991) Selective targeting of *p53* mutational hot spots in human cancers by etiologically defined carcinogens. *Cancer Res.* 51, 6185–6189

Sambrook, J., Fritsch, E.F. & Maniatis, T. (1989) *Molecular Cloning*. Cold Spring Harbor, NY, CSH Press

Saparbaev, M. & Laval, J. (1998) 3,N^4-Ethenocytosine, a highly mutagenic adduct, is a primary substrate for *Escherichia coli* double-stranded uracil-DNA glycosylase and human mismatch-specific thymine-DNA glycosylase. *Proc. Natl Acad. Sci. USA* 95, 8508–8513

Soini, Y., Welsh, J.A., Ishak, K.G. & Bennett, W.P. (1995) *p53* mutations in primary hepatic angiosarcomas not associated with vinyl chloride exposure. *Carcinogenesis* 16, 2879–2881

Soussi, T., Caron de Fromentel, C., Breugnot, C. & May, E. (1988) Nucleotide sequence of a cDNA encoding the rat p53 nuclear oncoprotein. *Nucleic Acids Res.* 16, 11384

Swenberg, J., Fedtke, N., Ciroussel, F., Barbin, A. & Bartsch, H., (1992) Etheno adducts formed in DNA of vinyl-chloride exposed rats are highly persistent in liver. *Carcinogenesis* 13, 727–729

Tornaletti, S. & Pfeifer, G.P. (1994) Slow repair of pyrimidine dimers at *p53* mutation hotspots in skin cancer. *Science* 263, 1436–1438

Tudek, B., Van Zeeland, A.A., Kuśmierek, J.T. & Laval, J. (1998) Activity of *Escherichia coli* DNA-glycosylases on DNA damaged by methylating and ethylating agents and influence of 3-substituted adenine derivatives. *Mutat. Res.* 407, 169–176

Wogan, G.N. (1992) Aflatoxins as risk factors for hepatocellular carcinoma in humans. *Cancer Res.* 52, 2114–2118

Yip, K.F. & Tsou, K.C. (1973) Synthesis of fluorescent adenosine derivatives. *Tetrahedron Lett.* 33, 3087–3090

Exocyclic DNA Adducts in Mutagenesis and Carcinogenesis
B. Singer & H. Bartsch
IARC Scientific Publications No. 150
International Agency for Research on Cancer, Lyon, 1999

Chapter VII. Mutagenesis and carcinogenesis

Cancer-prone oxyradical overload disease

S. Ambs, S.P. Hussain, A.J. Marrogi & C.C. Harris

Oxyradical overload disease develops in conditions involving chronic inflammation and may be of inherited etiology, e.g. haemochromatosis and Wilson disease, be acquired, e.g. infection with hepatitis B or C virus or *Helicobactor pylori*, or be chemically induced, e.g. acid reflux in Barrett oesophagus. Susceptibility to cancer is frequently a pathological consequence of extensive oxyradical damage that leads to a cycle of cell death and regeneration and causes mutations in cancer-related genes. In this brief review, we focus on the possible interactive effects of nitric oxide and the *p53* tumour suppressor gene in human carcinogenesis.

Nitric oxide synthases

Nitric oxide (NO) is an important bioactive agent and signalling molecule that mediates a variety of actions, such as vasodilatation, neurotransmission, host defense and iron metabolism; however, increased NO production may contribute to the pathogenesis of a variety of disorders, including cancer (Moncada *et al.*, 1991; Bredt & Snyder, 1994; Nathan & Xie, 1994; Hentze & Kuhn, 1996; Tamir & Tannenbaum, 1996; Ambs *et al.* 1997). NO is produced endogenously by a family of enzymes known as NO synthases (NOSs) (Marletta, 1993; Forstermann & Kleinert, 1995). Ca^{2+}-dependent isoforms (NOS1 and 3) were found to be expressed constitutively, while a Ca^{2+}-independent isoform required induction (iNOS or NOS2). It is now known that NOS1 and 3 can also be induced (Forstermann & Kleinert, 1995) and that NOS2 is expressed constitutively in some tissues, e.g. bronchus and ileum (Guo *et al.* 1995; Hoffman *et al.* 1997). Only the inducible isoform (NOS2 or Ca^{2+}-independent) produces sustained NO concentrations in the micromolar range, which is high when compared with the pico- to nanomolar concentrations produced by the neuronal (NOS1) and endothelial isoforms (NOS3), which are Ca^{2+}-dependent (Beckman *et al.* 1990; Malinski *et al.* 1993).

Recent studies have addressed the expression and activity of the three NOS isoforms in human cancer. Increased NOS expression and/or activity was observed in human gynaecological (Thomsen *et al.*, 1994), breast (Thomsen *et al.*, 1995) and central nervous system (Cobbs *et al.*, 1995) tumours. In gynaecological and breast cancers, the increased expression was inversely associated with the grade of differentiation of the tumour. Moreover, accumulation of nitrotyrosine in both the inflamed mucosa of patients with ulcerative colitis (Singer *et al.*, 1996) and in the stomachs of patients with *Helicobacter pylori* gastritis (Mannick *et al.*, 1996) indicates that NO production and the formation of peroxynitrite are involved in the pathogenesis of both diseases, which predispose individuals to cancer (Ohshima & Bartsch, 1994).

These observations indicate that NOS expression may contribute to tumour development or progression. Examples of chronic inflammatory and oxyradical overload diseases that have been associated with susceptibility to cancer are ulcerative colitis, viral hepatitis, Wilson disease, haemochromatosis, chronic gastritis, chronic pancreatitis and

Barrett oesophagus. NO has several properties that might enhance carcinogenesis. For example, it is an endothelial growth factor that specifically mediates tumour vascularization (Maeda *et al.*, 1994; Jenkins *et al.*, 1995) and tumour blood flow (Tozer *et al.*, 1997). Although high concentrations of NO induce apoptosis in susceptible cells (Nicotera *et al.*, 1997), low concentrations protect many cell types, including endothelial cells (Dimmeler *et al.*, 1997), from apoptosis. Because factors like cytokines and hypoxia synergistically induce NOS2 expression (Melillo *et al.*, 1995), the micro-environmental changes in premalignant and malignant tumour tissue may establish high, sustained NO production in a variety of tumour cells, thereby supporting clonal selection and tumour growth (Ambs *et al.*, 1998a,b).

Colon carcinogenesis

Genetic and epigenetic changes have been described in the multistage process of human colon carcinogenesis (Figure 1). Our recent data indicate that NOS2 expression coincides with COX2 and p53 overexpression in the progression from adenoma to carcinoma. We investigated NOS expression in human colon cancer with respect to tumour staging, NOS-expressing cell types, nitrotyrosine formation, inflammation and expression of vascular endothelial growth factor (Ambs *et al.*, 1998a). Ca^{2+}-dependent NOS activity was found in normal colon and in tumours but was significantly decreased in adenomas (p < 0.001) and carcinomas (Dukes stages A–D: p < 0.002). Ca^{2+}-independent NOS activity, indicating inducible NOS (NOS2), was markedly expressed in approximately 60% of human colon adenomas (p < 0.001 versus normal tissues) and in 20–25% of colon carcinomas (p < 0.01 versus normal tissues). Only low levels were found in the surrounding normal tissue. NOS2 was detected in tissue mononuclear cells, endothelium and tumour epithelium. There was a statistically significant

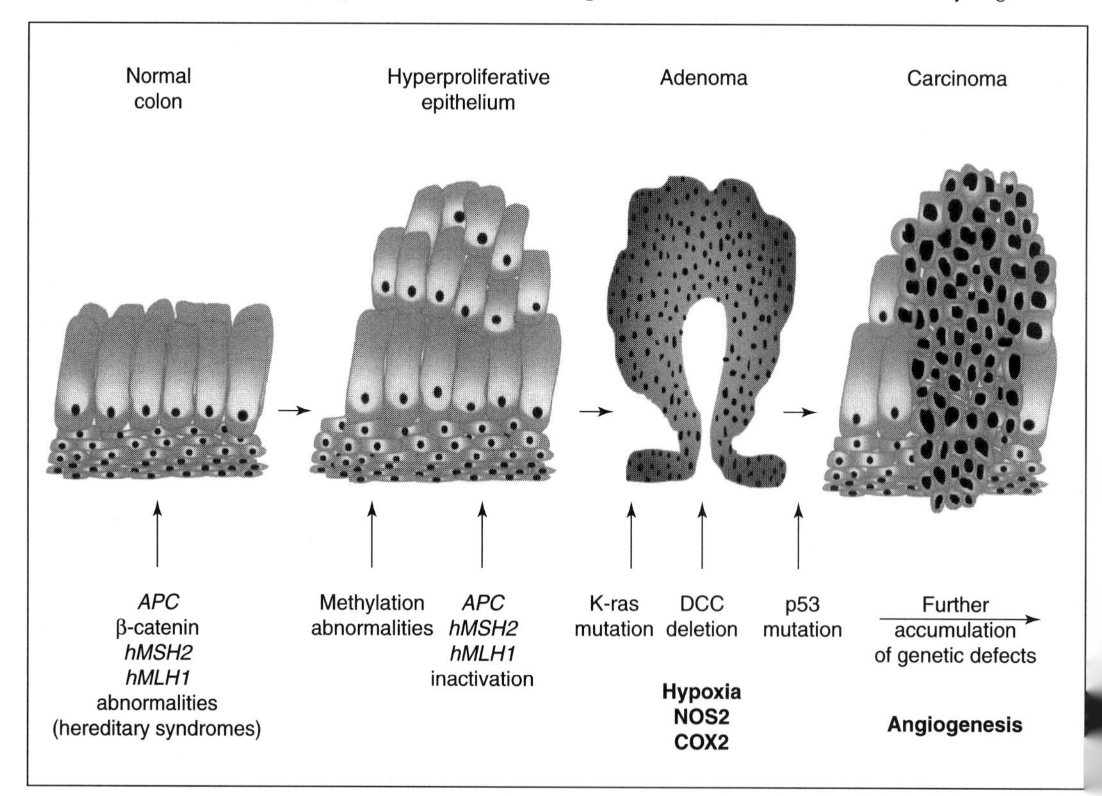

Figure 1. Genetic and epigenetic changes during multistage human colon carcinogenesis

correlation between NOS2 enzymatic activity and the level of protein detected by immunohistochemistry ($p < 0.01$). Western blot analysis of tumour extracts with Ca^{2+}-independent NOS activity showed up to three distinct NOS2 protein bands, at M_r 125 000–M_r 138 000. The same protein bands were heavily tyrosine-phosphorylated in some tumour tissues. Tissue mononuclear cells but not the tumour epithelium were immunopositive to a polyclonal anti-nitrotyrosine antibody; however, only a subset of the NOS2-expressing tissue mononuclear cells stained positively for nitrotyrosine, which is a marker of peroxynitrite formation. Furthermore, the expression of vascular endothelial growth factor was detected in adenomas that expressed NOS2. These data are consistent with the hypothesis that excessive NO production by NOS2 may contribute to the progression of colon cancer at the time of the transition of adenoma to carcinoma *in situ.*

We then investigated the hypothesis that NO generated by NOS mutates *p53* during human colon carcinogenesis (Ambs *et al.*, 1998c). We analysed 118 sporadic colon tumours for NOS2 expression and *p53* mutations and found high levels of NOS2 expression in various tumours throughout the right, left and sigmoid colon. The average NOS2 activity was highest in adenomas, declined with advancing tumour stage and was lowest in metastatic tumours, which confirmed our previous observations (Ambs *et al.*, 1998a). The decline in NOS2 activity with advancing tumour stage may be attributed to immunosuppression of NOS2 expression associated with tissue mononuclear cells in advanced tumours (Alleva *et al.*, 1994).

We next determined the *p53* mutation frequency and type in relation to NOS2 activity in colon tumours. We confined the *p53* mutation analysis to the evolutionarily conserved region, which contains about 90% of the mutations in this gene and all of the mutational hotspots at CpG sites (Hollstein *et al.*, 1991; Greenblatt *et al.*, 1994). We found 55 mutations in 26 adenomas (35% mutation frequency) and 92 Dukes A–D carcinomas (48% mutation frequency). There were 44 missense mutations, six nonsense mutations and five insertions, deletions or inversions. The predominant mutation was a G:C→A:T transition at CpG sites (62%), and a significant association was found between these transitions and increased NOS2 activity; the rates of all other mutations varied inversely with NOS2 activity.

Our investigation of primary human colon tumours established a strong positive relationship between the presence of NOS2 in tumours and the frequency of G:C→A:T transitions at CpG sites. These mutations are also common in lymphoid, oesophageal, head-and-neck, stomach, brain and breast cancers (Hollstein *et al.*, 1991; Levine *et al.*, 1991; Greenblatt *et al.*, 1994). Increased NOS2 expression has been demonstrated in four of these cancer types (Ellie *et al.*, 1995; Thomsen *et al.*, 1995; Ambs *et al.*, 1998a; Gallo *et al.*, 1998). Tumour-associated NO production may modify DNA directly, or it may inhibit DNA repair activity (Wink *et al.*, 1996); for instance, it may affect the recently described human thymine-DNA glycosylase, which has been shown to repair G:T mismatches at CpG sites (Sibghat-Ullah *et al.*, 1996). Because NO production also induces *p53* accumulation (Messmer & Brune, 1996; Forrester *et al.*, 1996), the resulting growth inhibition can provide a strong additional selection pressure for mutant *p53*. NO may therefore act as both an endogenous initiator and a promoter in human colon carcinogenesis. Specific inhibitors of NOS2, as demonstrated in an animal tumour model (Thomsen *et al.*, 1997), may have chemopreventive potential in human colorectal cancer.

Regulation of nitric oxide synthase 2 by *p53*

DNA damage triggers p53 protein accumulation (Kastan *et al.*, 1991; Lu & Lane, 1993), which produces either growth arrest (El Deiry *et al.*, 1993) or apoptosis (Lowe *et al.*, 1993a,b). Exposure of cells to high concentrations of NO causes DNA damage and apoptosis, and NO stimulates *p53* accumulation (Messmer *et al.*, 1994; Forrester *et al.*, 1996) and *p53*-mediated apoptosis (Fehsel *et al.*, 1995; Messmer & Brune, 1996; Ho *et al.*, 1997). Because *p53* is a transcription factor that down-regulates the promoters of *bcl-2* (Miyashita *et al.*, 1994) and *hsp70* (Agoff *et al.*, 1993), we investigated the possibility that *p53* also represses NOS2 promoter activity. We found that basal and cytokine-induced NOS2 promoter activity is down-regulated by *p53* in human cells *in vitro*, and we have located the region required for wild-type *p53*-mediated repression to approximately 400 base pairs upstream of the transcription start site (Forrester *et al.*, 1996).

To further characterize the feedback loop between NOS2 and *p53*, we investigated NO production, i.e. urinary nitrate plus nitrite excretion, and NOS2 expression in homozygous *p53* 'knockout' mice (Ambs *et al.*, 1998d). Untreated *p53* knockout mice excreted 70% more nitrite plus nitrate than mice with wild-type *p53*. NOS2 protein expression was constitutively detected in the spleen of untreated *p53* knockout mice, while it was undetectable in the spleen of wild-type *p53* controls. After treatment with heat-inactivated *Corynebacterium parvum*, urinary excretion of nitrite plus nitrate by *p53* knockout mice exceeded that of wild-type controls by approximately 200%. Treatment with *C. parvum* also induced the accumulation of *p53* in the liver. Splenectomy reduced the NO output of *C. parvum*-treated *p53* knockout mice but not that of wild-type *p53* controls. Although NO production and NOS2 protein expression were increased to a similar extent in knockout and wild-type *p53* mice 10 days after injection of *C. parvum*, NOS2 expression returned to baseline levels only in wild-type *p53* controls and remained up-regulated in *p53* knockout mice. These genetic and functional data indicate that *p53* is an important transrepressor of NOS2 expression *in vivo* and attenuates excessive NO production in a regulatory negative feedback loop (Figure 2).

Implication in human cancer progression

Normal tissue homeostasis is maintained by regulation of cell proliferation and death. Both external and internal signals can either initiate or inhibit cell proliferation and apoptosis. During carcinogenesis, genetic and epigenetic lesions accumulate in dysplastic and neoplastic cells and generate an imbalance between growth and death regulatory pathways (Figure 1). Eventually, clonal expansion gives rise to clinical cancer. In this scenario of tumour progression, *p53* muta-

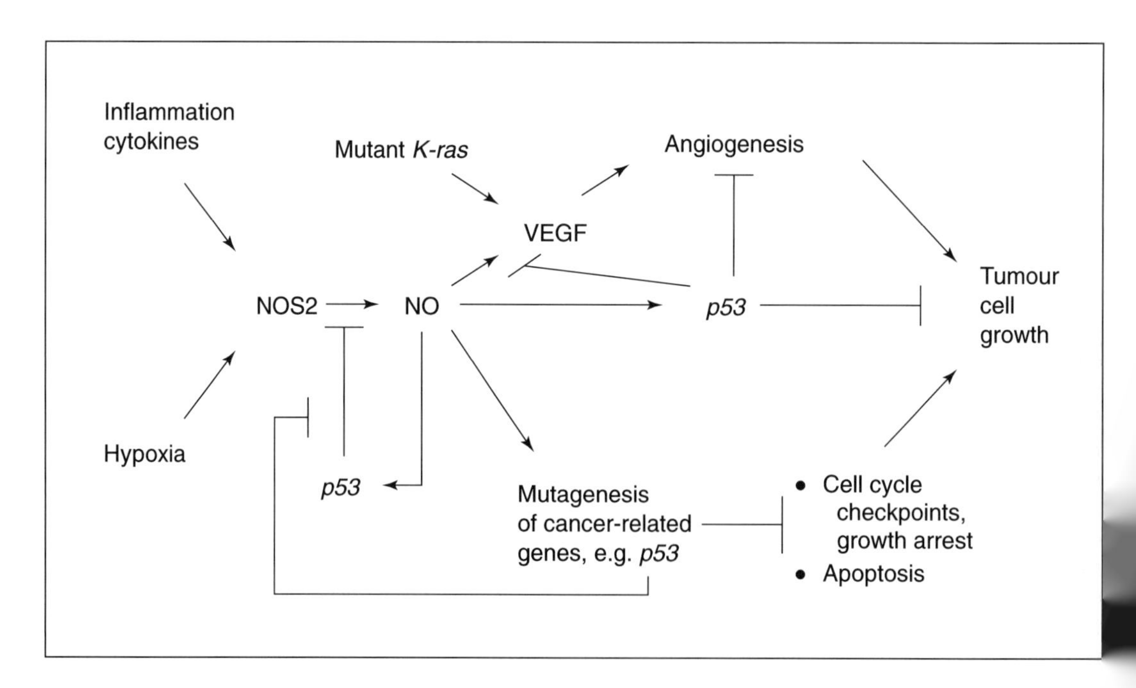

Figure 2. Regulation of inducible nitric oxide synthase (NOS2), production of nitric oxide (NO) and its interactive effects with the *p53* tumour suppressor gene during carcinogenesis and tumour progression

VEGF, vascular endothelial growth factor

tions frequently occur after the initiating events of carcinogenesis. Not surprisingly, hypoxia, which arises as tumours increase in size, has been found to select for mutant *p53* cells that are resistant to hypoxia-induced apoptosis (Graeber *et al.*, 1996). NO may select cells carrying mutant *p53* as it has both genotoxic and angiogenic properties. Furthermore, mutant *p53* cells may be less susceptible to NO-induced apoptosis, or low to moderate concentrations of NO may exert anti-apoptotic properties specifically in those cells. If hypoxia and/or cytokines induce NOS expression, only cell clones with nonfunctional *p53* would tolerate the stress by sustained NO production and escape growth arrest and apoptosis while tumour growth is supported. Furthermore, NO may act directly by stimulating endothelial cell growth or may induce the expression of angiogenic factors, such as vascular endothelial growth factor (Takahashi *et al.*, 1995; Hanahan & Folkman, 1996). Alternatively, NOS2 and vascular endothelial growth factor may have common induction pathways, and NO may synergistically enhance expression of this factor or increase its secretion into the cellular microenvironment. Additionally, mutant *p53* has been shown to synergize with protein kinase C in the induction of vascular endothelial growth factor (Kieser *et al.*, 1994).

Much of the information presented in this brief review is consistent with the hypothesis that sustained NO production in human tumours induces DNA damage and triggers *p53*-mediated growth arrest and apoptosis. As a result, NO production would contribute to human cancer progression by selecting against wild-type *p53* and for mutant *p53* cells (Figure 2). Further studies are required to test this hypothesis and to answer the following questions. Firstly, does NO cause *p53* mutations in human cells, and is there a positive correlation between NOS expression in premalignant and malignant tissues and an increased frequency of *p53* mutations in human cancers? Is there a preference for certain types of mutations indicating direct NO involvement? Does NO production in some chronic inflammatory diseases that predispose patients to cancer, e.g. ulcerative colitis and chronic active viral hepatitis, generate an increased *p53* mutation frequency in the inflamed portion of the affected tissue? The *p53* mutation frequency and type can be determined by a highly sensitive method, restriction fragment length polymorphism–polymerase chain reaction (Aguilar *et al.*, 1993). Does the hepatitis virus X protein, which interacts with and functionally inactivates *p53* (Wang *et al.*, 1994), permit excessive NO production by NOS2? Does hypoxia induce NOS2 expression in human cells and increase NO-induced micro-vascularization? A synergism has been found between hypoxia and interferon-γ in murine NOS2 expression (Melillo *et al.*, 1995). As NO induces neovascularization, does this pathway involve other angiogenic factors, e.g. vascular endothelial growth factor? (Ambs *et al.*, 1998a,b). These and other questions should be addressed to better understand the interactive effects of NO and *p53* in human carcinogenesis.

Acknowledgement

We thank Dorothea Dudek for her editorial and graphic assistance.

References

Agoff, S.N., Hou, J., Linzer, D.I. & Wu, B. (1993) Regulation of the human hsp70 promoter by p53. *Science* 259, 84–87

Aguilar, F., Hussain, S.P. & Cerutti, P. (1993) Aflatoxin B1 induces the transversion of G→T in codon 249 of the p53 tumor suppressor gene in human hepatocytes. *Proc. Natl Acad. Sci. USA* 90, 8586–8590

Alleva, D.G., Burger, C.J. & Elgert, K.D. (1994) Tumor-induced regulation of suppressor macrophage nitric oxide and TNF-alpha production. Role of tumor-derived IL-10, TGF-beta, and prostaglandin E2. *J. Immunol.* 153, 1674–1686

Ambs, S., Hussain, S.P. & Harris, C.C. (1997) Interactive effects of nitric oxide and the p53 tumor suppressor gene in carcinogenesis and tumor progression. *FASEB J.* 11, 443–448

Ambs, S., Merriam, W.G., Bennett, W.P., Felley-Bosco, E., Ogunfusika, M.O., Oser, S.M., Klein, S., Shields, P.G., Billiar, T.R. & Harris, C.C. (1998a) Frequent nitric oxide synthase-2 expression in human colon adenomas: Implication for tumor angiogenesis and colon cancer progression. *Cancer Res.* 58, 334–341

Ambs, S., Merriam, W.G., Ogunfusika, M.O., Bennett, W.P., Ishibe, N., Hussain, S.P., Tzeng, E.E., Geller, D.A., Billiar, T.R. & Harris, C.C. (1998b) *p53* and vascular endothelial growth factor regulate tumor growth of NOS2-expressing human carcinoma cells. *Nature Med.*, 4, 1371–1376

Ambs, S., Bennett, W.P., Merriam, W.G., Ogunfusika, M.O., Oser, S.M., Shields, P.G., Felley-Bosco, E., Hussain, S.P. & Harris, C.C. (1998c) Characteristic *p53* mutations correlate with inducible NO synthase expression in human colorectal cancer. *J. Natl Cancer Inst.*, 91, 86–88

Ambs, S., Ogunfusika, M.O., Merriam, W.G., Bennett, W.P., Billiar, T.R. & Harris, C.C. (1998d) Upregulation of NOS2 expression in cancer-prone p53 knockout mice. *Proc. Natl Acad. Sci. USA* 95, 8823–8828

Beckman, J.S., Beckman, T.W., Chen, J., Marshall, P.A. & Freeman, B.A. (1990) Apparent hydroxyl radical production by peroxynitrite: Implications for endothelial injury from nitric oxide and superoxide. *Proc. Natl Acad. Sci. USA* 87, 1620–1624

Bredt, D.S. & Snyder, S.H. (1994) Nitric oxide: A physiologic messenger molecule. *Annu. Rev. Biochem.* 63, 175-195

Cobbs, C.S., Brenman, J.E., Aldape, K.D., Bredt, D.S. & Israel, M.A. (1995) Expression of nitric oxide synthase in human central nervous system tumors. *Cancer Res.* 55, 727–730

Dimmeler, S., Haendeler, J., Nehls, M. & Zeiher, A.M. (1997) Suppression of apoptosis by nitric oxide via inhibition of interleukin-1beta-converting enzyme (ICE)-like and cysteine protease protein (CPP)-32-like proteases. *J. Exp. Med.* 185, 601–607

El-Deiry, W.S., Tokino, T., Velculescu, V.E., Levy, D.B., Parsons, R., Trent, J.M., Lin, D., Mercer, W.E., Kinzler, K.W. & Vogelstein, B. (1993) WAF1, a potential mediator of p53 tumor suppression. *Cell* 75, 817–825

Ellie, E., Loiseau, H., Lafond, F., Arsaut, J. & Demotes-Mainard, J. (1995) Differential expression of inducible nitric oxide synthase mRNA in human brain tumours. *Neuroreport* 7, 294–296

Fehsel, K., Kroncke, K.D., Meyer, K.L., Huber, H., Wahn, V. & Kolb-Bachofen, V. (1995) Nitric oxide induces apoptosis in mouse thymocytes. *J. Immunol.* 155, 2858–2865

Forrester, K., Ambs, S., Lupold, S.E., Kapust, R.B., Spillare, E.A., Weinberg, W.C., Felley-Bosco, E., Wang, X.W., Geller, D.A., Billiar, T.R. & Harris, C.C. (1996) Nitric oxide-induced p53 accumulation and regulation of inducible nitric oxide synthase (NOS2) expression by wild-type *p53*. *Proc. Natl Acad. Sci. USA* 93, 2442–2447

Forstermann, U. & Kleinert, H. (1995) Nitric oxide synthase: Expression and expressional control of the three isoforms. *Naunyn. Schmiedebergs Arch. Pharmacol.* 352, 351–364

Gallo, O., Masini, E., Morbidelli, L., Franchi, A., Fini-Storchi, I., Vergari, W.A. & Ziche, M. (1998) Role of nitric oxide in angiogenesis and tumor progression in head and neck cancer. *J. Natl Cancer Inst.* 90, 587–596

Graeber, T.G., Osmanian, C., Jacks, T., Housman, D.E., Koch, C.J., Lowe, S.W. & Glaccia, A.J. (1996) Hypoxia-mediated selection of cells with diminished apoptotic potential in solid tumours. *Nature* 379, 88–91

Greenblatt, M.S., Bennett, W.P., Hollstein, M. & Harris, C.C. (1994) Mutations in the *p53* tumor suppressor gene: Clues to cancer etiology and molecular pathogenesis. *Cancer Res.* 54, 4855–4878

Guo, F.H., De Raeve, H.R., Rice, T.W., Stuehr, D.J., Thunnissen, F.B. & Erzurum, S.C. (1995) Continuous nitric oxide synthesis by inducible nitric oxide synthase in normal human airway epithelium *in vivo*. *Proc. Natl Acad. Sci. USA* 92, 7809–7813

Hanahan, D. & Folkman, J. (1996) Patterns and emerging mechanisms of the angiogenic switch during tumorigenesis. *Cell* 86, 353–364

Hentze, M.W. & Kuhn, L.C. (1996) Molecular control of vertebrate iron metabolism: mRNA-based regulatory circuits operated by iron, nitric oxide, and oxidative stress. *Proc. Natl Acad. Sci. USA* 93, 8175–8182

Ho, Y.S., Lee, H.M., Mou, T.C., Wang, Y.J. & Lin, J.K. (1997) Suppression of nitric oxide-induced apoptosis by N-acetyl-L-cysteine through modulation of glutathione, bcl-2, and bax protein levels. *Mol. Carcinog.* 19, 101–113

Hoffman, R.A., Zhang, G., Nussler, N.C., Gleixner, S.L., Ford, H.R., Simmons, R.L. & Watkins, S.C. (1997) Constitutive expression of inducible nitric oxide synthase in the mouse ileal mucosa. *Am. J. Physiol.* 272, G383–392

Hollstein, M., Sidransky, D., Vogelstein, B. & Harris, C.C. (1991) *p53* mutations in human cancers. *Science* 253, 49–53

Jenkins, D.C., Charles, I.G., Thomsen, L.L., Moss, D.W., Holmes, L.S., Baylis, S.A., Rhodes, P., Westmore, K., Emson, P.C. & Moncada, S. (1995) Roles of nitric oxide in tumor growth. *Proc. Natl Acad. Sci. USA* 92, 4392–4396

Kastan, M.B., Onyekwere, O., Sidransky, D., Vogelstein, B. & Craig, R.W. (1991) Participation of p53 protein in the cellular response to DNA damage. *Cancer Res.* 51, 6304–6311

Kieser, A., Weich, H.A., Brandner, G., Marme, D. & Kolch, W. (1994) Mutant *p53* potentiates protein kinase C induction of vascular endothelial growth factor expression. *Oncogene* 9, 963–969

Levine, A.J., Momand, J. & Finlay, C.A. (1991) The *p53* tumour suppressor gene. *Nature* 351, 453–456

Lowe, S.W., Ruley, H.E., Jacks, T. & Housman, D.E. (1993a) *p53*-dependent apoptosis modulates the cytotoxicity of anticancer agents. *Cell* 74, 957–967

Lowe, S.W., Schmitt, E.M., Smith, S.W., Osborne, B.A. & Jacks, T. (1993b) *p53* is required for radiation-induced apoptosis in mouse thymocytes. *Nature* 362, 847–849

Lu, X. & Lane, D.P. (1993) Differential induction of transcriptionally active *p53* following UV or ionizing radiation: Defects in chromosome instability syndromes? *Cell* 75, 765–778

Maeda, H., Noguchi, Y., Sato, K. & Akaike, T. (1994) Enhanced vascular permeability in solid tumor is mediated by nitric oxide and inhibited by both new nitric oxide scavenger and nitric oxide synthase inhibitor. *Jpn. J. Cancer Res.* 85, 331–334

Malinski, T., Taha, Z., Grunfeld, S., Patton, S., Kapturczak, M. & Tomboulian, P. (1993) Diffusion of nitric oxide in the aorta wall monitored *in situ* by porphyrinic microsensors. *Biochem. Biophys. Res. Commun.* 193, 1076–1082

Mannick, E.E., Bravo, L.E., Zarama, G., Realpe, J.L., Zhang, X.J., Ruiz, B., Fontham, E.T., Mera, R., Miller, M.J. & Correa, P. (1996) Inducible nitric oxide synthase, nitrotyrosine, and apoptosis in *Helicobacter pylori* gastritis: Effect of antibiotics and antioxidants. *Cancer Res.* 56, 3238–3243

Marletta, M.A. (1993) Nitric oxide synthase structure and mechanism. *J. Biol. Chem.* 268, 12231–12234

Melillo, G., Musso, T., Sica, A., Taylor, L.S., Cox, G.W. & Varesio, L. (1995) A hypoxia-responsive element mediates a novel pathway of activation of the inducible nitric oxide synthase promoter. *J. Exp. Med.* 182, 1683–1693

Messmer, U.K. & Brune, B. (1996) Nitric oxide-induced apoptosis: *p53*-dependent and p53-independent signalling pathways. *Biochem. J.* 319, 299–305

Messmer, U.K., Ankarcrona, M., Nicotera, P. & Brune, B. (1994) *p53* expression in nitric oxide-induced apoptosis. *FEBS Lett.* 355, 23–26

Miyashita, T., Harigai, M., Hanada, M. & Reed, J.C. (1994) Identification of a *p53*-dependent negative response element in the *bcl-2* gene. *Cancer Res.* 54, 3131–3135

Moncada, S., Palmer, R.M. & Higgs, E.A. (1991) Nitric oxide: Physiology, pathophysiology, and pharmacology. *Pharmacol. Rev.* 43, 109–142

Nathan, C. & Xie, Q.W. (1994) Nitric oxide synthases: Roles, tolls, and controls. *Cell* 78, 915–918

Nicotera, P., Bonfoco, E. & Brune, B. (1997) Mechanisms for nitric oxide-induced cell death: Involvement of apoptosis. *Adv. Neuroimmunol.* 5, 411–420

Ohshima, H. & Bartsch, H. (1994) Chronic infections and inflammatory processes as cancer risk factors: Possible role of nitric oxide in carcinogenesis. *Mutat. Res.* 305, 253–264

Sibghat-Ullah, Gallinari, P., Xu, Y.Z., Goodman, M.F., Bloom, L.B., Jiricny, J. & Day, R.S., III (1996) Base analog and neighboring base effects on substrate specificity of recombinant human G:T mismatch-specific thymine DNA-glycosylase. *Biochemistry* 35, 12926–12932

Singer, I.I., Kawka, D.W., Scott, S., Weidner, J.R., Mumford, R.A., Riehl, T.E. & Stenson, W.F. (1996) Expression of inducible nitric oxide synthase and nitrotyrosine in colonic epithelium in inflammatory bowel disease. *Gastroenterology* 111, 871–885

Takahashi, Y., Kitadai, Y., Bucana, C.D., Cleary, K.R. & Ellis, L.M. (1995) Expression of vascular endothelial growth factor and its receptor, KDR, correlates with vascularity, metastasis, and proliferation of human colon cancer. *Cancer Res.* 55, 3964–3968

Tamir, S. & Tannenbaum, S.R. (1996) The role of nitric oxide (NO•) in the carcinogenic process. *Biochim. Biophys. Acta* 1288, F31–36

Thomsen, L.L., Lawton, F.G., Knowles, R.G., Beesley, J.E., Riveros-Moreno, V. & Moncada, S. (1994) Nitric oxide synthase activity in human gynecological cancer. *Cancer Res.* 54, 1352–1354

Thomsen, L.L., Miles, D.W., Happerfield, L., Bobrow, L.G., Knowles, R.G. & Moncada, S. (1995) Nitric oxide synthase activity in human breast cancer. *Br. J. Cancer* 72, 41–44

Thomsen, L.L., Scott, J.M., Topley, P., Knowles, R.G., Keerie, A.J. & Frend, A.J. (1997) Selective inhibition of inducible nitric oxide synthase inhibits tumor growth *in vivo*: Studies with 1400W, a novel inhibitor. *Cancer Res.* 57, 3300–3304

Tozer, G.M., Prise, V.E. & Chaplin, D.J. (1997) Inhibition of nitric oxide synthase induces a selective reduction in tumor blood flow that is reversible with L-arginine. *Cancer Res.* 57, 948–955

Wang, X.W., Forrester, K., Yeh, H., Feitelson, M.A., Gu, J.R. & Harris, C.C. (1994) Hepatitis B virus X protein inhibits *p53* sequence-specific DNA binding, transcriptional activity, and association with transcription factor ERCC3. *Proc. Natl Acad. Sci. USA* 91, 2230–2234

Wink, D.A., Hanbauer, I., Grisham, M.B., Laval, F., Nims, R.W., Laval, J., Cook, J., Pacelli, R., Liebmann, J., Krishna, M., Ford, P.C. & Mitchell, J.B. (1996) Chemical biology of nitric oxide: Regulation and protective and toxic mechanisms. *Curr. Top. Cell Regul.* 34, 159–187

Exocyclic DNA Adducts in Mutagenesis and Carcinogenesis
B. Singer & H. Bartsch
IARC Scientific Publications No. 150
International Agency for Research on Cancer, Lyon, 1999

Role of etheno DNA adducts in carcinogenesis induced by vinyl chloride in rats

A. Barbin

Vinyl chloride, a hepatocarcinogen in humans and rodents, can form promutagenic etheno bases in DNA after metabolic activation. The formation of 1,N^6-ethenoadenine (εA) and 3,N^4-ethenocytosine (εC) was measured in adult Sprague-Dawley rats by immunoaffinity purification and ^{32}P-postlabelling. A highly variable background was found in all tissues from untreated animals: the mean molar ratios of εA:A and εC:C in DNA ranged from 0.043 x 10^{-8} to 31.2 x 10^{-8} and from 0.062 x 10^{-8} to 20.4 x 10^{-8}, respectively. After exposure to 500 ppm vinyl chloride by inhalation (4 h/day, 5 days/week for 8 weeks), increased levels of εA were found in the liver, lung, circulating lymphocytes and testis, the mean (± SD) of induced levels (treated–control values) being (4.1 ± 1.5) x 10^{-8} for these tissues. No increase in the εA:A ratio was observed in kidney, brain or spleen. The levels of εC increased in all the tissues examined except the brain. The mean value of the induced εC:C ratios was (7.8 ± 1.2) x 10^{-8} for the liver, kidney, lymphocytes and spleen, and these ratios were higher in the lung (28 x 10^{-8}) and testis (19 x 10^{-8}). The results suggest a variable repair capacity for εA or εC in different tissues. The results are discussed in relation to published studies on the accumulation and persistence of etheno bases in the liver during and after exposure to vinyl chloride and on mutation spectra in the *ras* and *p53* genes in liver tumours induced by vinyl chloride. In addition, we show that the linear relationship established for monofunctional alkylating agents between their carcinogenic potency in rodents and their covalent binding index for promutagenic bases in hepatic DNA holds for vinyl chloride. It is concluded that etheno bases are critical lesions in hepatocarcinogenesis induced by vinyl chloride. For a better understanding of the mechanism of action of this compound, further work is needed on the role of DNA repair pathways and of endogenous lipid peroxidation products in the formation and persistence of etheno bases *in vivo.*

Introduction

Vinyl chloride was shown to be carcinogenic in rats by Viola *et al.* in 1971 and in humans exposed occupationally by Creech and Johnson in 1974. These discoveries prompted epidemiological and experimental studies on the mechanism of action of this compound (IARC, 1979; Barbin & Bartsch, 1986). The hallmark of vinyl chloride is its ability to induce angiosarcoma of the liver, a very rare cancer in humans (Simonato *et al.*, 1991). It also induces angiosarcoma in rodents, as well as hepatocellular carcinomas and tumours at other sites (mammary gland carcinomas, nephroblastomas, Zymbal gland carcinomas, neuroblastomas, lung adenomas; Maltoni *et al.*, 1984). In humans, vinyl chloride is suspected but not proven to induce tumours other than angiosarcoma of the liver (IARC, 1979; Simonato *et al.*, 1991), although there have been reports of hepatocellular carcinoma which cannot be attributed to any other known risk factor (see e.g. Saurin *et al.*, 1997). Vinyl chloride is mutagenic in a number of organisms, inducing mainly base-pair substitution mutations (reviewed by Giri, 1995). These biological activities of vinyl chloride require metabolic activation by cytochrome P450 2E1 into chloroethylene oxide (Barbin, *et al.*, 1975; Guengerich *et al.*, 1991; El Ghissassi *et al.*, 1998), which is thought to be the ultimate carcinogen species (Zajdela *et al.*, 1980). Chloroethylene oxide rearranges spontaneously into another electrophilic species, 2-chloroacetaldehyde (O'Neill *et al.*, 1986).

Because chloroacetaldehyde is known to react *in vitro* with nucleic acid bases to yield exocyclic etheno derivatives (Kochetkov *et al.*, 1971), it was proposed that DNA etheno bases might be formed after exposure to vinyl chloride and could be responsible for the mutagenic and carcinogenic effects of that substance (Barbin *et al.*, 1975). The formation of etheno adducts by vinyl chloride in the presence of liver microsomal fractions was demonstrated in 1975–78 (Barbin *et al.*, 1975; Laib & Bolt, 1977, 1978), and the first studies showing the promutagenic properties of etheno bases were reported in 1981 (Barbin *et al.*, 1981; Hall *et al.*, 1981; Spengler & Singer, 1981). Subsequently, it appeared that chloroethylene oxide could yield etheno adducts by direct reaction with nucleobases (Barbin, *et al.*, 1986; Guengerich, 1992) and that chloroethylene oxide, not chloroacetaldehyde, is the metabolite involved in the formation of etheno bases by vinyl chloride in the presence of a microsomal system (Guengerich *et al.*, 1981). In addition to etheno bases, chloroethylene oxide yields 7-(2-oxoethyl)guanine in reaction with DNA. Although this is the major DNA adduct formed by vinyl chloride *in vitro* and *in vivo* (see below), it may not play a critical role in its genotoxic effects because it does not have promutagenic properties (Barbin *et al.*, 1985).

In contrast, the promutagenic properties of etheno bases are well established. The four etheno bases induce mainly base-pair substitutions (reviewed by Bartsch *et al.*, 1994). In the initial studies on mutations, oligonucleotides were used that had been modified by either chloroacetaldehyde or chloroethylene oxide; however, these two metabolites of vinyl chloride modify the bases differently, and chloroacetaldehyde forms 1,N^6-ethenoadenine (εA) = 3,N^4-ethenocytosine (εC) > N^2,3-ethenoguanine (εG) >> 1,N^2-εG in DNA (Dosanjh *et al.*, 1994). Barbin *et al.* (1981), using high levels of modification of poly(dA–dT) or poly(dA), found that εA directed the incorporation of dG which results in AT→CG transversions. This was confirmed by Singer *et al.* (1984) with various polymerases. The mutagenicity of εC was also investigated by both groups, who found T to be the favoured misincorporation for generating CG→AT transversions. In a site-directed replication system, Zhang *et al.* (1995) and Shibutani *et al.* (1996), using various polymerases, reported that εC preferentially incorporates A and T. An exception is polymerase β, which incorporates C as well. The precursor of εC, the hydrated form εC·H_2O, does not miscode for any base (Singer *et al.*, 1983). N^2,3-εG differs from the other etheno derivatives in that it is a very efficient mutagen; with various replicating enzymes, it pairs two to four times more often with T than the analogous wobble pair, G·T. Only G→A transitions occur (Singer *et al.*, 1991). 1, N^2-εG, which is the least frequently formed etheno base, has been investigated primarily in the laboratory of Guengerich, who found that this derivative led to GC→TA and GC→CG transversions (Langouët *et al.*, 1998), in addition to very low levels of –1 and –2 base frameshift mutations (Langouët *et al.*, 1997).

The extent and type of mutations induced *in vitro* are not necessarily the same as those found *in vivo*. With εA in a bacterial system, Basu *et al.* (1993) reported a low percentage (0.1% of survivors) of A→G transitions. Pandya and Moriya (1996), using simian kidney cells, described a high frequency (63%) of εA→G transitions with minor amounts of A→T and A→C mutations. Palejwala *et al.* (1991, 1993) and Basu *et al.* (1993) used a bacterial system, M13, to investigate mutations induced by εC. Both groups reported mainly εC→T and some εC→A transversions, although the frequencies differed. In monkey kidney cells, Moriya *et al.* (1994) found that T, A and C were all inserted opposite εC, with an overall 80% frequency, leading to high levels of transitions and transversions. Only one experiment with N^2,3-εG *in vivo* has been reported: Cheng *et al.* (1991), using the same M13 system, found that N^2,3-εG produced only G→A transitions, as *in vitro* (Singer *et al.*, 1991).

Four vinyl chloride–DNA adducts have been identified *in vivo* in rodents exposed to vinyl chloride: 7-(2-oxoethyl)guanine, the major adduct, and three etheno bases, εA, εC and N^2,3-εG (reviewed by Barbin, 1998). Because etheno adducts are formed at low levels, with typical molar ratios of etheno base to parent base in DNA of about 10^{-8}–10^{-7}, their occurrence in DNA after exposure of animals to vinyl chloride is difficult to demonstrate. Two approaches were used initially. In the first approach, animals were exposed to ^{14}C-vinyl chloride, DNA deoxyribonucleosides were separated by high-performance liquid chromatography (HPLC) and the adducts were quantified by scintillation counting (Laib, 1986). Green and Hathway (1978) used HPLC or gas chromatographic separation of the etheno bases or etheno nucleosides, followed by characterization by mass spectrometry. Subsequently, more sensitive techniques were developed. Monoclonal antibodies were raised against εdAdo and εdCyd and used in competitive radioimmunoassays after HPLC pre-purification of the etheno adducts (Eberle *et al.*, 1989). In parallel, a method based on electrophore labelling followed by negative-ion chemical ionization mass spectrometry was developed to measure N^2,3-εG (Fedtke *et al.*, 1990). With these techniques, εA, εC and N^2,3-εG were quantified in the liver, lung, kidney and brain of preweanling or adult rats exposed to vinyl chloride by inhalation (Ciroussel *et al.*, 1990; Swenberg *et al.*, 1992). More recently, an ultra-sensitive immunoaffinity/^{32}P-postlabelling method has been developed which permits the detection of endogenous background levels of εA and εC in various tissues from unexposed rodents and humans (Guichard *et al.*, 1993; Nair *et al.*, 1995).

In addition to the four adducts detected in DNA *in vivo* after exposure to vinyl chloride, two more adducts have been identified *in vitro* after reaction of chloroethylene oxide or chloroacetaldehyde with DNA: 1,N^2-εG and 5,6,7,9-tetrahydro-7-hydroxy-9-oxoimidazo[1,2-a]-purine (Müller *et al.*, 1997). Several other lesions (monoadducts and cross-links) have been observed in reactions of chloroethylene oxide or chloroacetaldehyde with nucleosides or polynucleotides, but they have not yet been identified in DNA *in vivo* (Barbin *et al.*, 1986; Singer *et al.*, 1986).

In this report, we present our results on the formation of εA and εC in adult rats exposed to vinyl chloride, some of which were published previously (Guichard *et al.*, 1996), and discuss their putative role in vinyl chloride-induced carcinogenesis in the light of recent data on mutation spectra in vinyl chloride-induced tumours (see also Marion, this volume) and on quantitative structure–activity relationships.

Results and discussion

Formation of etheno adducts in adult Sprague-Dawley rats exposed to vinyl chloride

The formation of εA and εC in six-week-old male Sprague-Dawley rats exposed to 500 ppm vinyl chloride by inhalation, 4 h/day, 5 days/week, for 1, 2, 4 or 8 weeks, and in untreated control animals (Guichard *et al.*, 1996) was analysed by immunoaffinity/^{32}P-postlabelling (Guichard *et al.*, 1993; Nair *et al.*, 1995). The levels of εA and εC in liver, lung, kidney and circulating lymphocytes were reported previously (Guichard *et al.*, 1996). In the study reported here, we analysed additional tissues, including the brain, spleen and testis, from three untreated rats and three rats exposed for eight weeks to vinyl chloride and killed immediately after the end of treatment (Figures 1 and 2). The analyses were done in duplicate.

Background levels were found in all of the tissues examined (Figures 1 and 2), probably originating from *trans*-4-hydroxy-2-nonenal, a lipid peroxidation product, or from its epoxide 2,3-epoxynonanal (El Ghissassi *et al.*, 1995; Chung *et al.*, 1996). For εA:A, the mean values ranged from 0.043 x 10^{-8} in liver to 35.0 x 10^{-8} in brain, corresponding to a 700-fold variation. For εC:C, they ranged from 0.062 x 10^{-8} in liver to 20.4 x 10^{-8} in brain, a 300-fold variation. The background levels in the heart (not shown in figures) were found to be (2.8 ± 1.1) x 10^{-8} for εA:A and (24 ± 16) x 10^{-8} for εC:C. Except in the liver, the mean values (± SD) in the various tissues were (12.3 ± 12.9) x 10^{-8} for εA:A and (12.3 ± 5.7) x 10^{-8} for εC:C. Higher background levels of both εA and εC have been observed in hepatic DNA from young adult Sprague-Dawley rats in other experiments (unpublished data), probably reflecting different dietary conditions (Fernando *et al.*, 1996; Nair *et al.*, 1997; Barbin, 1998).

After an eight-week exposure to vinyl chloride, increased levels of εC and/or εA were measured in most tissues, with varying degrees of significance depending both on the number of analyses and the background values (Figures 1 and 2). The levels of etheno adducts did not increase significantly in the brain, whereas in the kidney and spleen, the levels of εC but not of εA increased after exposure. The mean ± SD of the induced levels of εA:A (treated values–control values) was (4.1 ± 1.5) x

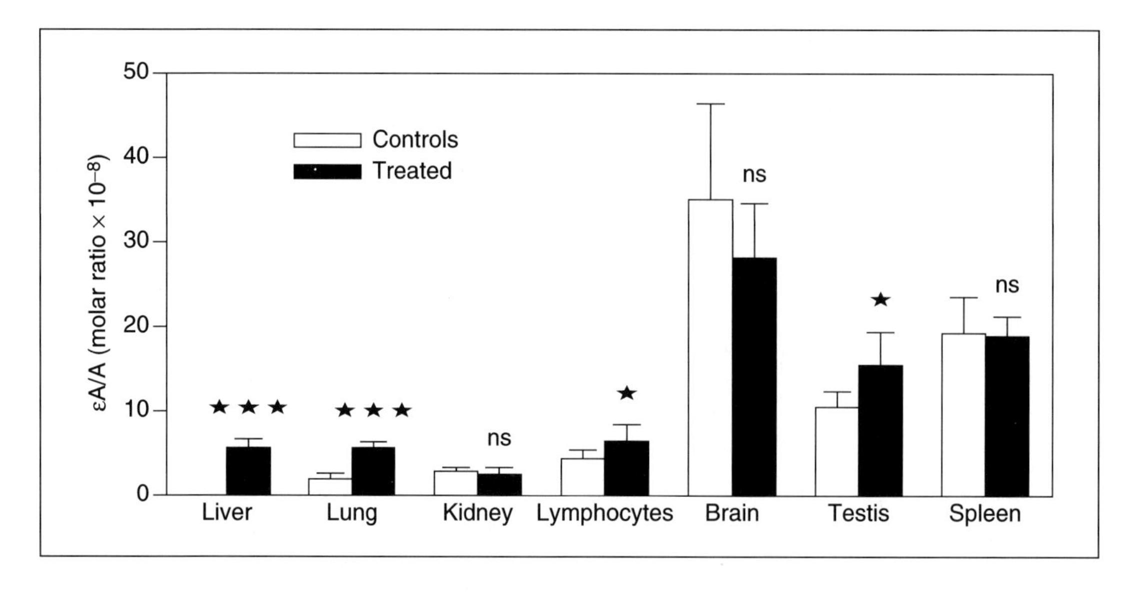

Figure 1. Formation of 1,N^6-ethenoadenine (εA) in DNA from adult Sprague-Dawley rats exposed to vinyl chloride for eight weeks

One-tailed *p* values (two-sample *t* test) for comparison of treated with control values: ns, not significant; *, $p < 0.05$; **, $p < 0.005$; ***, $p < 0.0001$

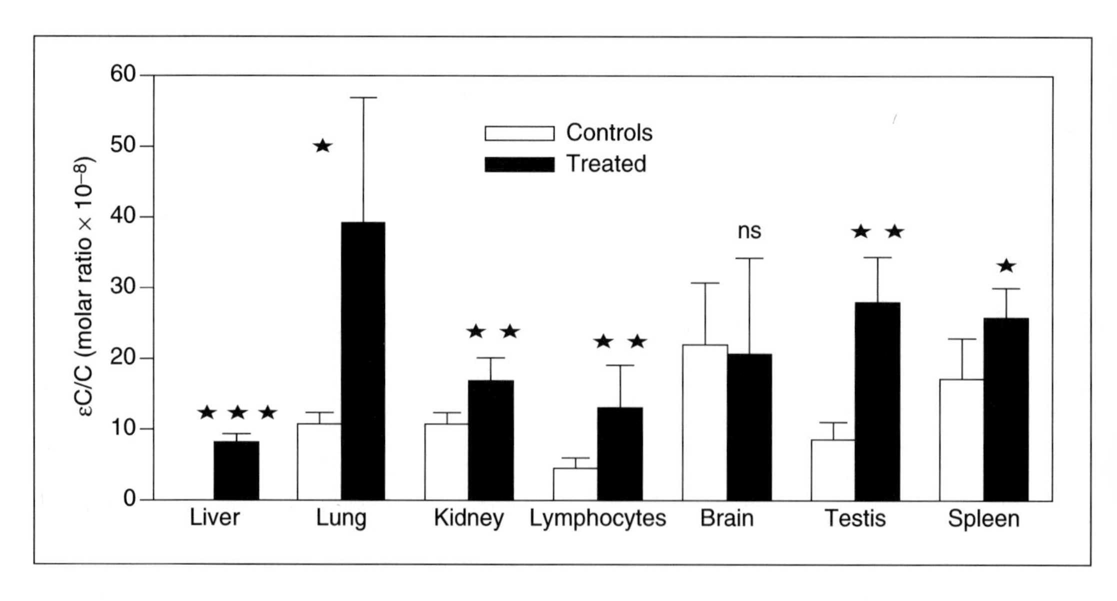

Figure 2. Formation of 3,N^4-ethenocytosine (εC) in DNA from adult Sprague-Dawley rats exposed to vinyl chloride for eight weeks

One-tailed *p* values (two-sample *t* test) for comparison of treated with control values: ns, not significant; *, $p < 0.05$; **, $p < 0.005$; ***, $p < 0.0001$

10^{-8} for liver, lung, lymphocytes and testis. The mean ± SD of the induced levels of εC:C was (7.8 ± 1.2) x 10^{-8} for liver, kidney, lymphocytes and spleen. The induced levels of εC:C were higher in the lung (mean value, 28 x 10^{-8}) and testis (mean value, 19 x 10^{-8}). There was no correlation between background levels and induced levels. The levels of vinyl chloride-induced and endogenous etheno adducts in various organs did not correlate with the target of vinyl chloride carcinogenesis: the liver, the major target organ, did not have higher levels of etheno adducts than the other tissues.

These data suggest a fairly homogeneous distribution among tissues in the induction of εA and εC by vinyl chloride, despite the large variations in the endogenous levels of these lesions. This might be explained by the fact that practically all inhaled vinyl chloride is activated in the liver (El Ghissassi *et al.*, 1998). Thus, alkylation of DNA in various tissues could be due mostly to generation of chloroethylene oxide in the liver and its distribution through the bloodstream. The levels measured after eight weeks of exposure, however, reflect different kinetics of formation and repair. It was shown previously that εA accumulates almost linearly in the liver during prolonged (1–8 weeks) exposure to vinyl chloride, whereas in the lung the level of εA increased during the first two weeks of exposure and remained at a plateau thereafter (Guichard *et al.*, 1996). The lack of any increase in εA levels in the kidney, brain and spleen after exposure to vinyl chloride may reflect efficient repair in those tissues. In contrast, εC may be poorly repaired in the lung and testis. As εA and εC can be repaired by the 3-methyladenine-DNA glycosylase (Saparbaev *et al.*, 1995; Hang *et al.*, 1996) and the thymine-DNA glycosylase (Saparbaev & Laval, 1998), respectively, it would be worthwhile examining the activities of these two enzymes in various tissues of rats, before and after exposure to vinyl chloride.

Spectra of mutations in liver tumours induced by vinyl chloride

Liver tumours obtained from Sprague-Dawley rats exposed to vinyl chloride by inhalation were analysed for mutations in the *ras* and *p53* genes (Froment *et al.*, 1994; Barbin *et al.*, 1997; Boivin *et al.*, unpublished data; see also Marion, this volume). In hepatocellular carcinomas, an activating AT→TA transversion was found at base 2 of codon 61 of the Ha-*ras* gene, but in angiosarcomas of the liver, no activating mutation was found in the Ha-*ras*, Ki-*ras* or N-*ras* genes in the 10 samples analysed (Froment *et al.*, 1994; Boivin *et al.*, unpublished data). *p53* gene mutations, mostly base-pair substitutions, were detected in 11/25 angiosarcomas, and one tumour had two point mutations, including a silent one. There were four AT→TA transversions (codons 160, 235, 253), two AT→GC transitions (codon 203), two AT→CG transversions (codon 235), three GC→AT transitions (codons 147, 152, 246) and one 12-base-pair deletion between codons 177 and 181. Codon 235 was mutated in three angiosarcomas. Two of the tumours had the same mutation in codon 203, and two others had the same mutation in codon 253. These data suggest the presence of 'hot spots' for *p53* mutations. Of eight heptocellular carcinomas analysed, only one contained a *p53* mutation, an AT→TA transversion in codon 283 (Barbin *et al.*, 1997). Thus, A:T base pairs are often involved in activating mutations of the Ha-*ras* gene in hepatocellular carcinoma and in mutations of the *p53* gene in angiosarcoma of the liver induced by vinyl chloride. This suggests that εA may be a critical lesion in hepatocarcinogenesis in rats exposed to vinyl chloride, a hypothesis that is compatible both with the promutagenic properties of εA (see Introduction) and with the accumulation of εA in hepatic DNA during prolonged exposure to vinyl chloride (Guichard *et al.*, 1996).

Correlation between the carcinogenic potency of vinyl chloride and its covalent binding index

On the basis of their chemical properties and genotoxic properties in *Drosophila*, etheno adduct-forming compounds such as vinyl chloride and urethane have been classified as a distinct group of alkylating agents, sharing the properties of monofunctional and cross-linking agents (Vogel *et al.*, 1996). A linear correlation was established between the carcinogenic potency of monofunctional carcinogens in rodents and their Swain-Scott constant, *s* (Barbin & Bartsch, 1989). Subsequently, this linear correlation was extended to monofunctional pro-carcinogens by use of the covalent binding index (Vogel *et al.*, 1996).

This index, originally proposed by Lutz (1979), is a measure of the covalent binding of a carcinogen to DNA in rodent liver after a single or a short-term treatment. Generally, total binding or the initial level of a major adduct, such as 7-alkylguanine, is considered. In a previous study on a series of 14 monofunctional alkylating agents, we estimated the covalent binding index for O^6-alkylguanine from that for N7-alkylguanine (Vogel *et al.*, 1996). We showed that the carcinogenic potency of these agents, expressed as the reciprocal of the TD_{50} value, is positively correlated with their covalent binding index for O^6-alkylguanine in liver DNA (Vogel *et al.*, 1996). This correlation has now been extended to 18 monofunctional alkylating agents (Figure 3).

To further understand the role of etheno adducts in carcinogenesis, adduct levels in liver can be expressed as a covalent binding index, thus allowing a quantitative comparison with the carcinogenic potency of vinyl chloride in rats. The carcinogenic potency of the compound can be calculated from the incidence of angiosarcoma of the liver observed in Sprague-Dawley rats exposed by inhalation to different concentrations (Maltoni *et al.*, 1984). When the dose (ppm in air) is transformed into the dose actually metabolized, a linear relationship is observed with tumour incidence (see, e.g. Chen & Blancato, 1989). From this relationship, we can calculate a TD_{50} value of 146 mmol/kg bw for vinyl chloride, corresponding to a carcinogenic potency ($1/TD_{50}$) of 0.00685 $mmol^{-1}$.kg bw. The covalent binding index for εA can be used for a comparison with carcinogenic potency, as this adduct appears to be a critical promutagenic lesion in the heptocarcinogenesis associated with exposure of Sprague-Dawley rats to vinyl chloride (see above). The εA levels induced in hepatic DNA after a five-day exposure to vinyl chloride were taken from the paper of Guichard *et al.* (1996), and the dose metabolized was calculated from the kinetic constants determined in the presence of hepatic microsomes from young adult Sprague-Dawley rats (El Ghissassi *et al.*, 1998). This yielded a covalent binding index of 0.00513 pmol.mg DNA^{-1}.$mmol^{-1}$.kg bw. As shown in Figure 3, vinyl chloride seems to fit the linear relationship established for monofunctional alkylating agents. This still holds true if εC and N^2,3-εG are included in the covalent binding index for vinyl chloride (data not shown). Thus, these data suggest that etheno bases are as efficient as O-alkylated bases in initiating carcinogenesis.

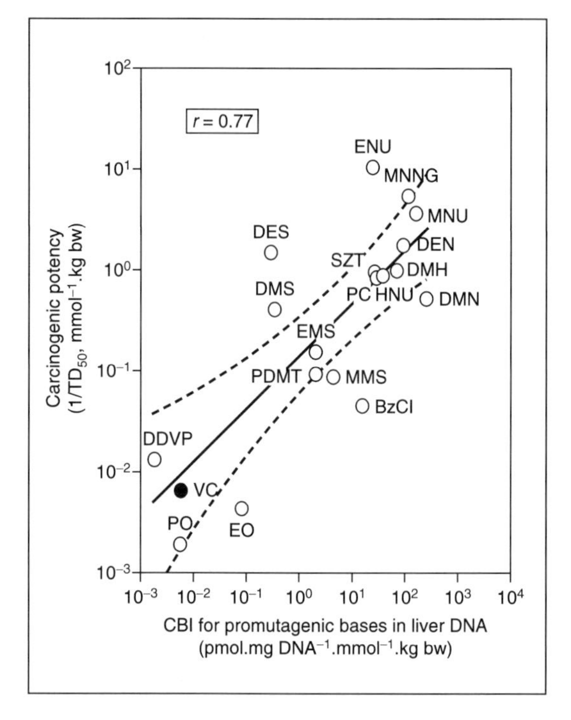

Figure 3. Carcinogenic potency (median $1/TD_{50}$ values in rodents) of monofunctional alkylating agents (open circles) and of vinyl chloride (VC) (filled circles) and comparison with covalent binding index (CBI) for promutagenic bases in liver DNA

Monofunctional agents: CBI for O^6-alkylguanine; VC, CBI for 1,N^6-ethenoadenine. The solid line represents the linear relationship for monofunctional carcinogens. $1/TD_{50}$ values and CBIs were calculated from published data, as described by Vogel *et al.* (1996).
ENU, *N*-ethyl-*N*-nitrosourea; MNNG, *N*-methyl-*N*′-nitro-*N*-nitrosoguanidine; MNU, *N*-methyl-*N*-nitrosourea; DEN, *N*-nitrosodiethylamine; DES, diethyl sulfate; SZT, streptozotocin; DMH, 1,2-dimethylhydrazine; HNU, *N*-hydroxyethyl-*N*-nitrosourea; PC, procarbazine; DMN, *N*-nitrosodimethylamine; DMS, dimethyl sulfate; EMS, ethyl methanesulfonate; PDMT, 1-phenyl-3,3-dimethyltriazene; MMS, methyl methanesulfonate; BzCl, benzyl chloride; DDVP, 2,2-dichlorovinyldimethyl phosphate; EO, ethylene oxide; PO, propylene oxide

Factors involved in the accumulation and persistence of etheno bases during exposure to vinyl chloride

As several factors may interact to modulate the formation and persistence of DNA etheno bases *in vivo* (Figure 4), interpretation of these measures, as shown in Figures 1 and 2, is challenging. Since DNA etheno bases are formed by the metabolite chloroethylene oxide during exposure to vinyl chloride, their levels in DNA are affected by the activities of cytochrome P450 2E1, detoxifying enzymes (epoxide hydrolases, glutathione transferases) and DNA repair proteins. In addition, there is a highly variable background of endogenous origin, probably resulting from reactive lipid peroxidation products (El Ghissassi *et al.,* 1995; Chung *et al.,* 1996). This background seems to be very sensitive to diet (Fernando *et al.,* 1996; Nair *et al.,* 1997; Barbin, 1998). Lipid peroxidation could be stimulated by exposure to vinyl chloride through the release of reactive oxygen species by CYP 2E1 (Ekström & Ingerman-Sundberg, 1989) and through glutathione depletion by chloroethylene oxide and chloroacetaldehyde. Chloroacetaldehyde has been shown to stimulate lipid peroxidation in isolated rat hepatocytes (Sood & O'Brien, 1993). Therefore, etheno bases are also formed through an indirect pathway (Figure 4), the contribution of which is probably variable and cannot be evaluated accurately. Stimulation of these endogenous processes could explain the paradox that, despite the fact that they are known to be repaired *in vitro* ((Saparbaev *et al.,* 1995; Hang *et al.,* 1996; Saparbaev & Laval, 1998), etheno adducts seem to accumulate in certain tissues during exposure to vinyl chloride and to persist in hepatic DNA after the end of treatment (Swenberg *et al.,* 1992; Guichard *et al.,* 1996).

DNA alkylation and the role of etheno bases in vinyl chloride-induced carcinogenesis have been investigated extensively; however, for a more comprehensive understanding of the mechanisms of action of this carcinogen, the repair pathways of etheno bases *in vivo* and the role of oxidative stress and lipid peroxidation must be further investigated.

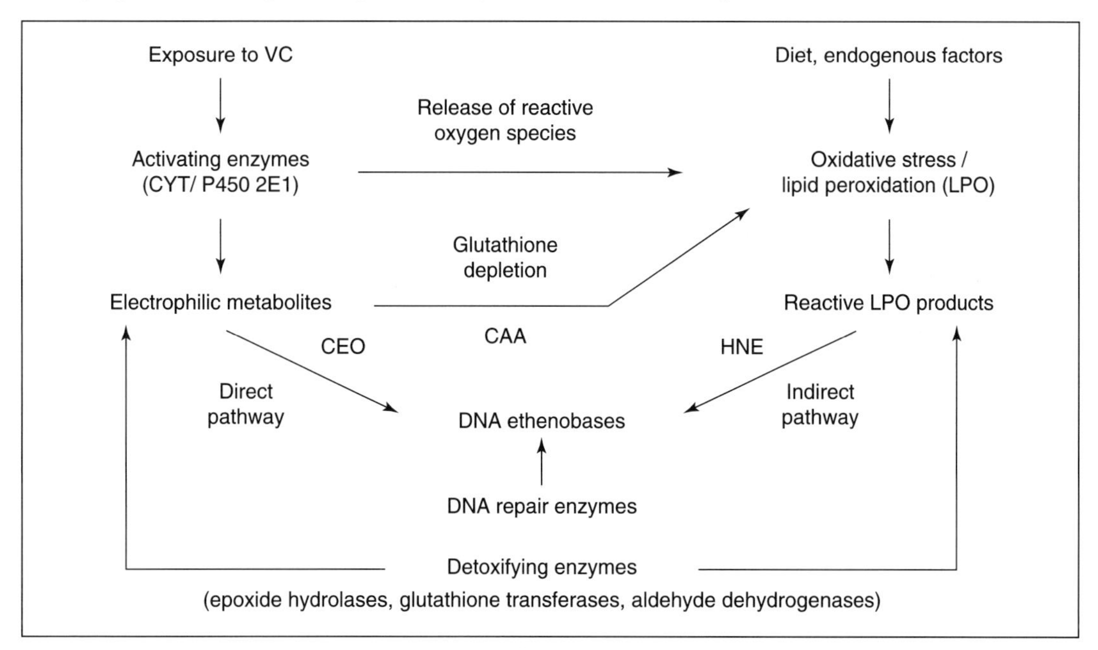

Figure 4. Factors possibly involved in the accumulation and persistence of DNA etheno bases during and after exposure to vinyl chloride (VC)

CEO, chloroethylene oxide; CAA, 2-chloroacetaldehyde; HNE, *trans*-4-hydroxy-2-nonenal.

Acknowledgements

We wish to thank Mrs G. Brun for her skilled technical assistance. This work was partly supported by the Environment and Climate Research Programme of the European Commission (Contract ENV4-CT97-0505).

References

Barbin, A. (1998) Formation of DNA etheno adducts in rodents and humans and their role in carcinogenesis. *Acta Biochem. Pol.* 45, 145–161

Barbin, A. & Bartsch, H. (1986) Mutagenic and promutagenic properties of DNA adducts formed by vinyl chloride metabolites. In: *The Role of Cyclic Nucleic Acid Adducts in Carcinogenesis and Mutagenesis* (Singer, B. & Bartsch, H., eds), IARC Scientific Publications No. 70. Lyon, IARC, pp. 345–358

Barbin, A. & Bartsch, H. (1989) Nucleophilic selectivity as a determinant of carcinogenic potency (TD_{50}) in rodents: A comparison of mono- and bifunctional alkylating agents and vinyl chloride metabolites. *Mutat. Res.* 215, 95–106

Barbin, A., Brésil, H., Croisy, A., Jacquignon, P., Malaveille, C., Montesano, R. & Bartsch, H. (1975) Liver-microsome-mediated formation of alkylating agents from vinyl bromide and vinyl chloride. *Biochem. Biophys. Res. Commun.* 67, 596–603

Barbin, A., Bartsch, H., Leconte, P. & Radman, M. (1981) Studies on the miscoding properties of 1,N^6-ethenoadenine and 3,N^4-ethenocytosine, DNA reaction products of vinyl chloride metabolites, during *in vitro* DNA synthesis. *Nucleic Acids Res.* 9, 375–387

Barbin, A., Laib, R.J. & Bartsch, H. (1985) Lack of miscoding properties of 7-(2-oxoethyl)guanine, the major vinyl chloride–DNA adduct. *Cancer Res.* 45, 2440–2444

Barbin, A., Friesen, M., O'Neill, I.K. & Bartsch, H. (1986) New, potentially miscoding adducts of chloroethylene oxide and chloroacetaldehyde with pyrimidine nucleotides. *Chem. Biol. Interactions* 59, 43–54.

Barbin, A., Froment, O., Boivin, S., Marion, M.J., Belpoggi, F., Maltoni, C. & Montesano, R. (1997) *p53* gene mutation pattern in rat liver tumours induced by vinyl chloride. *Cancer Res.* 57, 1695–1698

Bartsch, H., Barbin, A., Marion, M.J., Nair, J. & Guichard, Y. (1994) Formation, detection, and role in carcinogenesis of ethenobases in DNA. *Drug Metab. Rev.* 26, 349–371

Basu, A.K., Wood, M.L., Niedernhofer, L.J., Ramos, L.A. & Essigmann, J.M. (1993) Mutagenic and genotoxic effects of three vinyl chloride-induced DNA lesions: 1,N^6-Ethenoadenine, 3,N^4-ethenocytosine, and 4-amino-5-(imidazol-2-yl) imidazole. *Biochemistry* 32, 12793–12801

Chen, C.W. & Blancato, J.N. (1989) Incorporation of biological information in cancer risk assessment: Example—vinyl chloride. *Cell Biol. Toxicol.* 5, 417–444

Cheng, K.C., Preston, B.D., Cahill, D.S., Dosanjh, M.K., Singer, B. & Loeb, L.A. (1991) The vinyl chloride DNA derivative N^2,3-ethenoguanine produces G→A transitions in *Escherichia coli*. *Proc. Natl Acad. Sci. USA* 88, 9974–9978

Chung, F.-L., Chen, H.-J.C. & Nath, R.G. (1996) Lipid peroxidation as a potential endogenous source for the formation of exocyclic DNA adducts. *Carcinogenesis* 17, 2105–2111

Ciroussel, F., Barbin, A., Eberle, G. & Bartsch, H. (1990) Investigations on the relationship between DNA ethenobase adduct levels in several organs of vinyl chloride-exposed rats and cancer susceptibility. *Biochem. Pharmacol.* 39, 1109–1113

Creech, J.L., Jr & Johnson, M.N. (1974) Angiosarcoma of liver in the manufacture of polyvinyl chloride. *J. Occup. Med.* 16, 150–151

Dosanjh, M.K., Chenna, A., Kim, E., Fraenkel-Conrat, H., Samson, L. & Singer, B. (1994) All four known cyclic adducts formed in DNA by the vinyl chloride metabolite chloroacetaldehyde are released by a human DNA glycosylase. *Proc. Natl Acad. Sci. USA*, 91, 1024–1028

Eberle, G., Barbin, A., Laib, R.J., Ciroussel, F., Thomale, J., Bartsch, H. & Rajewsky, M.F. (1989) 1,N^6-Etheno-2′-deoxyadenosine and 3,N^4-etheno-2′-deoxycytidine detected by monoclonal antibodies in lung and liver DNA of rats exposed to vinyl chloride. *Carcinogenesis* 10, 209–212

Ekström, G. & Ingerman-Sundberg, M. (1989) Rat liver microsomal NADPH-supported oxidase activity and lipid peroxidation dependent on ethanol-inducible cytochrome P450 (P450 IIE1). *Biochem. Pharmacol.* 38, 1313–1319

El Ghissassi, F., Barbin, A., Nair, J. & Bartsch, H. (1995) Formation of 1,N^6-ethenoadenine and 3,N^4-ethenocytosine by lipid peroxidation products and nucleic acid bases. *Chem. Res. Toxicol.* 8, 278–283

El Ghissassi, F., Barbin, A. & Bartsch, H. (1998) Metabolic activation of vinyl chloride by rat liver microsomes: Low dose kinetics and involvement of cytochrome P450 2E1. *Biochem. Pharmacol.* 55, 1445–1452

Fedtke, N., Boucheron, J.A., Walker, V.E. & Swenberg, J.A. (1990) Vinyl chloride-induced DNA adducts. II: Formation and persistence of 7-(2′-oxoethyl)guanine and N^2,3-etheno-guanine in rat tissue DNA. *Carcinogenesis* 11, 1287–1292

Fernando, R.C., Nair, J., Barbin, A., Miller, J.A. & Bartsch, H. (1996) Detection of 1,N^6-ethenodeoxyadenosine and 3,N^4-ethenodeoxycytidine by immunoaffinity/^{32}P-postlabelling in liver and lung DNA of mice treated with ethyl carbamate (urethane) or its metabolites. *Carcinogenesis* 17, 1711–1718

Froment, O., Boivin, S., Barbin, A., Bancel, B., Trépo, C. & Marion, M.J. (1994) Mutagenesis of *ras* proto-oncogenes in rat liver tumors induced by vinyl chloride. *Cancer Res.* 54, 5340–5345

Giri, A.K. (1995) Genetic toxicology of vinyl chloride—A review. *Mutat. Res.* 339, 1–14

Green, T. & Hathway, D.E. (1978) Interactions of vinyl chloride with rat liver DNA *in vivo*. *Chem.-biol. Interactions* 22, 211–224

Guengerich, F.P. (1992) Roles of the vinyl chloride oxidation products 1-chlorooxirane and 2-chloroacetaldehyde in the *in vitro* formation of etheno adducts of nucleic acid bases. *Chem. Res. Toxicol.* 5, 2–5

Guengerich, F.P., Mason, P.S., Stott, W.T., Fox, T.R. & Watanabe, P.G. (1981) Roles of 2-haloethylene oxides and 2-haloacetaldehydes derived from vinyl bromide and vinyl chloride in irreversible binding to protein and DNA. *Cancer Res.* 41, 4391–4398

Guengerich, F.P., Kim, D.H. & Iwasaki, M. (1991) Role of human cytochrome P-450 IIE1 in the oxidation of many low molecular weight cancer suspects. *Chem. Res. Toxicol.* 4, 168–179

Guichard, Y., Nair, J., Barbin, A. & Bartsch, H. (1993) Immunoaffinity clean-up combined with ^{32}P-postlabelling analysis of 1,N^6-ethenoadenine and 3,N^4-ethenocytosine in DNA. In: *Postlabelling Methods for Detection of DNA Adducts* (Phillips, D.H., Castegnaro, M. & Bartsch, H., eds), IARC Scientific Publications No. 124. Lyon, IARC, pp. 263–269

Guichard, Y., El-Ghissassi, F., Nair, J., Bartsch, H. & Barbin, A. (1996) Formation and accumulation of DNA ethenobases in adult Sprague-Dawley rats exposed to vinyl chloride. *Carcinogenesis* 17, 1553–1559

Hall, J.A., Saffhill, R., Green, T. & Hathway, D.E. (1981) The induction of errors during *in vitro* DNA synthesis following chloroacetaldehyde-treatment of poly(dA-dT) and poly(dC-dG) templates. *Carcinogenesis* 2, 141–146

Hang, B., Chenna, A., Rao, S. & Singer, B. (1996) 1,N^6-Ethenoadenine and 3,N^4-ethenocytosine are excised by separate human DNA glycosylases. *Carcinogenesis* 17, 155–157

IARC (1979) *IARC Monographs on the Evaluation of Carcinogenic Risks to Humans*, Vol. 19. Lyon, IARC, pp. 377–438

Kochetkov, N.K., Shibaev, V.N. & Kost, A.A. (1971) New reaction of adenine and cytosine derivatives, potentially useful for nucleic acid modifications. *Tetrahedron Lett.* 22, 1993–1996

Laib, R.J. (1986) The role of cyclic base adducts in vinyl chloride-induced carcinogenesis: Studies on nucleic acid alkylation *in vivo*. In: *The Role of Cyclic Nucleic Acid Adducts in Carcinogenesis and Mutagenesis* (Singer, B. & Bartsch, H., eds), IARC Scientific Publications No. 70. Lyon, IARC, pp. 101–108

Laib, R.J. & Bolt, H.M. (1977) Alkylation of RNA by vinyl chloride metabolites *in vitro* and *in vivo*: Formation of 1,N^6-ethenoadenosine. *Toxicology* 8, 185–195

Laib, R.J. & Bolt, H.M. (1978) Formation of 3,N^4-ethenocytidine moieties in RNA by vinyl chloride metabolites *in vitro* and *in vivo*. *Arch. Toxicol.* 39, 235–240.

Lutz, W.K. (1979) *In vivo* covalent binding of organic chemicals to DNA as a quantitative indicator in the process of chemical carcinogenesis. *Mutat. Res.* 65, 289–356

Langouët, S., Müller, M. & Guengerich, F.P. (1997) Misincorporation of dNTPs opposite 1,N^2-ethenoguanine and 5,6,7,9-tetrahydro-7-hydroxy-9-oxoimidazo[1,2-α]purine in oligonucleotides by *Escherichia coli* polymerases I exo$^-$ and II exo$^-$, T7 polymerase exo$^-$, human immunodeficiency virus-1 reverse transcriptase, and rat polymerase β. *Biochemistry* 36, 6069–6079

Langouët, S., Mican, A.N., Müller, M., Fink, S.P., Marnett, L.J., Muhle, S.A. & Guengerich, F.P. (1998) Misincorporation of nucleotides opposite five-membered exocyclic ring guanine derivatives by *Escherichia coli* polymerases *in vitro* and *in vivo:* 1,N^2-Ethenoguanine, 5,6,7,9-tetrahydro-9-oxoimidazo-[1,2-α]purine, and 5,6,7,9-tetrahydro-7-hydroxy-9-oxoimidazo[1,2-α]purine. *Biochemistry* 37, 5184–5193

Maltoni, C., Lefemine, G., Ciliberti, A., Cotti, G. & Caretti, D. (1984) Experimental research on vinyl chloride carcinogenesis. In: *Archives of Research on Industrial Carcinogenesis* (Maltoni, C. & Mehlman, M.A., eds). New York, Princeton Scientific Publishers

Moriya, M., Zhang, W., Johnson, F. & Grollman, A.P. (1994) Mutagenic potency of exocylic DNA adducts: Marked differences between *Escherichia coli* and simian kidney cells. *Proc. Natl Acad. Sci. USA,* 91, 11899–11903

Müller, M., Belas, F.J., Blair, I.A. & Guengerich, F.P. (1997) Analysis of 1,N^2-etheno-guanine and 5,6,7,9-tetrahydro-7-hydroxy-9-oxoimidazo[1,2-α]purine in DNA treated with 2-chlorooxirane by high performance liquid chromatography/electrospray mass spectrometry and comparison of amounts to other DNA adducts. *Chem. Res. Toxicol.* 10, 242–247

Nair, J., Barbin, A., Guichard, Y. & Bartsch, H. (1995) 1,N^6-Ethenodeoxyadenosine and 3,N^4-ethenodeoxycytine in liver DNA from humans and untreated rodents detected by immunoaffinity/^{32}P-postlabeling. *Carcinogenesis* 16, 613–617

Nair, J., Vaca, C.E., Velic, I., Mutanen, M., Valsta, L.M. & Bartsch, H. (1997) High dietary ω-6 polyunsaturated fatty acids drastically increase the formation of etheno-DNA base adducts in white blood cells of female subjects. *Cancer Epidemiol. Biomarkers Prev.* 6, 597–601

O'Neill, I.K., Barbin, A., Friesen, M. & Bartsch, H. (1986) Reaction kinetics and cytosine adducts of chloroethylene oxide and chloroacetaldehyde: Direct observation of intermediates by FTNMR and GC–MS. In: *The Role of Cyclic Nucleic Acid Adducts in Carcinogenesis and Mutagenesis* (Singer, B. & Bartsch, H., eds), IARC Scientific Publications No. 70. Lyon, IARC, pp. 57–73

Palejwala, V.A., Simha, D. & Humayun, M.Z. (1991) Mechanisms of mutagenesis by exocylic DNA adducts. Transfection of M13 viral DNA bearing a site-specific adduct shows that ethenocytosine is a highly efficient RecA-independent mutagenic noninstructional lesion. *Biochemistry* 30, 8736–8743

Palejwala, V.A., Rzepka, R.W & Humayun, M.Z. (1993) UV irradiation of *Escherichia coli* modulates mutagenesis at a site-specific ethenocytosine residue on M13 DNA. Evidence for an inducible recA-independent effect. *Biochemistry* 32, 4112–4120

Pandya, G.A. & Moriya, M. (1996) 1,N^6-Ethenodeoxyadenosine, a DNA adduct highly mutagenic in mammalian cells. *Biochemistry* 35, 11487–11492

Saparbaev, M. and Laval, J. (1998) 3,N^4-Ethenocytosine, a highly mutagenic adduct, is a primary substrate for *Escherichia coli* double-stranded uracil-DNA glycosylase and human mismatch-specific thymine-DNA glycosylase. *Proc. Natl Acad. Sci. USA* 95, 8508–8513

Saparbaev, M., Kleibl, K. & Laval, J. (1995) *Escherichia coli, Saccharomyces cerevisiae,* rat and human 3-methyladenine DNA glycosylases repair 1,N^6-ethenoadenine when present in DNA. *Nucleic Acids Res.* 23, 3750–3755

Saurin, J.C., Tanière, P., Mion, F., Jacob, P., Partensky, C., Paliard, P. & Berger, F. (1997) Primary hepatocellular carcinoma in workers exposed to vinyl chloride: A report of two cases. *Cancer 79,* 1671–1677

Shibutani, S., Suzuki, N., Matsumoto, Y. & Grollman, A.P. (1996) Miscoding properties of 3,N^4-etheno-2'-deoxycytidine in reactions catalyzed by mammalian DNA polymerases. *Biochemistry* 35, 14992–14998

Simonato, L., L'Abbé, K.A., Andersen, A., Belli, S., Comba, P., Engholm, G., Ferro, G., Hagmar, L., Langård, S., Lundberg,I., Pirastu, R., Thomas, P., Winkelmann, R. & Saracci, R. (1991) A collaborative study of cancer incidence and mortality among vinyl chloride workers. *Scand. J. Work. Environ. Health* 17, 159–169

Singer, B., Kuśmierek, J.T. & Fraenkel-Conrat, H. (1983) *In vitro* discrimination of replicases acting on carcinogen-modified polynucleotide templates. *Proc. Natl Acad. Sci. USA,* 80, 969–972

Singer, B., Abbott, L.G. & Spengler, S.J. (1984) Assessment of mutagenic efficiency of two carcinogen-modified nucleosides, 1,N^6-ethenodeoxyadenosine and O^4-methyldeoxythymidine, using polymerases of varying fidelity. *Carcinogenesis* 5, 1165–1171

Singer, B., Holbrook, S.R. Fraenkel-Conrat, H. & Kuśmierek, J.T. (1986) Neutral reactions of haloacetaldehydes with polynucleotides: Mechanisms, monomer and polymer products. In: *The Role of Cyclic Nucleic Acid Adducts in Carcinogenesis and Mutagenesis* (Singer, B. & Bartsch, H., eds), IARC Scientific Publications No. 70. Lyon, IARC, pp. 45–56

Singer, B., Kuśmierek, J.T., Folkman, W., Chavez, F. & Dosanjh, M.K. (1991) Evidence for the mutagenic potential of the vinyl chloride induced adduct, N^2,3-ethenodeoxyguanosine, using a site-directed kinetic assay. *Carcinogenesis* 12, 745–747

Sood, C. & O'Brien, P.J. (1993) Molecular mechanisms of chloroacetaldehyde-induced cytotoxicity in isolated rat hepatocytes. *Biochem. Pharmacol.* 46, 1621–1626

Spengler, S. & Singer, B. (1981) Transcriptional errors and ambiguity resulting from the presence of 1,N^6-ethenoadenosine or 3,N^4-ethenocytidine in polyribonucleotides. *Nucleic Acids Res.* 9, 365–373

Swenberg, J.A., Fedtke, N., Ciroussel, F., Barbin, A. & Bartsch, H. (1992) Etheno adducts formed in DNA of vinyl chloride-exposed rats are highly persistent in liver. *Carcinogenesis* 13, 727–729

Viola, P.L., Bigotti, A. & Caputo, A. (1971) Oncogenic response of rat skin, lungs, and bones to vinyl chloride. *Cancer Res.* 31, 516–522

Vogel, E.W., Nivard, M.J.M., Ballering, L.A.B., Bartsch, H., Barbin, A., Nair, J., Comendador, M.A., Sierra, L.M., Aguirrezabalaga, I., Tosal, L., Ehrenberg, L., Fuchs, R.P.P., Janel-Bintz, R., Maenhaut-Michel, G., Montesano, R., Hall, J., Kang, H., Miele, M., Thomale, J., Bender, K., Engelbergs, J. & Rajewsky, M.F. (1996) DNA damage and repair in mutagenesis and carcinogenesis: Implications of structure–activity relationships for cross-species extrapolation. *Mutat. Res.* 113, 177–218

Zajdela, F., Croisy, A., Barbin, A., Malaveille, C., Tomatis, L. & Bartsch, H. (1980) Carcinogenicity of chloroethylene oxide, an ultimate reactive metabolite of vinyl chloride, and bis(chloromethyl)ether after subcutaneous administration and in initiation–promotion experiments in mice. *Cancer Res.* 40, 352–356

Zhang, W., Johnson, F., Grollman, A.P. & Shibutani, S. (1995) Miscoding by the exocylic and related DNA adducts 3,N^4-etheno-2'-deoxycytidine, 3,N^4-ethano-2'-deoxycytidine, and 3-(2-hydroxy-ethyl)-2'-deoxyuridine. *Chem. Res. Toxicol.* 8, 157–163.

Exocyclic DNA Adducts in Mutagenesis and Carcinogenesis
B. Singer & H. Bartsch
IARC Scientific Publications No. 150
International Agency for Research on Cancer, Lyon, 1999

Vinyl chloride-specific mutations in humans and animals

M.J. Marion & S. Boivin-Angele

Vinyl chloride is a potent hepatocarcinogen which reacts with DNA to generate etheno bases. In order to determine whether mutational patterns in target genes *in vivo* are characteristic of vinyl chloride and could be explained by the mutagenic properties of the etheno bases, human and rat liver tumours associated with exposure to vinyl chloride were analysed for point mutations in the *ras* and *p53* genes. In this paper, we review these data and report our latest results on animal tumours. Two alterations were found which could be attributed to a direct effect of vinyl chloride: a GC→AT transition which leads to a GGC→GAC mutation at codon 13 of the Ki-*ras* gene in human liver angiosarcomas, and lesions at AT base pairs, mostly AT→TA transversions, which lead to mutations in the *p53* gene in human and rat angiosarcomas and to a CAA→CTA mutation at codon 61 of the Ha-*ras* gene in rat hepatocellular carcinomas.

Introduction

Carcinogens are often mutagens, which are believed to initiate carcinogenesis by altering genes that are critical for the control of cell proliferation. They react directly with DNA bases by covalent binding. The modified bases, called adducts, may subsequently induce misincorporation during replication or transcription, leading to base-pair substitution in the resulting DNA or RNA. Notably on the basis of analyses of experimental tumours, it was assumed that a particular mutation found in a tumour could be a direct effect of a carcinogenic agent; however, it is difficult to establish unequivocally a link between the mutational spectrum in a tumour and a specific exposure, particularly for human tumours.

Vinyl chloride is a gas used in the plastics industry to produce polyvinyl chloride; it is also a potent hepatocarcinogen, as first shown in animals by Viola *et al.* in 1971 and confirmed by Maltoni and Lefemine (1974), who observed angiosarcomas of the liver, hepatocellular carcinomas and cholangiocarcinomas and other tumours in rats exposed to vinyl chloride. This compound was found to be carcinogenic in humans after the retrospective identification of three cases of angiosarcoma of the liver in the same polyvinyl chloride production plant in Louisville, Kentucky, USA (Creech & Johnson, 1974). The causal relationship between exposure to vinyl chloride and angiosarcoma has since been well established in numerous epidemiological studies (Simonato *et al.*, 1991); some cases of histologically confirmed hepatocellular carcinoma and cholangiocarcinoma have also been reported, alone or associated with angiosarcoma (Byren & Holmberg, 1975). Humans are exposed to vinyl chloride mainly by inhalation in plants where it is synthesized and especially in polymerization plants.

Many studies have been carried out to elucidate the metabolic pathways of vinyl chloride and to identify its DNA adducts and mechanism of mutagenicity. It is activated by the microsomal cytochrome P450 mono-oxygenase system into chloroacetaldehyde and chloroethylene oxide, which is thought to be the ultimate carcinogenic metabolite. These two compounds induce mutations by base-pair substitution in various test systems, suggesting that vinyl chloride acts by forming miscoding DNA adducts. The identification of these etheno adducts, their mutagenic properties *in vitro* and their formation and persistence in the liver and other organs of animals have been reviewed (Barbin, 1998).

Angiosarcomas of the liver are rare. For example, an epidemiological survey in the State of New York showed an annual incidence rate of 0.25 case per million, including angiosarcomas induced by radiation and chemicals. One spontaneous angiosarcoma was observed among 1420 Sprague-Dawley rats used in long-term toxicological studies (Zwicker *et al.*, 1995). Angiosarcomas occurring in humans or in Sprague-Dawley rats after exposure to vinyl chloride might be expected to exhibit mutations in target genes such as *ras* and *p53* caused by direct interaction with DNA. Genetic alterations possibly arising from the formation of vinyl chloride–DNA adducts have been identified in Ki-*ras* and *p53* genes in human cases of angiosarcoma of the liver associated with exposure to vinyl chloride (Marion *et al.*, 1991; Hollstein *et al.*, 1994), and mutations in the *p53* gene have been investigated in liver tumours induced in Sprague-Dawley rats (Barbin *et al.*, 1997).

In a preliminary study, five angiosarcomas of the liver and two hepatocellular carcinomas induced by long-term exposure to gaseous vinyl chloride were analysed for *ras* gene mutations by polymerase chain reaction (PCR) amplification, allele-specific amplification and sequencing (Froment *et al.*, 1994). The analysis was completed by analysing other liver tumours from the same experiment by PCR amplification, single-strand conformation polymorphism analysis and sequencing. We review all these results in this paper.

Materials and methods

Vinyl chloride-induced liver tumours in rats

The experimental conditions of exposure of Sprague-Dawley rats to vinyl chloride and the collection of the tissues have been described elsewhere (Froment *et al.*, 1994). In the 44 vinyl chloride-exposed rats, 66 hepatic lesions were identified, including nodular hyperplasia, endothelial cell hyperplasia, peliosis, adenomas, benign cholangiomas, angiosarcomas and hepatocellular carcinomas. In the work presented here, 11 angiosarcomas and eight hepatocellular carcinomas were analysed; these included the five angiosarcomas and the two hepatocellular carcinoma studied previously.

DNA extraction

Microdissection was used to select areas containing at least 10–20% of tumour cells particularly in the tumours in which the sinusoidal form predominated. DNA was extracted with proteinase K and phenol–chloroform (Marion *et al.*, 1991) from frozen tissue sections of all the tumours, except for three very small hepatocellular carcinomas for which paraffin sections were used.

Amplification of ras genes by polymerase chain reaction

The primers used for PCR, single-strand conformation polymorphism (SSCP) analysis and sequencing were those described by Verlaan-de Vries *et al.* (1986) and Fujimoto *et al.* (1991). DNA (0.5 µg or less) was subjected to 35 cycles of amplification, as described previously (Marion *et al.*, 1991), with a GeneAmp PCR system 2400 (Perkin Elmer, Applied Biosystems Division, Foster City, CA, USA). After a 5-min denaturation step at 94 °C, 33 cycles of 10 s at 94 °C, 10 s at 53 °C and 15 s at 72 °C were run, with a final extension step of 5 min at 72 °C. The amplification products were analysed by electrophoresis in 4% Nu-Sieve agarose gel (FMC Bioproducts, Rockland, ME, USA), the bands corresponding to the expected DNA fragment were excised under ultraviolet light, and the DNA was eluted in water.

Single-strand conformation polymorphism analysis

Two microlitres of the eluted DNA were subjected to 35 cycles of PCR as described above in 10 µl of the same medium containing 50 µmol/L dCTP and 2 µCi of [α-^{32}P]dCTP (specific activity, > 111 TBq/mmol) (ICN Biomedicals Inc., Costa Mesa, CA, USA). One microlitre of the radioactive PCR product was mixed with 4 µl of loading buffer containing 95% formamide, 20 mmol/L EDTA, 0.05% bromophenol blue and 0.05% xylene cyanole. Each sample was denaturated and loaded onto a 0.5% Hydrolink MDE™ gel (FMC Bioproducts, Rockland, ME, USA). Electrophoresis was performed under four conditions: in the presence of 0 or 5% glycerol, at 4 °C or at room temperature. The gel was dried on Whatman 3MM paper and exposed to Biomax MR film (Eastman Kodak Company, Rochester, NY, USA).

Direct sequencing

DNA eluted from the shifted bands of the SSCP gel was sequenced with the Sequenase™ Version 2.0 DNA Sequencing kit (United States Biochemicals, Cleveland, OH, USA), as described previously (Froment *et al.*, 1994).

Results

All of the angiosarcomas were investigated for mutations at exons 1 and 2 of the Ha-, Ki- and N-*ras* genes by PCR amplification and SSCP analysis. Similar patterns were obtained for liver DNA from a control Sprague-Dawley rat and tumour DNA from all of the angiosarcomas analysed by SSCP with the four experimental conditions (data not shown), demonstrating the absence of mutation in the DNA fragments amplified.

In eight hepatocellular carcinomas analysed by SSCP, no mutation was detected in exons 1 and 2 of the Ki- and N-*ras* genes or at exon 1 of the Ha-*ras* gene. However, a CAA→CTA mutation at codon 61 of Ha-*ras* gene was found by allele-specific oligonucleotide hybridization and confirmed by direct sequencing of the PCR product (Froment *et al.*, 1994). Figure 1 shows the results of analysis of these tumours with probes containing either the normal codon, CAA, or the mutated CTA codon and the DNA sequences of point mutations in two hepatocellular carcinomas. One of these tumours was subsequently used as a positive control, and six others were analysed by PCR and SSCP. Figure 2 shows the pattern obtained by analysis of the Ha-*ras* exon 2 in four of these tumours: it was similar to that of the positive tumour DNA. Shifted bands were observed, in addition to the main bands present in all the samples, which corresponded to the wild-type sequence. Shifted bands were sometimes also observed in tissue surrounding the tumour but at lower intensity. Sequencing of the shifted bands confirmed unambiguously the presence of a CAA→CTA mutation at codon 61 (Gln→Leu). Shifted bands were also observed in DNA extracted from paraffin sections of the three small tumours (data not shown), but the sequencing of these bands was inconclusive and the mutation could not be confirmed.

Discussion

Human tumours

Six angiosarcomas of the liver from workers exposed to vinyl chloride were analysed for mutations at codons 12–13 and 61 of Ha-, Ki- and N-*ras* genes by microdissection, PCR amplification with exonic primers and allele-specific oligonu-

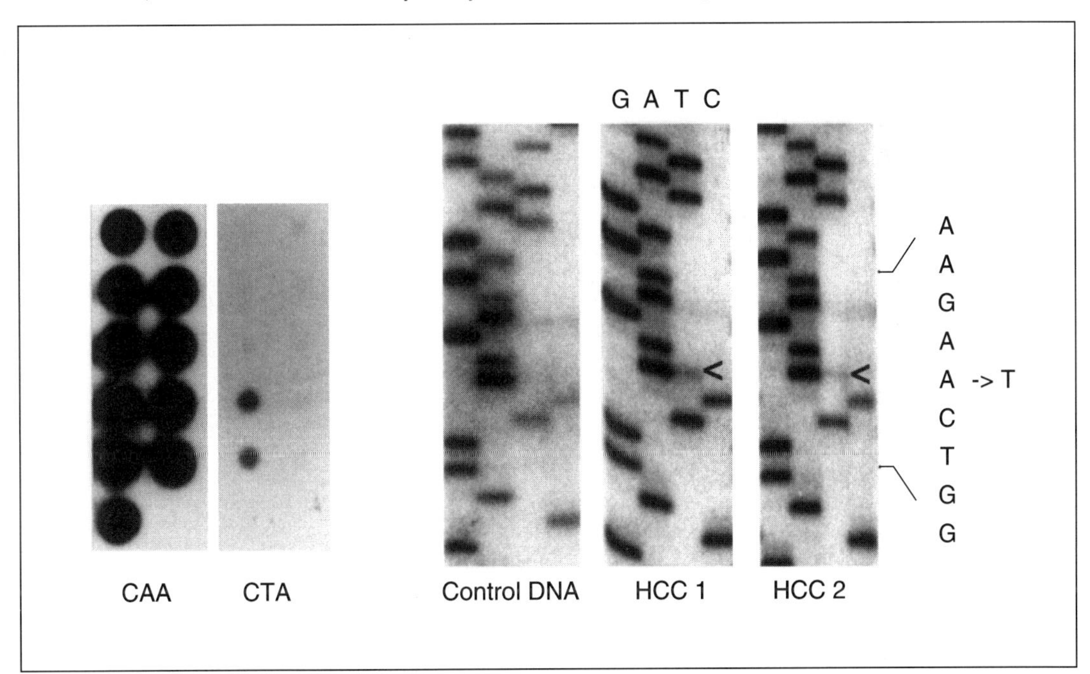

Figure 1. Detection by allele-specific oligonucleotide hybridization (left panel) and DNA sequencing (right panel) of point mutations in the second exon of the Ha-*ras* oncogene in hepatocellular carcinomas (HCC) induced by vinyl chloride in rats

cleotide hybridization. DNA from none of the six tumours hybridized with mutated probes for codons 12 and 61 of the three *ras* genes, nor for codon 13 of Ha- and N-*ras*. DNA from five tumours hybridized with a probe in which the normal codon 13 of the Ki-*ras* gene –GCC– (non-coding strand) had been replaced by –GTC–. DNA from normal tissue or from lymphocytes from the same patients had the normal sequence, indicating that the observed mutation was somatic (Marion *et al.*, 1991). The sequencing confirmed the presence of a GGC→GAC mutation at codon 13 (M.J. Marion, unpublished data).

Ha-, Ki- and N-*ras* genes are members of a family involved in the signal transduction pathways known to be activated in tumours by point mutations at 'hot-spot' codons, 12, 13 or 61, in about 40% of human tumours of all types and at higher frequency in some tumours in experimental animals. In humans, mutations at codon 13 are relatively rare; they occur, for example, in only 10% of lung and colon carcinomas carrying a mutated Ki-*ras* gene. Moreover, in spontaneous human tumours, the mutations may involve either the first or second base of the codon and can lead to the incorporation of various amino acids. Therefore, the fact that the same mutation was found in most of the tumours points to a specific mechanism of mutagenesis.

Table 1 shows the mutation patterns of *ras* genes in sporadic and vinyl chloride- and Thorotrast-induced angiosarcomas of the liver. Thorotrast is a colloidal solution of radioactive ThO_2 which was used as a radiographic contrast medium in the 1950s and which induces angiosarcomas of the liver, hepatocellular carcinomas and cholangiocarcinomas. Two types of mutation were observed in sporadic and Thorotrast-induced tumours at codon 12 of the Ki-*ras* gene, GGT→GAT and GGT→TGT; in some cases, both mutations were present in the same tumour (Przygodzki *et al.*, 1997). The mutation

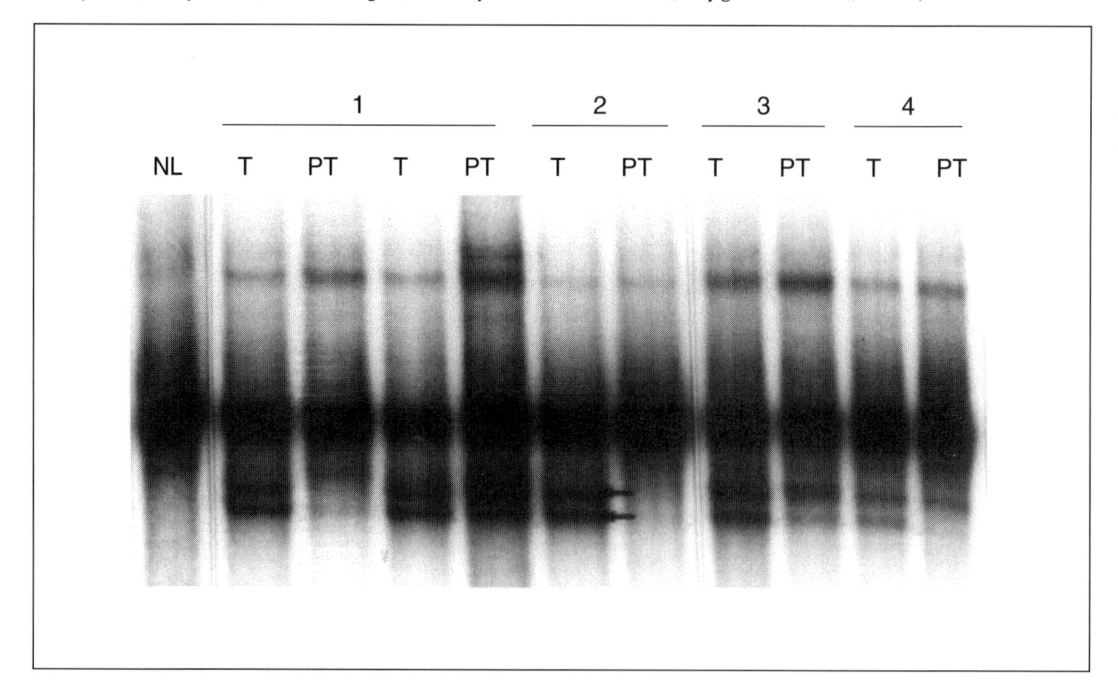

Figure 2. Detection by polymerase chain reaction and single-strand conformation polymorphism analysis of point mutations in the second exon of the Ha-*ras* oncogene in four hepatocellular carcinomas induced by vinyl chloride in rats

Samples were electrophoresed through an 0.5% Hydrolink MDE gel without glycerol at 4 °C. NL, DNA from normal liver tissue of an untreated rat; T, DNA extracted from tumoral tissue; PT, DNA extracted from peritumoral tissue

Table 1. Mutation spectra of the Ki-*ras* gene in human liver angiosarcomas of various etiology

Exposure	Frequency[a]	Mutation	Reference
Vinyl chloride	5/6	$GGC_{13} \rightarrow GAC$	Marion *et al.* (1991)
None or	2/19	$GGT_{12} \rightarrow GAT$	Przygodzki *et al.* (1997)
unknown	3/19	$GGT_{12} \rightarrow GAT + GGT_{12} \rightarrow TGT$	
	1/5	$GGT_{12} \rightarrow GAT$	
Thorotrast	1/5	$GGT_{12} \rightarrow GAT + GGT_{12} \rightarrow TGT$	

[a] Number of tumours with a mutation/number of tumours analysed

pattern in the Ki-*ras* gene in vinyl chloride-induced tumours was different, however, both in the frequency and in the nature of the mutation (Table 1). The mutation patterns in Thorotrast-induced and sporadic angiosarcomas, which are indistinguishable, have been attributed to lipid peroxidation (Przygodzki *et al.*, 1997). These *ras* mutation patterns thus suggest (i) that the Ki-*ras* gene is probably involved in the development of angiosarcomas of the liver, whatever their etiology, and (ii) that the GGC→GAC mutation in codon 13 is specific to vinyl chloride.

Comparison of the mutational patterns of the *p53* gene in human angiosarcomas also suggests that vinyl chloride induces specific mutations in this target gene. Table 2 shows the results of investigations of mutations in *p53* in sporadic and vinyl chloride- and Thorotrast-induced angiosarcomas. Four angiosarcomas of the liver and one hepatocellular carcinoma from workers exposed to vinyl chloride were investigated for the presence of mutations in exons 5–8 by microdissection, DNA amplification with intronic primers and sequencing. Two A→T missense

Table 2. Mutation spectra in the *p53* gene in human liver tumours induced by vinyl chloride or Thorotrast and in sporadic tumours

Tumour	Exposure	Frequency	Mutation	Source
Angiosarcoma	Vinyl chloride	2/4	$A_{255} \rightarrow T$, $A_{249} \rightarrow T$	Hollstein *et al.* (1994)
		1/1[a]	$A_{179} \rightarrow T$	Boivin *et al.* (1997)
	Thorotrast	0/5	None	Soini *et al.* (1995)
	None or unknown	2/16	$G_{141} \rightarrow A$, $C_{136} \rightarrow T$	
	Thorotrast	0/9	None	Andersson *et al.* (1995)
Hepatocellular	Vinyl chloride	0/1	None	Hollstein *et al.* (1994)
carcinoma	Thorotrast	1/18	$T_{176} \rightarrow G$	Andersson *et al.* (1995)
Cholangiocarcinoma	Thorotrast	0/9	None	Andersson *et al.* (1995)
		0/11		Hollstein *et al.* (1997)
		2/2[b]	$C_{138} \rightarrow A$, $G_{248} \rightarrow A$	Iwamoto *et al.* (1998)

[a] Myofibroblast-like cell line isolated from a vinyl chloride-induced angiosarcoma

[b] One patient with two different tumour types; two nodules were tested.

mutations were found, one at codon 249 (AGG→TGG, Arg→Trp) and one at codon 255 (ATC→TTC, Ile→Phe) (Hollstein *et al.*, 1994). Later, a third mutation, CAT→CTT (His→Leu) was identified at codon 179 in a myofibroblast-like cell line established from a liver angiosarcoma induced by vinyl chloride (Boivin *et al.*, 1997). Like the Ki-*ras* mutations, the *p53* mutations were tumour-specific, since they were not found in non-tumour tissues from the same patients. Mutations in the *p53* gene were observed at various codons, but in all cases they resulted from an AT→TA transversion. This mutation is relatively rare, occurring in only 5.5% of the human cell lines and tumours tested so far (Hainaut *et al.*, 1998). In contrast, in our small series of vinyl chloride-associated angiosarcomas, the incidence of this transversion was very high (3/5, 60%).

Tumours in rats

In our study, 15 of 44 rats (34%) exposed to vinyl chloride developed angiosarcoma of the liver, which contrasts well with the rarity of spontaneous tumours of this type in rats. Although the tumours in humans and rats are histologically similar (Popper *et al.*, 1977), none of the tumours in the rats contained point mutations in exon 1 or 2 of the three *ras* genes. This suggests that development of angiosarcomas in rats is *ras*-independent or involves another component of the *ras* cascade.

A mutation at the second base of codon 61 of the Ha-*ras* gene was found in five of eight hepatocellular carcinomas, indicating a specific mechanism of mutagenesis. This assumption is strengthened by the rarity of activated *ras* genes in chemically induced hepatocellular carcinomas, particularly by Ha-*ras* mutations at codon 61, at least in Wistar and Fischer rats. The presence of CAA→CTA and CAA→AAA mutations has been reported in preneoplastic lesions and hepatocellular carcinomas induced in Sprague-Dawley rats by *N*-nitrosomorpholine (Baba *et al.*, 1997). These data and our results are summarized in Table 3. In contrast, the Ha-*ras* gene has frequently been found to be activated in spontaneous liver tumours in B6C3F$_1$ and CD1 mice (for a review, see Maronpot *et al.*, 1995). The mutations detect-

Table 3. Frequency of mutations at codon 61 of the Ha-*ras* gene in rat liver tumours induced by various carcinogens

Rat strain	Carcinogen	Tumour	Frequency	Mutation	Source
Fischer	MNU, NDEA	Preneoplastic lesions Hepatocellular carcinoma	0/70 0/27		Li *et al.* (1990) Sakai & Ogawa (1990)
Wistar	NDELA,NDEA/PB, MNU, AAF, AAP, AAS	Preneoplastic lesions Hepatocellular carcinoma	0/23 0/80		Buchmann *et al.* (1991) Ostermayer *et al.* (1991) Bitsch *et al.* (1993)
Sprague-Dawley	NMOR	Preneoplastic lesions	4/8[a]	CAA→CTA 2/4 CAA→AAA 1/4	Baba *et al.* (1997)
		Hepatocellular carcinoma	7/19[a]	CAA→CTA 4/7 CAA→AAA 2/7	
	Vinyl chloride	Hepatocellular carcinoma	8/8[a]	CAA→CTA 5/8	Froment *et al.* (1994); this study

MNU, *N*-methyl-*N*-nitrosourea; NDELA, *N*-nitrosodiethanolamine; NDEA, *N*-nitrosodiethylamine; PB, phenobarbital; AAF, 2-acetylamoinofluorene, AAP, 2-acetylaminophenanthrene; AAS, *trans*-4-acetylaminostilbene; NMOR, *N*-nitrosomorpholine

[a] Some mutations could not be identified.

ed in these tumours were, in order of decreasing frequency, CAA→AAA, CAA→CGA and CAA→CTA. Mouse liver tumours induced by urethane or its metabolite, vinyl carbamate, also contained an activated Ha-*ras* gene, but in this case the predominant mutation was CAA→CTA, which has also been detected in vinyl chloride-induced hepatocellular carcinomas in rats. It is worth noting that both vinyl chloride and vinyl carbamate are oxidized by the cytochrome P450 isoenzyme 2E1 to the corresponding epoxides, and the two produce the same etheno adducts (Park *et al.*, 1993).

Out of 31 angiosarcomas and hepatocellular carcinomas induced by vinyl chloride in the study of Barbin *et al.* (1997), 12 had a mutated *p53* gene. The frequency of point mutations in the *p53* gene in rat hepatocellular carcinomas is generally very low, except in those induced by Tamoxifen, 50% of which have a mutation in the *p53* gene mostly involving A→G transitions, and those induced by *N*-nitroso-2-acetylaminofluorene and 2-acetylaminofluorene which show mutation frequencies of 19 and 31%, respectively, and most of which are G→A transitions (Vancutsem *et al.*, 1994; Ho *et al.*, 1995). In nine out of 11 vinyl chloride-induced tumours with a point-mutated *p53* gene, the lesion involved an AT base pair, as in human angiosarcomas.

Conclusions

The mutations identified in vinyl chloride-induced tumours in humans and in animals at the *ras* and *p53* genes are listed in Table 4: 68% of all the mutations are due to AT base-pair lesions, and 52% result in AT→TA transversions. Three etheno adducts were found in greater amounts in the liver and other organs of rats after exposure to vinyl chloride; these were 1,N^6-ethenoadenine, 3,N^4-ethenocytosine and N^2,3-ethenoguanine (Eberle *et al.*, 1989; Swenberg *et al.*, 1992; Guichard *et al.*, 1996). These DNA adducts are highly mutagenic both *in vitro* and *in vivo* (Barbin *et al.*, 1981; Cheng *et*

Table 4. *ras* and *p53* gene mutation spectra in human and rat liver tumours induced by vinyl chloride

Gene	Species	Tumour	Frequency	Mutation	Source
Ki-*ras* 2nd base codon 13	Human	Angiosarcoma	5/6	5 GC→AT	Marion *et al.* (1991)
Ha-*ras* 2nd base codon 61	Rat	Hepatocellular carcinoma	5/8	5 AT→TA	Froment *et al.* (1994; this study
p53	Human	Angiosarcoma	3/5	3 AT→TA	Hollstein *et al.* (1994)
		Hepatocellular carcinoma	0/1	None	Boivin *et al.* (1997)
p53	Rat	Angiosarcoma	11/25	4 AT→TA 2 AT→GC 2 AT→CG 3 GC→AT	Barbin *et al.* (1997)
		Hepatocellular carcinoma	1/8	1 AT→TA	

al., 1991; Singer *et al.*, 1991; Basu *et al.*,1993; Palejwala *et al.*, 1991; Moriya *et al.*, 1994; Pandya & Moriya, 1996). It is likely that the CAA→CTA transversion found in hepatocellular carcinomas induced in rats by vinyl chloride arose from the formation of 1,N^6-ethenoadenine adducts in hepatic DNA. The second most common lesion observed in these tumours is the GC→AT transversion (31%), which is consistent with the known mutational pattern induced by vinyl chloride in bacteria and human fibroblasts. In bacteria, *TrpA*$^+$ revertants induced by vinyl chloride were preferentially generated through GC→AT transitions (Barbin *et al.*, 1985). In human fibroblasts, Matsuda *et al.* (1995), using shuttle vectors, showed that 72% of the genetic lesions induced by chloroacetaldehyde were point mutations, 54% being GC→AT transitions. Such transitions could arise from the formation of 3,N^4-ethenocytosine or N^2,3-ethenoguanine, both of which are mutagenic (Cheng *et al.*, 1991; Palejwala *et al.*, 1991). Thus, the occurrence of particular mutations in hepatic tumours induced by vinyl chloride in both rats and humans is consistent with the known etiology of the tumours and with the mutagenic properties of the etheno bases, although this unproven correlation requires further investigation.

Acknowledgments

This work was supported in part by grants from the Association pour la Recherche sur le Cancer and the National Institute of Environmental Health Sciences, grant ESO5948. We wish to thank Lydie Lefrançois for skilled technical assistance.

References

Andersson, M., Jönsson, M., Nielsen, L.N., Vyberg, M., Visfeldt, K., Storm, H.H. & Wallin, H. (1995) Mutations in the tumour suppressor gene *p53* in human liver cancer induced by α-particles. *Cancer Epidemiol. Biomarkers Prev.* 4, 765–770

Baba, M., Yamamoto, R., Iishi, H. & Tatsuta, M. (1997) Ha-*ras* mutations in *N*-nitrosomorpholine-induced lesions and inhibition of hepatocarcinogenesis by antisense sequences in rat liver. *Int. J. Cancer* 72, 815–820

Barbin, A. (1998) Formation of DNA ethenoadducts in rodents and humans and their role in carcinogenesis. *Acta Biochim. Pol.* 45, 145–161

Barbin, A., Bartsch, H., Lecomte, P. & Radman, M. (1981) Studies on the miscoding properties of 1,N^6-ethenoadenine and 3,N^4-ethenocytosine, DNA reaction products of vinyl chloride metabolites, during *in vitro* DNA synthesis. *Nucleic Acids Res.* 9, 375–387

Barbin, A., Besson, F., Perrard, M.H., Béréziat, J.C., Kaldor, J., Michel, G. & Bartsch, H. (1985) Induction of specific base-pair substitutions in *E. coli trp* A mutants by chloroethylene oxide, a carcinogenic vinyl chloride metabolite. *Mutat. Res.* 152, 147–156

Barbin, A., Froment, O., Boivin, S., Marion, M.J., Belpoggi, F., Maltoni, C. & Montesano, R. (1997) *p53* gene mutation pattern in rat liver tumours induced by vinyl chloride. *Cancer Res.* 57, 1695–1698

Basu, A.K., Wood, M.L., Niedernhofer, L.J., Ramos, L.A. & Essigmann, J.M. (1993) Mutagenic and genotoxic effects of three vinyl chloride-induced DNA lesions: 1,N^6-ethenoadenine, 3,N^4-ethenocytosine and 4-amino-5-(imidazol-2-yl)imidazole. *Biochemistry* 32, 12793–12801

Bitsch, A., Röschlau, H., Deubelbeiss, C. & Neumann, H.G. (1993) The structure and function of the Ha-*ras* proto-oncogene are not altered in rat liver tumours initiated by 2-acetylaminofluorene, 2-acetylaminophenanthrene and *trans*-4-acetylaminostilbene. *Toxicol. Lett.* 67, 173–186

Boivin, S., Pedron, S., Bertrand, S., Desmouliere, A., Martel-Planche, G., Bancel, B., Montesano, R., Trépo, C. & Marion, M.J. (1997) Myofibroblast-like cells exhibiting a mutated p53 gene isolated from a liver human tumour of a worker exposed to vinyl chloride. In: *Cells of the Hepatic Sinusoids* (Wise, E., Knook, D.L. & Balabaud, C., eds). Leiden, The Kupffer Cell Foundation, pp. 449–451

Buchmann, A., Bauer-Hofmann, R., Mahr, J., Drinkwater, N.R., Luz, A. & Schwarz, M. (1991) Mutational activation of the c-Ha-*ras* gene in liver tumours of different rodent strains: Correlation with susceptibility to hepatocarcinogenesis. *Proc. Natl Acad. Sci. USA* 88, 911–915

Byren, D. & Holmberg, B. (1975) Two possible cases of angiosarcoma of the liver in a group of Swedish vinyl chloride–polivinyl chloride workers. *Ann. N.Y. Acad. Sci.* 246, 249–250

Cheng, K.C., Preston, B.D., Cahill, D.S., Dosanjh, M.K., Singer, B. & Loeb, L.A. (1991) The vinyl chloride DNA derivative N^2,3-etheno-guanine produces G→A transitions in *Escherichia coli*. *Proc. Natl Acad. Sci. USA* 88, 9974–9978

Creech, J.L. & Johnson, M.N. (1974) Angiosarcoma of the liver in the manufacture of polyvinyl chloride. *J. Occup. Med.* 16, 150–151

Eberle, G., Barbin, A., Laib, R.J., Ciroussel, F., Thomale, J., Bartsch, H. & Rajewsky, M. (1989) 1,N^6-Etheno-2′-deoxyadenosine and 3,N^4-etheno-2′-deoxycytidine detected by monoclonal antibodies in lung and liver DNA of rats exposed to vinyl chloride. *Carcinogenesis* 10, 209–212

Froment, O., Boivin, S., Barbin, A., Bancel, B., Trepo, C. & Marion, M.J. (1994) Mutagenesis of *ras* proto-oncogenes in rat liver tumours induced by vinyl chloride. *Cancer Res.* 54, 5340–5345

Fujimoto, Y., Ishizaka, Y., Tahira, T., Sone, H., Takahashi, H., Enomoto, K., Mori, M., Sugimura, T. & Nagao, M. (1991) Possible involvement of *c-myc* but not *ras* genes in hepatocellular carcinomas developing after spontaneous hepatitis in LEC rats. *Mol. Carcinog.* 4, 269–274

Guichard, Y., El-Ghissassi, F., Nair, J., Bartsch, H. & Barbin, A. (1996) Formation and accumulation of DNA ethenobases in adult Sprague-Dawley rats exposed to vinyl chloride. *Carcinogenesis* 17, 1553–1559

Hainaut, P., Hernadez, T., Robinson, A., Rodriguez-Tome, P., Flores, T., Hollstein, M., Harris, C.C. & Montesano, R. (1998) IARC database of *p53* gene mutations in human tumours and cell lines: Updated compilation, revised formats and new visualization tools. *Nucleic Acids Res.* 26, 205–213

Ho, Y.S., Cheng, H.T., Wang, Y.J. & Lin, J.K. (1995) *p53* gene mutational spectra in hepatocellular carcinomas induced by 2-acetylaminofluorene and *N*-nitroso-2-acetylaminofluorene in rats. *Mol. Carcinog.* 13, 182–190

Hollstein, M., Marion, M.J., Lehman, T., Welsh, J., Harris, C.C., Martel-Planche, G., Kusters, I & Montesano, R. (1994) *p53* mutations at A/T base pairs in angioarcomas of vinyl chloride-exposed factory workers. *Carcinogenesis* 15, 1–3

Hollstein, M., Bartsch, H., Wesch, H., Kure, E.H., Mustonen, R., Muhlbauer, K.R., Spiethoff, A., Wegener, K., Wiethege, T. & Muller, K.M. (1997) *p53* gene mutation analysis in tumours of patients exposed to alpha-particles. *Carcinogenesis* 18, 511–516

Iwamoto, K., Mizuno, T., Kurata, A., Masuzawa, M., Mori, T. & Seyama, T. (1998) Multiple, unique and common p53 mutations in a Thorotrast recipient with four primary cancers. *Hum. Pathol.* 29, 412–416

Li, H., Lee, G.H., Cui, L., Liu, J., Nomura, K., Ohtake, K. & Kitagawa, T. (1990) Absence of H-*ras* point mutations at codon 12 in *N*-methyl-*N*-nitrosourea-induced hepatocellular neoplasms in the rat. *J. Cancer Res. Clin. Oncol.* 116, 331–335

Maltoni, C. & Lefemine, G. (1974) Carcinogenicity bioassays of vinyl chloride: I. Research plan and early results. *Environ. Res.* 7, 387–405

Marion, M.J., Froment, O. & Trepo, C. (1991) Activation of Ki-*ras* gene by point mutation in human liver angiosarcoma associated with vinyl chloride exposure. *Mol. Carcinog.* 4, 450–454

Maronpot, R.R., Fox, T., Malarkey, D.E. & Goldsworthy, T.L. (1995) Mutations in the *ras* proto-oncogene: Clues to etiology and molecular pathogenesis of mouse liver tumours. *Toxicology* 101, 125–156

Matsuda, T., Yagi, T., Kawanish, M., Matsui, S. & Tabebe, H. (1995) Molecular analysis of mutations induced by 2-chloroacetaldehyde, the ultimate carcinogenic form of vinyl chloride in human cells using shuttle vectors. *Carcinogenesis* 16, 2389–2394

Moriya, M., Zhang, W., Johnson, F. & Grollman, A.P. (1994) Mutagenic potency of exocyclic DNA adducts: Marked differences between *Escherichia coli* and simian kidney cells. *Proc. Natl Acad. Sci. USA* 91, 11899–11903

Ostermayer, R., Suchy, B. & Rabes, H.M. (1991) Carcinogen-induced liver tumours of Wistar rats: Absence of activated *ras* genes and of N-*ras* C. *J. Cancer Res. Clin. Oncol.* 117, 381–384

Palejwala, V.A., Simha, D. & Humayun, M.Z. (1991) Mechanisms of mutagenesis by exocyclic DNA adducts. Transfection of M13 viral DNA bearing a site-specific adduct shows that ethenocytosine is a highly efficient RecA-independent mutagenic noninstructional lesion. *Biochemistry* 30, 8736–8743

Pandya, G.A. & Moriya, M. (1996) 1,N^6-Ethenodeoxyadenosine, a DNA adduct highly mutagenic in mammalian cells. *Biochemistry* 35, 11487–11492

Park, K.K., Liem, A., Stewart, B.C. & Miller, J.A. (1993) Vinyl carbamate epoxide, a major strong electrophilic, mutagenic and carcinogenic metabolite of vinyl carbamate and ethyl carbamate (urethane). *Carcinogenesis* 14, 441–450

Popper, H., Maltoni, C., Selikoff, I., Squire, R.A. & Thomas, L.B. (1977) Comparison of neoplastic hepatic lesions in man and experimental animals. In: *Origins of Human Cancer* (Hiatt, H.H., Watson, J.D. & Winsten, J.A., eds). Cold Spring Harbor, New York, CSH Press, pp. 1359–1382

Przygodzki, R.M., Finkelstein, S.D., Keohavong, P., Zhu, D., Bakker, A., Swalsky, P.A., Soini, Y., Ishak, K.G. & Bennett, W.P. (1997) Sporadic and Thorotrast-induced angiosarcomas of the liver manifest frequent and multiple point mutations in K-*ras*-2. *Lab. Invest.* 76, 153–159

Sakai, H. & Ogawa, K. (1990) Mutational activation of H-*ras* and K-*ras* genes is absent in *N*-nitroso-*N*-methylurea induced liver tumours in rats. *Jpn. J. Cancer Res.* 81, 437–439

Simonato, L., L'Abbe, K.A., Andersen, A., Belli, S., Comba, P., Engholm, G., Ferro, G., Hagmar, L., Langard, S., Lundberg, I., Pirastu, R., Thomas, P., Winkelmann, R. & Saracci, R.A. (1991) Collaborative study of cancer incidence and mortality among vinyl chloride workers. *Scand. J. Work Environ. Health* 17, 159–169

Singer, B., Kusmierek, J.T., Folkman, W., Chavez, F. & Dosanjh, M.K. (1991) Evidence for the mutagenic potential of the vinyl chloride induced adduct, N^2,3-etheno-deoxyguanosine, using a site-directed kinetic assay. *Carcinogenesis* 12, 745–747

Soini, Y., Welsh, J.A., Ishak, K.G. & Bennett, W.P. (1995) *p53* mutations in primary hepatic angiosarcomas not associated with vinyl chloride exposure. *Carcinogenesis* 16, 2879–2881

Swenberg, J.A., Fedtke, N., Ciroussel, F., Barbin, A. & Bartsch, H. (1992) Etheno adducts formed in DNA of vinyl chloride-exposed rats are highly persistent in liver. *Carcinogenesis* 13, 727–729

Vancutsem, P.M., Lazarus, P. & Williams, G.M. (1994) Frequent and specific mutations on the rat *p53* gene in hepatocarcinomas induced by tamoxifen. *Cancer Res.* 54, 3864–3867

Verlaan-de Vries, M., Bogaard, M.E., van den Elst, H., van Boom, J.H., van der Eb, A.J. & Bos, J.L. (1986) A dot–blot screening procedure for mutated *ras* oncogenes using synthetic oligodeoxynucleotides. *Gene* 50, 313–320

Viola, P.L., Bigotti, A. & Caputo, A. (1971) Oncogenic response of rat skin, lungs and bones to vinyl chloride. *Cancer Res.* 31, 516–522

Zwicker, G.M., Eyster R.C., Sells D.M. & Gass, J.H. (1995) Spontaneous vascular neoplams in aged Sprague-Dawley rats. *Toxicol. Pathol.* 23, 518–526

Exocyclic DNA Adducts in Mutagenesis and Carcinogenesis
B. Singer & H. Bartsch
IARC Scientific Publications No. 150
International Agency for Research on Cancer, Lyon, 1999

Solution conformation and mutagenic specificity of 1,N^6-ethenoadenine

A.K. Basu, J.M. McNulty & W.G. McGregor

Site-specific studies in several laboratories established that each of the three etheno adducts, 1,N^6-ethenoadenine (εA), 3,N^4-ethenocytosine (εC) and N^2,3-ethenoguanine (N^2,3-εG), is mutagenic. In *Escherichia coli*, εA is only weakly mutagenic in single-stranded DNA (mutation frequency, 0.1%), and εC is at least 20 times more mutagenic than εA. Prior treatment of host cells with ultraviolet irradiation enhaces the mutagenic frequency of εC by 30–60%, even when the *E. coli* is *recA*. Likewise, enhanced mutagenicity was observed when the host cells lacked 3′→5′ exonuclease activity of DNA polymerase III. εA induces all three base substitutions, but A→G predominates. εC induces εC→T and εC→A substitutions, but only the latter was enhanced after ultraviolet irradiation of host cells. In contrast to the results in bacteria, both εA and εC are potent mutagenic lesions in simian kidney cells, inducing 70 and 81% base substitutions, respectively. In simian kidney cells, εA exclusively induces εA→G transitions, whereas εC→A transversions are the major type of mutation induced by εC. Nuclear magnetic resonance (NMR) spectrometry of the four possible pairs containing εC indicated that both εC:G and εC:T pairs are stabilized by hydrogen bonds. Even though the latter forms the most stable pair containing εC, the etheno adduct is in *syn* alignment. DNA polymerase appears to continue DNA synthesis with a *syn*-orientated base only in the absence of proofreading exonuclease activity or when ultraviolet irradiation-inducible proteins are present. For εA, only εA:T and εA:G pairs have been studied by NMR, which showed that the former has no hydrogen bond whereas the latter maintains two hydrogen bonds with the etheno base in *syn* orientation. Determination of the relationship between a particular conformation of εA and its mutagenic activity must await further studies. In a site-specific study of εA with human cell extracts, an 11-mer oligonuclotide with a single εA was inserted into an M13 bacteriophage containing an SV40 origin of replication. This vector was replicated *in vitro* with human fibroblast cell extracts, and the replicated products were analysed. In this experiment, εA induced predominantly εA→G transitions but at a mutation frequency of 0.14%

Introduction

An interesting modification of adenine and cytosine derivatives was observed by Kochetkov and coworkers in 1971 when chloroacetaldehyde was allowed to react with the nucleic acid derivatives (Kochetkov *et al.*, 1971). These so-called etheno adducts received much attention in the 1970s, not only because of their structural resemblance to wyosine and related uncommon natural bases but also because of their intense fluorescence (Leonard, 1984). 1,N^6-Ethenoadenine (εA) nucleosides and nucleotides, the most fluorescent exocyclic derivatives in this group, have been used as fluorescent probes for investigations of the mechanism of action of enzymes and coenzymes and of t-RNA tertiary structures (for reviews, see Leonard, 1984, 1992). Subsequently, the biological effects of the etheno adducts were studied extensively, because these adducts are formed in mammals not only after exposure to a series of exogenous chemicals but also by endogenous lipid peroxidation (Figure 1). In the early 1980s, several research groups independently investigated the mispairing properties of εA *in vitro* (Barbin *et al.*,

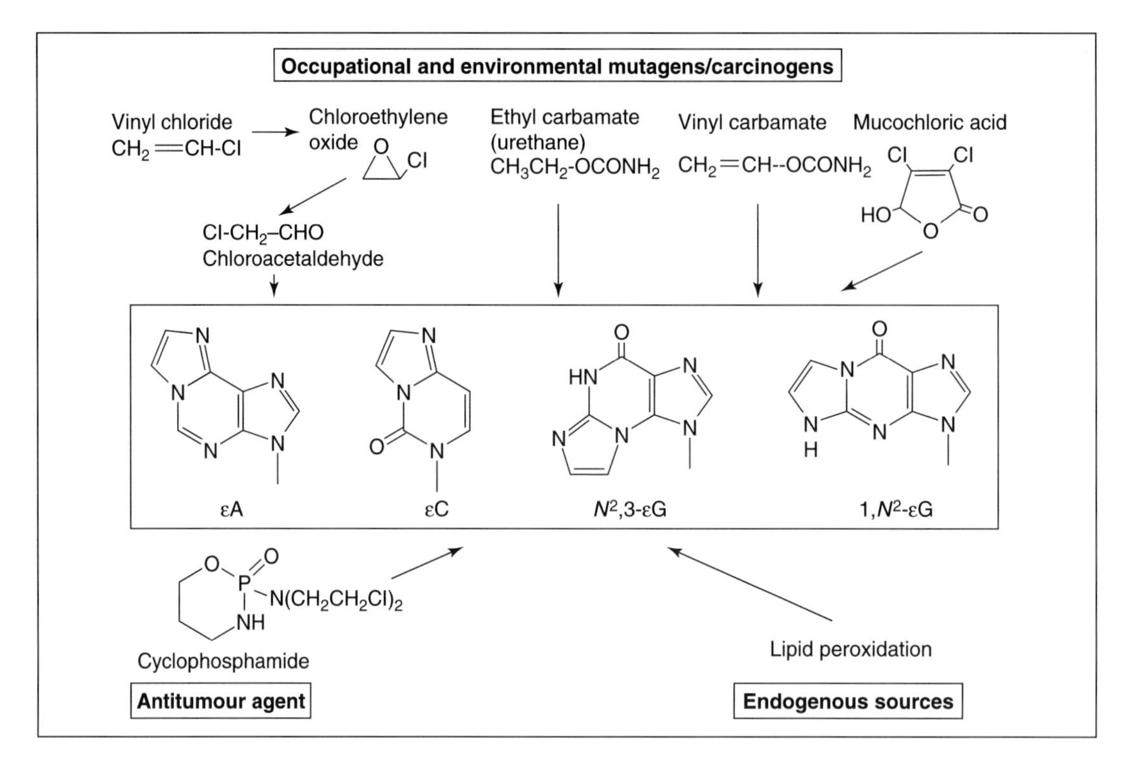

Figure 1. Chemicals that form the etheno adducts either directly or via metabolic intermediates

εA, 1,N^6-ethenoadenine ; εC, 3,N^4-ethenocytosine; N^2,3-εG, N^2,3-ethenoguanine; 1,N^2-εG, 1,N^2- ethenoguanine

1981; Hall *et al.*, 1981; Spengler & Singer, 1981; Singer *et al.*, 1984) and showed that misincorporation may depend on the DNA polymerase used. Incorporation of G opposite εA was found to be a major event, although C incorporation was also detected in one study (Hall *et al.*, 1981). Misincorporation of T and A opposite 3,N^4-ethenocytosine (εC) (Barbin *et al.*, 1981; Hall *et al.*, 1981; Spengler & Singer, 1981) and T opposite N^2,3-εG (Singer *et al.*, 1987) were observed *in vitro*. Since then, extensive *in-vivo* and biophysical studies have been performed, particularly with εA and εC, and unusual mutational patterns have been seen. It is intriguing that both these etheno adducts are recognized by repair enzymes in many organisms (reviewed by Singer & Hang, 1997).

Mutagenicity of etheno adducts

Mutagenicity in bacteria

When a partially double-stranded M13 bacteriophage DNA was treated with chloroacetaldehyde and replicated in *Escherichia coli*, most of the mutations in the single-stranded region were at C residues (Jacobsen *et al.*, 1989). Mutations were one order of magnitude less frequent at A and two orders of magnitude less frequent at G as compared with C. Of the mutations targeted at cytosines, 80% were C→T transitions and 20% were C→A transversions. Although N^2,3-ethenoguanine (N^2,3-εG) residues were not formed in the single-stranded region of the DNA, more εA residues were likely to be formed than εC under the experimental conditions. Therefore, εC was considered to be significantly more mutagenic than εA, and this was later confirmed in site-specific studies.

When bacteriophage M13 DNA containing a single εA or εC was replicated in repair-competent cells, the viability was unaffected when the adduct was placed in double-stranded DNA, whereas in single-stranded DNA εA and εC reduced survival to ~30% and ~10%, respectively (Basu *et al.*, 1993). Only 0.1% of the surviving progeny from εA-containing single stranded M13 had targeted mutations. Even though εA→G was the major type of mutation, all possible base substitutions were detected. εC was found to be nearly 20 times more mutagenic than εA, and it induced primarily εC→T transitions, with a lower frequency of εC→A transversions and targeted C deletions. Subsequent studies in other laboratories showed similar results for εC in uninduced repair-competent cells. Both Moriya *et al.* (1994) and Palejwala *et al.* (1994) showed that εC induces 2–4% mutations (more εC→T than εC→A), but that prior treatment of the host cells with ultraviolet irradiation (UV) enhances the mutagenicity to 30–60%. This increase was mainly in the frequency of εC→A transversions. Another interesting finding was that the UV-induced enhancement of mutagenicity was *recA*-independent, which suggests that, in addition to an SOS regulatory system, there may be additional UV-inducible mutagenic mechanism in *E. coli*. Mutagenesis of εC, and particularly εC→A transversion events, were also enhanced in a *mutD* strain in which the 3′→5′ exonucleolytic proofreading activity of the DNA polymerase is impaired.

Because of its instability, N^2,3-εG has not been studied as extensively as the other etheno adducts. In the only site-specific study in *E. coli*, N^2,3-εG was shown to be a potent mutagen in duplex DNA, inducing nearly 13% G→A transitions (Cheng *et al.*, 1991).

Mutagenicity in mammalian cells

In contrast to their weak constitutive mutagenicity in *E. coli*, both εA and εC are strongly mutagenic in simian kidney cells (Moriya *et al.*, 1994; Pandya & Moriya, 1996), with mutation frequencies of 81% for εC and 70% for εA when site-specifically modified single-stranded shuttle vectors were replicated in COS cells. Therefore. the ability of the mammalian DNA polymerase to incorporate the correct nucleotide opposite each of these lesions had been highly impaired. εC→A was the major type of mutation (~50% of the progeny), although εC→T occurred at a significant frequency (~29%). Indeed, the frequency of incorporation of the correct nucleotide (19%) was lower than that of either of the two major mutagenic events. In the case of εA, the major mutation was εA→G (63%), whereas nonmutagenic by-pass was detected in 31% of the progeny.

Relationship between conformation of the lesion and mutagenicity

The striking difference in mutation frequency between *E. coli* and COS cells leads to speculation about the mutational mechanism of etheno adducts in different organisms. Several structural studies have been performed with these lesions. NMR investigations provided evidence that each of the four pairs containing εC maintains a right-handed B-DNA helix. The nonmutagenic εC:G pair is stabilized by hydrogen bonding, and both εC and G are in *anti* orientation (Cullinan *et al.*, 1997). The two observed mutagenic intermediates, the εC:A and εC:T pairs, show some interesting differences. No hydrogen bonds were detected in the εC:A pair, in which glycosidic torsion angles of both bases are *anti* and the pair is stabilized by stacking interaction with the neighbouring bases (Korobka *et al.*, 1996). In contrast, the εC:T pair is stabilized both by hydrogen bonding and stacking, but the εC remains in *syn* alignment (Cullinan *et al.*, 1996). Thermodynamic data suggest that the latter pair causes less enthalpic damage (Gelfand *et al.*, 1998). Therefore, the εC:T pair may be more stable than both the nonmutagenic εC:G and the mutagenic εC:A pair; however, unlike the other two pairs, εC:T has increased stabilization at the expense of a *syn*-oriented εC.

It is still unclear how a DNA polymerase bypasses a lesion and what roles are played by hydrogen bonding, stacking and base rotation. On the basis of NMR and thermodynamic data on εC pairing with different bases, it is tempting to speculate that the *E. coli* DNA polymerase extends the εC:G pair efficiently in the absence of UV-inducible proteins, when the 3′→5′ exonuclease activity of the DNA polymerase is intact. After induction of certain proteins by exposure to UV light, the 3′→5′ exonuclease activity of the polymerase may be compromised and extension of the εC:T pair occurs more efficiently, resulting in the observed mutagenicity. This model is consistent with the observation that εC→A occurs at a high rate in strains that carry a mutation in the *mutD* (*dnaQ*) gene (Moriya *et al.*, 1994), which results in a defect in the 3′→5′ proofreading exonuclease activity of the polymerase.

Only two pairs of εA have been studied by NMR. Both bases adopt *anti* orientation in the nonmutagenic εA:T pair (Kouchakdjian *et al.*,

1991). They are directed into the interior of a right-handed helix without disruption of the flanking base pairs. To avoid a steric clash, εA and its partner T are not in the same plane, and there is no hydrogen bonding between them. This alignment is stabilized by stacking interaction between the T and the εA at the lesion site (Figure 2). In the εA:G pair, however, the εA adopts a *syn* glycosidic torsion angle, while all of the other bases, including the partner G, adopt an *anti* orientation (de los Santos *et al.*, 1991). The right-handed helix is stabilized at the lesion site by two hydrogen bonds from the NH1 and NH_2-2 of G (*anti*) to N9 and N1 of εA (*syn*) (Figure 2). Crystal structure analysis of the εA:G pair in a different sequence is consistent with this observation, except that it also showed a third, weak hydrogen bond between the O^6 of G (*anti*) and the C8–H of εA (*syn*) (Leonard *et al.*, 1994). In order to relate the mutagenic effects of εA to its conformation and base pairing, the other two pairs must also be studied. The εA:C pair, which models the premutagenic intermediate for the εA→G transition, will be the most interesting pair to investigate. It is important, however, that such studies take into account the influence of the flanking bases on polymerase by-pass of εA (Litinski *et al.*, 1997).

Mutagenicity of 1,N^6-ethenoadenine in human cell extracts

The objective of our current study is to determine whether εA is mutagenic in human cells. As discussed elsewhere in this volume, studies of human cells have shown that vinyl chloride-induced mutations occur in adenines. Also of interest is the observation that εA is repaired in mammalian cells (Hang *et al.*, 1996, 1997). It is

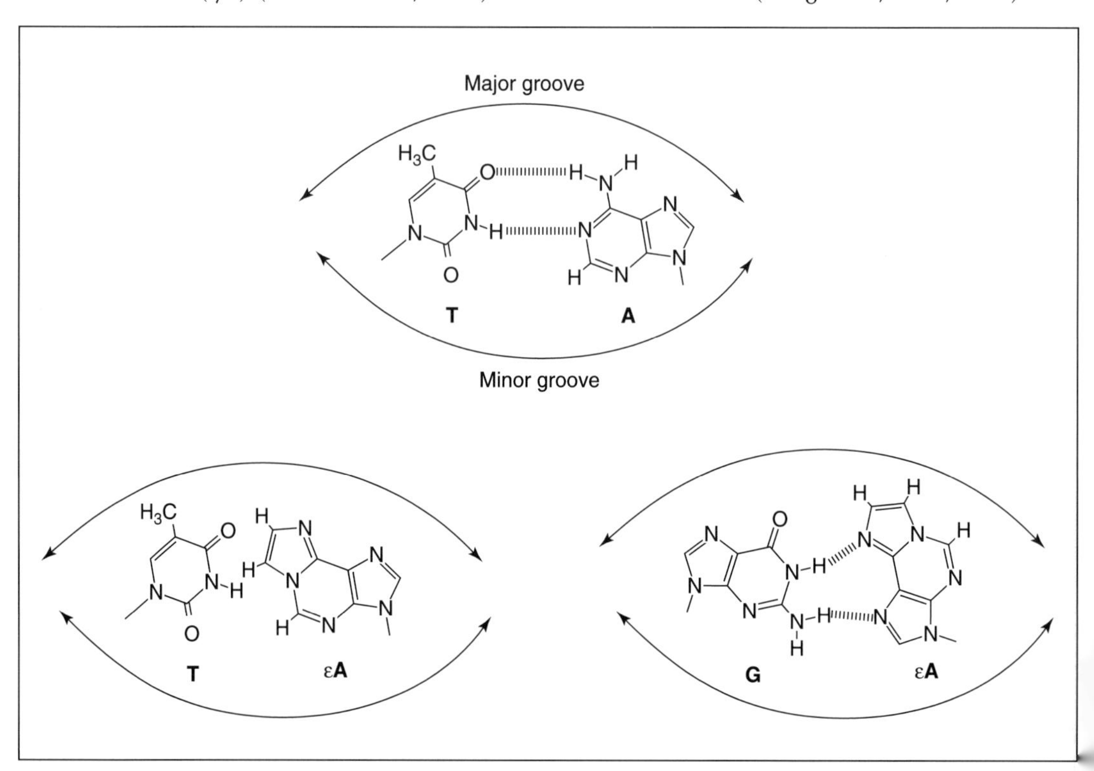

Figure 2. Base-pairing modes of a normal A:T pair compared with εA:T and εA:G pair as determined by nuclear magnetic resonance spectrometry

εA, 1,N^6-ethenoadenine

therefore important to determine whether low levels of εA in duplex DNA are mutagenic in human cells. We used a site-specific approach with a double-stranded vector. Construction of the vector involved two essenetial steps: first, an oligonucleotide was synthesized with a single εA at a particular site, and then the modified oligonucleotide was inserted into a gapped vector by recombinant DNA techniques.

Stability of 1,N^6-ethenoadenine in DNA

Although the purity of the lesion is important for any investigation, it is absolutely crucial for studying the mutagenicity of single adducts. A small amount of a mutagenic impurity can lead to erroneous data. Indeed, in earlier studies we found that εA-containing oligonucleotides are highly base-labile (Basu *et al.*, 1993). Even at neutral pH, aqueous solutions of oligonucleotides containing εA undergo conversion to a biimidazole derivative during prolonged storage (Figure 3) (Yip & Tsou, 1973). For short oligonucleotides, analysis by reversed-phase HPLC is usually sufficient (Basu *et al.*, 1993). For oligonucleotides longer than a decamer, a combination of polyacrylamide gel electrophoresis and fluorescence can be used to demonstrate their purity. For example, the εA 11-mer, 5′-CC(εA)TCGCTACC-3′, migrates faster than the 11-mer containing the biimidazole derivative III on a 20% polyacrylamide gel. Thus, we were able to determine that the 11-mer used was more than 97% pure. Since εA is intensely fluorescent (emission maximum, 410 nm) with a lifetime of 20 ns in water for the ribonucleotide (Barrio *et al.*, 1972), while the biimidazole derivative has no fluorescence, the purity can be confirmed by measuring the intensity of fluorescence of the oligonucleotides despite much quenching in DNA (Basu *et al.*, 1987). As shown in Figure 4, an 11-mer containing εA is intensely fluorescent (A), whereas of the two additional 11-mers that can be isolated after base treatment one has much weaker fluorescence (B) and the other one (C) is not fluorescent at all. We suspect that the weakly fluorescent 11-mer contains compound I or II, which could partially regenerate εA, whereas the nonfluorescent one contains the biimidazole derivative III. A similar observation was made with a hexanucleotide containing εA (Basu *et al.*, 1993).

Construction and replication of 1,N^6-ethenoadenine-containing vector

We inserted a single εA residue at a preselected site in an M13mp2 bacteriophage DNA containing an SV40 origin of replication at the *Ava* I site (abbreviated to M13mp2 SV). The general protocol for construction is shown in Figure 5. Briefly, a +11 mutant clone of M13mp2 SV was generated by inserting 5′-GGTAGCGATGG-3′ into the *Eco*R I site. A gapped heteroduplex was formed by denaturation, followed by renaturation of the single-stranded M13mp2 SV +11 with *Eco*R I-linearized M13 mp2 SV duplex DNA. Subsequently, 5′-phosphorylated 5′-CC(εA)TCGCTACC-3′ and its unmodified analogue were ligated in the presence of T4 DNA ligase to the gap to construct εA-containing and control vectors, respectively.

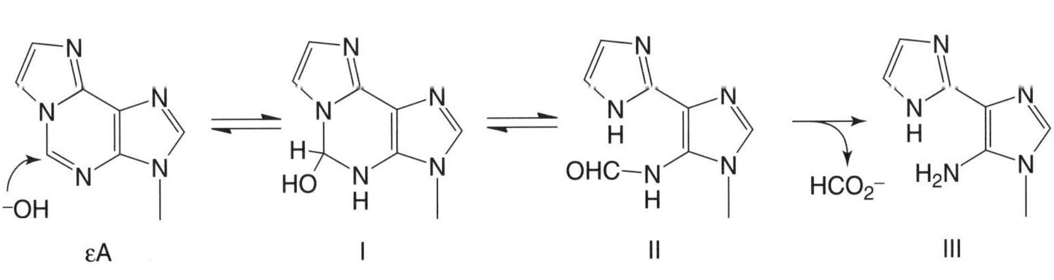

Figure 3. Conversion of 1,N^6-ethenoadenine (εA) to the biimidazole derivative under alkaline conditions

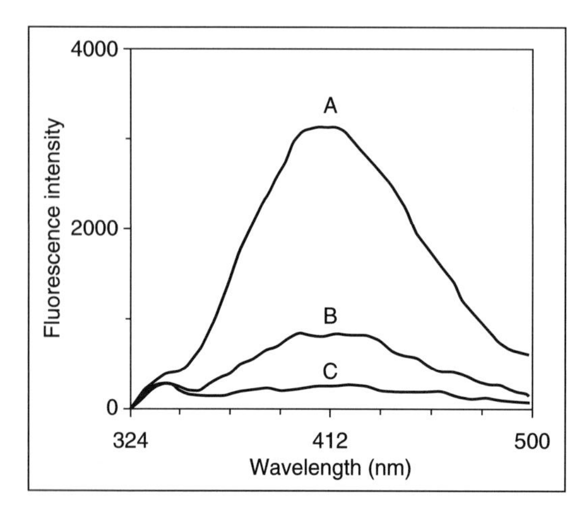

Figure 4. Fluorescence emission of three 11-mers containing 1,N^6-ethenoadenine (A), biimidazole derivative (C) and an intermediate (B) after excitation at 300 nm

The constructed genomes were purified and replicated *in vitro* in an extract of human fibroblasts in the presence of SV40 T antigen and dNTPs. The replicated DNA was purified and treated with *Dpn* I to inactivate any unreplicated templates and transfected into *E. coli* by electroporation. Mutants were scored by oligonucleotide hybridization, and the mutant phage DNA was subjected to DNA sequencing (Figure 6). After analysing 3512 plaques from the singly modified genome, we detected five mutants (mutation frequency, 0.14%). Four of the mutants had a εA:T→G:C transition, and one contained a C(εA) dinuclotide deletion. This result indicates that εA could be mutagenic in repair-competent human cells even when it is present in duplex DNA. Further studies on the mutagenicity and repair of εA in human cell extracts are in progress.

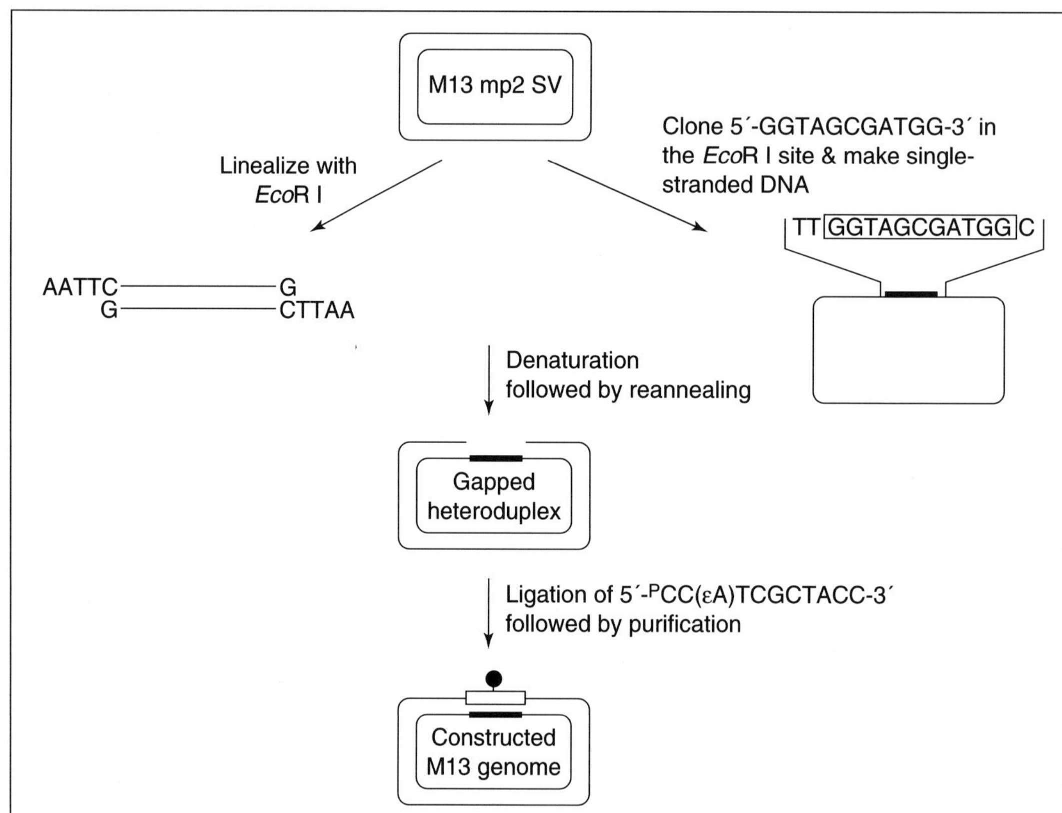

Figure 5. Construction scheme of M13mp2 SV genome containing a site-specific 1,N^6-ethenoadenine

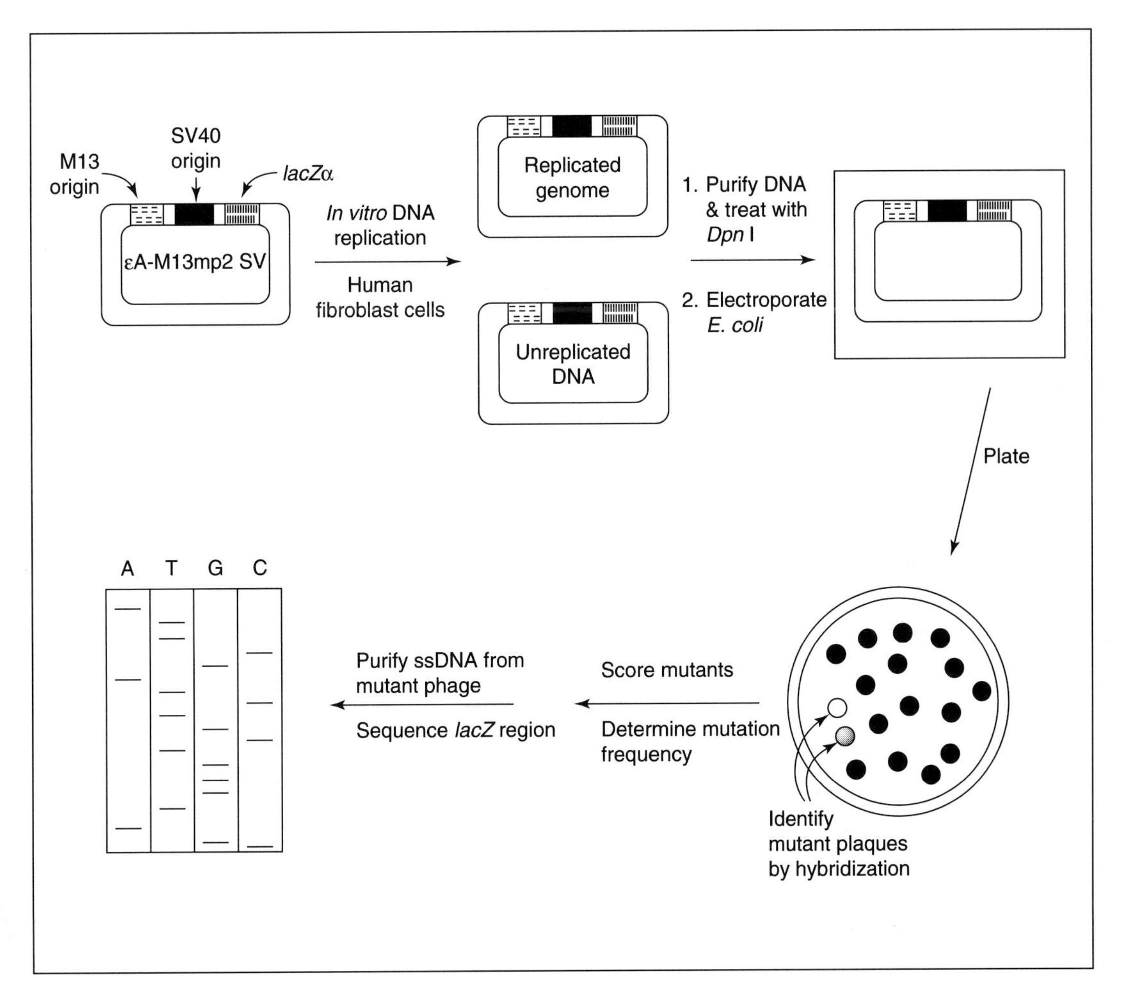

Figure 6. Replication *in vitro* of M13mp2 SV-containing a site-specific 1,N^6-ethenoadenine (εA) and determination of the types and frequency of mutations

ss, single-stranded

Acknowledgements

The project described was supported by the National Institute of Environmental Health Sciences, (grant ES07946 and ES09127) to A.K.B. and by the National Cancer Institute grant CA73984 to W.G.M. A.K.B. is a recipient of a Research Career Development Award from the National Institute of Environmental Health Sciences (grant 1K02 ES00318). J.M.M. was supported in part by a training grant (ES07163) from the National Institute of Environmental Health Sciences.

References

Barbin, A., Bartsch, H., Leconte, P. & Radman, M. (1981) Studies on the miscoding properties of 1,N^6-ethenoadenine and 3,N^4-ethenocytosine, DNA reaction products of vinyl chloride metabolites, during *in vitro* DNA syntheis. *Nucleic Acids Res.* 9, 375–387

Barrio, J.R., Secrist, J.A., III & Leonard, N.J. (1972) Fluorescent adenine and cytidine derivatives. *Biochem. Biophys. Res. Commun.* 46, 597–604

Basu, A.K., Niedernhofer, L.J. & Essigmann, J.M. (1987) Deoxyhexanucleotide containing a vinyl chloride-induced DNA lesion, 1,N^6-ethenoadenine: Synthesis, physical characterization and incorporation into a duplex bacteriophage M13 genome as part of an amber codon. *Biochemistry* 26, 5626–5635

Basu, A.K., Wood, M.L., Niedernhofer, L.J., Ramos, L.A., and Essigmann, J.M. (1993) Mutagenic and genotoxic effects of three vinyl chloride induced DNA lesions: 1,N^6-Ethenoadenine, 3,N^4-ethenocytosine, and 4amino-5-(imidazol-2-yl)-imidazole. *Biochemistry* 32, 12793–12801

Cheng, K.C., Preston, B.D., Cahill, D.S., Dosanjh, M.K., Singer, B. & Loeb, L.A. (1991) The vinyl chloride DNA derivative N^2,3-ethenoguanine produces G→A transitions in *Escherichia coli*. *Proc. Natl Acad. Sci. USA* 88, 9974–9978

Cullinan, D., Korobka, A., Grollman, A.P., Patel, D.J., Eisenberg, M. & de los Santos, C. (1996) NMR solution structure of an oligodeoxynucleotide duplex containing the exocyclic lesion 3,N^4-etheno-2′-deoxycytidine opposite thymidine: Comparison with the duplex containing deoxyadenosine opposite the adduct. *Biochemistry* 35, 13319–13327

Cullinan, D., Johnson, F., Grollman, A.P., Eisenberg, M. & de los Santos, C. (1997) Solution structure of a DNA duplex containing the exocyclic lesion 3,N^4-etheno-2′-deoxycytidine opposite 2′-deoxyguanosine. *Biochemistry* 36, 11933–11943

Gelfand, C.A., Plum, G.E., Grollman, A.P., Johnson, F. & Breslauer, K.J. (1998) The impact of an exocyclic cytosine adduct on DNA duplex properties: Significant thermodynamic consequences despite modest lesion-induced structural alterations. *Biochemstry* 37, 12507– 12512

Hall, J.A., Saffhill, R., Green, T. & Hathway, D.E. (1981) The induction of errors during *in vitro* DNA synthesis following chloroacetaldehyde-treatment of poly(dA–dT) and poly(dC–dG) templates. *Carcinogenesis* 2, 141–146

Hang, B., Chenna, A., Rao, S. & Singer, B. (1996) 1,N^6-Ethenoadenine and 3,N^4-ethenocytosine are excised by separate human DNA glycosylases. *Carcinogenesis* 17, 155–157

Hang, B., Singer, B., Margison, G.P. & Elder, R.H (1997) Targeted deletions of alkylpurine-DNA-N-glycosylase in mice eliminates repair of 1,N^6-ethenoadenine and hypoxanthine but not of 3,N^4-ethenocytosine or 8-oxoguanine. *Proc. Natl Acad. Sci. USA* 94, 12869–12874

Jacobsen, J.S., Perkins, C.P., Callahan, J.T., Sambamurti, K. & Humayun, M.Z. (1989) Mechanism of mutagenesis by chloroacetaldehyde *Genetics* 121, 213–222

Kochetkov, N.K., Shibaev, V.N. & Kost, A.A. (1971) New reactions of adenine and cytosine derivatives, potentially useful for nucleic acid modification *Tetrahedron Lett.* 22, 1993–1996

Korobka, A., Cullinan, D., Cosman, M., Grollman, A.P., Patel, D.J., Eisenberg, M. & de los Santos, C. (1996) Solution structure of an oligodeoxynucleotide duplex containing the exocyclic lesion 3,N^4-etheno-2′-deoxycytidine opposite 2′-deoxyadenosine, determined by NMR spectroscopy and restrained molecular dynamics. *Biochemistry* 35, 13310–13318

Kouchakdjian, M., Eisenberg, M., Yarema, K., Basu, A., Essigmann, J. & Patel, D.J. (1991) NMR studies of the exocyclic 1,N^6-ethenodeoxyadenosine adduct (εdA) opposite deoxythymidine in a DNA duplex. Non-planar alignment of εdA (anti) and dT (anti) at lesion site. *Biochemistry* 30, 1820–1828

Leonard, N.J. (1984) Etheno-substituted nucleotides and coenzymes: Fluorescence and biological activity. *CRC Crit. Rev. Biochem.* 15, 125–199

Leonard, N.J. (1992) Etheno-bridged nucleotides in structural diagnosis and carcinogenesis. *Biochem. Mol. Biol. Chemtracts* 3, 273–297

Leonard, G.A., McAuley-Hecht, K.E., Gibson, N.J., Brown, T., Watson, W.P. & Hunter, W.N. (1994) Guanine 1,N^6-ethenoadenine base pairs in the crystal structure of d(CGCGAATT(εdA)GCG). *Biochemistry* 33, 4755–4761

Litinski, V., Chenna, A., Sagi, J. & Singer, B. (1997) Sequence context is an important determinant in the mutagenic potential of 1,N^6-ethenodeoxyadenosine (εA): Formation of εA basepairs and elongation in defined templates. *Carcinogenesis* 18, 1609–1615

Moriya, M., Zhang, W., Johnson, F. & Grollman, A.P. (1994) Mutagenic potency of exocyclic DNA adducts: Marked difference between *Escherichia coli* and simian kidney cells. *Proc. Natl Acad. Sci. USA* 91, 11899–11903

Palejwala, V.A., Pandya, G.A., Bhanot, O.S., Solomon, J.J., Murphy, H.S., Dunman, P.M. & Humayun, M.Z. (1994) UVM, an ultraviolet-inducible recA-independent mutagenic phenomenon in *Escherichia coli. J. Biol. Chem.* 269, 27433–27440

Pandya, G.A. & Moriya, M. (1996) 1,N^6-Ethenodeoxyadenosine, a DNA adduct highly mutagenic in mammalian cells. *Biochemistry* 35, 11487–11492

de los Santos, C., Kouchakdjian, M., Yarema, K., Basu, A., Essigmann, J. & Patel, D.J. (1991) NMR studies of the exocyclic 1,N^6-ethenodeoxyadenosine adduct (εdA) opposite deoxyguanosine in a DNA duplex. εdA (syn): dG (anti) pairing at lesion site. *Biochemistry* 30, 1828–1835

Singer, B. & Hang, B. (1997) What structural features determine repair enzyme specificity and mechanism in chemically modified DNA? *Chem. Res. Toxicol.* 10, 713–732

Singer, B., Abbott, L.G., & Spengler, S.J. (1984) Assessment of mutagenic specificty of two carcinogen-modified nucleosides, 1,N^6-ethenoadenine and O^4-methyldeoxythymidine, using polymerases of varying fidelity. *Carcinogenesis* 5, 1165–1171

Singer, B. Spengler, S.J., Chavez, F. & Kuśmierek, J.T. (1987) The vinyl chloride-derived nucleotide, N^2,3-ehenoguanosine, is a highly efficient mutagen in transcription. *Carcinogenesis* 8, 745–747.

Spengler, S. & Singer, B. (1981) Transcriptional errors and ambiguity resulting from the presence of 1,N^6-ethenoadenosine and 3,N^4-ethenocytidine in polyribonucleotides. *Nucleic Acids Res.* 9, 365–373

Yip, K.F., & Tsou, K.C. (1973) Synthesis of fluorescent adenosine derivatives. *Tetrahedron Lett.* 33, 3087–3090

Exocyclic DNA Adducts in Mutagenesis and Carcinogenesis
B. Singer & H. Bartsch
IARC Scientific Publications No. 150
International Agency for Research on Cancer, Lyon, 1999

Genetic effects of exocyclic DNA adducts *in vivo*: Heritable genetic damage in comparison with loss of heterozygosity in somatic cells

M.J.M. Nivard & E.W. Vogel

Etheno adducts are promutagenic lesions which generate point mutations, deletions, homologous recombination and gross structural DNA aberrations. High ratios of chromosome loss to forward mutations characterize vinyl bromide, vinyl chloride, ethyl carbamate, vinyl carbamate and its epoxide as effective clastogens in postmeiotic germ cells of *Drosophila melanogaster*. Of the mutants induced by vinyl carbamate at the *vermilion* gene, 68% were intra-locus or multi-locus deletions. In view of the far-reaching concordance between mutation spectra in mice and *Drosophila* observed in specific-locus tests with genotoxic agents, etheno bases are expected to generate mainly deletions in male mammals in the postmeiotic germ-cell stages. Twenty-two of 23 base substitutions induced in the *vermilion* gene after treatment of postmeiotic stages with vinyl carbamate or vinyl bromide fall into four categories of mutations expected from etheno bases: GC→AT, AT→GC, GC→TA and AT→TA base-pair changes. These types of point mutations occurred in mutated proto-oncogenes of tumours induced in rodents by vinyl chloride, ethyl carbamate or their metabolites. Of interest is the ability of vinyl carbamate to produce persistent lesions in otherwise highly repair-active premeiotic cells of *Drosophila*, leading to mutations of yet unknown nature.

Etheno bases are also potent pro-clastogenic lesions in somatic cells *in vivo*. Strongly positive responses were reported for ethyl carbamate and vinyl carbamate in assays for micronucleus formation in mouse bone marrow and in the *Drosophila white⁺/white* eye mosaic test. Loss of heterozygosity in somatic cells of *Drosophila* was due primarily to ring-X chromosome loss, followed by homologous mitotic recombination. Particularly striking is the near failure of ethyl carbamate and vinyl carbamate to generate significant frequencies of intrachromosomal recombination. The overall genetic activity profiles of etheno adduct-forming chemicals in mice and in *Drosophila* support the hypothesis that vinyl carbamate is the proximate mutagen of ethyl carbamate, and vinyl carbamate epoxide is the ultimate electrophilic mutagen and carcinogen.

Introduction

Ethyl carbamate (urethane) was first reported to produce chromosomal rearrangements, recessive lethal and visible mutations in *Drosophila melanogaster* 50 years ago (Vogt, 1948). It was no less than 25 years later that ethyl carbamate and its structural analogues received general attention, by the demonstration that vinyl chloride induces cancer in humans exposed occupationally (reviewed by Bartsch *et al.*, 1994; Barbin, 1998). Mechanistic studies revealed that vinyl chloride, vinyl bromide, ethyl carbamate, vinyl carbamate and vinyl carbamate epoxide produce exocyclic etheno (ε) adducts with DNA and RNA (reviewed by Singer & Bartsch, 1986; Bartsch *et al.*, 1994; these proceedings).

The joint metabolic features of the entire group are that oxidation by CYP 2E1 (Guengerich & Kim, 1991; Guengerich *et al.*, 1991; Park *et al.*, 1993) of ethyl carbamate to vinyl carbamate and further oxidative conversion of vinyl carbamate, vinyl chloride and vinyl bromide yields the corresponding epoxides vinyl carbamate epoxide, chloroethylene oxide and bromoethylene oxide. Chloroethylene oxide and bromoethylene oxide rearrange spontaneously to the corresponding 2-haloacetaldehydes, 2-chloroacetaldehyde and 2-bromoacetaldehyde (reviewed by Singer & Bartsch, 1986). The consensus view is that the epoxides chloroethylene oxide, bromoethylene oxide and vinyl carbamate epoxide are the ultimate reactive electrophilic and carcinogenic metabolites of vinyl chloride, vinyl bromide and vinyl carbamate, respectively (Leithauser *et al.*, 1990; Guengerich & Kim, 1991; Park *et al.*, 1993; Bartsch *et al.*, 1994). Additional degradation products of the epoxides and aldehydes have been described. Park *et al.* (1993) proposed that acetyl chloride is another rearrangement product of chloroethylene oxide. A structural analogue is 2,2′-dichlorodiethyl ether, which releases chloroacetaldehyde (Gwinner *et al.*, 1983), glycolaldehyde and methyl carbamate intracellularly. Glycolaldehyde was identified as a degradation product of chloroethylene oxide and vinyl carbamate epoxide in water (O'Neill *et al.*, 1986) and has cross-linking activity with proteins (Acharya & Manning, 1983).

Few reports are available on the genotoxic effects of etheno adduct-forming chemicals in somatic mammalian cells *in vivo* (Williams *et al.*, 1998), and attempts to induce genetic damage in mouse germ cells with vinyl chloride and ethyl carbamate were unsuccessful (Anderson *et al.*, 1977; Russell *et al.*, 1987). Thus, it is pertinent to ask (i) which of the numerous metabolites of vinyl halides and mechanistic analogues have genotoxic properties *in vivo*, and (ii) which of the compounds capable of interacting with DNA are the most relevant intermediates for the induction of genotoxic damage. We have investigated the genetic activity profiles of these chemicals by estimating the forward mutation spectra, structural chromosomal aberrations and mitotic recombination, using germ cells and somatic cells from *Drosophila* (Ballering *et al.*, 1996; 1997). Here, we summarize the outcome of earlier studies and new results on the genotoxic effects of vinyl carbamate and vinyl carbamate epoxide in male germ cells and of ethyl carbamate and vinyl carbamate in a new assay for simultaneous measurement of mitotic recombination and clastogenicity. We also included the bifunctional agent bis(chloroethyl)nitrosourea (BCNU), because cyclic ethano adducts with adenine, guanine and cytosine appear as intermediates in the formation of cross-linked bases in DNA reacted with BCNU (reviewed by Eisenbrand *et al.*, 1994). Thus, a comparison of the genetic activity profiles of the vinyl chloride group with bifunctional and monofunctional nitrosoureas (which do not form ethano bases) is of interest. The comparison is restricted to genetic activity profiles *in vivo*.

Materials and methods

Chemicals

Ethyl carbamate (CAS 51-97-6) was obtained from Aldrich, Axel, The Netherlands; BCNU was kindly provided by Dr E. Eisenbrand (University of Kaiserslautern). Vinyl carbamate was synthesized by the method of Schaefgen (1968) and was characterized by nuclear magnetic resonance (NMR) analysis, which showed no impurities. Vinyl carbamate was used to prepare vinyl carbamate epoxide according to the procedure of Park *et al.* (1990); NMR analysis showed 10% contamination with vinyl carbamate. In the genetics experiments, vinyl carbamate was clearly less active than its epoxide, proving that the epoxide itself is strongly mutagenic. Vinyl carbamate was stored at 4 °C and its epoxide at –70 °C.

Genetic assays on postmeiotic male germ cells

Bis(chloroethyl)nitrosourea was dissolved in Tween 80:ethanol (1:3; final concentration, 4%) and further diluted in phosphate buffer (16.5 mmol/L Na_2HPO_4. 16.5 mmol/L KH_2PO_4, pH 6.8) containing 5% sucrose. Ethyl carbamate was dissolved in the buffer. Male flies were exposed to BCNU and ethyl carbamate by feeding on glass microfibre paper for 24 h. Vinyl carbamate epoxide was first dissolved in acetone and diluted with 0.7% NaCl to a final concentration of 1.5% acetone. Vinyl carbamate was diluted in 0.7% NaCl. Both chemicals were injected directly (± 0.15 µl) into the abdomen of adult males.

Tests for ring-X loss were conducted by mass mating with two broods (48 and 72 h, respectively) in bottles after exposure of males to vinyl carbamate or its epoxide. *R1(2) y B; B^S Y y^+* males were mated with *y w spl sn^3* females at a ratio of one male to four females. For the determination of recessive lethal mutations on the X chromosome, mutagen-treated Hikone-R males were mated individually with four *Inscy, y^{C4} v; bw* females. (For a description of the symbols, see Lindsley & Zimm, 1992.)

M_{NER-}/M_{NER+} hypermutability ratios were determined by a method based on the test for recessive lethal mutations in which forward mutations on the entire X chromosome of *D. melanogaster* are measured. Postmeiotic male germ-cell stages that have been exposed to the mutagen are transferred after development into mature spermatozoa to either repair-proficient (NER$^+$) or repair-deficient (NER$^-$) oocytes. The frequency of recessive lethal mutations (M_{NER-}) induced in NER$^-$ background (NER$^-$ female x NER$^+$ male) is divided by the mutant frequency (M_{NER+}) induced under the repair-proficient condition (NER$^+$ female x NER$^+$ male), providing the M_{NER-}/M_{NER+} ratio. A M_{NER-}/M_{NER+} ratio of ~1 is taken to indicate no potentiating effect of NER$^-$ on mutation fixation. For the test, Hikone-R males injected with vinyl carbamate epoxide were individually mated with either four *Inscy, y^{C4} v; bw* (NER$^+$) females or with four *Inscy, y^{C4} v; mus201^{D1} bw* (NER$^-$) females. For a more detailed description of the protocol, see Ballering *et al.* (1996).

vermilion mutants were isolated as described previously (Nivard *et al.*, 1992). Briefly, *brown* (*bw*) males were exposed to BCNU and mated in bottles with *Inscy, y^{C4} v; bw* virgin females at a sex ratio of 1:1 and at 48- and 72-h intervals, to yield broods 1A and 2A. Inseminated females from the 'A' broods were separated from the males and transferred to new bottles for 72 h, to yield the 'B' broods.

The F_1 female progeny and flies of each sex of the F_2 generation were screened for newly induced *vermilion* mutations. Strains were built up from transmissible mutants which were either homozygous for the new *vermilion* mutation or heterozygous (hemizygous lethal mutants). In the latter case, the X chromosome containing the *vermilion* mutation was balanced over an *Inscy, v* chromosome, which has a 15-base-pair deletion in its *vermilion* gene. Chromosomes from female-sterile mutants containing a *vermilion* mutation were also balanced opposite the *Inscy, v* chromosome. The DNA of mutant strains or single flies was isolated and subjected to polymerase chain reaction and sequencing procedures, as described previously (Nivard *et al.*, 1992).

Assays with premeiotic germ cells and somatic cells

Male larvae were treated in premeiotic germ-cell stages for analysis of recessive lethal mutations. Females were allowed to oviposit for 24 h and were then discarded. Forty-eight hours later, 2 ml of vinyl carbamate solution were applied to the surface of developing cultures. Emerging males were mated twice to groups of four females, at intervals of 48 and 72 h. Males used for the second chromosome assay of recessive lethal mutations had been made lethal-free before the experiments were begun. For details, see Table 5.

For parallel detection of recombination and clastogenicity in somatic cells, all crosses were set up as mass cultures with 30–40 pairs of flies per bottle. P_1 females heterozygous for a ring-X (rX) chromosome and a rod-X chromosome were mated with *y w f* males, generating four genotypes in the F_1 which can be distinguished by their phenotypes: *y$^+$ w$^+$ f$^+$/y w f* (phenotype wild-type, XX females), *y f* (rX)/*y w f* (yellow, forked rXX females), *y$^+$ w$^+$ f$^+$* (wild-type XY males), and *y f* (rX) (yellow, forked rXY males). P_1 flies were discarded after 48 h, and one day later developing larvae were treated by placing 3 ml of mutagen solution on top of the culture medium. F_1 flies were scored for mosaic *white* spots according to the technique previously described (Vogel & Nivard, 1993).

The following events lead to visible light spots in the red eyes of F_1 adult flies: homologous (inter-chromosomal) recombination between the centromere and the *white* locus of the two X chromosomes of XX females, sister chromatid exchanges and DNA strand breaks in rXX females, loss of the *w$^+$* function by point mutations, small deletions or chromosome loss, and intrachromosomal recombination. We reasoned previously that the contribution of mutations and deletions to total clone formation can be neglected, except for strong point mutagens such as *N*-ethyl-*N*-nitrosourea (ENU), because a 40–375-fold increase in mutation frequency at the *white* locus would only double the frequency of spots in the *white/white$^+$* system (Vogel, 1992).

Results and Discussion

Mutagenesis of exocyclic DNA adduct-forming chemicals in germ cells

Drosophila: High ratios of chromosome loss to recessive lethal mutation of 2.7–6.9 identified vinyl bromide, vinyl chloride, vinyl carbamate and ethyl carbamate as potent clastogens in postmeiotic *Drosophila* germ cells (Ballering *et al.*, 1996), and new results for vinyl carbamate and its epoxide confirm this conclusion (Table 1). These ratios are of the same magnitude as those estimated for cross-linking halonitrosoureas (fotemustine and BCNU), nitrogen mustards, aziridines, alkane sulfonates and epoxides (Vogel *et al.*, 1998). This parameter thus allows no distinction between different classes of strong clastogens; however, they are differentiated from monofunctional agents, which generally have chromosome loss to recessive lethal mutation ratios < 1. These differences between mono- and bifunctional agents are illustrated in Figure 1 for five etheno adduct-forming chemicals (vinyl bromide, vinyl chloride, ethyl carbamate, vinyl carbamate and vinyl carbamate epoxide), two bifunctional nitrosoureas (BCNU and fotemustine) and two monofunctional nitrosoureas (ENU and *N*-methyl-*N*-nitrosourea (MNU)). There is a noticeable step-wise increase in genotoxic activity from the parent ethyl carbamate via the proximate metabolite.

Ratios of chromosome loss to recessive lethal mutation > 2 indicate the occurrence of deletions (> 50%) among the specific-locus mutations induced in postmeiotic stages (Figure 2). Of 13 and 19 *vermilion* mutants induced by BCNU (Table 2) and vinyl carbamate (Ballering *et al.*, 1997), respectively, 12 and 13 were intra-locus or multi-locus deletions. The latter are seen more frequently with bifunctional than with any monofunctional agent, as shown in Figure 2 for BCNU, vinyl carbamate, methyl methane sulfonate and ENU. None of 28 *vermilion* mutants isolated from ENU-treated flies carried a deletion mutation (Pastink *et al.*, 1989).

Twenty-two of the 23 base substitutions induced in the *vermilion* gene by vinyl carbamate or vinyl bromide fell into four categories of mutations expected from etheno adducts (Table 3). Eight were transitions (three GC→AT, five AT→GC), seven GC→TA and seven AT→TA transversions. Only one GC→CG but no AT→CG transversion was found. The miscoding or ambiguous base-pairing properties of etheno bases were shown in assays for replication and transcription fidelity and oligo- or polynucleotides modified by these adducts (Barbin *et al.*, 1981; Hall *et al.*, 1981, Spengler & Singer, 1981). Their promutagenic potential for mutation induction was further shown in site-specific mutagenesis studies with bacterial and mammalian hosts. Thus, AT→GC transitions and AT→TA and AT→CG transversions can be caused by 1,N^6-ethenoadenine (Basu *et al.*, 1993; Pandya & Moriya, 1996), GC→AT transitions by 3,N^4-ethenocytosine and N^2,3-ethenoguanine (Cheng *et al.*, 1991; Singer *et al.*, 1991; Palejwala *et al.*, 1993), GC→TA transversions by 3,N^4-ethenocyto-

Table 1. Determination of chromosome loss (CL) to recessive lethal (RL) mutation for vinyl carbamate (VCA) and vinyl carbamate epoxide (VCO) in postmeiotic male germ cell-stages of *Drosophila*; broods 1 and 2 pooled for this comparison

Chemical		VCA (0.1 mmol/L)	VCA (1.0 mmol/L)	VCO (1.0 mmol/L)	Control
CL					
	N	2256	691	1880	991
	% CL	3.55	9.41	13.40	1.82
	% $Cl_{induced}$	1.73	7.59	11.58	–
RL					
	N	1302	868	435	18 239[a]
	% RL	0.31	1.96	3.22	0.12
	% $Rl_{induced}$	0.19	1.84	3.10	–
CL:RL ratio		(9.1)	4.1	3.7	(15)

[a] From Vogel & Nivard (1997)

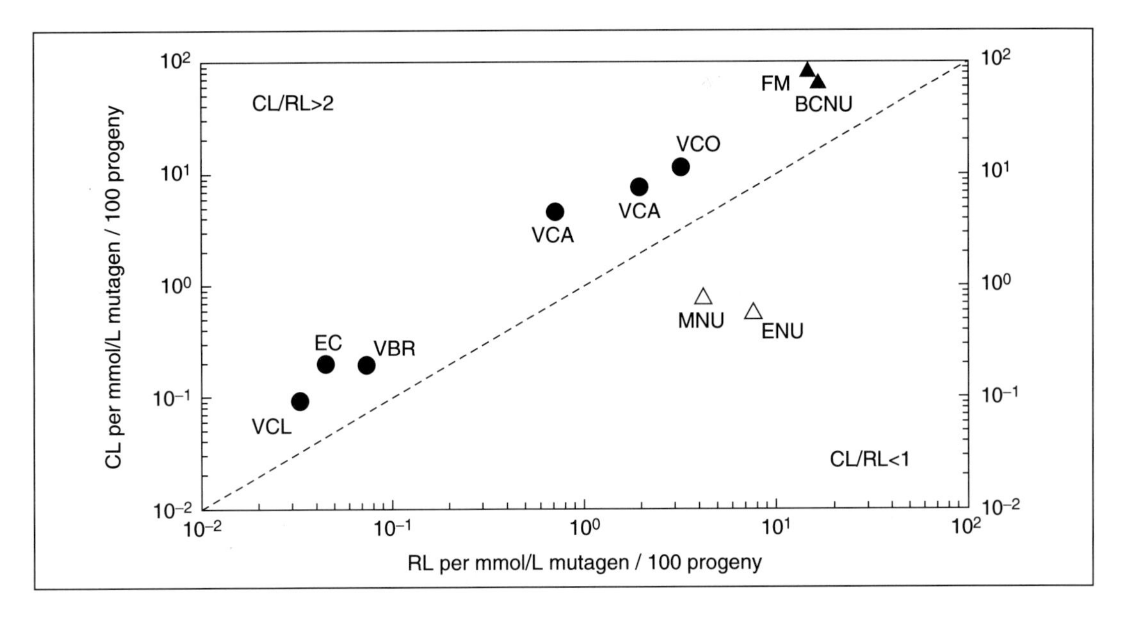

Figure 1. Induced frequencies of ring-X loss (CL) *versus* those of recessive lethal (RL) mutations in *Drosophila* for five etheno adduct-forming chemicals: ethyl carbamate (EC), vinyl bromide (VBR), vinyl carbamate (VCA), vinyl chloride (VCL) and vinyl carbamate epoxide (VCO); two monofunctional nitrosoureas: *N*-ethyl-*N*-nitrosourea (ENU) and *N*-methyl-*N*-nitrosourea (MNU) and two bifunctional nitrosoureas: bis(chloroethyl)nitrosourea (BCNU) and fotemustine (FM)

Chromosome loss (CL)/recessive lethal (RL) mutation ratios are for postmeiotic male germ cells (broods 1 and 2 pooled for this comparison). The stippled line indicates a CL/RL ratio of 1. Activities for vinyl bromide and vinyl chloride are per 24 000 ppm.h inhalation (Ballering *et al.*, 1996). References: ethyl carbamate, vinyl bromide, vinyl chloride, vinyl carbamate (Ballering *et al.*, 1996); BCNU, ENU and MNU (Vogel *et al.*, 1993); vinyl carbamate and vinyl carbamate epoxide (this paper); fotemustine (Ashby *et al.*, 1993)

sine and 1,N^2- ethenoguanine (Langouët *et al.*, 1997; Palejwala *et al.*, 1993) and GC→CG by 1,N^2-ethenoguanine (Langouët *et al.*, 1997).

Mutations induced by vinyl carbamate, vinyl chloride and ethyl carbamate have been observed in activated proto-oncogenes. Marion *et al.* (1991) reported a GC→AT transition in codon 13 of the Ki-*ras* gene in a human liver angiosarcoma associated with exposure to vinyl chloride. Hollstein *et al.* (1994) and Trivers *et al.* (1995) reported AT→TA transversions in the *p53* tumour suppressor gene in liver angiosarcomas from two patients with a history of exposure to vinyl chloride. In proto-oncogenes from rat and mice tumours induced by vinyl chloride, ethyl carbamate or its metabolites, primarily AT→TA, AT→GC, AT→CG and GC→AT base changes and occasional GC→CG transversions seemed to arise from the formation of etheno bases (reviewed by Barbin, 1998).

M_{NER-}/M_{NER+} ratios provide information on the efficiency of DNA repair which takes place in the egg after fertilization and on the type of DNA adduct(s) carried by mutagen-damaged male germ cells (Vogel, 1989). Figure 3 shows the M_{NER-}/M_{NER+} ratios for three cross-linking nitrosoureas (BCNU, fotemustine and chlorozotocin) and for seven chemicals that form etheno bases with DNA. 2-Chloroacetaldehyde and 2-bromoacetaldehyde are not included in the second group. 2-Chloroacetaldehyde, a borderline case in *Drosophila*, is approximately 50 times less active in inducing recessive lethal mutations than chloroethylene oxide, and 2-bromoacetaldehyde was not mutagenic at all (Vogel, 1989; Ballering *et al.*, 1996). Glycidaldehyde, which has been reported to form cyclic products in reactions with deoxyguanosine in calf thymus DNA (Van Duuren & Lowengart, 1977), showed the strongest response in the repair assay.

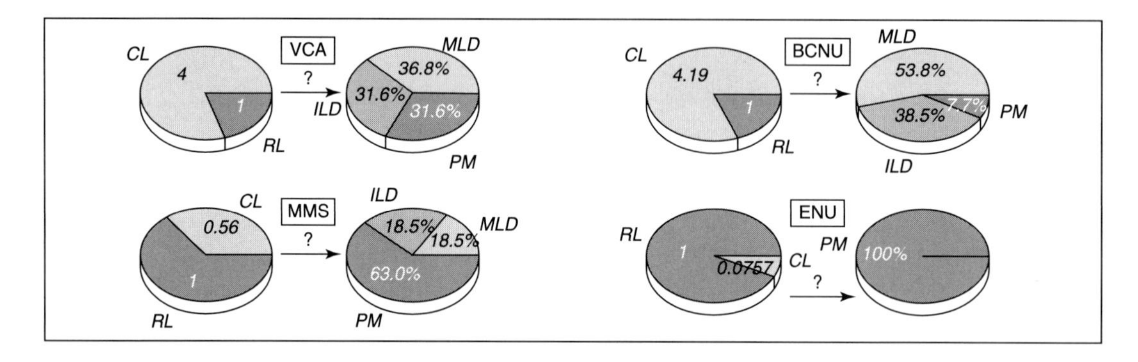

Figure 2. High chromosome loss (CL)/recessive lethal (RL) mutation ratios (left-hand pies) have predictive value for the occurrence of intra-locus (ILD) and multi-locus (MLD) deletions among visible mutations (right-hand pies) induced at the *vermilion* gene of *Drosophila*

VCA, vinyl carbamate; BCNU, bis(chloroethyl)nitrosourea; MMS, methyl methane sulfonate; ENU, *N*-ethyl-*N*-nitrosourea. CL:RL ratios are 0.076 (ENU), 0.56 (MMS), 4.1 (VCA), 4.2 (BCNU). Forward mutations were measured as recessive lethal mutations. Point mutations (PM; base substitutions) in the *vermilion* gene. The sum of ILD and MLD is 0% for ENU (Pastink *et al.*, 1989), 37% for MMS (Nivard *et al*, 1992), 68% for VCA (Ballering *et al.*, 1997) and 92% for BCNU (Table 2). References for CL:RL ratios: BCNU, ENU, MMS (Vogel *et al.*, 1993); VCA (this paper)

Table 2. Classification of *vermilion* mutants induced by 1.0 mmol/L bis(chloroethyl)nitrosourea in postmeiotic male germ cells of *Drosophila*. Cross *bw* males x *Inscy, y^{C4} v; bw* females[a]

Mutant	Brood	Position	Genetic change	Target sequences (5´–3´)[b]
Sterile mutants				
B-2	F_1-2A		Multi-locus deletion	
B-3	F_1-1B		Multi-locus deletion	
B-4	F_1-1B		Multi-locus deletion	
B-6	F_1-1C		No mutation found	
B-7	F_1-2C		Multi-locus deletion	
B-8	F_1-1A		Multi-locus deletion	
Male-lethal mutants				
B-1	F_1-2A		Multi-locus deletion	Df(1) 9B-10A1
B-9	F_1-1B		Multi-locus deletion	
Male-viable mutants				
B-5	F_1-1B	1168	GC→CG trp→cys	AAGTG G AGATgt
B-10	F_1-1C	169–186	Deletion/insertion	CACGAT-**gattcggcggtgccatt**-AACCACø CACGAT-**CGTTTG**-AACCAC
B-16	F_1-1C	496–498	Deletion/insertion	GAGTG-**gt**-TCTGAT ø GAGTG-**AACCGCTAAAGAG**-TCTGAT
B-18	F_1-2A	878–915	Deletion	CAGGAG-**ag**---**ca**-GAGCG ø CAGGAGCG
B-13	F_1-2C	234–256	Deletion/insertion	ACAA-**actgctggatgccagtgtatg**-CTCTCC ø ACAA-**TGCCTC**-CTCTCC
B-11	F_2-2A	29–249	Deletion/insertion	TTC-**gctgggg-----ggtca**-TTGATG ø TTC-**TGG**-TTGATG

[a] Offspring scored for *vermilion* mutants: F_1: 13 mutants in 61 880 females, mutation frequency = 2.1 x 10^{-4}; F_2 (mass cultures): 1 mutant in 14 827 flies, mutation frequency = 0.7 x 10^{-4}

[b] Sequence changes: deletions, bold small letters; insertions, bold capital letters; small letters, intron sequences; capital letters, exon sequences

Table 3. Types of single base changes induced by vinyl carbamate and vinyl bromide in postmeiotic germ cells of *Drosophila*

Mutation	Vinyl carbamate		Vinyl bromide (only NER⁻)	Total number of mutations	Suggested mutational specificity[a]
	NER⁺	NER⁻			
Transitions					
GC→AT	1		2	3	3,N^4-Ethenocytosine and N^2,3-ethenoguanine
AT→GC	2	3		5	1,N^6-Ethenoadenine
Transversions					
GC→TA		2	5	7	3,N^4-Ethenocytosine and 1,N^2,3-ethenoguanine
AT→TA	2	1	4	7	1,N^6-Ethenoadenine
GC→CG	1			1	1,N^2-Ethenoguanine
AT→CG				0	1,N^6-Ethenoadenine
				23	

From Ballering *et al.* (1997)
[a] For references, see text

M_{NER-}/M_{NER+} ratios are also clearly enhanced for vinyl bromide, vinyl chloride, ethyl carbamate and vinyl carbamate (Ballering *et al.*, 1996) and also for vinyl carbamate epoxide (Table 4), but surprisingly not for chloroethylene oxide. The lack of a potentiating effect of the NER⁻ condition on chloroethylene oxide-induced DNA damage is in contradiction with the general view that this compound is the ultimate DNA reactive intermediate of vinyl chloride (reviewed by Bartsch *et al.*, 1994). Our data suggest that vinyl chloride produces some DNA modification(s) that are not caused by chloroethylene oxide. One possibility could be the involvement of lipid peroxidation products generated by chloroacetaldehyde, as it has been shown to stimulate lipid peroxidation in isolated hepatocytes. In rats exposed to vinyl chloride, accumulation and persistence of etheno bases in DNA has been proposed to result from a direct pathway involving chloroethylene oxide and an indirect pathway involving lipid peroxidation products (Guichard *et al.*, 1996, and references therein).

The nitrosoureas are a separate subclass in that they generate adducts at oxygens in DNA (Singer & Bartsch, 1986; Eisenbrand *et al*, 1994). With BCNU, ~3% of all the products identified are O^6-2-hydroxyethylguanine. BCNU was also shown to form cross-linked adducts between two guanine bases and between guanine and cytosine (Eisenbrand *et al.*, 1994). The absence of significant hypermutability effects with BCNU, chlorozotocin and fotemustine indicates that they make no measurable contribution to the overall induction of mutation by nitrogen alkylation, because nitrogen mustards that transfer $Cl.CH_2.CH_2$-groups to ring nitrogens in DNA show strongly enhanced M_{NER-}/M_{NER+} ratios. The monofunctional 2-chloroethylamine, for instance, a borderline case in NER⁺ genotypes, becomes a powerful mutagen (33-fold increase in mutation frequency) in a NER⁻ background. Cross-linking analogues of 2-chloroethylamine also gave some positive responses in an assay for DNA repair, which were attributed to unrepaired mono-adducts at ring nitrogens (Vogel *et al.*, 1998).

In both *Drosophila* and mice, effective DNA repair is the mechanism by which premeiotic stages show high resistance to damage by monofunctional alkylating agents interacting with ring nitrogens in DNA (reviewed by Vogel *et al.*, 1998). It is therefore remarkable that etheno bases are

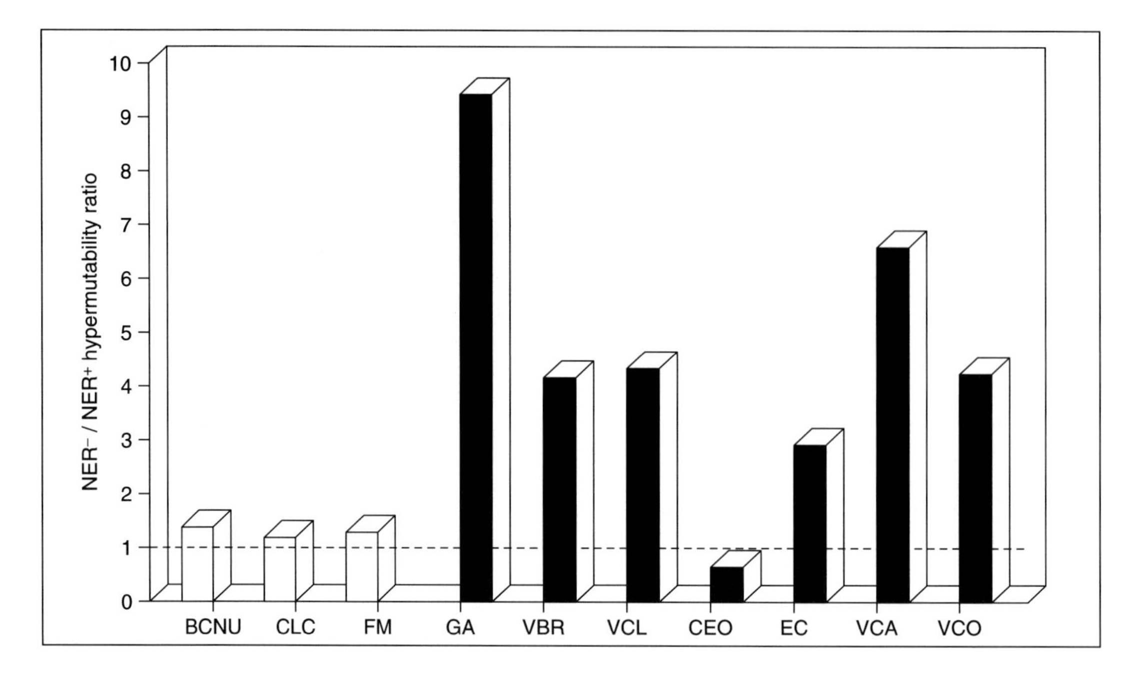

Figure 3. M_{NER-}/M_{NER+} hypermutability ratios in *Drosophila* for three halonitrosoureas (white columns) and seven chemicals that form etheno bases with DNA

BCNU, bis(chloroethyl)nitrosourea; CLC, chlorozotocin; FM, fotemustine; GA, glycidaldehyde; VBR, vinyl bromide; VCL, vinyl chloride; CEO, chloroethylene oxide; EC, ethyl carbamate; VCA, vinyl carbamate; VCO, vinyl carbamate epoxide. M_{NER-}/M_{NER+} ratios are for postmeiotic male germ cells (broods 1 and 2 pooled for this comparison). References: BCNU, glycidaldehyde and chloroethylene oxide (Vogel, 1989); vinyl bromide, vinyl chloride, ethyl carbamate, vinyl carbamate (Ballering *et al.*, 1996); vinyl carbamate epoxide (Table 3); chlorozotocin and fotemustine (E.W. Vogel & M.J.M. Nivard, unpublished data)

Table 4. The mutator effect of suppressed excision repair in the egg of *Drosophila* on mutation fixation of premutagenic lesions induced by 0.5 mmol/L vinyl carbamate epoxide during spermatogenesis

Female genotype	Mating interval (days)	Chromosomes tested	Mutations		M_{NER-}/M_{NER+} [a]
			n	%	
Wild-type	0–2	1068	21	1.97 ± 0.43	
	2–5	772	17	2.20 ± 0.53	
	0–5	1840	38	2.07 ± 0.33	
mus201	0–2	973	81	8.32 ± 0.89	4.1
	2–5	596	52	8.72 ± 1.16	4.1
	0–5	1569	133	8.48 ± 0.70	4.1

[a] Corrected for spontaneous mutations; control mutation frequencies are 0.09 and 0.14% (wild-type), 0.58 and 0.34 (*mus201*) for brood 1 and 2, respectively. Pooled mutation frequencies are 0.12% (wild-type) and 0.47% (*mus201*). Data from Vogel & Nivard (1997)

efficient promutagenic lesions in these cells (Table 5). After larvae were exposed to 2.0 mmol/L of vinyl carbamate, the ratio of the frequencies of recessive lethal mutations on the X and the second chromosome was about 8.5, which is considerably larger than the 2:1 ratio expected on the basis of the differences in their length (Lindsley & Zimm, 1992). The high yield of mutations in the second chromosome might be an indication of the occurrence of deletion mutations, which cause cell death before meiosis in hemizygous (as in XY cells of males) but not in homozygous conditions (as in II/II cells). A comparison of the mutation spectra in X-linked and autosomal reporter genes should clarify this most interesting question.

The finding that vinyl carbamate-induced etheno bases are mutagenic in *Drosophila* spermatogonia adds a further genotoxic action principle to the previously restricted group of DNA lesions shown to generate heritable damage in premeiotic cells. The other lesions are DNA *O*-alkyl adducts at pyrimidines induced by ENU and MNU, and DNA–DNA cross-links (reviewed by Vogel *et al.*, 1998).

Male mouse germ cells: Attempts to induce heritable genetic damage in male mouse germ cells by chemicals that form etheno bases have been largely unsuccessful. This is presumably a matter of metabolism. Procarbazine appears to be the only procarcinogen reported to have induced mutation in male mouse germ cells (reviewed by Vogel *et al.*, 1998). Dominant lethal mutations were not induced by vinyl chloride (Anderson *et al.*, 1977), and ethyl carbamate did not induce gene mutations or clastogenic damage (Russell *et al.*, 1987 and references therein).

Alkyl nitrosoureas are effective mutagens in male mice. For post-spermatogonial stages, largely concordant results have been reported for one monofunctional (MNU) and two bifunctional (BCNU and 1-(2-chloroethyl)-3-cyclohexyl-1-nitrosourea (CCNU)) nitrosoureas in tests for dominant lethal and specific-locus mutations. Positive results for both end-points were obtained with MNU at all germ-cell stages except spermatozoa (Ehling & Neuhäuser-Klaus, 1991). With 75 mg/kg of body weight MNU, an extremely high frequency of 90.8 x 10^{-5} mutations per locus was found in progeny conceived in matings 36–42 days (differentiating spermatogonia) after treatment (Russell & Hunsicker, 1983). The corresponding mutation frequency in a study with 70 mg/kg of body weight MNU (Ehling & Neuhäuser-Klaus, 1991) was lower, 29.8 x 10^{-5}, for the mating interval 33–40 days, whereas a second peak (mutation frequency, 21.8 x 10^{-5}) was seen in spermatids that had not been reported in the first study. Both BCNU and CCNU induced dominant lethals in spermatids and spermatocytes and specific-locus mutations in post-spermatogonial germ-cell stages of mice. In addition, CCNU increased the specific-locus

Table 5. Mutant frequencies (recessive lethal mutations) in premeiotic male germ cells from *Drosophila* exposed to vinyl carbamate

Treatment (mmol/L)	X-Chromosome[a] tested	Recessive lethals		2nd chromosome[b] tested	Recessive lethals	
		n	% ± SD		*n*	% ± SD
Control[c]	18 239[c]	21	0.12 ± 0.03[c]			
LA, 25.0	2161	13	0.60 ± 0.17			
LS, 0.25	651	4	0.61 ± 0.31			
LS, 2.0	5235	98	1.87 ± 0.19	475	72	15.2 ± 1.65
LS, 4.0	5402	154	2.85 ± 0.23			

L, larval treatment; LA, acute; LS, surface treatment
[a] Crosses: Hikone-R male x *Inscy, y^{c4} v; bw* female
[b] Crosses: Hikone male x *In(2LR)SM5, al^2 Cy ltv sn^2 sp^2* female
[c] From Vogel & Nivard (1997); broods 1 and 2 pooled for this comparison

mutation frequency in stem-cell spermatogonia in two of three experiments (Ehling *et al.*, 1997). ENU and *N*-propyl-*N*-nitrosourea also induced specific-locus mutations in stem-cell spermatogonia, ENU being the most potent point mutagen in mice (Favor *et al.*, 1990, and references therein; Murota & Shibuya, 1991). All five alkyl nitrosoureas induced specific-locus mutations in repair-active premeiotic cells, although differences were observed in the maximum yield of mutations induced in specific cell stages.

Mutagenesis of exocyclic DNA adduct-forming chemicals in somatic cells

Drosophila: The standard *white*$^{+}$/*white* eye mosaic assay is used predominantly to measure homologous recombination between the X chromosomes of XX cells. In this assay, only marginal genotoxic activity was observed with acetyl chloride, glycolaldehyde and 2,2′-dichlorodiethyl ether, whereas methyl carbamate was inactive (Ballering *et al.*, 1996). These low genotoxic activities do not suggest a significant role of acetyl chloride or glycolaldehyde in the genotoxicity of vinyl chloride and ethyl carbamate, which are clearly recombinagenic in somatic tissues of *Drosophila* (Frölich & Würgler, 1990; Vogel & Nivard, 1993).

Recently, we developed a system (see Methods) for the parallel detection of homologous and intrachromosomal recombination, chromosome loss (as a measure of clastogenicity) and gene mutation. In this new four-end-point assay, both vinyl carbamate and ethyl carbamate were extraordinarily active in generating ring-X loss in XrX female and rXY male genotypes (Figure 4). Ring chromosome loss may result from chromosome breakage or sister chromatid exchange (Ashburner, 1989). Studies with methyl methane sulfonate, ENU and cisplatin had shown that the frequency of mosaic *white* clones is not significantly reduced in inversion heterozygotes (InXrX), if a ring-chromosome is placed opposite an inverted X chromosome (InX) (E.W. Vogel &

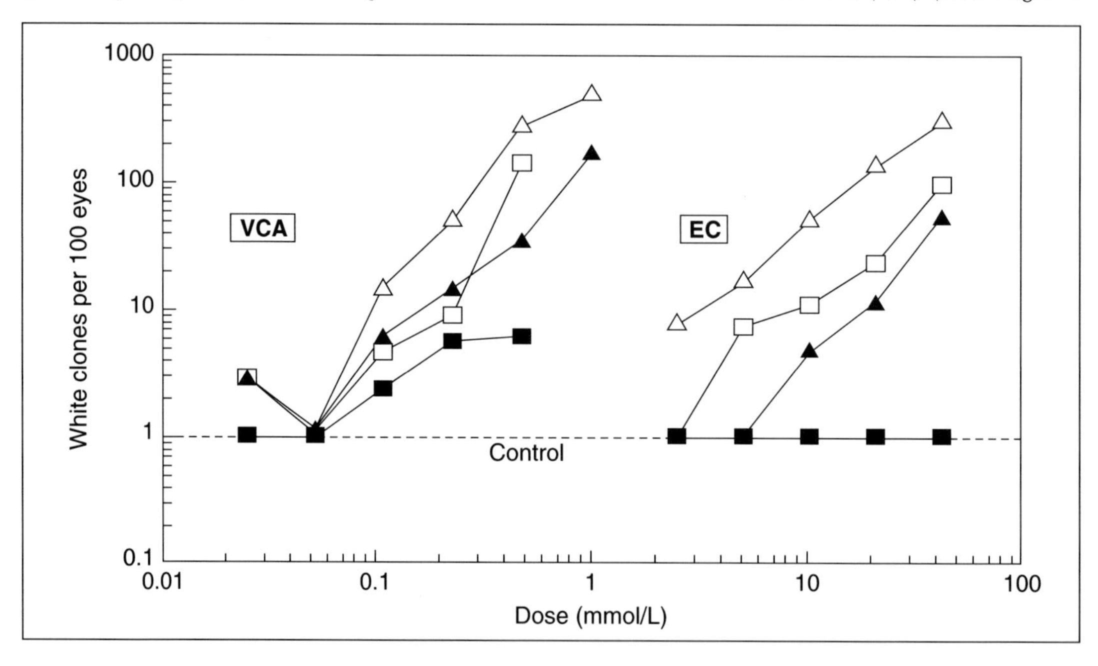

Figure 4. Activity of ethyl carbamate (EC) and vinyl carbamate (VCA) in a modified *w/w*$^{+}$ eye mosaic assay in *Drosophila*

Number of eyes scored per dose for *white* clones: ethyl carbamate, 200 eyes with 2.5–20 mmol/L, 100 with 40 mmol/L; vinyl carbamate, 200 eyes with 0.025–0.20 mmol/L, 30–190 eyes with 0.50 and 1.0 mmol/L. (open triangles, ring-X loss in XrX cells; filled triangles, recombination in XX cells; open squares, ring-X loss in rXY cells; filled squares, intrachromosomal change in XY cells

M.J.M. Nivard, unpublished data). However, in female genotypes carrying one normal and one inverted X chromosome (XInX), clone formation is strongly reduced or even suppressed, depending on the type of mutagen, showing that the predominant event measured in XX cells is indeed homologous mitotic recombination (Vogel, 1992).

In comparison with their efficiency in inducing clastogenic damage, vinyl carbamate and ethyl carbamate are relatively weak recombinagens: in female genotypes, both chemicals are approximately 10-fold more clastogenic than recombinagenic. Both chemicals also cause a high incidence of ring-X loss in rXY cells but do not induce significant frequencies of intrachromosomal recombination. Thus, the genetic action spectra of ethyl carbamate and vinyl carbamate are similar, supporting the general view that they act through the same ultimate electrophilic species, most likely the vinyl carbamate epoxide.

The genetic activity profiles of BCNU and ENU show significant differences from those of ethyl carbamate and vinyl carbamate, and they also differ from each other in various respects (Figure 5). Ring-X loss is the dominant type of damage found after exposure to BCNU, followed by homologous recombination and a low incidence of intra-chromosomal recombination. ENU provides an example that high frequencies of intra-chromosomal changes can be induced by chemical mutagens. With ENU, all four endpoints are induced at high frequency.

Since ring-X loss in XrX cells is the most frequent class of clone induced by all four mutagens, we investigated whether other chemicals might produce a higher yield of homologous recombination than chromosome loss. We have so far identified one genotoxic agent, 7,12-dimethylbenz[*a*]anthracene, that is clearly more recombinagenic than clastogenic (E.W. Vogel & M.J.M. Nivard, paper submitted).

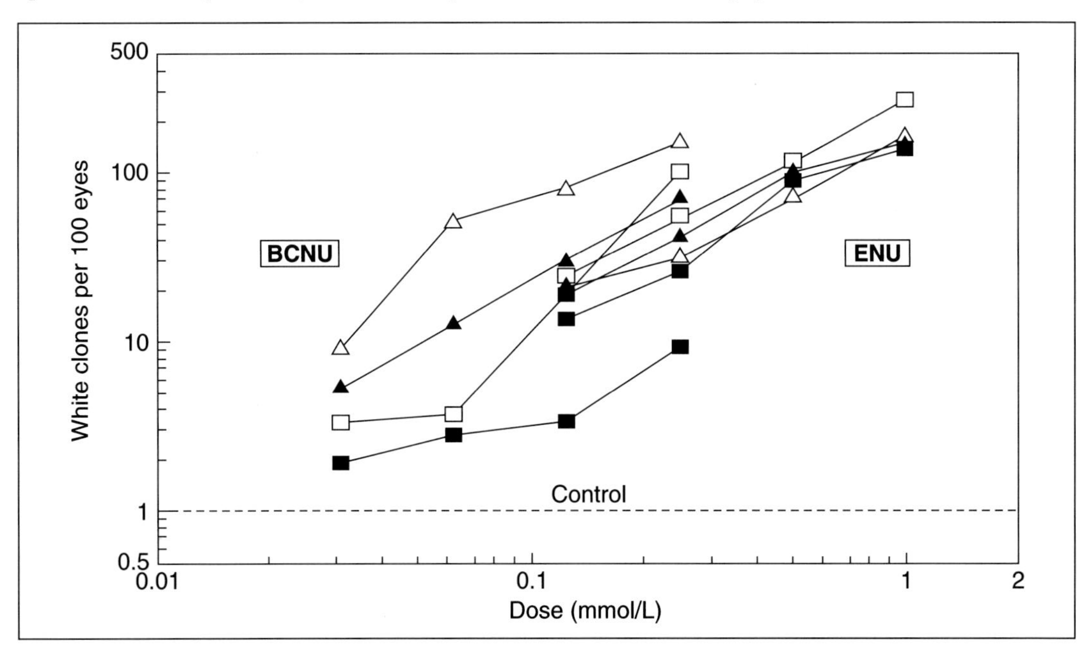

Figure 5. Activity of *bis*(chloroethyl)nitrosourea (BCNU) and *N*-ethyl-*N*-nitrosourea (ENU) in a modified *w/w*$^+$ eye mosaic assay in *Drosophila*

Number of eyes scored per dose for *white* clones: ENU, 150–200 eyes; BCNU, 170–200 eyes. open triangles, ring-X loss in XrX cells; filled triangles, recombination in XX cells; open squares, ring-X loss in rXY cells; filled squares, intrachromosomal change in XY cells

Mouse: Several studies in mice have established that ethyl carbamate is genotoxic in the bone marrow (Ashby *et al.*, 1990, and references therein). After repeated or single intraperitoneal injections of 300–1200 mg/kg of body weight, ethyl carbamate induced a high frequency of micronuclei in mouse bone marrow (Ashby *et al.*, 1990). In the *lacZ*⁻ transgenic mouse mutation assay (Muta™Mouse), a single dose of 900 mg/kg of body weight ethyl carbamate induced statistically significant (about twofold) increases in mutation frequency in the lung and liver and weaker effects in the spleen and bone marrow (Williams *et al.*, 1998). Concomitant assays for micronucleus formation in bone marrow confirmed the earlier report of the strong clastogenic effects of ethyl carbamate. Vinyl carbamate was about 40 times more potent than ethyl carbamate in inducing micronuclei in mouse bone marrow (Wild, 1991). Methyl carbamate, which cannot undergo transformation to an epoxide, was not active in this test (Shelby & Tice, 1991).

In conclusion, the overall genetic activity profiles seen for etheno adduct-forming chemicals in mice and *Drosophila* support the hypothesis that vinyl carbamate is the proximate mutagen of ethyl carbamate, and vinyl carbamate epoxide is the ultimate electrophilic mutagen and carcinogen. The experimental data available on BCNU and other bifunctional halonitrosoureas indicate a dominant role of DNA cross-links in mutagenesis.

Acknowledgements

We are very grateful to Sharda Bisoen, Corrie van Veen and Ineke Bogerd for their careful technical assistance. The authors gratefully acknowledge financial support from the 'Environment' Programme, contracts EV5V-CT940409, and the 'Environment & Climate' Programme, contract ENV4-CT97-0505, of the European Union.

References

Acharya, S.A. & Manning, J.M. (1983) Reaction of glycolaldehyde with proteins: Latent crosslinking potential of α-hydroxyaldehydes. *Proc. Natl Acad. Sci. USA* 80, 3590–3594

Anderson, D., Hodge, M.C.E. & Purchase, I.F.H. (1977) Dominant lethal studies with the halogenated olefins vinyl chloride and vinylidene dichloride in male CD-1 mice, *Environ. Health Perspectives* 21, 71–78

Ashburner, M. (1989) *Drosophila, A Laboratory Handbook*. Cold Spring Harbor, New York, CSH Press

Ashby, J., Tinwell, H. & Callander, R.D. (1990) Activity of urethane and N,N-dimethylurethane in the mouse bone-marrow micronucleus assay: Equivalence of oral and intraperitoneal routes of exposure. *Mutat. Res.* 245, 227–230

Ashby, J., Vogel, E.W., Tinwell, H., Callander, R.D. & Shuker, D.E.G. (1993) Mutagenicity to *Salmonella, Drosophila* and the mouse bone marrow of the human antineoplastic agent fotemustine: Prediction of carcinogenic potency. *Mutat. Res.* 286, 101–109

Barbin, A. (1998) Formation of DNA etheno adducts in rodents and humans and their role in carcinogenesis. *Acta Biochim. Pol.* 45, 145–161

Barbin, A., Bartsch, H., Leconte, P. & Radman, M. (1981) Studies of the miscoding properties of 1,N^6-ethenoadenine and 3,N^4-ethenocytosine, DNA reaction products of vinyl chloride metabolites during in vitro synthesis. *Nucleic Acid Res.* 9, 375–387

Ballering, L.A.P., Nivard, M.J.M. & Vogel, E.W. (1996) Characterization by two-endpoint comparisons of the genetic toxicity profiles of vinyl chloride and related etheno-adduct forming carcinogens in *Drosophila*. *Carcinogenesis* 17, 1083–1092

Ballering L.A.P., Nivard, M.J.M. & Vogel, E.W. (1997) Preferential formation of deletions following *in vivo* exposure of postmeiotic *Drosophila* germ cells to the DNA ethenoadduct forming carcinogen vinyl carbamate. *Environ. Mol. Mutag.* 30, 321–329

Bartsch, H., Barbin, A., Marion, M.J., Nair, J. & Guichard, Y. (1994) Formation detection, and role in carcinogenesis of ethenobases in DNA. *Drug. Metab. Rev.* 26, 349–371

Basu A.K., Wood M.L., Niedernhofer L.J., Ramos L.A. & Essigmann, J.M. (1993) Mutagenic and genotoxic effects of three vinylchloride induced DNA lesions: 1,N^6 ethenoadenine, 3,N^4 ethenocytidine and 4-amino-5-(imidazol-2- yl)imidazole. *Biochemistry* 32, 12793–12801

Cheng, K.C., Preston, B.D., Cahill, D.S., Dosanjh, M.K., Singer, B. & Loeb, L.A. (1991) The vinyl chloride derivative, N^2,3-ethenoguanine, produces G→A transitions in *Escherichia coli*. *Proc. Natl Acad. Sci. USA* 88, 9974–9978

Ehling, U.H. & Neuhäuser-Klaus A. (1991) Induction of specific-locus mutations and dominant lethal mutations in male mice by 1-methyl-1-nitrosourea (MNU). *Mutat. Res.* 250, 447–456

Ehling, U.H., Adler, I.-D., Favor, J. & Neuhäuser-Klaus, A. (1997) Induction of specific-locus and dominant lethal mutations in male mice by 1,3-bis(2-chloroethyl)-1-nitrosourea (BCNU) and 1-(2-chloroethyl)-3-cyclohexyl-1-nitrosourea (CCNU). *Mutat. Res.* 379, 211–218

Eisenbrand, G., Pfeiffer, C. & Tang, W. (1994) DNA adducts of *N*-nitrosoureas. In: *DNA Adducts, Identification and Biological Significance* (Hemminki, K., Dipple, A., Shuker, D.E.G., Kadlubar, F.F., Segerbäck, D. & Bartsch, H., eds), IARC Scientific Publications No. 124. Lyon, IARC, pp. 277–293

Favor, J., Sund, M., Neuhäuser-Klaus, A. & Ehling, U.H. (1990) A dose–response analysis of ethylnitrosourea-induced recessive specific-locus mutations in treated spermatogonia of the mouse. *Mutat. Res.* 231, 47–54

Frölich, A. & Würgler, F.E. (1990) Genotoxicity of ethyl carbamate in the *Drosophila* wing spot test: Dependence on genotype-controlled metabolic capacity. *Mutat. Res.* 244, 201–208.

Guengerich, F.P., Kim, D.-H. & Iwasaki, M. (1991) Role of human cytochrome P450 IIEI in the oxidation of many low molecular weight cancer suspects. *Chem. Res. Toxicol.* 4, 168–179

Guengerich, F.P. & Kim, D.-H. (1991) Enzymatic oxydation of ethyl carbamate to vinyl carbamate and its role as an intermediate in the formation of 1,N^6-etheno adenosine. *Chem. Res. Toxicol.* 4, 413–421

Guichard, Y., El Ghissassi, E., Nair, J., Bartsch, H. & Barbin, A. (1996) Formation and accumulation of DNA ethenobases in adult Sprague-Dawley rats exposed to vinyl chloride. *Carcinogenesis* 17, 1553–1559

Gwinner, L.M., Laib, R.J., Filser, J.G. & Bolt, H.M. (1983) Evidence of chloroethylene oxide being the reactive metaboite of vinyl chloride towards DNA: Comparative study with 2,2′-dichlorodiethyl ether. *Carcinogenesis* 4, 1483– 1486

Hall, J.A., Saffhill, R., Green, T. & Hathway, D.E. (1981) The induction of errors during *in vitro* DNA synthesis following chloroacetaldehyde-treatment of poly (dA–dT) and poly (dC–dG) templates. *Carcinogenesis* 2, 141–146

Hollstein, M., Marion, J.-M., Lehman, T., Welsh, T., Harris, C.C., Martel-Planche, G., Kusters, I. & Montesano, R. (1994) *p53* mutations at A:T base pairs in angiosarcomas of vinyl chloride-exposed factory workers. *Carcinogenesis* 15, 1–3

Langouët, S., Müller, M., & Guengerich, F.P. (1997) Misincorporation of dNTPs opposite 1,N^2-ethenoguanine and 5,6,7,9-tetrahydro-7-hydroxy-9-oxoimidazol[1,2-α]purine in oligonucleotides by *Escherichia coli* polymerases I exo$^-$ and II exo$^-$, T7 polymerase exo$^-$, human immunodeficiency virus-1 reverse transcriptase, and rat polymerase β. *Biochemistry* 36, 6069–6079

Leithauser, M.T., Liem, A., Steward, B.A., Miller, E.C. & Miller, J.A. (1990) 1,N^6-Ethenoadenosine formation, mutagenicity and murine tumor induction as indicators of the generation of an electrophilic epoxide metabolite of the closely related carcinogens ethyl carbamate (urethane) and vinyl carbamate. *Carcinogenesis* 11, 463–473

Lindsley, D.L. & Zimm, G. G. (1992) *The Genome of Drosophila melanogaster*. New York, Academic Press

Marion, M.J., Froment, O. & Trépo, C. (1991) Activation of Ki-*ras* gene by point mutation in human liver angiosarcomas associated with vinyl chloride exposure. *Mol. Carcinog.* 4, 450–454

Murota, T. & Shibuya, T. (1991) The induction of specific-locus mutations with *N*-propyl-*N*-nitrosourea in stem-cell spermatogonia of mice. *Mutat. Res.* 264, 235–240

Nivard, M.J.M., Pastink, A. & Vogel, E.W. (1992) Molecular analysis of mutations in the *vermilion* gene of *Drosophila melanogaster* by methyl methanesulfonate. *Genetics* 131, 673–682

O'Neill, I., Barbin, A., Friesen, M. & Bartsch, H. (1986) Reaction kinetics and cytosine adducts of chloroethylene oxide and chloro acetaldehyde: Direct observation of intermediates by FTNMR and GC–MS. In: *The Role of Cyclic Nucleic Acid Adducts in Carcinogenesis and Mutagenesis* (Singer, B. & Bartsch, H., eds), IARC Scientific Publications No. 70. Lyon, IARC, pp. 57–73

Palejwala, V.A., Rzepka, R.W., Simha, D. & Humayun, M.Z. (1993) Quantitative multiplex sequence analysis of mutational hot spots. Frequency and specificity of mutations induced by a site-specific ethenocytosine in M13 viral DNA. *Biochemistry* 32, 4105–4111

Pandya, G.A. & Moriya, M. (1996) 1,N^6-Ethenodeoxyadenosine, a DNA adduct highly mutagenic in mammalian cells. *Biochemistry* 35, 11487–11492

Park, K.K., Surh, Y.J., Stewart, B.C. & Miller, J.A. (1990) Synthesis and properties of vinyl carbamate epoxide, a possible ultimate electrophilic and carcinogenic metabolite of vinyl carbamate and ethylcarbamate. *Biochem. Biophys. Res. Commun.*169, 1094–1098

Park, K.K., Liem, M., Steward, B.C. & Miller, J.A. (1993) Vinyl carbamate epoxide, a major strong electrophilic, mutagenic and carcinogenic metabolite of vinyl carbamate and ethyl carbamate (urethane). *Carcinogenesis* 14, 441–450

Pastink, A., Vreeken, C., Nivard, M.J.M., Searles, L. & Vogel, E.W. (1989) Sequence analysis of N-ethyl-N-nitrosourea induced *vermilion* mutations on *Drosophila melanogaster*. *Genetics* 123, 123–129

Russell, W.L. & Hunsicker. P.R. (1983) Extreme sensitivity of one particular germ-cell stage in male mice to induction of specific-locus mutations by methylnitrosourea. *Environ. Mutagen* 5, 498

Russell, L.B., Hunsicker, P.R., Oakberg, E.F., Cummings, C.C. & Schmoyer, R.L. (1987) Tests for urethane induction of germ-cell mutations and germ-cell killing in the mouse. *Mutat. Res.* 188, 335–342

Schaefgen, J.R. (1968) Poly(vinyl chloroformate) and derivatives: Preparation and properties. *J. Polymer Sci.*, Part C, No. 24, 75–88

Shelby, M.D. & Tice, R.R. (1991) Methyl carbamate: Negative results in mouse bone-marrow micronucleus test. *Mutat. Res.* 260, 311

Singer, B. & Bartsch, H., eds (1986) *The Role of Cyclic Nucleic Acid Adducts in Carcinogenesis and Mutagenesis*, IARC Scientific Publications No. 70. Lyon, IARC

Singer, B., Kusmierek, J.T., Folkman, W., Chavez F. & Dosanjh, M.K. (1991) Evidence for the mutagenic potential of the vinyl chloride-induced adduct N^2,3-ethenodeoxyguanine, using a site directed kinetic assay. *Carcinogenesis* 12, 745– 747

Spengler, S. & Singer, B. (1981) Transcriptional errors and ambiguity resulting from the presence of 1N^6-ethenoadenosine or 3,N^4-ethenocytidine in polyribonucleotides. *Nucleic Acids Res.* 9, 365–373

Trivers, G.E., Cawley, H.L., DeBenedetti, V.M., Hollstein, M., Marion, M.J., Bennett, W.P., Hoover, M.L., Prives, C.C., Tamburro, C.C. & Harris, C.C. (1995) Anti-p53 antibodies in sera of workers occupationally exposed to vinyl chloride. *J. Natl Cancer Inst.* 87, 1400–1407

Van Duuren, B.L. & Loewengart, G. (1977) Reaction of DNA with glycidaldehyde. *J. Biol. Chem.* 252, 5370–5371

Vogel, E.W. (1989) Nucleophilic selectivity of carcinogens as a determinant of enhanced mutational response in excision repair-defective strains in *Drosophila*: Effects of 30 carcinogens. *Carcinogenesis* 10, 2093–2106

Vogel, E.W. (1992) Tests for recombinagens in somatic cells of *Drosophila*. *Mutat. Res.* 284, 159–175

Vogel, E.W. & Nivard M.J.M. (1993) Performance of 181 chemicals in a *Drosophila* assay predominantly monitoring interchromosomal mitotic recombination. *Mutagenesis* 8, 57–81

Vogel, E.W. &. Nivard, M.J.M (1997) The response of germ cells to ethylene oxide, propylene oxide, propylene imine and methyl methanesulfonate is a matter of cell stage-related DNA repair. *Environ. Mol. Mutag.* 29, 124–135

Vogel, E.W., Zijlstra, J.A. & Nivard, M.J.M. (1993) A genetic method for classification of genotoxins into monofunctional or cross-linking agents. *Environ. Mol. Mutag.*, 21, 319–331

Vogel, E.W., Barbin, A., Nivard, M.J.M., Stack, H.F., Waters, M.D. & Lohman, P.H.M. (1998) Heritable and cancer risks of exposure to anticancer drugs: Inter-species comparisons of covalent deoxyribonucleic acid-binding agents. *Mutat. Res.* 400, 509–540

Vogt, M. (1948) Mutationsauslösung bei *Drosophila* durch Äthylurethan. *Experientia* 4, 68–69

Wild, D. (1991) Micronucleus induction in bone marrow by vinyl carbamate, a hypothetical metabolite of the carcinogen urethane (ethyl carbamate). *Mutat. Res.* 260, 309–310

Williams, C.V., Fletcher, K., Tinwell, H. & Ashby, J. (1998) Mutagenicity of ethyl carbamate to *lacZ*$^{-}$ transgenic mice. *Mutagenesis* 13, 133–137.

Exocyclic DNA Adducts in Mutagenesis and Carcinogenesis
B. Singer & H. Bartsch
IARC Scientific Publications No. 150
International Agency for Research on Cancer, Lyon, 1999

Appendix to papers by Chenna & Singer, Sági & Singer and Singer & Hang

Distortion of the double helix by *para*-benzoquinone adducts

Geometrically optimized structures of *para*-benzoquinone (pBQ)-C (Figure 2), pBQ-A (Figure 3) and pBQ-G (Figure 4) adduct-containing duplexes, as compared with an unmodified duplex (Figure 1), show differential distortions of the helix at the benzoquinone adduct (shown in green). For optimization, the same 25-mer sequences that were used in biochemical and biophysical studies (Sági & Singer, this volume) were built as models in canonical B conformation with HyperChem (version 4.5 for SGI; Hypercube, Inc., Gainsville, Florida, USA). Exocyclic adducts were formed manually on the selected bases. Molecular mechanics were calculated by AMBER3 force field with distance-dependent dielectric constant *in vacuo*. The steepest descent method and then the Polak-Ribiere conjugate gradient method were used to reach an RMS gradient of < 0.1 kcal/A mol. The geometrically optimized structures, representing the minimum energy of the system, are used only to visualize structural distortion caused by the adducts in a 10-base-pair structure.

Carbon atoms are represented in blue, nitrogen in purple, oxygen in red, phosphorus in yellow and hydrogen in white.

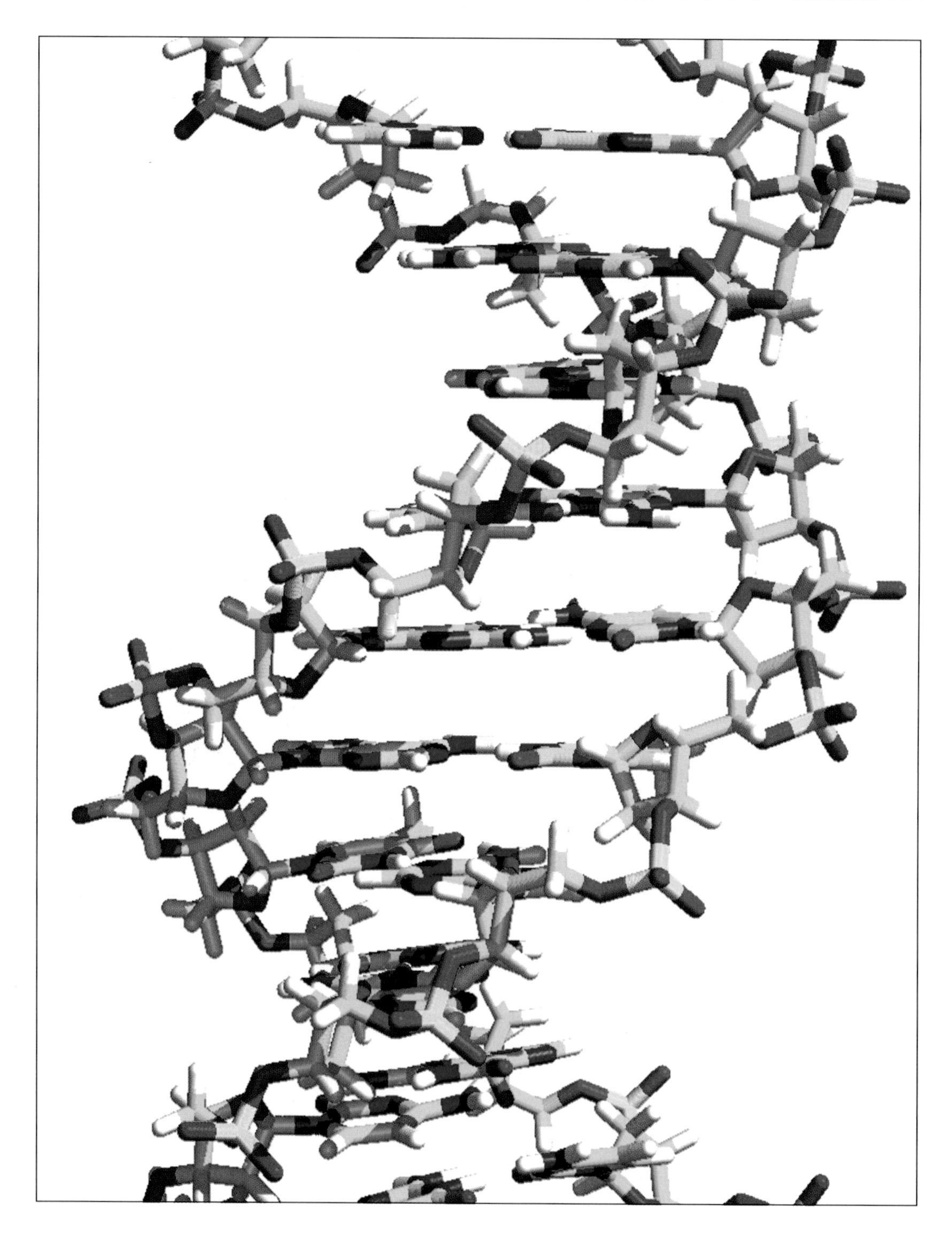

Figure 1.

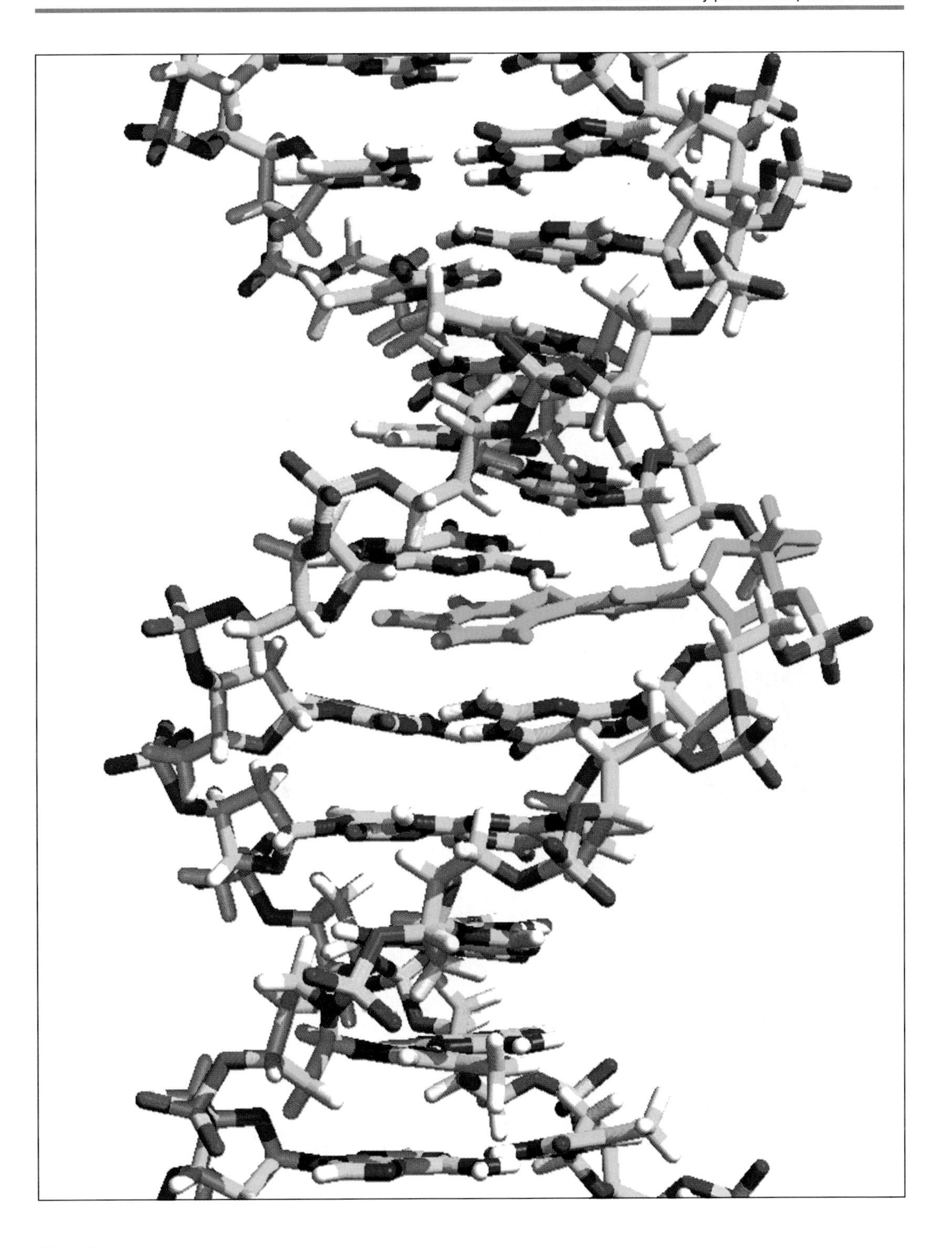

Figure 2.

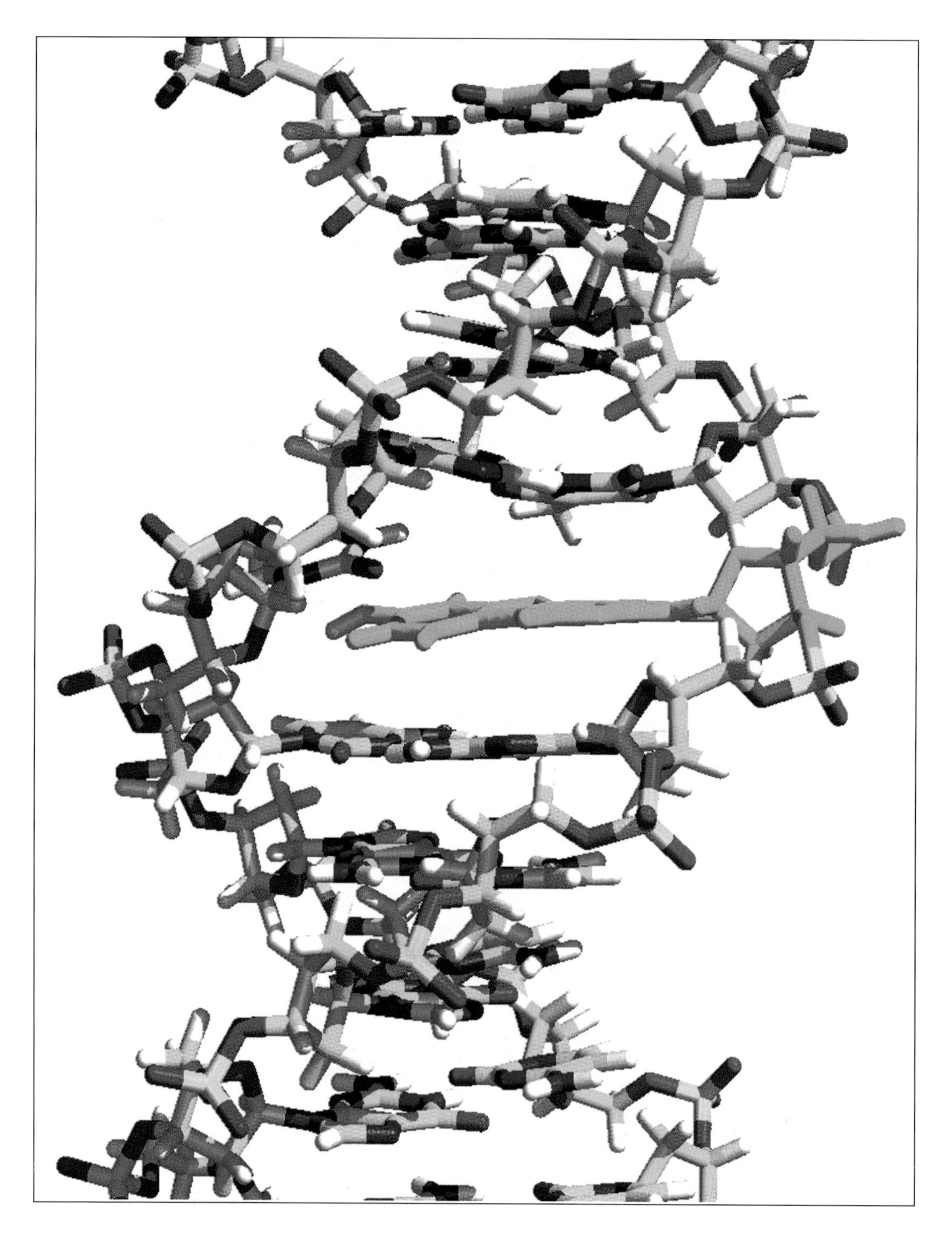

Figure 3.

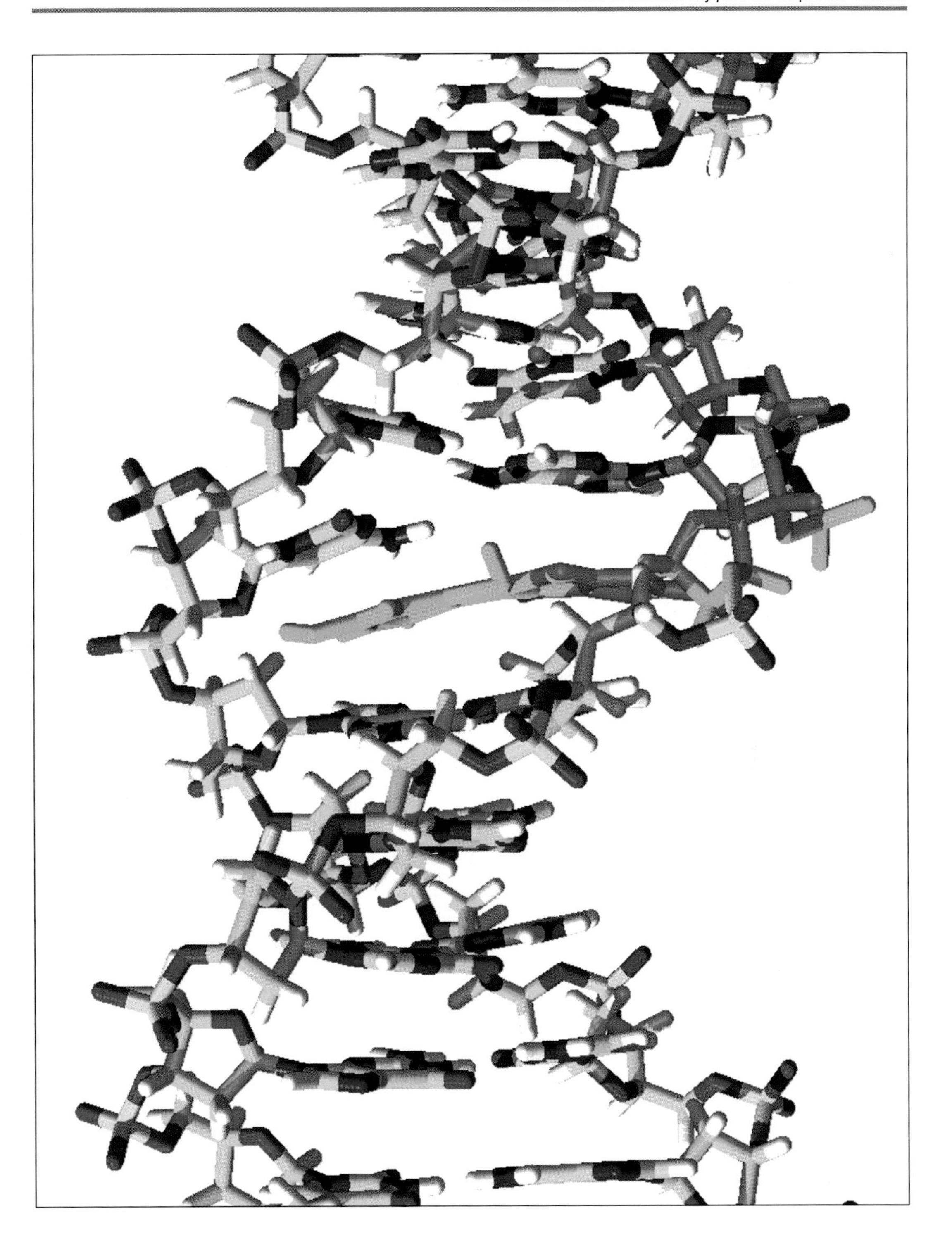

Figure 4.

Subject Index

Achevé d'imprimer sur les presses de l'imprimerie Darantiere
à Dijon-Quetigny, en octobre 1999

Dépôt légal : 4e trimestre 1999 - N° d'impression : 99-0457